The Fundamentals of Horticulture
Theory and Practice

Essential reading for all studying horticulture and keen gardeners. This clear introduction to the principles underlying the practical applications of horticulture opens up the excitement of growing plants and garden development without readers wading through complex information. Written by a team of highly motivated and experienced horticultural tutors, the text supports the newly restructured RHS Level 2 qualifications with related Level 3 topics in boxes and sign-posting to Level 4 topics, together with other horticultural qualifications at these levels.

Full colour images tied closely to the text and practical case study boxes inspire readers by making topics relevant to their own horticultural experiences. A comprehensive glossary helps build confidence in the use of classical horticulture language as well as new developing terms, and end-of-chapter questions encourage readers to apply what they have learned. Extensive online supporting material includes mind maps showing the relationship of topics and aiding students in revision, and can be found at www.cambridge.org/horticulture.

Chris Bird has over 25 years of horticultural teaching experience in guiding learners through the complete suite of Royal Horticultural Society (RHS) qualifications, together with other horticultural qualifications including the BTEC National Diploma and City & Guilds Subsidiary, Diploma and Extended Diploma. He has co-ordinated Sparsholt College's Chelsea Flower Show educational gardens for the past 14 years, winning 14 RHS show medals including five at gold level, and has written articles for a number of gardening magazines.

Royal Horticultural Society

Sharing the best in Gardening

The RHS is the UK's leading gardening charity, dedicated to advancing horticulture and promoting good gardening. Its work includes training the next generation of gardeners and creating hands-on opportunities for children to grow their own plants. For more information visit www.rhs.org.uk or call 0845 130 4646.

The Fundamentals of Horticulture
Theory and Practice

Edited by **Chris Bird**
Sparsholt College, Winchester

CAMBRIDGE
UNIVERSITY PRESS

CAMBRIDGE
UNIVERSITY PRESS

University Printing House, Cambridge CB2 8BS, United Kingdom

Cambridge University Press is part of the University of Cambridge.

It furthers the University's mission by disseminating knowledge in the pursuit of
education, learning, and research at the highest international levels of excellence.

www.cambridge.org
Information on this title: www.cambridge.org/horticulture

First published 2014

Printed in Spain by Grafos SA, Arte sobre papel

A catalogue record for this publication is available from the British Library

Library of Congress Cataloguing in Publication data
The fundamentals of horticulture : theory and practice / edited by Chris Bird,
Sparsholt College, Winchester.
    pages   cm
Includes bibliographical references and index.
ISBN 978-0-521-70739-8 (alk. paper)
1. Horticulture.   I. Bird, Chris, 1960–
SB318.F86   2014
635–dc23
                                                            2014001826

ISBN 978-0-521-70739-8 Paperback

Additional resources for this publication at www.cambridge.org/horticulture

CONTENTS

CONTRIBUTORS

Chris Allen Editor of the online horticultural magazine: thegardeningtimes.com. Formerly of Kingston Maurward College, Dorchester.

Chris Bird Sparsholt College.

Michael Buffin Gardens and Parks Advisor, National Trust.

Neil Helyer IPM Specialist, Fargro Ltd.

Kelvin Mason Sparsholt College.

Aaron Mills Sparsholt College.

Jenny Shukman RHS Professional Associate, formerly of Sparsholt College.

Tim Upson Cambridge University Botanic Garden.

Daphne Vince-Prue Formerly of the University of Reading (now retired).

Rosie Yeomans Sparsholt College.

PREFACE

WHY ANOTHER HORTICULTURAL TEXTBOOK?

Well, the world of horticulture is constantly changing, and the Royal Horticultural Society, one of the major guardians of excellence in this area, has rethought and restructured the information required to obtain its highly regarded qualifications. These changes make the transition through the levels of knowledge more logical and smoother for anyone wishing to know more about plant selection, propagation, growing, use and design.

This is the first textbook written to take account of these latest adjustments and provide sequential development over a wide range of topics, arranged in three broad parts, allowing quick access to an appropriate starting point for readers to move forward in their quest for horticultural knowledge and understanding.

Most people start learning about plants through their families, especially grandparents: the techniques and dates of practical operations have been handed down from one generation to the next, over time, but the reasoning and science behind these events has become confused or totally lost. Therefore with changes such as the advent of climate change, the expanded plant palette at our disposal, the re-awakening of interest in the natural world and the joy of feeding our families by home production of fruit and vegetables, we need to revisit our timings and methods and ensure maximum success with minimum inputs of resources, including the efficient use of valuable time.

Also the mental and spiritual wellbeing found within the activities of plant growing and use can not be understated at this time, with the continued rapid development of horticulture as a successfully method of therapy: an antidote to busy, sometimes stressful, lives. Many people are looking to enhance, or produce, their own green oasis; the chapters on the selection and use of plants will assist this process.

This textbook is not only for students wishing to study for examinations leading to valuable qualifications, but also for all those who have any interest, or an emerging puzzlement, of how plants grow, develop and can be used.

Structured in three parts, Part 1 The foundations, Part 2 The adjustments and Part 3 The applications, the book provides a clear development of the material provided by an experienced team who have between them amassed over 250 years of horticultural and 150 years of teaching experience; and are still learning.

STRUCTURE OF THE BOOK

With the resurgence of interest in the 'green agenda' and grow-your-own campaigns, such as the National Trust's 'Plot to Plate', the Royal Horticultural Society's 'Grow your Own' and a number of major supermarkets leading national campaigns (also related to encouraging uptake by school children and their wider families) there has never been a better opportunity to spend some time exposed to the **foundations** of growing; being reminded of the **adjustments** that can be undertaken and introduced to the **applications** that can be achieved within your own community. Also, with the publication of a new syllabus emerging from the RHS and other horticultural qualifications, we have taken a fresh look at the fundamentals of horticulture, remembering to include an emphasis on the first syllable of fundamentals.

One of the current buzz-phrases of education is 'distance travelled'. Think of this also within a garden context. How far have you travelled on these subjects?

- ✿ Environmental concerns
- ✿ Sustainable developments
- ✿ Reducing waste
- ✿ Offsetting climate change
- ✿ Promoting the balance of wildlife

This book is divided into three parts: developing a sequential progression of interest and knowledge. Part 1, entitled **The foundations**, as the name suggests provides an introduction to the background techniques and topics that underpin the whole process of the successful growing of plants. Part 2, **The adjustments**, consists of the developmental topics that having mastered, or at least understood, the foundations lead on to.

Finally, Part 3, **The applications**, shows the diversity that horticulture encompasses, including chapters on landscape design; plant use; pest, diseases and disorders; and commercial production. These provide an increased understanding of your plot and build on its history, providing a balance of conditions to aid your improved success, whatever your starting point.

HOW TO USE THIS BOOK

Do you wish to enjoy or endure your studies? This new, full colour, textbook has been developed to balance the need for an understanding of technical language with full descriptions of traditional and modern, fully illustrated, practical techniques, all designed to enhance the joy of study.

Each chapter begins by listing the key concepts that the chapter covers and features an opening case study showing the practical applications of the theory covered.

The main text is supported by a comprehensive glossary with key terms clearly highlighted in **bold** in the text. Within the chapters themselves, given within boxed sections, are extra details or further examples, which provide information to support RHS Level 3 qualifications, and boxed case studies show the real-life applications of the theory.

Chapters are rounded off by summaries, further reading lists and some revision questions for you to confirm your knowledge or to highlight where further study time is required.

All the major horticultural syllabuses have been considered and the main text is aimed at the RHS Level 2 learner, but will also inform the keen gardener or those seeking some information to get started. Clear signposting to the RHS Level 3 areas of information are highlighted within boxes at the end of each chapter, helping students to organise their learning. However, it should be noted that some topics have been re-allocated into Level 4 and these are covered in Chapters 9 and 15. Where comprehensive tables have been developed these are provided as appendices to maintain the smooth flow of the text: they make excellent revision notes. The text also supports NPTC (National Proficiency Test Council) City & Guilds horticultural qualifications, such as Extended Diploma, Diploma and Subsidiary Diploma in Horticulture at Level 3 and Level 2, together with BTEC National Diploma and Certificate in Horticulture.

The book has been designed so that you can dip in and out of the different parts and chapters as you wish, to experience new information or remind you of forgotten facts, methods or solutions. Enjoy your time becoming informed and guided around the fantastic role we can play in the growing and caring for the natural world, and especially the parts we may own or have responsibility for and can influence.

ONLINE RESOURCES

To continue your studies supporting material is available from www.cambridge.org/horticulture.
This includes:

- ✿ an extended glossary of common horticultural terms
- ✿ mind maps serving as study aids to show the structure and relationship of topics
- ✿ all figures from the book as JPEG files and PowerPoint slides
- ✿ plant photographs for identification purposes
- ✿ exam techniques for those taking the RHS exams
- ✿ example questions with outline answers
- ✿ video demonstration of chip budding technique.

ACKNOWLEDGEMENTS

Thank you to all our families who have borne the brunt of our trials and tribulations in the development of this work, especially to my wife Wendy; tea and cake always came at the right time.

We wish to acknowledge the assistance provided by the staff at RHS Qualifications, the staff at Sparsholt College, and particularly Alan and Susan Clarke and Martin and Sally Burr for their unstinting support over the years. A special mention to the remaining members of the original team: David Ingram and Daphne Vince-Prue, for their sheer staying power and guidance.

Our thanks go to Claire Eudall and her team at Cambridge University Press and Rae Spencer-Jones at RHS Publishing for their patience, guidance and continued support throughout the whole project.

Also my particular thanks to the team for their agreement to contribute knowledge, support and commitment to this publication.

PART 1
The foundations

INTRODUCTION

As with every complex endeavour it is always best to learn, or at least appreciate, the foundations of that subject or topic. Horticulture provides one of the most challenging but rewarding mixtures of endeavours, encompassing, but not limited to: art, chemistry, design, faith, frustration, health, history, languages, patience, physical effort, relaxation, religion, science, social development, therapy and wildlife.

This part comprises the chapters that relate to the background of growing and using plants. Answering those perennial horticultural questions: How does a plant work? What situation should it grow in? Where can I use it for maximum effect? Why was it so good last year/month/week?

Please use it to introduce, re-acquaint or remind yourself of the wonders that are found in the natural world and that you can tap into to provide a lasting and satisfying result: be it food production for the family, a wildlife haven, or a green oasis away from the busy world of today.

Starting with the range and development of the 'five kingdoms' classification and naming of plants, this leads on to the structure of plants in their many forms, and the new and developing language, such as eudicots, providing valuable technical updating for all those with an interest in plants. Also covered are the basic

environmental conditions required for successful plant establishment and growth, this is encompassed within the requirement and effect of light, water and its importance for plants, and the ever-changing and most talked about topic: climate, weather and seasonal effects. This must be one of the most challenging aspects of modern gardening. With no growing season being the same as any other, and the range of temperatures within a 24-hour period being so wide, this continues to focus the mind of everyone growing plants.

Underpinning all of this is an understanding of soils and plant nutrients, covering the usual questions: How are soils formed? What type of soil have I got? What can I use, if not soil? What are nutrients, and how can I supply them without harming the environment?

By studying all of these aspects, in balance, the maximum reward can be obtained from your situation and whatever plant palette you personally wish to develop, using whatever timescale you wish, resulting in successful production of flowers, seeds and fruits: how to aid in their formation, storage and germination.

Please also use this part as a quick reference, when required, for the clarification of lapsed memories or positive re-inforcement of mis-remembered facts and details.

Also remember that plants are very forgiving and usually possess unlimited fortitude and patience, even if we do not.

CASE STUDY A living fossil

The Maidenhair tree, *Ginkgo biloba*, is regarded as a 'living fossil' and is the only living example of *Ginkgoales*, which is a plant order. This plant was thought to be extinct in the wild for many years (surviving wild plants are now known in SE China) but the species was kept alive by Buddhist Monks in their monasteries as a sacred tree particularly in China and Japan. Increasingly it is valued as a long-lived street tree in urban situations due to its abilities to prosper in high temperatures, elevated levels of atmospheric pollution and resistance to pests. It is also regarded as an international symbol of peace because of established specimens growing next to the epicentre of the atom bomb site at Hiroshima, Japan survived the bomb and were the first living things to re-grow afterwards.

Another feature is its use as a medicinal plant to promote the flow of blood over the cranium thereby improving memory. However, without this plant being found, recognised, valued and reintroduced all these features would not be available for us to use.

The fallen leaves and fruits of a female Maidenhair tree, *Ginkgo biloba*. One of the oldest living lineages of seed plants alive today.

Chapter 1

Plant diversity

Michael Buffin and Tim Upson

INTRODUCTION: DIVERSITY AND EVOLUTION

In planting a garden, we celebrate diversity. There are the plants that we deliberately cultivate for our own benefit, but also the vast and often unseen array of microscopic organisms as well as the obvious birds and butterflies that we may actively encourage.

If you study nature, or are just intrigued by the diversity in the garden, it can perhaps at first be puzzling. Looking more closely at the range of diversity you soon realise there are patterns that when pieced together help to create a picture of life, which we can classify into a system that can be communicated. This pattern is the result of almost 4 billion years of organic evolution on Earth from a common ancestor that we all ultimately share. In sharing common ancestry, whether from millions of years ago or more recently, we have shared characteristics, such as the details of our cellular structure and chemistry or the form of a flower. These shared characteristics are the raw data that enable us to discover these patterns and ultimately build a classification. This provides a structure within which to name organisms and recognise their evolutionary relationships, a system that can be understood worldwide and without ambiguity. This is the science of systematics, which helps bring order and sense to this diversity, and an understanding of it is key knowledge in horticulture.

Diversity exists at a number of different levels, not just among the organisms that we may identify and name around us. Differences exist between individuals: the genetic diversity that gives the variation providing new garden plants. This is also the raw material for breeding or finding resistance to pests and diseases. It also exists in a wider sense beyond the garden, into the urban landscape and countryside beyond, and the underlying soil, geology and prevailing climate, which all ultimately dictate what can be successfully cultivated.

This diversity does not exist in isolation but has evolved together creating interactive and often complex relationships between organisms, which are further influenced by the physical environment around them – the ecosystem. Gardens are ecosystems in their own right, albeit ones artificially created and manipulated by their creators.

Key concepts

✿ Patterns and nature of diversity on Earth

✿ The major groups of plants

✿ Organising, naming and communicating diversity

✿ Interaction of diversity in the garden

✿ Plant collecting: how our diversity of garden plants arose

✿ Understanding diversity through different types of plant collections

The Fundamentals of Horticulture: Theory and Practice, ed. C. Bird. Published by Cambridge University Press. © The Royal Horticultural Society 2014.

THE FIVE KINGDOMS – THE DIVERSITY OF LIFE

The vast diversity of life that has evolved on Earth can be perplexing, but major groupings of organisms can be recognised. Initially split into animals and plants, we now recognise other groupings or kingdoms that reflect both the early forms of life and progress through to the more complex groups that have evolved. These five kingdoms are bacteria, representing the earliest forms of life, protists, fungi, plants and animals.

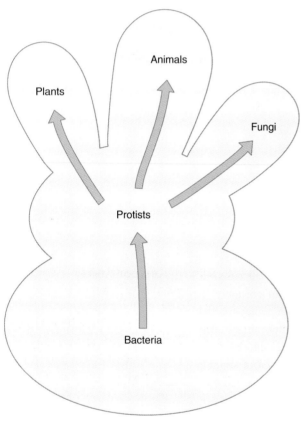

Figure 1.1 Diagrammatic representation of the five kingdoms.

Within these five kingdoms the most fundamental split is in the single-celled organisms, which lack a nucleus – the prokaryotes, meaning 'first kernals', and form the kingdom bacteria. The other four kingdoms contain organisms with more complex cells, containing a nucleus and other membrane-bound organelles – they are called the eukaryotes, meaning 'true kernals'. This division of prokaryotic and eukaryotic cells represent two distinct levels of cellular organisation and a fundamental distinction.

Kingdom bacteria

The most basic form of life, the kingdom bacteria comprises all organisms with prokaryotic cells and is further divided into the true bacteria (*Eubacteria*) and the *Archaebacteria*. The *Archaebacteria* represent the oldest living organisms on Earth today and are typically found in some of the most inhospitable environments, including oceanic volcanic vents and salt pans, reflecting their early origin when the Earth's atmosphere consisted of poisonous gases, was very hot and lacked oxygen.

The true bacteria are more complex and common, and are found everywhere around us. Many are familiar to our everyday lives as they can cause disease, are used to help ferment milk and form important relationships with both plants and animals. Nitrogen-fixing bacteria convert nitrogen in the atmosphere to ammonia, the form of nitrogen that can be used by plants.

One of the important groups, the cyanobacteria (once called the blue-green algae), were one of the first organisms to produce energy through photosynthesis: by harnessing the energy of the sun to produce sugars and releasing oxygen. Over time the gradual build up of oxygen changed the early atmosphere and led to the near extinction of oxygen-intolerant organisms including the *Archaebacteria*.

Kingdom *Protista* (the protists)

The protists are one of the most diverse kingdoms, a rather loose grouping of different lineages that have a relatively simple organisation, and if multicellular show no differentiation into distinct tissues. All protists evolved from a symbiosis between at least two different kinds of bacteria. Endosymbiosis is the term given when an organism lives within another: cellular organelles such as chloroplasts and mitochondria were originally free-living bacteria that became engulfed within other cells.

Protists include a range of organisms with which we are familiar, including many seaweeds. The various groups of algae also belong here, including the green algae (*Chlorophyta*) although some authors include them with plants. Others, such as the diatoms, are often studied by botanists and known for their beautifully sculpted hard coats and are commonly found in ponds and lakes. Perhaps one of the most puzzling organisms encountered in gardens, the cellular slime moulds belong here, and are encountered in the autumn where they live in damp soils and rotting vegetation. They live as independent feeding and dividing amoebas and only become visible when they aggregate into a slimy mass puzzled over by many.

For the horticulturist, the protists include some of the most serious disease-causing organisms. These include the genus *Plasmodiophora*, which live within plants and are the cause of club root in cabbages, *Brassica oleracea* (Capitata Group), and powdery scab of potato, *Solanum tuberosum*. Another group are the Oomyceta, known as the water moulds and once included with the fungi. These organisms extend fungus-like threads into plant tissues causing white rusts and downy mildews. *Phytophthora* causes some of the most serious plant diseases including late blight of potatoes due to *P. infestans*, which most famously caused the Irish potato blight, famine and emigration of the population. Currently sudden oak death, caused by *P. ramorum* and others, is reshaping landscapes and is a threat to major timber crops such as *Larix* (larch).

Kingdom fungi

The fungi are an essential group of organisms in the garden, often unseen until betrayed by their fruiting bodies – mushrooms and toadstools. The study of fungi, known as mycology, has often been undertaken by botanists as they were once considered to be primitive or degenerative plants that lacked chlorophyll. They are now recognised as being a distinct life form more closely related to animals – containing chitin in their cell walls as do some animals (arthropods). Over 100,000 fungi have been described and it is estimated that the total number of species may exceed 1.5 million, placing them second only to insects in their diversity.

Fungi consist of small filamentous structures known as hyphae, which collectively form a mass called a mycelium that can be extensive in spread. They are primary decomposers using enzymes to break down organic compounds from other organisms, which are absorbed for nutrition, the hyphae spreading to seek new food supplies. A few fungi familiar to us, such as the single-celled yeasts, obtain energy by fermentation and are most notably used in bakers' and brewers' yeast – to make bread rise or to produce alcoholic drinks.

Fungi reproduce through the formation of microscopic spores that can be produced sexually or asexually. Dry and very small, they are easily dispersed through air currents or in free-flowing water and rain splash. In some of the major groups of fungi sexual reproduction leads to the formation of reproductive structures, the familiar mushrooms, which consist of tightly packed mycelium and a remarkable range of forms.

Important for plants are the mycorrhizal fungi, literally fungus roots. These are of major ecological importance, forming a mutualistic relationship with the roots of vascular plants. This association provides the fungi with sugars produced by the plant; the plant in return benefits from the higher absorptive capacity of fungi for water and minerals, due to the large surface area of mycelium. This association is key for good plant growth and to increase crop yields. While only a small proportion of mycorrhizal fungi have been described, they are widespread. A huge number are yet to be named and they form an association with 95% of those vascular plant families examined. They can be important in allowing plants to colonise nutrient-poor soils.

One of the largest groups of fungi are the ascomycetes, or sac fungi, most familiar in the garden for the cup- or sac-like mushrooms and moulds that cause food spoilage, powdery mildews and several devastating diseases, including Dutch elm disease, *Ophiostoma novo-ulmi*, which devastated and reshaped the English countryside in the 1970s. Others produce the healing penicillin from *Penicilliums* – the first antibiotics that were effective against serious diseases. The most distinct and familiar group of fungi are the basidiomycetes, which include mushrooms, toadstools, puffballs and bracket fungi. These reproductive structures are spore-producing structures or basidia. Members of this group are particularly important in the decomposition of plant litter and the recycling of nutrients. In trees they form part of the natural cycle in decomposing wood, but in the garden this can be problematic if it leads to structural weakness. Plant pathogens, most notably the rusts and smuts, are included here.

Lichens – Included within the kingdom are the lichens, a mutualistic partnership between a fungal partner (the mycobiont) and a population of a photosynthetic algae or cyanobacteria (the photobiont). The fungi receive carbohydrates and nitrogen from their photosynthetic partners while providing a place to grow.

Lichens are able to live in some of the harshest places on Earth, and hence in gardens they are able to colonise rocks, roofs and the trunks of trees and shrubs, where their presence can cause concern – but they are completely harmless. One aspect of their ability to survive inhospitable places is their capacity to dry out very quickly, cease photosynthesising and enter a state of suspended animation. They also produce lichen acids that play a role in the weathering of rock and consequently soil formation. Lichens are also useful as environmental indicators. They are unable to secrete elements absorbed so are sensitive to toxic compounds, particularly sulphur dioxide found in polluted air and are used to monitor atmospheric pollution in cities.

Kingdom *Animalia* (the animals)

Animals form one of the best known and recognised kingdoms, a diverse assemblage including the most complex organisms on Earth. Most animals are multicellular, with a definite body plan and tissues that form organs and the systems that sustain them. They are all motile, able to move spontaneously and independently, even if this is only at certain stages in their life. Unable to make their own food, they generally ingest nutrients into a digestive chamber and are thus reliant on other kingdoms, especially plants, as a food source. They are also distinguished from plants, protists and fungi by their development, which is progressive and can involve different stages in their life cycle. Nearly all animals undergo some form of sexual reproduction with motile sperm and a larger, non-motile egg that fuse to form zygotes and develop into new individuals.

One of the basic and fundamental divisions within animals is whether they possess an external skeleton (invertebrates) or an internal skeleton (vertebrates). Those with an internal skeleton (vertebrates) have a backbone, which is made up of a column of vertebrae. During development, the internal skeleton forms a relatively flexible framework upon which cells can move about and be re-organised, making complex structures possible.

The most numerous and diverse animals found in gardens are the invertebrates, these are important within the soil biota, as key pollinators of flowers and often encouraged for their interest and beauty. They can also be destructive pests and a major cause of plant losses. The nematodes or eelworms are usually tiny soil-living species found almost everywhere in the world, but also include large animal parasites such as hookworms. Some are serious pests invading the root systems of plants and include potato cyst nematode, *Globodera pallida*, stem and bulb nematodes affecting daffodils, *Narcissus* spp., onions, *Allium cepa*, and beans, e.g. *Phaseolus coccineus*. Others are vectors for viruses, while predatory nematodes are increasingly being used as biological controls attacking soil-dwelling pests such as wireworm, *Agriotes* spp., and vine weevil, *Otiorhynchus sulcatus*.

The annelids (or segmented worms) include leeches, possible inhabitants of ponds, but also one of the most important garden animals, the earthworms. There are many different species ranging from those living in the leaf litter in composting bins to those that enrich the soil by mixing and allowing air and water to penetrate. In this respect they are perhaps one of the most important garden organisms ensuring a healthy soil, essential for good plant growth. They are also key prey species for birds, shrews, e.g. *Sorex araneus*, and badgers, *Meles meles*.

While the gastropods are most diverse in marine environments and include clams, mussels and other seashells, in the garden they are principally represented by the terrestrial slugs and snails. Both feed on plant material using their file-like tongue or radula, causing large amounts of damage, moving by means of a large muscular foot and leaving a tell-tale slimy trail. Slugs are one of the most serious pests in the garden, feeding on plant material including roots, tubers and bulbs in the soil and are a major reason for the failure of plants. Snails differ in bearing a protective shell and often feed on dead organic matter, but can also be problematic, particularly for young seedlings and some crop plants.

Perhaps the most important group are the arthropods, characterised by their jointed limbs and hard, protective exoskeleton or cuticle. This hard cuticle is inflexible and thus prevents growth, meaning they need to moult their skin to grow. Included in this large and diverse group are many familiar aquatic organisms such as crabs and shrimps, but they are equally diverse in the garden environment. Familiar garden arthropods include woodlice, millipedes and centipedes, which can occasionally cause damage. Mites can be found on plants where they cause mottling and distortion, while soil mites are generally beneficial to the soil biota. Spiders also belong here, not always welcomed by some, but important in helping to control pest levels with their webs providing autumnal interest.

By far the most diverse and important group of arthropods in the garden are the insects, recognised by their six legs and jointed body plan divided into three parts – head, thorax and abdomen. Many are pest species either sucking on plant sap or are chewing insects capable of attacking all plant parts including stored seeds. Others are beneficial, for example wasps, which feed on other pest species. While their presence in the garden in summer and autumn is unwelcome, as they search for sugary foods, they are the gardener's friend in spring, preying on larvae using their sting to paralyse the prey used to feed their own grubs. Others such as butterflies and the related moths, *Lepidoptera*, may be actively attracted into gardens with specific plantings of nectar-rich flowers, yet their larvae may actively feed on our garden plants. Perhaps the greatest benefit in the garden is pollination, one of the most important relationships between flowering plants and insects. Pollinating insects, such as the various species of bees and hoverflies, are essential for ensuring seed set and without which many of our cultivated fruit plants would be sterile.

Vertebrates are among the most familiar animals in gardens. Those most obviously associated with water bodies are fish. Water is also required by amphibians such as toads, frogs and newts to breed, their egg-containing spawn is an indication of spring; at other times they seek damp moist places, under logs or stones, and can be useful predators of garden pests. Of the reptiles, the most likely to be encountered is the grass snake, *Natrix natrix*, often seen swimming in water or utilising the warmth of compost heaps to incubate their eggs.

One group often deliberately encouraged are birds, for which gardens have become increasingly important habitats. They are important predators, particularly in spring, eating vast amounts of larvae to feed their young. A good population of birds can be vital in controlling numbers of pest species and reducing or even eliminating the need for chemical controls. They can, however, also be pests: wood pigeons, *Columba palmus*, being notorious for stripping leaves from plants and devastating crops.

Mammals are perhaps the largest animals found in gardens and, although rarely encouraged, are enjoyed by many despite the fact that some can cause immense damage. Common native garden mammals include hedgehogs, *Erinaceus europaeus*, which help control slugs and insects, and foxes, *Vulpes vulpes*, now frequent in urban gardens. Less welcome can be mice, eating young plants and seeds, tunnelling moles, *Talpa europaea*, which can ruin lawns with piles of earth, and badgers, *Meles meles*, which can damage borders and lawns in search of worms. Introduced mammals such as grey squirrels, *Sciurus carolinensis*, can be particularly problematic barking trees, eating shoots and even predating young nesting birds. Large herbivores, such as various species of deer, can cause devastation: in extreme cases requiring often expensive fencing to exclude them. These animals might seem to cause endless damage but this can often be at tolerable levels, and sharing our space with them can equally bring endless joy and interest.

Viruses

The organisms classified within the five kingdoms are all cellular but there is one group that does not fit this description, the viruses. They are composed of just DNA (deoxyribonucleic acid) or RNA (ribonucleic acid) enclosed in a protein coat and much smaller than any cell. Viruses replicate but can only do so by entering a cell and using its living mechanism. Outside the cell viruses cannot reproduce, feed or grow but can be incredibly tough and able to survive for long periods in extreme conditions. For the horticulturist they are, however, an important disease-producing organism causing yield loss in crops and even plant death.

THE DIVERSITY OF PLANTS IN THE GARDEN

Over the last 500 million years, plants have undertaken an evolutionary journey that has ultimately altered the planet to one dominated by green plants that are fundamental to supporting life on Earth today. Many of these major plant groups or evolutionary lineages are part of the cultivated diversity in gardens, either deliberately grown or natural colonisers. The diversity of plants described here represents many different evolutionary lineages: from the earliest plant groups to colonise land, which are still alive today, to the flowering plants that now dominate the Earth and our gardens.

Land plants evolved from green algae, at one time grouped with them but now placed in the kingdom *Protista* (the protists), although still often studied by botanists. Most green algae are aquatic, a key part of the food chain but sometimes problematic, such as the common blanket weed, e.g. *Cladophora* spp., that can turn water green in summer. Elsewhere in the garden algae are the 'green dusting' on tree trunks most evident in winter, or can be found colonising natural stone paving or rocks making them slippery. Of the several lineages of green algae, it is now widely agreed that the closest relatives to terrestrial land plants are the charophytes or stoneworts. Frequent in various water bodies, they have stems and whorls of leaf-like structures and might occasionally be found in garden ponds and lakes.

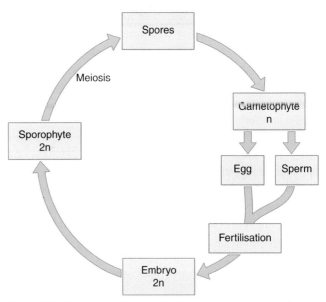

Figure 1.2 Diagram to illustrate basic principle (non-specific).

THE LIFE CYCLE IN PLANTS: ALTERNATION OF GENERATIONS

The plant life cycle is different from most other organisms in having two different generations: the gametophyte, which produces the gametes (eggs and sperms), and the sporophyte, which gives rise to a spore-producing phase. Spores are reproductive cells capable of growing into a new organism, they are resistant to dessication and small to allow dispersal on air currents. Alternation of generations refers to these two distinct phases. The critical stages are cell division through meiosis causing the change from the sporophyte to the gametophyte, which as a result is haploid (*n*), with a single set of chromosomes. Fertilisation, the fusion of the eggs and sperm to create an embryo, causes the change from the gametophyte to sporophyte, which is diploid (*2n*) with a full set of chromosomes. The sporophyte produces spores by meiosis starting a new life cycle with the gametophyte generation.

The balance and the dominance of each phase shifts between different groups of plants with a trend through evolutionary history for the gametophyte generation to get smaller and for the sporophyte to become the larger, dominant and longer living phase. This change in dominance reflects the gradual adaptation to terrestrial life. Early plant lineages require free water during the gametophyte generation for the sperm to swim to the egg and effect fertilisation. The sporophyte generation evolved a more protective cuticle to prevent drying out, vascular tissue to move water through the plants and eventually enclosed and protected the gametophyte stage, ultimately removing the reliance on water for fertilisation. This allowed the later-evolving lineages of plants with a dominant sporophyte to colonise every **niche** on Earth.

THE BRYOPHYTES: NON-VASCULAR PLANTS

The earliest lineages of terrestrial plants alive today are the bryophytes, a collective term for three distantly related lineages: liverworts, mosses and hornworts. Bryophytes lack a complex vascular system, instead absorbing water through the leaves and stems. This limits their size and also means the water content of the plant is closely related to the environment and one reason why bryophytes are most often associated with damp environments. They can also survive severe desiccation and come into growth again under wetter conditions, enabling them to survive dry periods in the year. The word bryophyte, is derived from a Greek term meaning to 'swell on hydration'. Bryophytes are reliant on water to complete their life cycle: rain splash or a film of water to allow the free-swimming sperm to reach and fertilise the egg.

In the garden the liverworts and mosses are frequently encountered, although they are rarely deliberately cultivated. Reflecting their early origins they are pioneers, able to colonise bare ground, rocks and man-made surfaces including roofs, tarmac and paving when conditions allow. The hornworts (*Antherocerotophyta*), which take their name from their elongated horn-like structure, are a small group of 200 species found worldwide, but rarely in gardens.

The liverworts (*Marchantiophyta*) are typically prostrate and flat, and they consist of a ribbon-like thallus, frequently forking and anchored by thread-like rhizoids. The liverworts are most likely to be encountered in the garden growing outside in damp places, or as a weed carpeting pot plants or capillary matting. Liverworts are limited in size by their lack of conducting tissues, their form is thought to be most reminiscent of the first plants to colonise dry land, and they are often referred to as the simplest true plants. Today there are estimated to be about 9,000 species of liverworts found worldwide.

Sexual reproduction occurs with the production of small chambers on the surface of the thallus, which are either male antheridia producing sperm or female archegonia containing the egg. Variation does occur and in the common *Marchantia polymorpha* often encountered in gardens, the antheridia and archegonia are borne on umbrella-like structures, the receptacles. Following fertilisation and formation of the embryo, the sporophyte develops from it and is a club-like structure from which spores are eventually released. Liverworts also reproduce asexually by gemmae, little groups of cells formed in small cup-like structures, the cupules also borne on the thallus. They are usually dispersed from the cupules by rain splash and in the right conditions they develop into new liverwort plants, a very successful way of spreading.

Figure 1.3 A typical moss with the capsules of the mature sporophyte emerging from the gametophyte.

The mosses (*Bryophyta*) are small clump- or mat-forming plants typically 1 to 10 cm (1/3 to 4 inches) tall with leaves arranged around a central axis. Mosses are extremely diverse with over 12,000 species found from the coldest polar regions to tropical forests. In gardens they are most commonly found in damp or shady locations but given their ability to survive drying out are frequent on roofs, gutters, rocks and paving.

The leafy moss gametophyte is the dominant phase in the life cycle, producing archegonia containing eggs and antheridia producing sperm. Following fertilisation, the resultant embryo gives rise to the sporophyte, which remains attached and reliant on the gametophyte. The mature sporophyte typically consists of a capsule, containing the spores, borne on a thin stalk. The lid of the capsule breaks to allow the spores to disperse. They can also reproduce asexually through fragmentation, where pieces break off and are then able to regrow into a new moss gametophyte.

Mosses are major components of some wetland ecosystems, where acid conditions and low nutrients limit other plant growth. These areas are the source of peat, primarily composed of decayed moss and controversially used in horticulture in growing media and for soil improvement. Mining of peat can have a detrimental effect on these unique habitats with moral pressure to find alternatives.

THE SPORE-PRODUCING VASCULAR PLANTS

The development of a vascular system to transport water and nutrients around the plant body enabled these plants to attain a much greater size, and in the geological past they formed large forests of giant plants. While many of these are now extinct, their remains form the coal deposits we now mine for energy. Those lineages that have survived through to the present day, such as club mosses, horsetails, true ferns and their allies, represent a small fraction of this previous diversity.

These groups also show a switch in the dominance of the sporophyte generation over the gametophyte: this general trend is illustrated by the life cycle in ferns. The spores, on germinating, give rise to the gametophyte, which is typically a small heart-shaped tissue, the prothallus. This contains the male antheridium and female archegonium, which following fertilisation gives rise to the sporophyte. This is initially reliant on the gametophyte but as the sporophyte matures and becomes rooted in the ground the gametophyte gradually disintegrates, leaving the dominant and long-lived sporophyte, the plant most familiar to us. On the fronds of the mature sporophyte are the sporangia, spore-producing structures that give rise to the next generation. However, the reliance on water for successful fertilisation to complete the life cycle remains a restriction to their distribution.

The club mosses (*Lycopodiopsida*) have small, scale-like leaves borne on forking stems and cone-like sporangia producing spores; they are believed to be structurally similar to the earliest vascular plants. They are a small group today of about 1,200 species and outside of plant collections are occasional houseplants such as the spike mosses (*Selaginella*) grown for their foliage and ability to survive shade.

The horsetails (*Equisetopsida*) – once diverse and reaching tree-like proportions, living horsetails now form a single family and genus *Equisetum* of 15 species. Typically they bear distinctive ribbed stems, 1 to 2 m (3 ft 3 in to 6 ft 6 in) high with obvious internodes, that bear tiny leaf-like scales. The spores are borne in small cone-like strobili on top of the stems. The common name, scouring rushes, refers to their use for cleaning utensils due to the gritty silica found in their cells. In the garden some are grown for their architectural form, but most are notorious as pernicious weeds that are extremely difficult to control.

True ferns (*Polypodiopsida*) are familiar in the garden for their attractive fronds that are usually further divided into smaller leaflets that unfurl from attractive curled crosiers. There are over 9,000 species of true ferns and their diversity of form is enormous: from small floating ferns (*Azolla*) just a centimetre in size through to tree ferns (*Dicksonia*) with distinct trunks and attractive rosettes of large fronds. We usually associate them with moist, shady places but some species can survive dry, sunny positions and are even epiphytes on trees.

In gardens they are grown for their architectural form, attractive in their own right but also a foil for other plants. The ability of many to thrive in shade makes them useful in the garden, although others are able to survive drought and full sun.

Figure 1.4 *Matteuccia struthiopteris*, ostrich fern, illustrating the unfurling fronds and curled crosiers.

THE SEED PLANTS

The seed was a great innovation during the evolution of vascular plants: providing protection for the embryonic plant contained within, and stored food to aid the critical stages of germination and establishment. Seeds are the perfect structure for dispersal through time and space. This gave the seed plants a great selective advantage over their spore-producing ancestors and gradually they became the dominant group of plants today.

In the life cycle of seed plants the sporophyte is totally dominant and fully free living; the gametophyte is highly reduced to a few cells and protected within the tissues of the sporophyte. Seed plants are heterosporous, meaning the sporophyte produces two different types of spores that give rise to male or female gametophytes. In the case of the female gametophyte they are contained and protected in a structure called the ovule and produce the egg cells. The male gametophyte is protected within a resistant structure, the pollen grain, enabling it to be dispersed over long distances. On fertilisation the ovule develops and matures into the seed.

THE GYMNOSPERMS: CYCADS, GINKGOS AND CONIFERS

The gymnosperms, meaning 'naked seeds', represent several distinct lineages that do not enclose their ovules in additional structures. Common garden plants included here are the cycads, ginkgos and conifers and also the *Gnetophytes*. Although rarely encountered in gardens they include joint firs (*Ephedra*), characterised by their mass of green jointed leafless stems, and the strange *Welwitschia*, which just produces two strap-like leaves growing continuously from a woody base.

Gymnosperms have not fully eliminated the need for water in their life cycle. The ovule produces a 'pollen drop' at the apex so providing a liquid medium for the sperm to reach the egg. In the case of cycads and ginkgos the pollen grain contains free-swimming sperm, a character shared in common with earlier plant lineages. In conifers a pollen tube germinates and grows towards the female gametophyte allowing the cells of the male gametophyte to effect pollination.

Cycads (*Cycadopsida*) reached the pinnacle of their diversity during the age of the dinosaurs, only 210 species survive today in the tropical and subtropical parts of the world. Their large ornamental frond-like leaves, usually borne on a distinct trunk, make them highly desirable as architectural garden plants, large specimens being expensive and sometimes leading to their illegal collection from the wild. Beetles are common pollinators, making cycads one of the first lineages to evolve insect pollination.

Ginkgo (*Ginkgoopsida*) – there is only one species that survives today, the maidenhair tree (*Ginkgo biloba*) – although it is well known from the fossil record due to its distinctive and attractive fan-shaped leaves. They are popular trees as garden specimens but also in urban environments as they are resistant to pollution and pests.

Conifers (*Pinopsida*) – the conifers are the largest and most diverse group of living gymnosperms, with over 600 species ranging from shrub-like junipers (*Juniperus*) to the largest organisms by volume, the giant redwood, *Sequoiadendron giganteum*, from North America. Most conifers bear the ovules on woody scales that collectively form the readily recognised cones, although some like junipers, bear fleshy cones. Worldwide in their distribution, their ability to survive cold, dry conditions means they dominate large tracts of the Earth's surface in northern latitudes. Most conifers are evergreen but a few are deciduous, including the larch, *Larix*, and the dawn redwood, *Metasequoia*, giving good autumn colour. With relatively fast growth many are important timber trees and widely planted as forestry trees.

Conifers are often planted as specimen trees or windbreaks where space allows, and include majestic trees such as silver firs,

Abies, cedars, *Cedrus*, spruces, *Picea*, and pines, *Pinus*. For smaller spaces and rock gardens slow-growing conifers are suitable providing year-round interest. Other cultivated conifers have been selected for their coloured foliage, usually glaucous blue or yellow, most notably in the Lawson cypress, *Chamaecyparis lawsoniana*. Some are fine hedging plants, particularly the yew *Taxus baccata*, although the large fast-growing Leyland cypress, × *Cuprocyparis leylandii* has become notorious for casting heavy shade when sited in unsuitable positions.

Figure 1.5 *Pinus wallichiana*, the Bhutan pine, illustrating typical needle-like foliage and woody cones.

THE ANGIOSPERMS (*MAGNOLIOPSIDA*): THE FLOWERING PLANTS

The flowering plants are the most abundant and visible group of plants on Earth today with approximately 350,000 species (estimates vary from 250,000 to 450,000). Their success is due to the combined advantages of the seed and evolutionary innovations of the flower and fruit. The flower is the reproductive structure that ultimately gives rise to the fruit and seeds. Many flowers have evolved to be attractive to animals and use them as a vector for the transfer of pollen. A typical flower consists of four whorls, although there is much variation from this basic pattern. The outermost whorl consists of the sepals that collectively form the calyx, typically green and enclosing the flower in bud. The second whorl is composed of petals that collectively form the corolla, which is often highly coloured and modified to attract pollinators and aid pollination. The next whorl is fertile and contains the male parts or androecium, which consists of the stamens, composed of the stalk-like filaments that bear the pollen-producing anthers. The female part or gynoecium includes one of the defining structures of flowering plants: the carpel. This structure is sometimes singular but typically consists of several fused together to form a hollow structure, the ovary. Protected within are the one to many ovules, which develop into seeds on fertilisation, while the carpel itself develops into the fruit wall. Rising from the ovary is the style terminating in the pollen-receiving stigma.

Pollen that lands on the receiving stigma produces a pollen tube that grows down the style to deliver sperm to the ovules and eggs within. The stigma and style are adapted so that only compatible pollen grains germinate, to encourage outcrossing (fusion of sperm and eggs from different individuals) or allow selfing (fusion of sperm and egg from the same flower).

Angiosperms have a unique double fertilisation: one sperm cell fertilising the egg to produce the embryo, another giving rise to the endosperm, a food store. On fertilisation the ovules develop into seeds and the ovary into a fruit, sometimes with additional structures. By developing a wide range of fruits, angiosperms are able to disperse seeds efficiently and widely whether by wind or animal vectors.

ANGIOSPERM DIVERSITY

The angiosperms have traditionally been divided into two major groups: monocotyledons and **dicotyledons**. Recent advances using molecular techniques have changed our understanding and we now recognise many more lineages of angiosperms, some the earliest to evolve and represented by just a few species today. We can have confidence in this new classification as it is based on objective scientific techniques and represents consensus under the umbrella of the Angiosperm Phylogeny Group (APG). For convenience and clarity the angiosperms are broadly grouped here and further major lineages are highlighted within.

The basal angiosperms

This group represents the ancestral radiation of the angiosperms and can be unequivocally traced back to about 135 million years ago. The oldest living angiosperm is *Amborella*, from the island of New Caledonia, but more familiar are the *Nymphaeaceae* (water-lilies), probably the first family to achieve a worldwide distribution.

Later groups within the basal angiosperms are the *Magnoliidae*, which includes some familiar garden plants such as Dutchman's pipe, *Aristolochiaceae*, bay laurels, *Lauraceae*, and magnolias, *Magnoliaceae*. These old lineages represent less than 2% of angiosperm diversity today. Magnolias and water-lilies illustrate some of the general features of these early lineages: many carpels that are free (not fused), numerous stamens, spiral arrangement of flower parts and large colourful petals that are free.

The monocotyledons

An early lineage to evolve, the monocotyledons remain diverse today, representing nearly a quarter of angiosperms. Over half of this diversity is found in two families, the *Orchidaceae* (orchids) and *Poaceae* (grasses). Other important groups in horticulture and for ethnobotanical uses include the *Arecaceae* (palms), *Liliaceae* (lilies), *Iridaceae* (iris), *Amaryllidaceae* (daffodils), *Asparagaceae* (asparagus) and the exotic *Zingiberaceae* (gingers) and *Musaceae* (bananas). The monocotyledons are recognised by their flower

Figure 1.6 *Magnolia × loebneri* 'Leonard Messel', illustrating free colourful tepals, numerous stamens and many carpels in the centre of the flower.

Figure 1.7 *Lilium martagon*, Turk's cap lily, a monocotyledon with flower parts in multiples of threes.

parts in threes or multiples thereof, a single cotyledon or seed leaf, parallel-veined leaves and are mainly herbaceous.

The eudicots (the true dicotyledons)

The largest and most successful group of angiosperms, they represent nearly three quarters of today's species, and are the major group of plants cultivated in gardens. The eudicots still contain the majority of the families previously included in the dicotyledons. The key feature that distinguishes them from the other angiosperm lineages are the pollen grains, which are tri-aperturate (with three openings) compared to the monoaperturate (with a single opening) in the earlier lineages. The eudicots bear two cotyledons, flower parts are borne in multiples of four or five, the leaves typically have highly branched veins, true secondary thickening is commonly present and hence trees and shrubs are frequent.

There are a number of more informal groupings recognised and these include important garden plants. The early-diverging eudicots include familiar garden plants including the *Proteaceae* (proteas), *Berberidaceae* (barberrys), *Papaveraceae* (poppies) and the *Ranunculaceae* (buttercups). The rest form the core eudicots, major groups including the Caryophyllales: the *Caryophyllaceae* (pink family), *Cactaceae* (cacti), *Polygonaceae* (docks and rhubarb) and a number of carnivorous plants such as the *Droseraceae* (sundews) and *Nepenthaceae* (pitcher plants).

The rosids form an enormous group that includes 140 families representing about one third of angiosperm species. Important families included are: *Rosaceae* (rose family); *Leguminosae* (pea family), *Urticaceae* (nettle family), *Curcubitaceae* (cucumbers), *Brassicaceae* (cabbages) and *Malvaceae* (mallow family). The asterids include major families such as the *Labiatae* (mint family), *Solanaceae* (tomato and potato family), *Asteraceae* (daisy), *Boraginaceae* (comfrey family), *Apiaceae* (carrot family) and *Dipsacaceae* (teasel family).

Figure 1.8 *Dahlia coccinea*, a member of the *Asteraceae*, one of the most diverse and advanced families within the eudicots.

PRINCIPLES OF EVOLUTION AND RELATIONSHIPS

Evolution is the theory that best describes how the diversity of life originated, from a common ancestor existing nearly 4 billion years ago, to the diversity of organisms today including ultimately ourselves. Evolution is the process of changes in the inherited characteristics within individuals and populations that usually takes place over successive generations. Evolutionary processes give rise to diversity at every level from the DNA, individuals and populations, through to species.

The theory of evolution by means of natural selection was arrived at by Charles Darwin and independently by Alfred Wallace. Evolution by natural selection is a process that can be inferred from three key processes: that more offspring are produced than can possibly survive; that characteristics vary among individuals, leading to different rates of survival and reproduction; and these different characteristics are heritable.

Those individuals that are best adapted to survive and reproduce in their environment are favoured: this is the process of natural selection or survival of the fittest. This can result in the gradual accrual of changes. Extinction is also an important and natural part of the evolutionary process. Individuals and populations that are less well adapted to their environment may become extinct, their remains captured in the fossil record, the ancestors of today's diversity.

Evolution by natural selection is not the only process that can give rise to changes. Mutations, changes in the coding genes, can result in dramatic changes. It is often these that are selected in horticulture as improvements or novelties. Examples include changes in flower colour: a result of mutations controlling the biochemical pathway synthesising flower pigments. Mutations can also be artificially induced through chemicals or radiation, which has frequently been used in breeding programmes. Hybridisation is also an important evolutionary process in plants (less so in animals), a remixing of genetic material, which may give rise to progeny that if better adapted to the environment than the parents can persist through time. Many plant species have probably arisen through hybridisation. Not all hybrids may be better adapted as many can be sterile and hence are fleeting and unlikely to persist in the wild. Under the care of horticulturists they are often brought into cultivation and survive.

DATABASES AND DATABASE MANAGEMENT

There are generally three methods of maintaining living plant collection records and they are an accession ledger or book, a card index system or a computerised plant record system. Computerised plant record systems can be as simple as a word processed document, a Microsoft Excel or Access database, or a specialised plant record database such as Plant Heritage's Demeter® database, or *BG-BASE*™, which is a widely used plant records and management system in botanic gardens, zoos and other organisations maintaining biological collections.

The Royal Horticultural Society uses *BG-BASE*™ to manage its living collection, its herbarium and it is also the system behind its horticultural database and online version of the *PlantFinder*. Generally these systems are either standalone, as in the case of Demeter®, or held on a central server so they are available across a single site. However, the National Trust's Gardens and Parks Plant Database is an intranet-based system that anyone with a National Trust account can log into and view. All of these databases are compatible with the International Transfer Format for Botanic Garden Park Records (ITF), which is a set of fields of data in an agreed format that enables data to be electronically shared across institutions. ITF Version 01.00 was established in 1987 and agreed 33 data fields to ensure that core plant record data is of a consistent standard. The fields are a fixed length and the main categories are those associated with the plant name, its origin, form and status of individual accessions. The new ITF version 2.0

did not supersede, but supports, Version 01.00 and includes a greater range of fields and more flexibility to include additional data sets. Computerised records have many advantages over paper-based records as they can easily re-sort and search accession data within the system. Botanic garden databases can be used to generate detailed reports to show the level of plants verified within a living collection; they can highlight the range of plants from particular regions of the world or generate lists of plants that require labelling.

Modern computerised plant record systems can be linked to global positioning system (GPS) mapping software, used in an interpretive and public or educational purpose, and linked to a public interface via a smartphone application ('app') using QR code technology (quick response codes) to provide greater information on a plant collection, which is instantly accessible.

GENETIC DIVERSITY

Organisms vary in their appearance and a large part of this is due to variation in the controlling genetic code between individuals. Sow seed from a plant that has been cross-pollinated and the resulting offspring may look similar but vary in their detailed characteristics – they show genetic diversity. This diversity is important for the survival of a species, enabling it to advantageously adapt to changes in the environment. In the garden, genetic diversity may show itself in differences in size, flower colour, ability to withstand cold, through to immunity to pests and diseases. These are the kinds of characteristics that form the raw material from which the horticulturists can select superior forms and breed new plants. Sometimes changes in genes give rise to unusual and unique variants: different flower colour being a prime example. These might provide novelty for the garden, although being unlikely to attract natural pollinators in the wild they rarely persist. Variation in plants can also be due to environmental factors: those growing in windy exposed positions might be dwarfed while those growing in shelter are much taller, so-called phenotypic variation. Grow these plants under the same environmental conditions and they would be a similar size.

Genetic diversity in plants can be furthered by hybridisation, the crossing between two different taxa. Unlike animals this is far more frequent in plants and can be particularly so in cultivation. In the wild, spatial barriers or different flowering times separate out species, but in gardens these barriers are often removed meaning the likelihood of hybrids occurring is more frequent. The hybrids share characteristics of both parents, a mixing of genetic diversity, and the resultant offspring often have greater vigour. Many of our garden plants have arisen through hybridisation in gardens, creating a unique resource of genetic diversity.

Sometimes plants propagate by means that do not encourage genetic diversity, through self-pollination, which reduces the genetic mixing and thus diversity. This can be common in annuals or plants from harsh environments when the opportunity for cross-pollination is reduced and hence self-pollination is an insurance to ensure the future generation. Plants can also reproduce themselves clonally. Horticulturists take advantage of this through cuttings or grafting plants to provide a way of ensuring each individual has the same genetic make-up as the parents, a way of capturing individuals with superior genetic diversity and maintaining this through time.

DIVERSITY OF HABITATS FOR PLANTS

Principles of ecology and phytosociology

Plants do not exist in isolation but interact with other organisms and the environment around them. Ecology is the science that studies these relationships. Horticulturists need to use ecological principles and knowledge if they are to be successful in cultivating plants. No plant species occurs everywhere in the world, each has its own ecological tolerances that determine its distribution. The various ways we cultivate plants or manage gardens illustrate some of the basic ecological principles.

Plants are adapted to the environment in which they naturally grow and have their own individual requirements related to their ecology in the wild. We need to understand and manipulate these requirements, matching our own garden conditions with plants that naturally grow in similar places, or adapting our own local ecology to suit the plants. For acid-loving *Rhododendron* species, whole gardens are devoted to them where the local ecology suits their cultivation. In other gardens we may change local soil conditions to make them suitably acidic, or perhaps grow them in a pot so the right soil can easily be provided. Working against the local ecology can be time consuming and expensive – most modern gardeners will try and match plants to their garden's condition and ecology, a more sustainable approach.

Different life forms are recognised in plant ecology, adaptations that will allow plants to survive in certain conditions. This can be demonstrated by 'alpine plants', as if you visit any collection you will find a range of life forms. These include bulbs, plants with underground storage organs that enable them to avoid seasonal conditions unfavourable for plant growth – from winter cold to summer drought. Collections will also include 'true alpines', those plants from high altitudes or latitudes, growing in harsh conditions with little competition from other plants. These plants are adapted to a specific niche, the functional position of that organism in its environment. In the case of cushion plants, their niche is an extreme environment where other plants lack the adaptations to survive. In this case, the horticulturist needs to provide the right niche in the garden, a pot of free-draining compost in an alpine frame. Such cushion plants are an example of plants that are stress tolerators able to survive in places where most plants would die. However, they are susceptible to competition from other plants, though able to survive in the most extreme conditions they are easily overgrown and smothered by other more competitive plants when grown under normal garden conditions. The idea of niches is most powerfully demonstrated by plants from oceanic islands. These are islands that have arisen through volcanic activity from the ocean floor. Plants and animals have to disperse from adjacent land masses to colonise them. Those that do successfully colonise find that islands offer numerous different niches, from dry lowland areas and wetter uplands to cliffs and forest floors. This offers the opportunity for plants to evolve and occupy these unfilled niches, which can lead to the evolution of numerous species on such islands. A classic example is offered by *Aeonium* species, these are well-known succulent plants with attractive rosettes of leaves and thus widely grown, but have differentiated from mainland ancestors into over 35 species in the Canary Islands.

Plants can also be associated by the communities in which they are found, this is called phytosociology. Different vegetation types can be recognised by their specific composition of plant species. The British Isles, for example, has been mapped and the various plant communities described, ranging from specialist maritime plant communities along the coast to a wide range of grasslands – the exact community is influenced by factors including soil type, altitude and hydrology. Most gardens are a mix of plants from different communities but that are nevertheless able to grow in the common conditions created. Increasingly ideas about plant communities are being used to create garden displays, most obviously by sowing wild flower meadows, sometimes based on our native plant communities but increasingly around exotic mixes representing North American prairies.

Succession is an important concept in ecology, the way an ecosystem moves from one state to another over time. In managing a garden we are often artificially influencing and controlling succession. The mowing of meadows in a garden is an important management technique, but also one that prevents succession. Without cutting (or grazing with animals) woody shrubs and trees would naturally seed in, eventually suppressing the meadow and its herbaceous plants. If we leave areas uncultivated, fast-growing woody plants like brambles, *Rubus* spp., or strong growing herbs such as stinging nettles, *Urtica dioica*, will quickly invade unless we intervene to prevent this. Ideas of succession are vividly demonstrated by plants from Mediterranean areas where fire is an important ecological process. This natural process might occur naturally every 15 to 20 years, resetting the succession each time. This importantly encourages diversity through time and space: bulbs and those plants reproducing by seed such as lavenders, *Lavandula*, and sun roses, *Cistus*, germinate following a fire and thrive over the initial years until gradually replaced by more woody species such as oaks, *Quercus*, and strawberry tree, *Arbutus*. Hence, plants such as *Lavandula* and *Cistus* naturally become unattractive over time unless we prune them or regenerate through cuttings as needed. This illustrates how dynamic ecology is – a process rather than an unchanging state.

Plant geography: major vegetation and ecosystem types

Scientists divide the Earth into vegetation regions based on factors such as the heat and cold tolerance of plants and drought resistance. In general terms the Earth's climate tends to be moist and warm around the equator and gradually gets colder and drier as we move towards the poles. Other factors such as mountain ranges, which cause increased rainfall on the windward side and dry areas on the leeward side (the rain shadow effect), give further diversity to vegetation patterns

In the hottest and wettest parts of the Earth around the equator, which receive the highest amount of solar energy, a band of tropical rainforests is found. The rainforest vegetation is multilayered with large canopy trees, a continuous mid-layer, ground storey and further niches for epiphytic plants or climbing lianas. Important for its immense diversity and in the Earth's ecological processes, plants from these regions are rarely suitable for wider cultivation in the UK except in specialist glasshouses and a few selected as houseplants.

Under drier conditions this gradually gives way to tropical and subtropical deciduous forest and eventually savannah (or tropical grassland). Distinctive Mediterranean ecosystems occur at around 40° latitude, close to the tropics and are characterised by hot dry summers and cool moist winters. The vegetation is dominated by evergreen shrubs and trees, and while occupying only a small land area is extremely diverse. These regions occur around the Mediterranean basin, California in South West USA, central Chile, the Cape region of South Africa and in the South and South Western parts of Australia. These areas have given us some of the most desirable garden plants, increasingly popular for their drought tolerance but sometimes limited to warmer parts of the country by lack of winter cold tolerance.

Desert areas are defined by low rainfall, typically with average annual precipitation of less than 25 cm (10 inches). They also have high day-time and low night-time temperatures, often below freezing. They tend to be concentrated in areas between 15° and 30° latitude and cover about one third of the Earth's land surface. They include areas from the Sahara and Kalahari in Africa, the Arabian Peninsula, central Australia and Atacama in South America. Plant life in such areas is highly specialised to survive long periods of drought. These include drought resistors, for example succulent plants able to store water in their tissues, including popular horticultural groups such as the cacti. Woody plants such as *Acacia* are drought enduring, typically with small leaves, physical defences such as spines and a deep root system able to reach underground water. Drought avoiders include annuals that lie dormant as seeds in the soil, growing and making the desert bloom after sufficient rain.

At mid-latitude regions with warm moist summers and mild winters, deciduous broad-leaved forest develops typical of eastern North America, Europe and Eastern Asia. These forests have a rich understorey of plants, flowering before the trees break leaf in spring and producing attractive autumn colours before the trees become dormant over winter. In warmer parts of Asia evergreen broad-leaved forests are found while in particularly wet regions such as Western North America the climate favours forests dominated by certain conifers. These regions have primarily given us the diversity of trees cultivated in our gardens.

At higher latitudes and altitudes, where the climate is cool and low temperatures limit the seasonal growth, coniferous forest dominates. These form large circumpolar belts in the northern hemisphere or distinctive altitudinal zones on mountains. The reduced leaves help trees survive both summer drought and winter physiological drought due to freezing, while evergreen leaves allow trees to photosynthesise as soon as growing conditions allow. Making leaves in such harsh conditions is expensive for the plant, so evergreens dominate. Trees from these regions are widely cultivated and many are important forestry trees.

Where seasonal climate such as low rainfall and cold favour perennial grasses over trees, temperate grassland occurs. Typically these are located in the central regions of continents from the prairies of North America, the steppes of Eurasia, high plateau grassveld's of Southern Africa to the pampas of South America. They have given us many of the herbaceous perennials and grasses grown in garden borders and inspiration to recreate them as meadow plantings in gardens.

Tundra occurs where tree growth becomes difficult towards the polar regions and on high mountains. Here the lack of summer warmth and short seasons limit the vegetation to small shrubs, herbs, mosses and lichens. Many of our cultivated alpine plants come from these areas, particularly the high mountains of the European Alps or the Himalaya in Asia. Other tundra vegetation occurs towards the poles as the high latitude again prevents the growth of trees. Tundra eventually gives way to permanent ice sheet towards the poles or snow cover on mountains.

Other important vegetation types are the freshwater wetlands, which can be very extensive, particularly in Scandinavia, Canada and Russia. These occur where the soil remains saturated with water for most of the year. In cool, moist climates and at higher latitudes, bogs and fens form distinguished by the origin and water chemistry that feeds them. Bogs are fed by the rain that falls directly upon them, are nutrient poor and usually acidic – characterised by mosses and a few small shrubs. Fens are fed by water from the surrounding area and tend to be more nutrient rich, dominated by sedges and grasses. Those communities where trees have developed over waterlogged ground are termed swamps and include the famous Everglades in the Southern USA.

Freshwater ecosystems encompass lakes and river channels through which precipitation is returned to the ocean as part of the hydrological cycle. Here specialist plants adapted for aquatic life are found including horticulturally important groups such as water-lilies (*Nymphaea* spp.).

Coastal zones and ecosystems such as dune systems, salt marshes and, in tropical areas, mangroves, are important ecological buffer zones providing natural protection from storms and inundation from the sea. Those plants found in dunes and on shingle beaches can make excellent garden subjects for dry sunny areas due to their ability to survive drought. The effect of exposure is also powerfully illustrated in exposed coastal areas, often by trees reduced in size with their branches bent horizontal in the direction of the prevailing wind. Tall-growing shrubs such as common rosemary, *Rosmarinus officinalis*, or common juniper, *Juniperus communis*, are often found as prostrate shrubs, hugging the ground away from the wind. Careful selection of plants that can survive salt exposure is needed in coastal areas. Many gardens in such areas need to be planted with high hedges or windbreaks of trees that are able to withstand wind and survive salt exposure. This creates a sheltered area in what are often very mild areas suited for the cultivation of a wide range of plants that cannot easily be cultivated elsewhere, and exemplified by the rich diversity of Cornish gardens.

CLASSIFYING DIVERSITY: THE NAMING OF PLANTS

Systematics describes the diversity of organisms around us, organises this into a classification system and, perhaps most importantly for horticulture, provides a framework and system for accurately naming plants, nomenclature. Taxonomy is often used interchangeably to describe this area of study, although strictly refers to the naming and description of organisms. Horticulture is about cultivating diversity so the science of systematics is a fundamental discipline to understand.

The naming and classifying of objects is a natural activity for humans. Shopping in a market we might group cabbages, *Brassica oleracea* (Capitata Group), carrots, *Daucus carota*, and aubergines, *Solanum melongena*, together as vegetables for practical purposes. We also use common names to describe them, understood in that language but one that is not internationally recognised and open to ambiguity. Here scientific names provide a universal system for naming plants accurately and without ambiguity. This need is highlighted by the example of the common name 'bluebell', which usually refers to *Hyacinthoides non-scripta* from Western Europe but is also the common name for at least 15 unrelated plants, *Campanula rotundifolia* in Scotland or the bluebell creeper of Western Australia, *Sollya heterophylla*.

The taxonomic hierarchy
A key concept is the taxonomic hierarchy in which organisms are grouped into larger and larger related groups. Ranks are given to the various levels in this hierarchy. Individuals that are related form populations, populations can be grouped into species, related species into genera and related genera into families and so forth into larger groups and ultimately the highest level of kingdom. The ranks of genera and species are the everyday language of the horticulturist, while families are useful to help understand the wider diversity in our gardens.

This hierarchy also extends below the level of species and recognises further variation. This includes subspecies (abbreviated to subsp. or ssp.), which recognises populations within a species that vary and have a distinct geographic distribution. Variety (var.) refers to populations that vary within the range of the species, while *forma* (f.) refers to individuals that vary and is

mostly commonly used for different flower colours, such as individuals with white flowers. Such plants are often of interest to horticulturists selecting novel variants. A useful term that can refer to a plant at any rank is taxon or taxa (singular).

Family	*Lamiaceae*
Genus	*Lavandula*
Species	*angustifolia*
Subspecies	*angustifolia* (typical variant from the Alps, rather than subsp. *pyrenaica* from the Pyrenees)
Cultivar	'Hidcote' (a variant with dark blue calyces and flower)

Plants can be grouped together because they share characters in common. This highlights another key concept, that of predictivity in classifications. This means related plants will have characters in common. For example, the genus *Digitalis* (foxgloves) bear attractive tubular flowers that are readily recognised and have a common biochemistry. This means you can often assign an unknown plant to a genus on the basis of these shared characters.

The hierarchy also extends further to include cultivated plants. Many will have originated as chance seedlings and selections and if these are superior in their characteristics may be named as cultivars, derived from **culti**vated **vari**ety. Naming cultivated plants can be complex both due to their mixed origins and the need to market plants or give breeders protection. A trade designation or selling name can sometimes be given. The correct cultivar name for the popular yellow-leaved *Choisya ternata* is 'Lich' but the plant is marketed in the UK as *C. ternata* 'Sundance'. A new plant may be legally protected by plant breeders rights (PBR) so that the owner or their representatives have sole rights to propagate and receive royalties on sales. Horticultural taxonomy deals with the accurate naming of cultivated plants and the particular issues associated with this. A useful resource for finding more information on cultivated plants can be found through Hortax, the Horticultural Taxonomy Group (see www.hortax.org.uk).

Naming wild and cultivated plants
The naming of plants is governed by codes, which include the *International Code of Nomenclature for Algae, Fungi, and Plants*, most recently revised in 2011 (the Melbourne Code; McNeil *et al*, 2012), and the *International Code of Nomenclature for Cultivated Plants* (Brickell *et al*, 2009), which deals with some of the specific issues in horticultural taxonomy. These codes set out the rules that ensure names are published in a standard form, in an appropriate place such as a book or scientific paper and with herbarium specimens that act as a reference point, known as the type. The codes evolve over time by international agreement, most recent changes for example allowing the electronic publication of names and for descriptions to be in English rather than Latin. Essentially anyone can name a plant as long as it complies with the appropriate code and can be

demonstrated to be different from others that have been previously named.

In some cases different authors have published names for the same taxa, so we have conflicting names. Systematists aim to ensure that each taxon has a unique name to avoid confusion. In the case of competing names the codes of nomenclature have the principle of priority. This says the earliest validly published name has priority and should be the accepted name. The other names are superseded and known as synonyms. In monographic treatments of plants, these synonyms are often listed below the accepted name and in some cases can be quite extensive for widespread species studied by different botanists.

These codes also govern the conventions for writing names. The genus, species and any lower ranks of wild species are written in italics, the genus starting with an upper case letter and the species with a lower case letter. Increasingly family names are also being italicised in texts. A person's name follows the scientific name and this is the author or authority, so the person or people who first described and published the name. It is often abbreviated and depending on the type of publication is used once when the scientific name is first used but not repeated thereafter. Cultivars' names are usually enclosed in quotes and not written in italics and usually do not bear an authority.

PHYLOGENETIC STUDIES AND DATA SOURCES

Classifications aim to reflect the evolutionary relationships, so related plants are placed together. This has traditionally been based on the morphology of plants and their shared features. Often this is a good guide but sometimes it can be misleading. Similar features can evolve in unrelated plants groups because they have adapted to a similar need. For example, many different plant families bear red tubular flowers – a common adaptation to hummingbird pollination. Carnivorous plants have also evolved in several unrelated plant groups, again a common adaptation to nutrient-poor environments. Systematists also use other data sources such as pollen, chemistry and chromosome numbers to help discover relationships. In the last 20 years molecular techniques have been used to investigate relationships by comparing the DNA sequences. This has proved a very powerful data source, ultimately coding for the plant features and less likely to be influenced by environmental factors. These molecular techniques have given new insights, confirming long-standing classifications, provided new evidence to support conflicting ideas or suggest completely new relationships.

These advances have also been made possible due to the use of a new analytical technique – cladistics. This method explicitly sets out to find shared characters in common, and build hypotheses of relationships, which are expressed as a branched diagram or tree. These are hypotheses of relationships that can then be tested with statistical measures giving a measure of confidence in the classification. Systematists also look for congruence between different data sources and when they suggest the same relationships we can have further confidence. Perhaps the most powerful means of analysing is to combine different data sources, from gene sequences to morphology, using all the available data to build a classification. In essence this has brought more scientific rigour to building classifications and replaced the traditional 'expert' approach – classifications based on a person's interpretation of the data.

A common criticism of systematists is that they change names and classifications too often. This can be inconvenient to users and destabilising, especially when this is due to obscure rules of nomenclature. Today name changes are avoided if possible, the code of nomenclature now allows for the conservation of names. Changes to names and classification still occurs when science gives us new insights, and has recently given some of the most exciting changes to our understanding of plant families. One of the most fundamental new discoveries has been new relationships within flowering plant families. The new classification, developed and published by the Angiosperm Phylogeny Group (APG), has confirmed many previous ideas of relationships but has also shown new relationships. Some familiar garden plants such as *Acer* and *Aesculus* (horse chestnuts), traditionally placed in the *Aceraceae* and *Hippocastanaceae*, respectively, have now been found to be embedded in a large, chiefly tropical family the *Sapindaceae*, a treatment that has now been widely adopted. Other plant families have changed their relationships. The *Urticaceae* (stinging nettles) is now recognised to be most closely related to the *Rosaceae* (rose family) having previously been placed close to the *Moraceae* (mulberry family). In evolutionary terms it is essentially a relative of roses that has evolved wind pollination and hence small flowers with much reduced petals.

Other families have totally changed their circumscription – a prime example being the *Scrophulariaceae* (foxglove family), which includes numerous garden plants.

CASE STUDY *Scrophulariaceae* (foxglove family) and *Proteaceae* (protea family)

Here molecular data and cladistics analysis has given a new view of this medium-sized and once-diverse family that has now been split into seven families. The *Scrophulariaceae* itself is now smaller in size and includes *Scrophularia* (figworts), *Verbascum* (mulleins) and *Buddleja*, which itself was once placed in its own family. Familiar garden plants such as *Digitalis* (foxgloves), *Antirrhinums* (snapdragons) and *Veronica* have been transferred to the *Plantaginaceae* (plantains). We now view this group as having developed a range of pollination syndromes, which is reflected in the diversity of different flower types.

One of the most surprising new relationships revealed is between the families *Proteaceae* (proteas), the aquatic *Nelumbonaceae* (*Nelumbo*) and *Platanaceae* (plane trees). These families differ greatly in floral morphology and habit but are each other's closest relatives. They evolved from an ancient ancestor and have diversified in their morphology in different parts of the world. The morphology may not suggest this relationship but the clues remain locked in the plants' DNA.

As this new classification is based on scientific rigour and strong data supported by statistics and consensus among the systematic community through the APG we can have confidence this new classification is unlikely to change in the future. Ultimately, with the aim that classifications should reflect evolutionary relationships, we will reach stability.

COLLECTIONS AND THEIR IMPORTANCE

Figure 1.9 Diversity of a living collection held within a botanic garden.

Living collections
The relatively mild and varied climate of the UK means plants from all around the world can be cultivated here – we hold a truly unique living collection of plant diversity. These collections are living libraries of plant diversity, a resource for our futures.

A living collection of a botanic garden, arboretum, public or private garden may have one or more purposes, which may be related to public enjoyment, landscape setting, education, or of systematic or thematic collections. Systematic collections are groupings of plants based on a taxon to demonstrate evolutionary development or taxonomic variation and these may be laid out in the form of order beds. They are a reference source that can be used to compare the range of taxonomic and genetic diversity in a group for comparative research. Thematic collections are often habitat based and include plants from certain regions of the world, plants of economic importance, or herbal or medicinal plants based on ethnobotanical themes. Diverse and labelled collections are a great resource for formal and informal learning, this being a key function of many gardens today.

A variety of organisations hold collections used for these purposes. Some of the best known are the national botanic gardens: the Royal Botanic Gardens, Kew and Edinburgh; and the more

recently founded National Botanic Garden of Wales. Others are the many university botanic gardens such as those at Cambridge, Oxford, Ness, Durham and Dundee. Equally diverse and important collections are held by other organisations ranging from the National Trust and NT for Scotland, to the Forestry Commission and the Royal Horticultural Society. These collections are networked through the charity PlantNetwork, which promotes botanical collections in Britain and Ireland as a national resource for research, conservation and education. A directory of collections is maintained on their website: www.plantnetwork.org.

Collections are often about growing plants for reference but another key purpose is for conservation. Plant Heritage (formerly the National Council for the Conservation of Plants and Gardens, NCCPG) is dedicated to this: its mission is to encourage the conservation of cultivated plants in the British Isles, primarily through the National Plant Collections® scheme. Currently there are nearly 650 collections (Plant Heritage, 2012) some held by organisations but many by growers and private individuals, both amateur and professional. Collections are listed in a directory published each year or available via a database on their website, www.nccpg.com. This is not just a living resource but one researched by their holders interested in the origins, and the historical and cultural importance of their plants. The organisation has also pioneered the Threatened Plant Project to identify all cultivars that are threatened and to ultimately produce a red-list of cultivated plants. This will also be a major contribution to fulfilling the UK's obligation under the Aichi Biodiversity Target 13 for conserving plant diversity, which highlights the cultural and socioeconomic value of cultivated plants. Many National Collection holders maintain detailed computerised, herbarium and photographical records of their collections and undertake and publish their own taxonomic research. Many National Plant Collections are unique as they maintain a wide range of genetic material from both wild-collected populations and cultivated garden varieties, and these can be invaluable when undertaking taxonomic research, as the collection holder will often collect a very broad range of plants from within their chosen taxon.

We usually think of living collections as growing plants. However, they can also take the form of seed banks, using nature's own means of ensuring plant survival through time. Under normal ambient conditions seeds will survive for a few years depending on the species. However, dry the seeds and freeze them to $-20°C$ $(-0.4°F)$ and they can survive for centuries, and being relatively small many thousands can be stored in a small space. This is the technology behind major projects to conserve genetic diversity for all our futures, one of the leading being the Millennium Seed Bank run by the Royal Botanic Gardens, Kew – the largest ex situ plant conservation project in the world. A similar project is the Svalbard Global Seed Vault situated in the Norwegian Arctic, the mission of which is to provide a safety net against accidental loss in traditional seed banks; it is sometimes referred to as the Doomsday Seed Vault. More simply, the life span of seeds can be extended by drying seed over rice, *Oryza*, in an airproof plastic container and storing in a fridge.

Each botanical institution will have its own accession policy that determines the make-up and diversity of its living collection. An accession policy can have a multidisciplined focus that is linked to a strategic document or policy that will guide future introductions or removals (de-accessions) to and from the living collection. These strategic aims may be based around how the living collection is used: for science and research, public enjoyment or education. This may then break down into individual projects focused on a distinct region of the world and its flora: a group of plants, a taxon that researchers are working on, or a group of plants of horticultural merit.

CASE STUDY Worldwide links

Herbariums

Herbariums are libraries or repositories of plant diversity; the specimens are preserved in a flat dry state and mounted onto sheets for easy handling and to record information about their origin. Herbarium specimens ideally display all of the various parts that make up the plant, such as the leaves, flowers, stems and roots. They are created by compressing the plant between sheets of paper in a plant press and drying the specimen as quickly as possible. Specimens that are too fleshy to press easily can be artificially dried or may be preserved in spirits.

Specimens are usually stored in cabinets and arranged according to a classification. Herbaria in this sense are self-classifying as new specimens are simply placed in the appropriate place in the classification. These are resources to understand plant diversity – key for identifying unknown plants and to write monographic treatments and floras. Each specimen should have details of its collection location, habitat, and growing conditions and include a description of the flower colour, size and habit – characters not preserved by the collecting and drying process. Large herbaria, such as the Royal Botanic Gardens, Kew, hold over 7 million specimens providing a historic record of the flora growing in a particular location at a certain time in history, and plants that have long since become extinct in their native habitats or lost from cultivation in our gardens. Among the most important collections are the 350,000 type specimens that are maintained (Royal Botanic Gardens Kew, 2012). Type specimens are the unique reference point to which a name and its application is formally attached. This anchors the name and the defining features of the particular taxon.

Although many herbaria concentrate on wild plants, they are equally important for studying and preserving cultivated plants, often an important historical record as to when plants were introduced to cultivation, bred or selected. They can be particularly important for cultivars, acting as a definitive reference to interpret the name of a cultivar – especially if its true identity has become confused in cultivation. The Royal Horticultural Society has developed nomenclatural standard specimens, which include photographs, colour records, habit

and size – a complete record for future reference. They hold an important herbarium of these nomenclatural standards at Wisley Gardens, in Surrey, where nearly 5,000 nomenclatural standards are maintained (Royal Horticultural Society, 2012) along with other cultivated plants.

DOCUMENTING PLANT COLLECTIONS

Plant collections are of most value, particularly for research, if they are underpinned by information. It is useful to know exactly which plants are in a collection but information on their origin extends its usefulness, particularly if the plants are of wild origin. Information on the locality of the collection, the local environment, as well as details of the actual collectors provides essential background information for the researcher and the horticulturist. Details of the altitude are a potential indication of hardiness. Such information has historically been recorded in handwritten ledgers or on card-index systems, but today searchable databases provide a powerful and more widely accessible resource.

The correct and accurate labelling of each plant is not just useful and the first point of interpretation but provides a link to the data held. Ideally each plant in a collection should have a unique number, the accession number, a link from the plant in the ground to the information in the database. In botanic gardens it is common practice to give an eight-digit number, the first four digits referring to the year, followed by a sequential number given to each accession as they come into the garden through the year. There are several databases developed to manage plant collections including BG-BASE (www.bg-base.com), Iris database (www.irisbg.com) and BRAHMS (www.herbaria.plants.ox.ac.uk/bol). The National Trust's Gardens and Parks Plant Database operates an intranet-based system that anyone with a National Trust account can log into and view. Modern computerised plant record systems can be linked to global positioning system (GPS) mapping software, used as an interpretive and public or educational purpose, and linked to a public interface via a smartphone application ('app') using QR code technology (quick response codes) to provide greater information on a plant collection, which is instantly accessible. However, for smaller collections basic information can also be held using proprietary spreadsheet software – certainly suitable for recording information for a few hundred taxa.

These databases are compatible with the International Transfer Format (ITF) for Botanic Garden Records, which is a set of fields of data in an agreed format that enables data to be electronically shared across institutions. ITF Version 01.00 was established in 1987 and agreed 33 data fields to ensure that core plant record data is of a consistent standard. The fields are a fixed length and the main categories are those associated with the plant name, its origin, form and status of individual accessions. The new ITF Version 2.0 did not supersede, but supports, Version 01.00 and includes a greater range of fields and more flexibility to include additional data sets. Computerised records have many advantages over paper-based records as they can easily re-sort and search accession data within the system. Botanic garden databases can be used to generate detailed reports to show the level of plants verified within a living collection; they can highlight the range of plants from particular regions of the world or generate lists of plants that require labelling.

Accessing such information has been made easier by the internet and there are several valuable sources of information that can be interrogated online. Principal online databases include the garden and plant search facilities available via the website of Botanic Garden Conservation International (BGCI), which gives access to collections worldwide that have uploaded their data (www.bgci. org) and the multisite search site for gardens using BG-BASE (www.bg-base.com). The Electronic Plant Information Centre (ePIC) available through the website of the Royal Botanic Gardens, Kew not only gives access to the living plant collections (www.epic.kew.org) but also other databases including The Plant List – a working list of all known plant species. This is a response to target 1 of the Global Strategy for Plant Conservation (www.cbd.int/gspc) and for the first time lists all known vascular plants and bryophytes (www.theplantlist.org). For cultivated plant names, the RHS plays a leading role in maintaining lists of the correct names for cultivated plants. Updated names are published annually in the RHS Plant Finder, which can also be accessed online (www.rhs.org.uk/rhsplantfinder).

CHRONOLOGY OF PLANT DIVERSITY

Our gardens both public and private are made up of a vast array of plants from all around the world. A very small percentage of these are native to the United Kingdom and Ireland, and many have been introduced from Europe, Japan, China, North America, Australasia and South Africa by a diverse range of plant collectors who have risked their lives in search of new garden plants.

According to the Joint Nature Conservation Council's report published in collaboration with the Royal Botanic Gardens, Kew, and PlantLife (PlantLife, 2012) and reproduced on the Department for, Environment, Food and Rural Affairs (DEFRA) website (http://jncc.defra.gov.uk/page-1739) the native vascular flora for the United Kingdom can be divided into four distinct categories. The first category is the truly native species that arrived in the UK without the intervention of man. The second group consists of plants that became naturalised in the UK prior to AD 1500 and are termed archaeophytes, the penultimate group are made up of species naturalised since AD 1500 and these are termed neophytes. The final group are called casuals and are made up of non-native and non-naturalised plants. Using these definitions to establish the native flowering plant flora of the UK produces a figure in the region of 2,951 species.

Although we can establish how many native plants we have, it is far more difficult to determine how many non-native or exotic plants are currently growing in private and public gardens across the UK,

as a complete inventory of cultivated garden varieties has never been attempted. However, each year the Royal Horticultural Society publishes the *Plant Finder* and in 2012 it recorded the names of 70,000 plants available for sale from over 800 nurseries in the UK (Cubey & Merrick, 2012). According to the National Trust (Calnan, 2009) there are over 300,000 species of cultivated plants grown in UK gardens and these plants represent a living library of over 500 years of gardening history and plant introductions.

The first officially recorded plant introduction dates from 1495 BC (Musgrave *et al.*, 2000) when Queen Hatshepsut (1479–1457 BC) of Egypt sent explorers to Somalia to collect incense trees, *Commiphora myrrha*. Many of the early introductions made by the Roman Empire in Britain seem well established now and form part of our cultural history, which makes it hard to believe that plants such as the carrot, *Daucus carota*, and apples, *Malus pumila*, and pears, *Pyrus communis*, were introduced as a stable crop during the rule of the Roman Empire. According to Campbell-Culver (2001) plant introductions into the UK can be separated out into the following main periods of activity.

The period between 1000 and 1560 was dominated by plant introductions returning from continental Europe due to the Norman invasion, the crusades and monastic influences. Many herbs made their way to Britain during this time along with plants such as the wild carnation, *Dianthus caryophyllus*, the wallflower, *Erysimum cheiri*, and the Gallic rose, *Rosa gallica* var. *officinalis*. The Romans and Greeks have long cultivated roses as they were grown widely in Europe. Many fruit trees such as the black mulberry, *Morus nigra*, quince, *Cydonia oblonga*, and peach, *Prunus persica*, as well as other more exotic fruits such as the pomegranate, *Punicea granatum*, were also cultivated during this time.

Also during this time period plant introductions were heavily influenced by Europe, to which Britain was still a part, and the quest for herbal and medicinal plants remained at the forefront of many of the new introductions.

From 1560 to 1620 plants began to arrive from the Near East, West Asia and the Balkans. John Gerard compiled a list of plants that grew in his garden in Exeter and this list would eventually form part of the basis of his great Herbal. The quest for new plants continued and among these were plants such as the fringed and French lavenders, *Lavandula dentata* and *L. stoechas*, the white mulberry, *Morus alba*, the umbrella pine, *Pinus pinea*, the crown imperial, *Fritillaria imperialis*, the bay and cherry laurels, *Laurus nobilis* and *Prunus laurocerasus*.

The period from 1620 to 1662 was dominated by the Tradescant explorations to Russia and the North Africa Coast (John Tradescant the Elder) and his son's three expeditions to North America. In 1621, Oxford Physic Garden, the first garden of its kind was founded by the First Earl of Danby (Henry Danvers, 1573–1643) and this greatly encouraged the study of botany. In 1634 the Catalogue of Plants was published (*Plantarum in Horto*, John Tradescant the Elder); the second edition was produced in 1656 by John Tradescant the Younger. This period saw the continued introduction of plants from

Europe, including the horse chestnut, *Aesculus hippocastanum*, and the cedar of Lebanon, *Cedrus libani*. Plant exploration continued in America and Europe, and plant breeding and gardening was continuing at pace. The white sweet pea, *Lathyrus odoratus*, arrived from Sicily and was soon being hybridised and new colour forms were quickly available. Many new plants flooded in from North and South America and also from Asia with the arrival of the lemon, lime and citron, *Citrus limon*, *C. aurantifolia* and *C. medica* respectively.

The period from 1700 to 1820 saw the exploration of Australia and the southern hemisphere by Captain Cook (1728–1779) and Joseph Banks (1743–1820) while George Vancouver and Archibald Menzies explored the Pacific coast of North America. In 1731 the *Gardener's Dictionary* was published by Philip Millar (1691–1771). Millar was head gardener at the Chelsea Physic Garden and the book covered the vast array of new North American plants that found their way into Britain via the plant exploration undertaken by John Bartram in Virginia. During this time Britain gained access to vast new territories after the conclusion of the Seven Year War (1754–1763), and this brought great botanical wealth to Britain. In 1735 *System Naturae*, the binomial classification published by Carl Linné (Linnaeus) received a mixed review and formed the basis of the system we use today. During this time the maidenhair tree, *Ginkgo biloba*, arrived in Britain in 1750 after its original Dutch introduction in 1730. A plethora of plants arrived from North America via Bartram's collection, funded mainly by a group of English businessmen that included Peter Collinson (1694–1768). Probably the most significant event during this time period was the discovery of Australia and the new plants brought back by Cook and Banks following the epic voyage that concluded in 1771. There were so many notable and new introductions of plants from all around the world as new territories were explored: plants such as the monkey puzzle, *Araucaria araucana*, various *Fuchsias* from Chile, and *Leptospermums* and *Banksias* from Australia.

From 1800 to 1890 there were plant introductions from California and the west coast of North America, and the first seed arrived from expeditions to India and the Himalayas. In 1804 the Horticultural Society was founded and later become the Royal Horticultural Society. In 1856 China opened its doors to European trade, and for the first time far greater access was permitted to Europeans into the Chinese interior. The creation of the Horticultural Society placed a far greater emphasis on the introduction of 'good' garden plants. Also many of the great nurseries such as Veitch were funding plant-hunting expeditions. The East India Company undertook much of the exploration and trade, and funded various expeditions. In 1772, Francis Mason was sent by the Royal Botanic Gardens, Kew to the Cape Province of South Africa, from where almost a thousand new introductions were made. In 1810 there was Joseph Hooker's expedition to the Himalaya and the first introduction of the blood red flowering rhododendron, *Rhododendron arboretum*. While in 1844, Robert Fortune set off for the first of five expeditions to China on behalf of the Horticultural Society and the East India Company, and

subsequently introduced many new and highly ornamental plants. Fortune also produced copious notes on the cultivation of tea, *Camellia sinensis*, in China, and was later involved in moving thousands of tea plants from China to the Himalaya. In 1824, David Douglas set off to explore the West coast of North America where he introduced the grand fir and the noble fir, *Abies grandis* and *A. procera*. The Royal Botanic Gardens, Kew remained at the centre of many of the new introductions and also established a series of well-executed crop introductions into the colonies; the breadfruit, *Artocarpus altilis*, was introduced into the West Indies from the Pacific Islands as it was a staple diet in South-east Asia and needed to support the slave industry for sugar production.

The list of collectors working tirelessly during this period for Veitch's Nursery, the Royal Botanic Gardens, Kew or the Royal Society is staggering and the list of plant introductions is also overwhelming and led to a massive influx of new plants from the Orient.

The years from 1890 to 1930 were dominated by plants arriving from China and Japan. During this period many stories of the rich botanical treasures in China and Japan had made their way back from the East. During this period many of the most celebrated plant hunters brought back a rich array of new introductions, and in such quantities that these could be circulated to the British public via now well-established nurseries. Probably the most influential and celebrated plant hunter of this period was Ernest Wilson (1876–1930) who on behalf of Veitch's Nursery of Chelsea, London and later the Arnold Arboretum in the USA introduced over a thousand new species of plants from China and Japan. His many introductions included such botanical treasures as the pocket-handkerchief tree, *Davidia involucrata*, the regal lily, *Lilium regale*, and the chinese paperbark maple, *Acer griseum*. Meanwhile, George Forrest (1873–1932) collected for the Royal Botanic Garden, Edinburgh, private individuals and the Rhododendron Society in Yunnan and Tibet. Forrest pretty much lived in the Himalaya from 1904 until his death in 1932. He collected a vast range of plants from rhododendrons, magnolias, camellias and also alpine plants never before seen in the West. Frank Kingdon-Ward (1885–1958) led 20 expeditions spanning over 45 years to Burma, Tibet and Assam and was responsible for introducing the blue poppy, *Meconopsis betonicifolia*. He also introduced the yellow flowering rhododendron, *Rhododendron wardii*, and this led to a yellow colour break in rhododendron breeding. This period saw the greatest influx of new plants into Britain.

From 1930 to the modern day there has been a continued interest in plant introductions from around the world to support both plant conservation and the continuing search for new plants. The World Wars severely limited the opportunities for plant hunting, and the resetting of national boundaries and the rise of Communism restricted the access to a number of countries.

Modern-day travel has made exploring great distances far easier, as aeroplane travel, improved road infrastructure, four-wheel drive vehicles, GPS equipment, and satellite and mobile phones have improved exploration and communication making it far easier to access more difficult terrain. However, an increasing awareness of the importance of native flora and the medicinal value of plants and the alarming rate of extinction of both plants and animals saw the introduction of the Convention on Biological Diversity (www.cbd.int/convention). This was the first international treaty to protect biodiversity; to ensure the sustainable use of its components; and introduce the fair and equitable sharing of benefits arising from genetic resources. As a response to this, Botanic Gardens Conservation International in 1999 spearheaded the Global Strategy for Plant Conservation (www.bgci.org/ourwork) to encourage signatories to the CBD to establish clear targets to slow the rate of extinction of wild plants. Although these aim to conserve wild plants, they also place conditions on access and use of wild plant populations, and in some cases have had a negative effect on plant introductions. To overcome this many modern plant hunters are now working in collaboration with overseas institutions and are sharing their skills and knowledge base. Conservation work on Mexican oaks has encouraged collaborative work between Allen Coombes, the former botanist at the Sir Harold Hillier Gardens and the University Botanic Garden Pueblo, Mexico to introduce a new range of evergreen oaks into the UK, with many of these being endangered in their native habitat. Chris Chadwell has undertaken over 20 trips to the Himalaya in search of new plants and re-introductions of existing plants from Tibet and Bhutan. Roy Lancaster has travelled the world in search of new plants, and this has taken him to China, South America, Nepal and Chile. One of the most prolific of modern partnerships must be the husband and wife team from Crûg Farm Plants Nursery in North Wales. Bleddyn and Sue Wynn-Jones started the nursery more than 20 years ago and have been plant hunting at least once a year since the early 1990s, visiting the Far East, Middle East, South America and, in particular, Asia. Many of their introductions are new to cultivation in the UK and find their way into gardens as they are offered for sale via their nursery.

This continuing influx of new material into the UK has continued the rich tradition of a nation of gardeners seeking new plants from around the world and has added to the rich biodiversity of our gardens.

Summary

This chapter illustrates that diversity in gardens is more than the plants we deliberately grow and includes many other organisms that occur naturally or we may encourage. This diversity is found at various levels from the kingdoms representing the main life forms, the living plant lineages that reflect their evolutionary history, to the species that we choose to grow and the genetic diversity within them that provides new material for breeding. The systems and tools used for structuring the different levels of classification, and subsequent naming, are also highlighted. Together with an introduction to the language used within this changing area of continued emerging knowledge; enlightenment assisted by the use of DNA.

Plants have been moved around the world for centuries and we continue to bring them together in new ways in our gardens. Never before has the choice of plant palette been wider. Also, our understanding of their relationships, both between themselves and the wider ecosystem, is constantly being expanded and redefined. This should be considered together with the role that we can play: will it be a passive or an active one? Only by harnessing and encouraging a healthy garden ecosystem and the many interactions between organisms can we create interesting and sustainable environments around us and successfully cultivate plants whether for our pleasure or food.

Understanding that these plants and other organisms interact together and have their own niches is essential knowledge for the horticulturist, to ensure we grow plants in the right place and that the garden ecosystems we create are healthy and sustainable. Get this aspect of gardening right and then the establishment, maintenance and successful growth of plants follows.

While we might consider our own evolutionary journey remarkable, perhaps the most important has been those of plants. From early aquatic origins they colonised terrestrial ecosystems altering the very nature of our planet. We are a green planet dominated by plants, on which almost all other organisms rely. Through photosynthesis plants power the planet and make it habitable and at least for us provide everyday essentials from food and shelter to medicines. It is this diversity of plants and their importance to us that brings people into horticulture and can maintain a life-long interest and passion.

Revision questions

1. What are the **major** evolutionary lineages of non-vascular, vascular and seed plants?
2. Describe how the dominance of the gametophyte and sporophyte phases in the alternation of generations has **changed** through evolutionary time.
3. List the **key** features of the major groups of flowering plants.
4. Describe the **key** phases for the introduction of plant diversity into our gardens.
5. For a **favourite** plant, construct its taxonomic hierarchy.
6. List the Earth's **major** vegetation types and sketch the main areas of their occurrence.

REFERENCES

Brickell, C. D. *et al.* (2009) *International Code of Nomenclature for Cultivated Plants, Proceedings of the meeting of the I.U.B.S Commission for the Nomenclature of Cultivated Plants, Eighth Edition.* Acta Horticulurae.

Calnan, M. (2009). *Space to Grow: Why People Need Plants.* Park Lane Press.

Campbell-Culver, M. (2001). *The Origin of Plants: the People and Plants that Have Shaped Britain's Garden History Since the Year 1000.* Headline Book Publishing.

Cubey, J. & Merrick, J. (2012), *RHS Plant Finder 2012–2013.* The Royal Horticultural Society.

McNeil, J., Barrie, F. R., Buck, W. R. *et al.* (2012) *International Code of Nomenclature for Algae, Fungi, and Plants (Melbourne Code).* Koeltz Scientific Books.

Musgrave, T., Gardener, C. & Musgrave, W. (2000) *The Plant Hunters: Two Hundred Years of Adventure and Discovery Around the World.* The Orion Publishing Group.

Plant Heritage (2012) *The National Plant Collections Directory*. The National Council for the Conservation of Plants and Gardens.

PlantLife (2012) www.plantlife.org.uk/publications/plant_diversity_challenge_the_uks_response_to_the_global_strategy_for.

Royal Botanic Gardens Kew (2012) www.kew.org/collections/herbcol.html.

Royal Horticultural Society (2012) www.rhs.org.uk/Plants/Plant-science/RHS-Herbarium/Nomenclature.

FURTHER READING

Andrews, S., Leslie, A. C. & Alexander, C. (2000). *Taxonomy of Cultivated Plants: Third International Symposium*. Royal Botanic Garden.

Benson, W. (2012). *Kingdom of Plants: A Journey Through their Evolution*. Collins.

Bridson, D. & Forman, L. (eds) (2000). *The Herbarium Handbook,* 3rd edn. Kew Publishing.

Cox, P. & Hutchinson, P. (2008). *Seeds of Adventure: In Search of Plants*. Antique Collectors Club.

Cullen, J. (2006). *Practical Plant Identification: Including a Key to Native and Cultivated Flowering Plants in North Temperate Regions*. Cambridge University Press.

Fry, C. (2012). *The Plant Hunters: The Adventures of the World's Greatest Botanical Explorers.*, Andre Deutsch Ltd.

Gupta, R. (2012). *Plant Taxonomy: Past, Present, and Future*. The Energy and Resources Institute.

Heywood, V. H., Brummitt, R. K., Culham, A. & Seberg, O. (2007). *Flowering Plant Families of the World*. Kew.

Hickey, M. & King, C. (1997). *Common Families of Flowering Plants*. Cambridge University Press.

Mabberley, D. J. (2008). *Mabberley's Plant-Book. A Portable Dictionary of Plants, their Classification and Uses,* 3rd edn. Cambridge University Press.

Mekonnen, G. & Dessalegn, Y. (2012). *Plant Taxonomy and Systematics: Concepts, Sources, Botanical Nomenclature, Plant Collecting and Documentation, Herbaria and Data Information Systems*. Lambert Academic Publishing.

Oldfield, S. & BGCI (2010). *Botanic Gardens: Modern Day Arks*. New Holland Publishers Ltd.

Pandey, H. P. (2010). *Principles of Plant Systematics: With Special Reference to Current Trends in Plant Taxonomy*. Lambert Academic Publishing.

Postan, C. (1996). *The Rhododendron Story: 200 Years of Plant Hunting and Garden Cultivation*. The Royal Horticultural Society.

Short, P. (2004). *In Pursuit of Plants: Experiences of Nineteenth and Early Twentieth Century Plant Collectors*. Timber Press.

Stacey, R. & Hay, A. (2004). *Herbarium*. Cambridge University Press.

CASE STUDY Plant hormones

Knowledge of plant hormones led to the development of rooting hormone substances, this made it possible to propagate plants that were impossible or very difficult.

The growing of 'dwarf' pot plants such as poinsettias, *Euphorbia pulcherrima*, and chrysanthemums, *Chrysanthemum morifolium*, was made possible by the development of plant growth regulators. The plants are treated with a growth regulator to keep them a certain size for the pot plant market.

A recent development that is now commercially available is a product that is used to reduce the length of grass leaves. This results in a shorter but wider leaf blade that provides a better surface for sports play. As well as reducing the amount of mowing required, which saves time, fuel and carbon dioxide production, it improves the plant's habit of growth making it more suitable for sports use.

Photograph of *Fritillaria imperialis*.

Chapter 2
Plant structure

Kelvin Mason

INTRODUCTION

It is important to understand the structure and workings of plants if you are to grow them to their full potential. Knowledge about a plant's life cycle indicates when it should be sown or planted, grown on and harvested, with each of the different categories named, defined and discussed, together with what treatments the plant may require to get the best from them, this knowledge includes their durability, economic use, and decorative or structural value.

Knowing about a plant's external and internal structure and how they work is very useful for propagation, this leads on to such decisions as the type of material to use related to the time of year or the plant's ability to produce adventitious roots on a wide range of different parts. A selection of species within a plant genus can have very different propagation and maintenance requirements. Also this information has a role to play in solving physiological problems that may occur with plants, especially those related to nutritional imbalances.

However, please note that there is still much to learn about the internal workings of plants and research is still ongoing. This greater and developing understanding will be useful in the future to help improve plant growth and yields to feed an ever-growing population. To aid in the gaining of this knowledge you will need to first develop a basic understanding of how plants grow, and hopefully this chapter will give you this introductory understanding.

The external features have been used as vital keys to plant identification and here each term is considered in turn, including floral formulas. This is a shorthand method of recording the floral detail and how the features relate to each other, making a contribution to an accurate plant identification result, especially in the field.

Key concepts

Once you have read this chapter you should be able to:

✿ Name and describe the various plant life cycles and geophyte plants

✿ Describe the external and internal structure of a plant; and name and state the function of the plant external and internal parts

✿ Name and describe the plant organs

✿ Describe photosynthesis, respiration and transpiration

✿ Name and describe the function of the main plant auxins

The Fundamentals of Horticulture: Theory and Practice, ed. C. Bird. Published by Cambridge University Press. © The Royal Horticultural Society 2014.

PLANT LIFE CYCLES

Ephemerals

Ephemerals are plants that can complete their life cycle in less than a year. They grow from seed, produce leaves, flowers and seed in a few weeks or months. How long they take depends on the species and time of year. Many plants growing in deserts are ephemerals, this enables them to germinate, grow and flower during the short periods when water is available.

Some common weeds are ephemerals: including groundsel, *Senecio vulgaris*, hairy bitter cress, *Cardamine hirsuta*, and annual meadow grass, *Poa annua*. This is one of the reasons they are such successful weeds as they complete their life cycle in a short time and produce large amounts of seed each generation.

Figure 2.2 Photograph of an example of an annual *Tropaeolum* cultivar.

Figure 2.1 Photograph of annual meadow grass, *Poa annua*, a common ephemeral.

Annuals

A true annual completes its life cycle in one year in its native environment. The seed will germinate outdoors in the spring and the plant will produce leaves and flowers during the spring or summer. Seed is then produced in the autumn, which is dispersed, and the cycle repeats itself the following year. Annuals include *Calendula officinalis*, *Godetia grandiflora* and *Limnanthes douglasii*.

CASE STUDY | **Annual plants**

Horticulturalists have developed techniques to grow annuals in areas that, and at times when, the plant does not naturally grow. Half-hardy annuals (HHAs) are an example; these are plants that naturally grow in warmer climates, such as the Mediterranean or South Africa. In their native habitat they are annuals, but in the UK the growing season is not long enough and warm enough for them to complete their life cycle.

Therefore HHAs are sown in greenhouses or polytunnels, grown on, hardened off and then planted outside to complete their life cycle. Examples include: snapdragon, *Antirrhinum majus*, aster, *Callistephus chinensis*, and French marigold, *Tagetes patula*.

Some really hardy annuals can be sown outdoors during September. They will overwinter as small plants and then flower during the early summer the next year. Although this means the plant has lived through two calendar years it is still classed as an annual. Plants that can be treated like this include:

- pot marigold, *Calendula officinalis*
- love-in-a-mist, *Nigella damascena*
- poached egg plant, *Limnanthes douglasii*
- larkspur, *Consolida ajacis*

Biennials

These are plants that germinate during the summer of year one then flower and seed during year two. During year one they usually produce a small rosette of leaves that overwinter. They produce a flower stem and seeds in the summer of the second year. Some also produce a storage organ (tap root) to store food over winter, for example carrot, *Daucus carota*, parsnip, *Pastinaca sativa*, and beetroot, *Beta vulgaris*.

Some biennials are used for spring bedding displays in parks and gardens, although many spring bedding plants are short-lived perennials. Examples include wallflower, *Erysimum cheiri*, and forget-me-not, *Myosotis sylvatica*.

Figure 2.3 Photograph of an example of a biennial, *Digitalis purpurea*.

Perennials: herbaceous types

Hardy herbaceous perennials such as *Hosta, Crocosmia* and *Helenium* grow from a perennial root each spring. They produce stems and leaves through the spring and then flower in summer or autumn before dying back to the perennial rootstock. Hardy perennials do not require any protection in the UK.

Figure 2.4 A typical herbaceous perennial, *Echinacea purpurea*.

Most hardy perennials die back to a rootstock for the winter but a few are evergreen, such as *Bergenia cordifolia* and *Helleborus niger*.

Half-hardy perennials (HHPs): these are plants that are perennials in their native habitat but are not hardy enough to survive outdoors in the UK. The frost and cold of a normal winter would

Figure 2.5 Photograph of a perennial that holds its leaves over winter, *Bergenia cordifolia*.

kill them. Many plants used for summer bedding displays are examples of this group, such as *Petunia, Pelargonium* and *Canna*.

Growing half-hardy perennials – To ensure the plants have a sufficiently long growing season of warmer temperatures the plants are started in greenhouses or polytunnels in the early spring. Once the last frost has gone the plants are planted out after being hardened off. This type of plant can be propagated by seed or vegetative methods.

Tender perennials are plants from even warmer climates such as tropical areas of the world. These need to be grown indoors all year or may be put outside during the warm summer months. Examples include bananas, *Musa* spp., and *Canna indica*.

Herbaceous perennials do not usually produce woody growth although some can have leaves and stems all year round.

Perennials: woody plants

These are plants that produce growth of a woody nature, such as trees and shrubs. A woody plant is one that produces cells that become lignified and form a permanent structure.

Shrubs are perennial plants with a low branching woody structure. Unless they have been trained they do not usually have a single trunk, but produce many branches from just above soil level.

Trees are woody perennial plants usually with a single stem that branches a metre or more above soil level. Some trees can be multi-stemmed, particularly if the stem is damaged or cut back when very young. Nurserymen usually train trees to produce a clear stem (trunk) up to a height of 1.8 m before allowing it to branch and form a crown. This style of growing is called a standard.

Conifers tend to have a natural habit of growing a main trunk with radiating branches.

Trees and shrubs can be subdivided into the following categories:

- ✿ deciduous
- ✿ semi-evergreen
- ✿ evergreen
- ✿ broad-leaved
- ✿ conifers.

Deciduous plants – shed their leaves in the autumn, when the day-length shortens and temperature cools, and then produce fresh leaves each spring. Typical examples are sycamore, *Acer pseudoplatanus*, *Forsythia* and *Philadelphus*.

Semi-evergreens – are plants that in the UK can retain their leaves for most of the winter, but may lose them during really cold spells. In mild winters or areas they may retain the leaves all winter as an evergreen plant. Semi-evergreens include garden privet, *Ligustrum ovalifolium*, and some azaleas, *Rhododendron* spp.

Evergreens – are plants that hold their leaves all year. They do shed leaves but not all at once; they are shed evenly during most of the year. Plants such as holly, *Ilex aquifolium*, ivy, *Hedera helix*, and rhododendrons are evergreens.

Broad-leaved plants – are plants with broad leaves as opposed to needles or grass-like leaves. This is a bit of a loose description covering a wide range of plants that have wide or broad foliage.

Conifers – are plants in the gymnosperms group, such as pines, *Pinus* spp., Christmas tree, *Picea abies*, and Lawson's cypress, *Chamaecyparis lawsoniana*. They produce their seed in cones hence the name conifer. The vast majority are evergreen but a few are deciduous, such as the larch, *Larix decidua*, dawn redwood, *Metasequoia glyptostroboides*, and swamp cypress, *Taxodium distichum*.

Perennial plants – include a wide and diverse collection of plants that have successfully evolved to grow in many parts of the world coping with widely varying climates and habitats. They grow from the Arctic tundra to the tropics. Some plants have evolved (become specialised) to grow in particular habitats, these include alpines and aquatic plants.

Alpines – these have adapted to grow in a very hostile environment, which can be very windy, cold with high light levels and a very dry atmosphere. They are often buried under snow for part of the year. Some alpines have very hairy leaves that both reduce water loss and reflect light. They often have a low rosette type of habit, which reduces wind damage and provides some protection from grazing animals.

Aquatic plants – have evolved to grow in water or similar environments. Plants actually growing in water do not need woody support as the water will support them. Also if they are flexible they are less likely to be damaged by the water currents than if they had a woody stem.

GEOPHYTES

Geophytes are plants that have evolved part of the plant to adapt for a specialist function, normally for the storage of food. Different parts of the plant have evolved including leaves, stems and roots. Common types of geophytes are set out below.

Bulbs

A bulb consists of a compressed stem or basal root plate with scale leaves attached to it. The modified scale leaves act as the food storage organ and the outer leaves provide protection and prevent desiccation, and in the centre of the bulb is the main stem with flower initials. Examples of bulbs are tulip, *Tulipa* spp., onion, *Allium cepa*, and lily, *Lilium* spp.

Figure 2.6 Photograph of a bulb – onion, *Allium cepa*.

Corms

A corm is made up of a modified stem on a basal plate that consists of solid storage tissue. Plants grown from corms include *Crocus* spp., gladioli, *Gladiolus* spp., and *Freesia* spp.

Stem tubers

A stem tuber is a thickened underground stem that stores food. The potato, *Solanum tuberosum*, is a good example. They differ from corms in not having a basal plate and have buds around the outside. *Caladium* spp. are another example of stem tubers.

Rhizomes

These are stems that grow horizontally through or on the soil and some are capable of storing food such as *Iris germanica*. The rhizomes

Figure 2.7 Photograph of a corm – gladioli, *Gladiolus* cvs.

Figure 2.9 Photograph of a root tuber – dahlia, *Dahlia* cultivar.

Figure 2.8 Photograph of a stem tuber – potato, *Solanum tuberosum*, clearly visible are the eyes (buds).

Figure 2.10 Photograph of an enlarged hypocotyl – *Cyclamen* spp.

will have nodes, internodes and buds; some may have scale leaves attached at the nodes. Examples include lily of the valley, *Convallaria majalis*, calla lily, *Zantedeschia aethiopica*, and *Achimenes* spp.

Root tubers

These are swollen fleshy roots that have evolved to store food. The tubers have buds at the centre of the crown and these grow to produce stems, leaves and flowers. The dahlia, *Dahlia* spp., and *Anemone blanda, are typical plants with root tubers.*

Enlarged hypocotyls

The hypocotyl is the portion of the stem between the cotyledon and root; this can become enlarged and act as a storage organ. Many plants in this group were considered to be tubers but are now classed as enlarged hypocotyls. These plants include *Cyclamen* spp., gloxinia, *Sinningia speciosa*, and the tuberous rooted begonias, *Begonia tuberhybrida.*

External plant structure

When looking at a plant there are four main parts that are immediately obvious. These are the flower (see Chapter 7 for further details), stems, leaves and, although not often visible, the roots. In the next section we will look at the external parts in detail starting with the roots.

ROOT SYSTEM

The main functions of the roots are:

1. To anchor the plant in the soil/growing media and provide support for the stems.
2. To absorb water and minerals from the soil or growing media.

3. Some roots store food, e.g. carrot, *Daucus carota*.
4. To produce hormones.
5. Vegetative reproduction.

There are two main types of root system, the fibrous system typical of the many grass plants and the tap root type typical of many dicotyledons. If a plant is carefully dug up and the roots washed it can be seen that the roots are made up of a main root (or tap root) and many laterals roots.

(a)

Main root – this usually grows straight down and is thicker than the other roots. This root grows first from the seed as it develops from the radicle (also see Chapter 7). The tap root will store food if the plant has evolved to do this.

Lateral root – these branch from the main root and spread out in search of water and minerals. The lateral roots are multibranched and form a network of roots.

Root hairs – are near the tips of most roots and consist of tufts of very fine white hairs. It is these roots that absorb the bulk of the water and minerals.

Fibrous roots – grasses and some other plants form a fibrous mass of roots of a similar size. These form a massive network of roots spreading out from the plant base. There is no main or tap root.

Root modifications

Adventitious roots – these are roots that originate from other parts of the plant, such as the stems or leaves. When plants are propagated by vegetative means from stems or leaves the roots produced are **adventitious**. These grow on to produce a normal root system.

Another type of adventitious root is produced by climbing plants to cling to a support. Ivy, *Hedera helix*, produces short roots that cling to a wall or to tree bark, which enables the ivy to climb. The plant does not need a thick stem to support it, but uses other plants or structures.

(b)

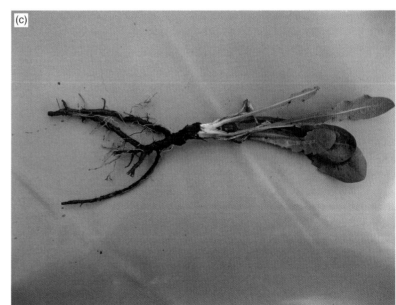

(c)

Figure 2.11 Photograph of root systems of (a) a shrub, *Rhamnus* spp., (b) annual meadow grass, *Poa annua*, and (c) dandelion, *Taraxacum officinale*, showing some of the different types of root systems.

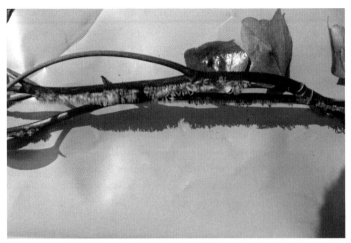

Figure 2.12 Photograph of ivy, *Hedera helix*, stem showing adventitious roots used for climbing.

Storage roots – carbohydrates are stored in many perennial roots (e.g. rhubarb, *Rheum × hybridum*) and some biennials, (e.g. carrot, *Daucus carota*) over winter. The carbohydrates are used to produce the stems and leaves the next spring. The food is stored in parenchymatous cells in the cortex of the root.

Figure 2.13 Photograph of a method of root storage –beetroot, *Beta vulgaris*, radish, *Raphanus sativus*, and carrot, *Daucus carota*.

Prop roots – Some plants have the ability to produce prop roots, which are adventitious roots that grow from the base of the stem. They grow out through the air and turn down into the soil. The roots then function as normal roots and absorb water and minerals. Prop roots give the plants extra stability as they act like guy wires holding up a tree or tall pole.

In mangroves the prop roots give stability in the swamps, especially in moving water. They also have a high number of lenticels on the aerial parts of the root, which enable gaseous exchange to take place. This allows in oxygen for the roots to use in respiration and the release of carbon dioxide.

Other plants that have prop roots include sweet corn, *Zea mays*, banyan tree, *Ficus benghalensis*, and screw pine, *Pandanus* spp.

Aerial roots – many plants that grow in the rainforests are epiphytes, these plants live on the bark of large trees. The aerial roots absorb rain and water that collects on the roots in the very humid conditions. The roots also attach the plant to the tree, but they do not damage or live off the tree. The Swiss cheese plant, *Monstera deliciosa*, well known to many as a common house plant, is a plant showing this feature.

Contractile roots – plants that grow from corms often produce contractile roots. These are roots that grow out from the base of the corm straight down. Once they are mature the upper part of the root contracts pulling down the corm, gladioli, *Gladiolus* spp., are a classic example. This helps to keep the plant at the same level each year, even though a new corm is produced on top of the old one.

STEM SYSTEM

The main functions of the stem are:

1. To support the upper parts of the plant.
2. To hold the leaves to receive maximum sunlight for photosynthesis.
3. To hold the flowers for pollination.
4. To transport water, minerals and food between the roots and leaves.

At the tips and along the stems are buds, these may be flower buds or stem/leaf buds. Buds are compressed leaves: an underdeveloped stem that will grow out to produce new stems and leaves. If a flower bud it will consist of compressed flower parts.

Apical buds – these are at the ends of the stem and will provide the extension growth to make the plant taller/larger, they are also called terminal buds. If the apical bud is a flower bud then stem growth will cease in that direction. The production of a terminal flower or flower head always stops extension growth on that stem.

Lateral buds – also called axillary buds. These are spaced out along the stem at the point where leaves are attached. This is called the node; the section of stem between the nodes is called the internode. Lateral buds grow out so the plant can bush out and become larger and wider. This enables it to compete with other plants by suppressing them and also gives it the maximum area for photosynthesis.

As the plant grows larger some of the axillary buds will grow, but not all, the others remain dormant. They may never grow but if a terminal bud is damaged or removed a dormant axillary bud will grow. This is why the tips of some plants are removed to make them bush out as this encourages the axillary buds to grow. Dormant buds can also be stimulated into growth by pollarding or coppicing trees such as poplar, *Populus* spp., and willows, *Salix* spp.

Stem modifications

Over millions of years plants have evolved to grow in various environments and situations. In order for them to do this the plant stems have modified.

Twining stems – e.g. runner beans, *Phaseolus coccineus*. These plants have long internodes between the nodes. The stem is very thin in relation to the height of the plant but it gains its support by twining around other plants or structures with the stems turning around fairly tightly and always in the same direction.

Stem tuber – e.g. potato, *Solanum tuberosum*. It can be proved that a potato is a stem by looking at it closely and noting the 'eyes', which are in effect small buds. These develop into stems and leaves when the tuber starts to grow. The stem tuber has evolved to store food to enable the plant to overwinter, with the food stored in cells as carbohydrate; humans have made use of this as a major food supply.

Corm – e.g. crocus, *Crocus* spp., and gladioli, *Gladiolus* spp. The stem has been modified as a very short squat round food store;

once again this is to enable the plant to overwinter. The swollen corm is surrounded by leaf scales that protect it, but if these are removed the central buds on the top and some of axillary buds can be seen. The leaves and flower stalk emerge from the buds on top of the corm in the spring, with adventitious roots produced from the base of the stem.

Rhizomes – e.g. *Iris germanica* and Solomon's seal, *Polygonatum multiflorum*. Rhizomes grow horizontally on or just below the soil surface, they have nodes with buds and small scale leaves. At the nodes the plant produces adventitious roots. If a bud grows up to the surface it will produce a normal stem and leaves. Rhizomes are a method for the plant to reproduce and spread as well as overwintering.

Runners – e.g. creeping buttercup, *Ranunculus repens*, and wild strawberry, *Fragaria vesca*. Runners spread above ground, they creep along the soil and will produce adventitious roots at nodes where they are in contact with the soil. Usually where a node touches the soil the runner will produce a new plant from an axillary bud.

Stolons – e.g. blackberry, *Rubus fruticosus*. These are very similar to runners but are normally long arching branches that bend down and come into contact with the soil. Where a node touches the soil adventitious roots are produced, which grow into the soil to secure the plant and provide it with water and minerals, with an axillary bud developing that will subsequently produce a new plant. This process can be encouraged by the gardener and is called layering (see Chapter 8 for further details).

Figure 2.14 Photograph of twining stems of *Lonicera* spp.

(a)

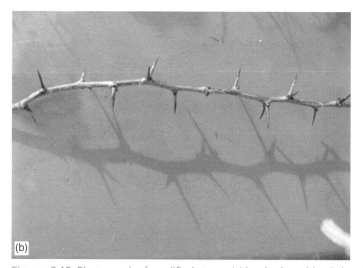

(b)

(c)

Figures 2.15 Photograph of modified stems. (a) hooks, bramble, *Rubus fruticosus*; (b) thorns, hawthorn, *Crataegus* spp., and (c) tendrils, white bryony, *Bryonia dioica*.

Offsets – e.g. *Echeveria* spp. and *Sempervivium* spp. These are like very short squat runners with many internodes, at the end of each offset a large bud is present that can develop into a plant, which if separated from the parent plant will produce adventitious roots and grow. This will be very familiar to anyone who has owned these types of plant.

Thorn – hawthorn, *Crataegus monogyna*, is a classic example: its thorns are produced by the plant as a form of protection and it helps to reduce grazing by animals. Thorns grow from axillary buds and some may have nodes, the whole part of the short stem has evolved to become a thorn.

Hooks – e.g. rose, *Rosa canina*, and bramble, *Rubus fruticosus*. Unlike a thorn a hook is where only part of a stem has become modified. The outer tissue (epidermis) and part of the cortex have evolved to produce the hook. This is used by the plant as an aid to grip and help it climb supports, note that the hooks point backwards to aid this function, and it also gives it some protection from grazing.

Tendrils – both leaves and stems can produce tendrils; however, if they are produced by the stem they are called stem tendrils as in white bryony, *Bryonia dioica*, with the tendril being a modified axillary branch.

Phylloclade – e.g. butchers broom, *Ruscus aculeatus*. The stems have become very flattened and green, looking like lines of stiff leaves, and in fact carry out photosynthesis. The actual leaves are reduced to small scale leaves found in the centre of the phylloclade.

LEAVES

The leaf can be divided into two main parts:

> petiole – the botanical name for the leaf stalk
> lamina – the botanical name for the leaf blade.

Petiole – the leaf stalk connects the leaf blade to the plant's stem at the node. Its function is to transport water and minerals to the leaf and carbohydrates to the stem for circulation to the rest of the plant, and it also holds the leaf in the best position to gain maximum light for photosynthesis. The length of the petioles varies between different plants.

Also at the base of the petiole some plants produce small outgrowths called stipules that look like very small leaves. These can be seen in the pea, *Pisum sativum*, and zonal pelargonium, *Pelargonium zonale*.

There are two main types of vein network:

1. Net or reticulate – this is a much-branched network of veins that extends to all parts of the leaf, common in most dicotyledon plants.
2. Parallel – these veins run parallel to the edges of the leaf and run in straight lines to the leaf tip. This type of vein is common in grasses (monocotyledons).

Figure 2.16 Photograph of pea, *Pisum sativum*, stipules.

Figure 2.17 Photographs of reticulate leaf-vein patterns. (a) *Geranium* spp.; (b) *Alchemilla mollis*.

If the leaf is joined straight to the stem with no petiole it is called sessile. *Zinnia* spp. are a group of plants with sessile leaves.

Lamina – this is the main flattened blade of the leaf and forms the bulk of it. It is within the lamina that photosynthesis and transpiration take place: this is the powerhouse of the plant.

In dicotyledonous plants the petiole runs straight through into the mid rib of the leaf blade. The mid rib is the main vein, which runs the length of the lamina and has many veins coming off it that form a venial network covering the whole leaf. The term reticulate is used to describe this, as the arrangement is net-like.

For the lamina to carry out these functions it needs a network of veins to supply it with water and minerals, and also to move the food away that is made by photosynthesis.

Leaves are usually classified into two groups: simple or compound.

Simple leaves – these are leaves that have a single leaf blade on the petiole, e.g. garden privet, *Ligustrum ovalifolium*.

Figure 2.18 Photograph of simple leaves.

Compound leaves – these consist of three or more leaflets, e.g. mountain ash, *Sorbus aucuparia*, ash, *Fraxinus excelsior*, and rose, *Rosa* spp.

Figure 2.19 Photograph of compound leaves – lupin, *Lupinus* spp.

The easiest way of telling whether a leaf is a single leaf or a number of leaflets, as in a compound leaf, is to check at the base of the leaf where it joins the stem. A true leaf will have a bud at the base; a leaflet will not, it will be attached to a leaf stalk, which is called a **rachis**.

Figure 2.20 Photograph of alternate leaf arrangement – beech, *Fagus sylvatica*.

Leaf arrangements

Leaves on plants are arranged in a particular layout on the stems. This arrangement varies between different species and can help in the identification of plants.

The arrangement should hold the leaves so that they receive the maximum amount of light for photosynthesis – thus allowing the plant to photosynthesise to its maximum potential. It is important that leaves at the top of the stem do not shade leaves lower down. As the leaves grow they are arranged in a helix pattern around the stem. This can be either clockwise or anti-clockwise depending on the species. The angles between the leaves follow a regular pattern and numerical sequence called a Fibonacci sequence, which is 1/2, 2/3, 3/5, 5/8, 8/13, 13/21, 21/34, 34/55, 55/89. The numerator (first number) indicates the number of complete turns of the stem; the denominator (second number) indicates the number of leaves in the arrangement. This can be shown by the following example:

Alternate leaves with an angle of:

180° or 1/2 – two leaves in each complete circle of the stem
120° or 1/3 –three leaves in each complete circle of the stem
144° or 2/5 –five leaves in two complete circles of the stem
135° or 3/8 – eight leaves in three complete circles of the stem.

The circles of the stem are called gyres.

There are three main leaf arrangements: alternate, opposite and whorled.

Alternate – single leaves are arranged on alternate sides of the stem as in beech, *Fagus sylvatica*, and hornbeam, *Carpinus betulus*.

Opposite – leaves are arranged on opposite sides of the stem as in coleus, *Solenostemon* spp.

Figure 2.22 Photograph of *Fritillaria imperialis* showing a plant with a whorled leaf arrangement.

Figure 2.21 Photograph of opposite leaf arrangement – *Euonymus fortunei* 'Emerald 'n' Gold'.

Figure 2.23 Photograph of *Verbascum* spp. showing a rosette habit just as the flower stem starts to grow in the spring.

Whorled – the leaves are arranged in a whorl (ring) around the stem, as in *Fritillaria imperialis*.

A fourth arrangement is called rosulate where the leaves form a rosette – this is often seen in plants that overwinter as biennials.

Leaf shapes

There are many different leaf shapes and edges that vary widely, and this is a useful means of identifying plants.

A few common shapes are illustrated below, there are many more but space is too limited to show them all.

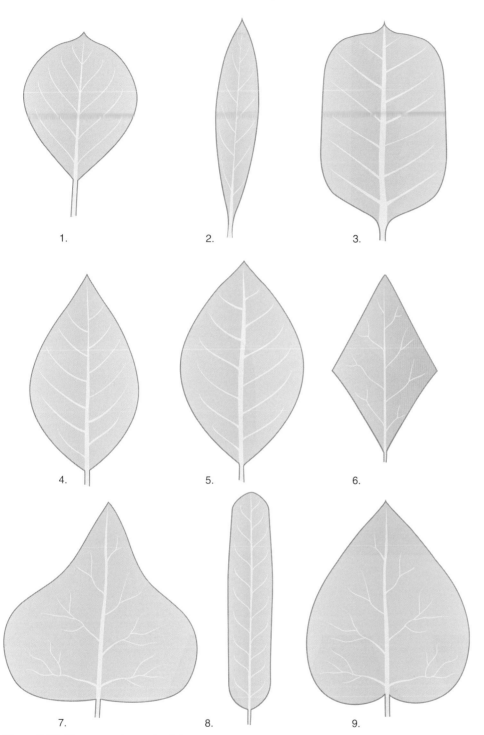

Figure 2.24 Diagrams of some leaf shapes.

(1) *Obovate* – inversely ovate.

(2) *Lanceolate* – lance shaped, widening above the base, long and tapering to the apex.

(3) *Oblong* – longer than broad, with nearly parallel sides.

(4) *Ovate* – egg shaped, wider at the base than the apex.

(5) *Elliptic* – two to three times longer than wide, widest in the middle and narrowing equally at both ends.

(6) *Rhomboidal* – diamond shaped.

(7) *Deltoid* – triangular shaped.

(8) *Linear* – long and narrow with parallel sides.

(9) *Cordate* – heart shaped, tapering to an acute apex. A notch at the base where attached to the petiole.

Leaf margins

These also aid identification. There is a wide range of leaf margins and a few are illustrated below.

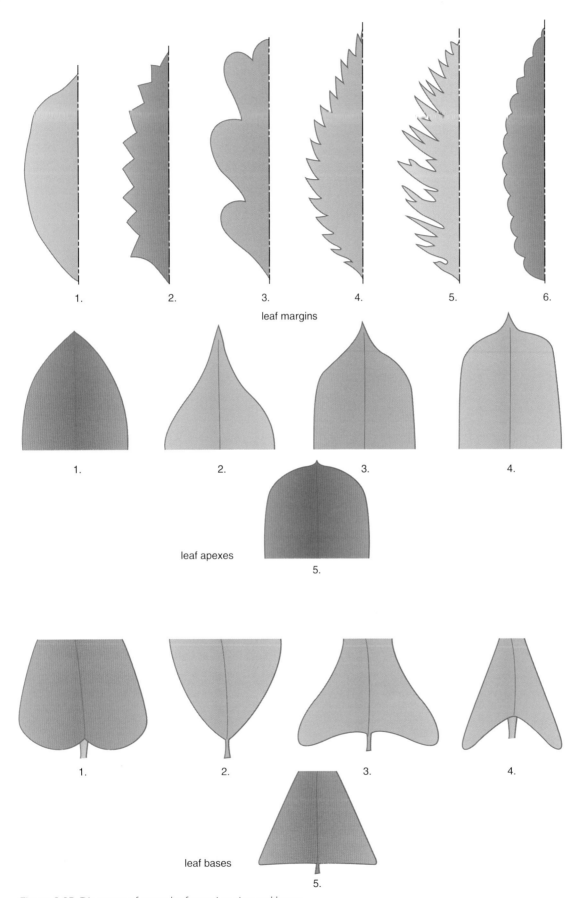

leaf margins

leaf apexes

leaf bases

Figure 2.25 Diagrams of some leaf margins, tips and bases.

Margins

(1) *Entire* – leaves have a smooth, undivided margin.
(2) *Dentate* – the edges have little teeth facing outward.
(3) *Lobed* – the edge has lobes that cut less than half way to the centre of the leaf.
(4) *Serrate* – small forward pointed teeth, like a saw blade.
(5) *Incised* – sharply, deeply and irregularly cut.
(6) *Crenate* – scalloped shaped edges.

Leaf apexes (tips)

(1) *Acute* –sharp pointed.
(2) *Acuminate* –tapering at the end with a long point.
(3) *Cuspidate* – a sharp, elongated rigid tip.
(4) *Mucronate*– ends abruptly with a sharp tip.
(5) *Obtuse* – rounded or blunt.

Leaf bases

(1) *Cordate* – heart shaped with a notch towards the petiole.
(2) *Cuneate* – wedge shaped.
(3) *Hastate* – shaped like a halberd with the basal lobes pointing outwards.
(4) *Sagittate* – shaped like an arrow head.
(5) *Truncate* – a flat end that looks as if cut off.

Leaf colour

The vast majority of leaves are green as the chlorophyll in the leaves reflects the green light and absorbs mainly in the blue/red spectrum (see Chapter 3 for further details). The shade of green can vary from very light to very dark, the shade can depend on a number of factors including how much chlorophyll *a* and *b* is present, whether the plants have evolved to grow in sun or shade, nutrient availability and genetic factors. Leaves show a wide variation in colour including reds, purples, yellows and variegated (green plus one other colour). Most of these have been bred or selected by humans.

Leaf modifications

Succulent leaves – these store water and food for use during periods of drought. Succulent plants usually grow in deserts or similar areas, and include *Echeveria* spp., *Sedum* spp. and *Aloe* spp. The food stored in the leaves can be used in crassulacean acid metabolism (CAM) photosynthesis, which is discussed further in Chapter 3.

Carnivorous leaves – some carnivorous plants have evolved leaves that trap insects, which are digested by the surface of the leaf and the nutrients are extracted for use by the plant.

Bulbs – see section on geophytes.

Bracts – the leaves colour up and help to show off the flower, which may be quite insignificant. The result can be highly decorative and long lasting. Poinsettia, *Euphorbia pulcherrima*, is a good example of this feature.

Figure 2.26 Photograph of succulent leaves – *Crassula* spp.

Figure 2.27 Photograph of *Euphorbia polychroma* showing bracts.

Tendrils – the leaves evolve into tendrils that enable the plant to climb, e.g. pea, *Pisum sativum*.

Spines – some cacti have evolved spines to protect the plant from browsing animals. The spines grow where a plant would

Figure 2.28 Photograph of poinsettia, *Euphorbia pulcherrima*, showing bracts.

Figure 2.30 Photograph of cacti showing spines.

Figure 2.29 Photograph of perennial sweet pea, *Lathyrus latifolius*, showing leaf tendrils.

Figure 2.31 Photograph of *Stachys byzantina* showing an example of hairy leaves.

normally produce leaves, e.g. *Mammillaria* spp. This configuration also reduces water loss from the plant.

Hairs – more correctly called trichomes, are produced on the surface of many leaves. Their function is to reduce water loss and trap surface humidity around the leaves in dry climates; they also help to create a large boundary layer around the leaf.

INTERNAL STRUCTURE OF PLANTS

Roots

It is important that the plant is anchored securely to grow and remain upright; it is the roots that provide this support. Roots have a cylindrical structure, which enables them to absorb water and minerals, and transport these to the stem and leaves efficiently. A root system has a large surface to volume ratio, which is ideal to absorb large amounts of water and then move it up to the stem.

Roots are also involved in the production of several hormones including cytokinins. It is thought that this may be the way the root controls plant size, by keeping shoot growth in step with root growth. It helps prevent the leaves transpiring more water than the roots can absorb.

Food is stored in some roots for example carrot, *Daucus carota*, parsnip, *Pastinaca sativa*, beetroot, *Beta vulgaris*, and radish, *Raphanus sativus*.

Root structure

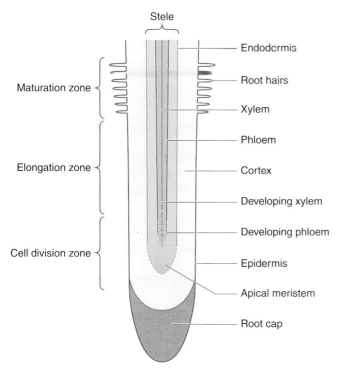

Figure 2.32 Diagram of longitudinal section of root tip.

Roots grow by the tip extending out from the plant centre, with the tip of the root called the root cap. Its function is to protect the apical meristem, which is just behind it. The root cap is cone shaped and as well as protecting the meristem it secretes mucilage. This helps to lubricate the root cap allowing it to penetrate the soil easier. It also encourages the growth of bacteria and fungi that assist with absorbing minerals and give the roots better contact with the soil particles. As the root cap pushes through the soil it is worn away by soil particles; the lost and damaged cells are replaced by the meristem.

The apical meristem is where cell division takes place. The cells in the meristem divide into two to produce two identical daughter cells. These cells then divide again and this process continues. As the root cap moves forward the first cells to divide start to expand. This is known as the region of elongation and it is this part of the root that actually grows longer. It is at this point that the cells start to differentiate into the various tissues.

The roots differentiates first into the epidermis, which is the outer skin of the root, then the cortex is formed just inside the epidermis and finally the stele is formed, which carries out the transporting function of the root. The cells in the stele differentiate into endodermis, phloem and xylem.

At this point the root hairs are formed and start to grow, and they absorb the vast majority of the water and minerals required by the plant. The root hairs are formed by epidermal cells growing into very narrow hairs that are unicellular; they only live for four to five days. New root hairs are being produced all the time the root is growing. The root hairs massively increase the surface area allowing the root to absorb the water the plant requires. They are also covered in mucilage, which improves water and mineral absorption.

The function of the root tissues are set out below:

Epidermis – as well as producing the root hairs the epidermis protects the inner tissues of the root. Even after the root hairs have died they provide protection for the young root. As the root ages the epidermis changes, this will be covered later in the chapter.

Cortex – the cortex forms a large part of the young root. It is made up of loosely packed cells with many cellular spaces to allow for the movement of water and oxygen. It is also the cortex that stores food in the root.

Endodermis – this is on the inside of the cortex and surrounds the pericycle. The endodermis is a single layer of cells and its function is to regulate the flow of water into the vascular tissues. The endodermis contains **suberin** and lignum, which helps to control water flow. The endodermis and everything inside it form the inner core of the root and this is called the stele.

Casparian strip – this is within the endodermic cells, although this is a single layer, sandwiched between the cells like cement between bricks in a wall. The Casparian strip is the part that controls the flow of water and minerals into vascular tissues. The minerals have to flow through a plasma membrane, which controls the molecules that are allowed through.

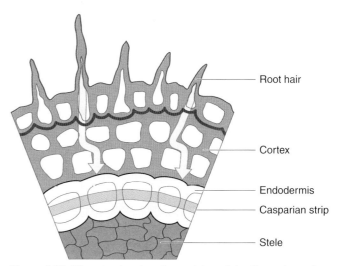

Figure 2.33 Diagram showing the position of the Casparian strip.

Pericycle – this is on the inside of the endodermis and consists of a single row of cells. Its function is to produce branch roots, which are initiated here and then push their way out of the root to emerge into the soil. This allows the plant to produce a branching network of roots.

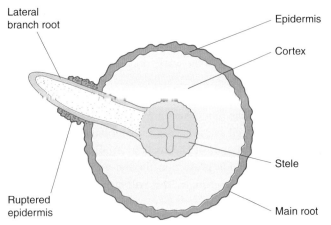

Figure 2.34 Diagram showing how new roots grow from the pericycle.

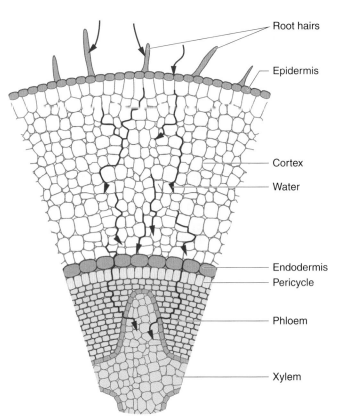

Figure 2.35 Diagram showing the pathway of water through the root.

Vascular tissue – this is inside the pericycle and the arrangement depends on the plant species. Vascular tissue consists of primary phloem, primary xylem and vascular cambium, as described below.

>   *Primary phloem* – are bundles of cells on the inside of the pericycle. Their function is to deliver food from the leaves and stem to the roots.
>   *Primary xylem* –these are a group of large cells that make up the central core and transport the water and minerals up to the stem and leaves.
>   *Vascular cambium* –this is between the phloem and xylem and usually consists of a single row of cells. Its function is to produce new phloem and xylem cells as the root grows.

How roots work

Water passes into the epidermis of the root mainly by osmosis. The water passes through the cell membrane to try to equalise the concentration on both sides of the membrane. If one side of the membrane has a high salt concentration (plant cell) and the other side has a low salt concentration (soil), water will pass from the low side to the high to try to equal it out. This will continue until the salt concentration is the same on both sides, see Chapter 4 for further details.

Once inside the root the water is passed through the cortex. Therefore the concentration at the outer cell is never of equal strength to that of the soil, and so the process continues.

From the epidermis the water passes between the cortex cells (intercellular spaces) to the endodermis, which controls the flow into the xylem. The water is then pushed up the xylem tubes to the stem and into the leaves.

Within the water solution that passes into the epidermis are the minerals the plant requires. The plant cells have adapted to allow some of these minerals into the root.

The mineral salts (nutrients) are taken in by the roots in a dilute solution and pass through the epidermis and between the cortex cells to the endodermis. Once at the endodermis they will only be let through if required by the plant. When in the transpiration stream of the xylem, they are moved to the point of use.

To assist in the uptake of minerals, roots have adapted and evolved a relationship with various fungi and bacteria.

Mycorrhizal fungi

Many species of plant have developed a symbiotic relationship with soil-inhabiting fungi. This is to the mutual benefit of both plants. There are two main types of mycorrhizal fungi:

1. **Ectomycorrhizal fungi** – occur in many woody plants, such as forest trees. The fungal hyphae penetrate the epidermis and outer cortex, but never actually invade any plant cells.
2. **Endomycorrhizal fungi** – penetrate the root cortex as far as the endodermis to the Casparian strip. They invade the cells but do not cause any internal damage as they do not penetrate the plasma or vacuole membranes.

The mycorrhizae take in phosphate and other nutrients and these are passed into the root and released for use by the plant. In return the plant allows the fungus use of some carbohydrates. Both the plant and the fungus need each other to thrive.

Rhizobium bacteria

These are a species of bacteria that live on the roots of certain plants including many members of the pea family, *Papilionaceae*, and other plants such as the alders, *Alnus* spp.

The bacteria have the ability to absorb nitrogen from the air (78% of the atmosphere is nitrogen). They convert the nitrogen into nitrogenous compounds that are released into the plant roots for growth. In return for this the plant releases sugars for the bacteria to use.

The bacteria enter the plant's roots through the root hairs and then send a thread into the cortex of the root. A root nodule is formed in the cortex and grows out of the epidermis, so is visible on the roots. (See also Chapter 6 for further details on nitrogen cycling.)

Secondary thickening in roots

This is similar to stems (see below) in that two cambiums are formed and these produce the secondary tissue. The cambium in the stele links up to form a ring of cambium. This cambium produces xylem on the inside and phloem on the outside, as in the stem. As the plant continues to grow the root circumference increases. This results in the tissues to the outside of the pericycle (epidermis and cortex) being pushed out and eventually shed. The pericycle now becomes meristematic and this produces the new cork cambium. This protects the root and helps to prevent water loss, but does not have the ability to absorb much water.

STEMS

The stem provides the transport structure to move water and minerals to the leaves and food from the leaves down to the roots and other plant parts. The stems also hold the leaves in an ideal position for maximum photosynthesis and flowers for pollination.

If the stem is green it may have the ability to photosynthesise, this is important in plants with few or very small leaves like the brooms, *Cytisus* spp., or gorse, *Ulex europaeus*.

Some woody stems have lenticels in the bark that enable the plant to take in oxygen for respiration in the stem. These are small, round, usually slightly raised spots on the young bark. This saves the plant having to move oxygen long distances around the plant.

The stem originates from the plumule that emerges when the seed germinates. It elongates in a similar way to the root although it is more complex.

Cells divide in the apical meristem at the tip of the stem in the terminal bud, which is protected from damage by bud scales. When the bud starts to grow the cells in the meristem actively divide, as in the root, with the mother cell producing two daughter cells.

The apical meristem forms a dome on top of the stem, on either side are two swellings, which are the leaf primordia. These fold over the meristem to protect it and are the first stage of leaf production. At the base of each of the leaf primordia is the axillary bud primordium, which remains dormant until activated by the plant. As the stem is formed it is divided up into sections of internodes and nodes.

Just behind the meristem is the zone of elongation where the internode cells elongate and the stem increases in length.

As the daughter cells expand and replicate the shoot gets longer. The cells start to differentiate in the region just behind the meristem. These cells form the epidermis, cortex and vascular tissues.

STEM STRUCTURE

Young or herbaceous stem

Epidermis – this is the outer layer of the stem and consists of parenchyma cells. They protect the inner parts of the plant and prevent excessive water loss from plant cells. The epidermis is often covered with **cutin** to prevent cell dehydration. In very hot, dry climates the cuticle may be covered with a wax-like substance to further reduce water loss. These two materials are also the plants first line of defence against pathogens such as fungi and bacteria.

To allow gaseous exchange to take place young stems have small swellings on the epidermis called lenticels. These allow oxygen in for respiration and waste carbon dioxide out.

Cortex– on the inside of the epidermis is a ring of tissue called the cortex. This ring varies in thickness depending on the species and age of the plant. If the stem of the plant is green this area is capable of photosynthesis and the cortex will contain chloroplasts. Set within the cortex of dicotyledons, usually in a ring, are the vascular bundles.

Vascular bundles – these are made up of three types of tissue: phloem, cambium and xylem.

Phloem – this transports food from the leaves, made by photosynthesis, around the plant, this is called translocation. Phloem consists of parenchyma and is divided up into two main parts: sieve tubes and companion cells. The sieve tube is where the food is moved, it comprises elongated tubes joined together. The end walls of the tubes have plates, which have many small holes in them to allow the liquid food to pass through. The plates are called sieve plates and have large **plasmodesmata**. The sieve tubes do not have a nucleus and have very few vacuoles.

Adjacent to the sieve tubes are companion cells, these carry out all of the cellular functions of the sieve tubes. The cytoplasm of the companion cell is connected to the sieve tube by the plasmodesmata. Without the companion cells the sieve tubes could not survive.

The movement of phloem sap is bidirectional and is caused by turgor pressure. The sap is water based but has a high amount of sugars dissolved in it.

Xylem – conducts water and minerals from the roots through the stem to the leaves and other parts of the plant. Most xylem tissues are dead; the water just passes through them.

There are two types of xylem cell: tracheids and vessel elements. Tracheids are long, tapered cells with angled end plates that join them together. These types of cells are common in conifers. Vessel elements are shorter tubes and much wider

Characteristic	Xylem	Phloem
Condition	Dead tissue	Living tissue
Essential tissue	Tracheids and vessel elements	Sieve tubes and companion cells
Associated tissues	Xylem fibres and parenchyma	Phloem fibres and parenchyma
Living tissues	Only xylem parenchyma	Sieve tubes, companion cells and phloem parenchyma
Function	Conduction of water	Conduction of food and other organic substances

Table 2.1 Main characteristics and differences between xylem and phloem tissues.

than tracheids, they do not have any end plates so water flows freely.

Cambium – this tissue is between the phloem and the xylem. Cambium becomes very important when secondary thickening starts. The cambium consists of a single layer of highly meristematic cells that divide and produce new phloem to the outside and new xylem to the inside.

Pith – in the centre of young stems is the pith, this is made up of large-celled parenchymatous tissue.

Some plants have hollow stems, for example *Leycesteria* spp. and *Forsythia* spp.

Secondary thickening in stems

This occurs in all woody plants such as trees and shrubs. It only occurs in a few monocots such as *Yucca, Dracaena* and similar plants.

Annuals and herbaceous perennials do not have secondary thickening as their stems die down at the end of each season. As trees and shrubs get larger the stems need to grow in circumference to be able to support the weight of the larger and longer growth. To enable the stem/trunk to expand in circumference, secondary thickening takes place in the cambium in the vascular bundles. This cambium becomes meristematic and the cells start dividing as they do in the apical meristem. The cambium cells divide in three directions: on the inside, outside and sideways.

Any new tissue on the inside becomes xylem; any produced on the outside becomes phloem. The sideways division enables the cambium in each bundle to link up and become a continuous circle between the xylem and phloem. As the circumference becomes larger the cambium expands to a complete ring. It forms a dividing line between the xylem wood and the phloem. The cambium continues to divide on both sides and produces new xylem and phloem. These form the secondary tissue and will continue to expand as long as the plant is alive.

Bark – in woody stems cork replaces the epidermis and forms a protective layer of outer tissue. Bark has a number of layers of cells and as the tree grows larger these expand outwards and are replaced by new inner bark. Some of the parenchyma cells become meristematic and form the cork cambium or **phellogen**. The cork

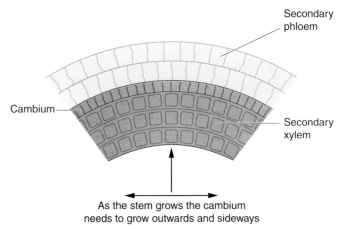

As the stem grows the cambium needs to grow outwards and sideways

Figure 2.36 Diagram showing the direction of cambium growth.

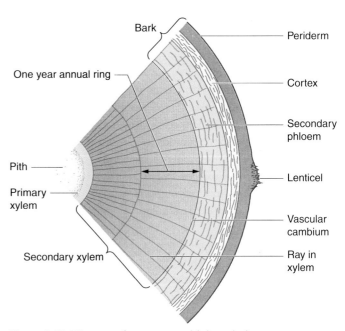

Figure 2.37 Diagram of a two-year-old dicotyledonous stem.

cambium produces cells on its inside and outside as does the vascular cambium. The cells on the inside form secondary cortex (called phelloderm) while the cells on the outside form cork (phellem). This whole area is called the periderm.

The bark of the cork oak, *Quercus suber*, is harvested and used by the drinks industry for corking wine etc.

Xylem – as a tree gets taller and the crown larger it needs more support. The plant therefore tends to produce more xylem heartwood than phloem as it is the lignified heartwood that gives it strength. The heartwood (secondary xylem) is always larger than the phloem and bark.

The xylem cells die but continue to transport water and minerals up the plant. Even though the cells are dead they are still capable of moving water by root pressure, transpiration pull and capillary action.

The xylem becomes darker and waste products from within the plant are often stored within the xylem cells. As the xylem ages it becomes less efficient at transporting water and forms mainly a support function.

Phloem – as the tree becomes larger it has a greater requirement to transport food. The circumference of the tree increases giving it a larger phloem volume for the movement of food etc.

Cambium – this continues to produce xylem and phloem for as long as the branch/trunk is alive.

CASE STUDY   **Monocotyledon trunk structures**

As well as only having one cotyledon at germination (hence the name monocotyledon), their internal structure is different to dicotyledons. The main difference is that the vascular bundles are scattered throughout the stem and not in a regular pattern around the stem as in dicotyledons. There are not distinct tissues like cortex and pith in monocots, but a 'ground tissue' that is surrounded by the epidermis.

Generally monocotyledons do not have a vascular cambium so cannot develop secondary thickening. The vast majority of monocots are small herbaceous plants so do not require it. Plants such as coconut palms, *Cocos nucifera*, and bamboos, *Arundinaria* spp., grow to tree-like size by primary thickening. The primary growth increases in width before the elongation has completed and the primary meristem is still productive. The trunk may still continue to grow in size in later years but this is because of the cells in the ground tissue increasing in size.

Plants such as *Dracaena* spp. can produce a cambium on the outside of the ground tissue and therefore produce secondary thickening similar to dicotyledons.

LEAVES

Function of leaves

The main function of leaves is to carry out photosynthesis and transpiration. The process of photosynthesis produces the food the plant requires to grow using light energy.

Most of the water taken up by the plant is lost by transpiration; a small amount is used in photosynthesis and other metabolism.

Internal structure

The vascular bundles from the stem enter the petioles and run through to the lamina. They are usually in a ring- or U-shaped formation in the petiole. They then split down and go into the veins.

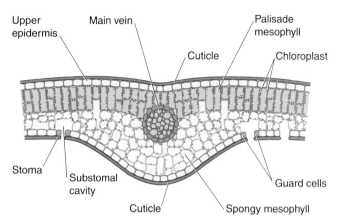

Figure 2.38 Diagram of a transverse section of part of a leaf.

Epidermis – this is a single layer of cells that forms the outer surface of the leaf, its main function is protection. It may be covered with a layer of waxy cutin, which helps to reduce water loss. Some plants have adapted by producing hairs for the same purpose e.g. lamb's ears, *Stachys byzantina*. The stinging nettle, *Urtica dioica*, has evolved hairs for protection; it is these hairs that sting you when touching the plant.

The epidermis covers the entire leaf and set within the epidermis are the stoma (plural stomata) – these are mainly on the lower surface of the leaf.

Palisade tissue – directly below the upper epidermis is the palisade tissue. This is where most of the photosynthesis takes place. Within the palisade cells are chloroplasts containing chlorophyll where photosynthesis occurs. The palisade tissue is between one and three cells thick.

Spongy tissue – below the palisade tissue is the spongy tissue. This also contains some chloroplasts but fewer than in the palisade tissue. Set within the spongy tissue are air spaces (cavities). This is where water and oxygen are released from the plant cells and pass out through the stomata into the atmosphere. It is also where carbon dioxide enters the plant cells. These air spaces are very humid; they provide an ideal environment for gases to be absorbed into the plant cells.

Stoma (plural stomata) – these arise in the leaf where an epidermal cell has divided into two sausage-shaped guard cells.

The guard cells control the outflow of water and oxygen and the inflow of carbon dioxide. The guard cells open during daylight hours providing the plant has a sufficient supply of water. When the guard cells are open water can escape into the atmosphere. Carbon dioxide can enter the stomata and passes into the cavities just inside; it is then available for use in photosynthesis.

If the plant is short of water the stomata will close to reduce water loss. However, this reduces the intake of carbon dioxide, which therefore reduces photosynthesis and the plant's food manufacturing.

The leaf tissue has a number of intercellular spaces, which allow the free movement and exchange of gases within the leaf.

Leaf abscission

Deciduous plants drop their leaves every autumn. This is a response to lower temperatures and shorter days.

At the base of the petiole where it joins the stem is an abscission zone. In the autumn this area releases enzymes that weaken the cell walls to start the abscission process. During windy or frosty weather the cell walls break away and the leaf drops off. Cells on the stem side now form a protective scar tissue to protect the stem. This is the leaf scar tissue seen on the twigs of plants.

PLANT ORGANS

Cells – plant cells come in a range of shapes and sizes. Their internal structure varies depending on the function they carry out. Typical functions include transporting water, photosynthesis, reproduction and protection.

The cell is the smallest living unit from which plants and animals are made.

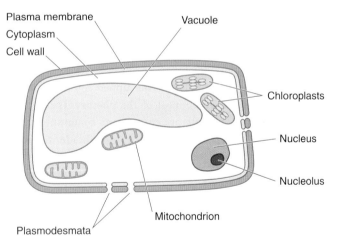

Figure 2.39 Diagram of a typical plant cell.

The diagram shows a basic plant cell and its main component parts. The plant cell is made up of various parts, as described below.

Cell wall – this holds the cell together and forms its shape. On the inside of the wall and in close contact is the plasma membrane: this is a sac that contains the rest of the cell. Everything contained within the plasma membrane is called the symplast; everything on the outside of it is called the apoplast.

The cell wall contains thin strands of cytoplasm that pass through it and connect to adjacent cells. The strands are called plasmodesmata, their function is to allow water and minerals to pass between cells.

Protoplasm – this is inside the plasma membrane and is the living part of the cell, which is made up of two parts: the nucleus and cytoplasm.

Nucleus – this is where the chromosomes containing the **DNA** are located, they hold all the genetic information for the cell.

Cytoplasm – this contains most of the organelles and is where most of the cell's metabolism takes place. The cytoplasm is a liquid jelly-like substance and is contained within a cytoplasm membrane. The liquid jelly contains the organelles that carry out the various functions of the cell.

One type of organelle is a chloroplast within which photosynthesis takes place; the chloroplasts contain the green pigment chlorophyll.

The mitochondria are another type of organelle. Cell respiration takes place in the mitochondria, extracting the energy required by the plant.

Vacuole – this is a sac filled with fluid. Vacuoles may contain water stored for the plant's later use and may also store excess minerals and any waste products from cell metabolism. Vacuoles tend to act as storage areas within the plant tissues.

Plant tissue – a tissue is a collection of plant cells that have differentiated into different cells with a common function. A simple tissue is made up of one cell type; whereas a complex tissue is made up of several different cell types.

These tissues have evolved to carry out different functions within the plant. An organ is a collection of tissues with a specific function, e.g. flower, leaves and stem.

Plant tissue can be living, as in parenchyma, or dead, as in mature sclerenchyma.

Living tissue can store food, transport materials around the plant and reproduce. The tissues that make up potatoes and apples are parenchyma tissue used to store carbohydrates. The phloem tissue that transports food around the plant is living parenchyma tissue.

THE MAIN INTERNAL PLANT FUNCTIONS

PHOTOSYNTHESIS

Photosynthesis is a complex process by which carbon dioxide (CO_2) and water (H_2O) are made into carbohydrates for the plant's use. Photosynthesis converts light energy into chemical energy. Using photosynthesis plants are able to manufacture complex food materials from simple inorganic raw materials. Below is a simplified description of photosynthesis, sufficient for Level 2. Students wishing to gain a greater understanding should look at

TYPES OF PLANT TISSUE

Parenchyma – this is the most common tissue type found in plants and makes up the bulk of the organs, such as the leaves, young stems and roots. Parenchyma cells have a thin, flexible cell wall. Their shape can vary depending on their function and some tissues have large intercellular spaces in them (leaf stoma cavities).

Parenchyma cells have large vacuoles and often store starch, water and other food reserves. The parenchyma cells in the leaf are where photosynthesis takes place.

Parenchyma cells often retain the ability to divide by mitosis. This can be used to form **callus tissue** if the plant is wounded and helps the wound to heal.

Collenchyma – this is similar to parenchyma but is not as common. It is made up of living cells, which are stronger and have thickened cell walls. Collenchyma is used to provide strength and support and is often found under the epidermis in stems. Cotton is made up of collenchyma cells.

Sclerenchyma – this is a supporting and strengthening tissue made up of dead lignified cells. Sclerenchyma forms two types of tissue:

1. *Sclerids* – these are small groups of cells and form a tough wood-type tissue. The stones in plums, *Prunus domestica*, and cherries, *Prunus avium*, are made of sclerid tissue.
2. *Fibres* – these are long strands of elongated cells. The fibres in New Zealand flax, *Phormium tenax*, are fibre tissue.

The tissues that make up the xylem are usually sclerenchyma tissue. These cells have been lignified to give them strength and basically form tubes that transport water up the plant. Sclerenchyma consists of dead, thick-walled cells.

Chapter 4 and consult books on plant biology (see the further reading list).

Photosynthesis takes place in chloroplasts in the leaves and stems. It has two main phases: the light reaction and the dark reaction (also called stroma reaction).

Phase 1

In the light reaction chlorophyll *b*, carotene and xanthophylls absorb light energy. This is channelled to chlorophyll *a*, which is 'charged' to a high energy potential. In this state the chlorophyll *a* electrons extract and store the energy for use in the synthesis of sugar. The electrons in chlorophyll *a* are used to split water into hydrogen and oxygen. So the first phase in photosynthesis splits water into hydrogen and oxygen. The oxygen is released into the atmosphere via the stomata. The hydrogen is retained for the next phase.

Phase 2

The second phase of photosynthesis is the fixation of carbon dioxide from the atmosphere. The carbon dioxide enters the plant via the stomata and diffuses through the air spaces in the leaf. It is dissolved in moisture on the cell walls that are adjacent to the air spaces. It then diffuses through the cell walls into the palisade cells and chloroplasts. The carbon dioxide is united with the hydrogen from the first phase. This forms molecules that come together to form the sugars glucose and fructose.

Photosynthesis equation

$$6\,CO_2 + 6\,H_2O + \text{Light Energy} = C_6H_{12}O_6 + 6\,O_2$$

The sugars produced by photosynthesis are converted by enzymes into sucrose and starch. The starch can then be transported to other parts of the plant for use or stored. Starch is stored in the plant and can be used in respiration to produce energy for growth.

RESPIRATION

This is the reverse of photosynthesis: respiration burns or oxidises glucose produced by photosynthesis and provides usable energy for both plants and animals. Respiration is carried out in every living cell; to provide the energy it uses carbohydrates and breaks them down into glucose molecules. It then uses the glucose to produce energy-rich ATP molecules.

There are three main phases of respiration:

1. Glycolysis
2. Krebs cycle
3. Electron transport chain

Glycolysis

This takes place in the cytosol of the cell and uses some energy to break down the carbohydrate into glucose molecules. Enzymes then break the molecule apart and form pyruvate.

Krebs cycle

The pyruvate is moved from the cytosol to a mitochondrion where enzymes further break down the pyruvate. The Krebs cycle now starts and the molecules are oxidised to produce carbon dioxide and energy-yielding molecules (6 NADH, 2 ATP and 2 FADH$_2$ molecules).

Electron transport chain

The Krebs cycle produces only a small amount of energy; the vast amount is produced during the electron transport phase. This takes place in the inner mitochondrion where proteins transfer electrons down a chain. At the end of this process between 32 and 34 ATP molecules are produced.

Respiration equation

$$C_6H_{12}O_6 + 6\,O_2 = 6\,CO_2 + 6\,H_2O + \text{Energy (ATP)}$$

The energy produced from respiration is used for plant growth and other metabolic functions within the plant.

TRANSPIRATION

This is the loss of water vapour from the plant, mostly through the stomata on the leaves. There is some water loss through the cuticle (covering the epidermis) and lenticels (on the stems). The transpiration stream is the flow of water from the root hairs through the vascular system to the leaf surface where the water evaporates. Up to 98% of the water taken in by the roots is lost through transpiration.

Transpiration serves two functions:

1. Water lost through the leaves cools the plant preventing cell damage caused by overheating on hot days.
2. It moves minerals dissolved in the water up the plant to be used in various parts of the plant.

Water is taken into the plant by osmosis, see Chapter 4 for further details. As explained earlier, in the section on roots, the water passes through the epidermis and cortex and on to the endodermis. At the endodermis the water is 'pumped' up the xylem and starts its movement up the plant to the leaves. Once in the xylem the water moves up by a combination of:

1. Root pressure – the endodermis pushes the water up the xylem tubes. This can be seen when a plant is cut down and sap oozes out of the cut surface.
2. Transpiration pull – as water is transpired from the leaves it pulls more water up from the xylem in the leaf, which in turn pulls water up the xylem in the stem. Also see Chapter 4 for further details on this function.
3. Capillary action – this is a result of water molecules having a tendency to stick together and to adhere to the sides of narrow tubes. This effect helps the water move up narrow tubes; water molecules tend to cling to the molecules above them and this pulls the water up the xylem tube.

Transpiration is what enables water to rise to the tops of tall trees in large quantities, especially when it is hot and sunny and the tree is losing vast amounts of water to keep cool.

The water moves up the xylem in the stem and is distributed around the plant to the various stems/branches. It enters the leaves via the petioles and diffuses out of the xylem into intercellular spaces in the spongy mesophyll. It then diffuses out of the leaf via the stomata. Some water is also lost through the epidermis.

The stomata are normally open during daylight and close at night. When a plant is short of water the stomata will close to try to prevent water loss, if this happens photosynthesis will virtually stop as carbon dioxide cannot enter the plant.

The transpiration rate depends on:

✿ humidity of the air surrounding the leaf/plant
✿ wind speed
✿ moisture level in the soil
✿ light energy/temperature.

The lower the humidity surrounding the plant the higher the transpiration rate, as water vapour is easily absorbed by the surrounding air. If the relative humidity is high there is less water lost from the plant as the atmosphere has difficulty holding any more water vapour. If the temperature increases, the amount of water the air will hold will also increase, which will increase the transpiration rate.

On still days the humidity surrounding the plant will stay fairly constant, but as the wind increases the air around the plant moves away from the plant taking the moisture with it. This allows more water to evaporate from the leaves. Windy days will therefore increase the rate of transpiration.

Transpiration will continue as long as there is available water in the soil. If the soil becomes dry the plant cannot absorb water as easily and therefore the transpiration rate will slow down. In drought conditions or if the soil freezes and the plant cannot absorb water, transpiration will stop and the plant will wilt.

On hot days a plant may wilt temporary during the day because water is lost at a higher rate than it can be replaced, but the plant will usually recover overnight when the transpiration rate is lower.

Plants growing in the desert and in other hostile conditions have adapted to reduce their transpiration rate. These plants are called xeromorphic and have the following adaptions:

✿ thicker waxy cuticles
✿ smaller or hairy leaves
✿ no leaves – as in cacti
✿ fewer stomata on their leaves, which may be sunken into the epidermis to reduce water loss.

Guttation

This is caused by root pressure pushing water up the xylem to the stem and leaves. During the night the stomata are closed so little water can escape, so to prevent damage to the cell structure the water is allowed to escape out of openings called **hydrathodes**. These allow droplets of water to escape and during the morning these evaporate from the leaf. The hydrathodes are positioned at the edges and tips of leaves. In grasses they are at the tips, and in *Fuchsia × hybrida* and *Alchemilla mollis* they are positioned along the edges. The droplets of water can easily be seen early in the morning along the leaf edges, although you should not confuse them with dew.

PLANT AUXINS

These are also called plant hormones and plant growth regulators (PGRs). They are chemicals that control the growth of plants in some way. Different chemicals have different responses, some will make the plant smaller, others larger and some will encourage flowering. The synthetic chemicals produced by humans usually mimic natural chemicals produced by the plant. The natural plant auxins occur in the plant in very low concentrations; these are produced in various parts of the plant and have different effects on the plant. Some of the effects on plant growth include:

- flower initiation, opening and whether it is male or female if the plant is monoecious
- bud dormancy growth
- ripening of fruit
- growth of stems, roots and leaves
- direction of growth.

There are five major classes of plant hormone at present, although this may increase in the future as new discoveries are made. The five classes are based on the chemical structure and the effect the hormone has on the plant.

These are:

- abscisic acid (ABA)
- auxins
- cytokinins
- gibberellins
- ethylene.

Abscisic acid

This was first identified in 1963 as a naturally occurring compound in plants. It is produced in plastids, including chloroplasts. Abscisic acid is produced as a response to stresses in the plant, such as water loss, drought or low temperatures. It can be moved around the plant in any direction.

Functions of abscisic acid – Abscisic acid encourages the closing of stomata (caused by water stress). It inhibits shoot growth, but does not seem to affect root growth. It also inhibits the effect of gibberellins and has some effect on the induction and maintenance of dormancy.

Auxins

The term auxin is used to cover a number of compounds that are similar to indole-3-acetic acid (IAA), which was the first auxin isolated from plants. Auxins were first discovered in the 1880s and Darwin carried out some experiments to determine the effects of auxins.

At present, research indicates that most auxins are produced in the tip of the stem and moved down to the site required. Research is still being carried out and other sites and effects may be discovered in the future.

Functions of auxins – Auxins stimulate cell elongation and division; stimulate the differentiation of phloem and xylem; and stimulate root initiation on stem cuttings. They mediate the effects of the tropisms of light and gravity, and suppress lateral bud development. They can induce fruit set and growth in some plants, and can delay fruit ripening. They stimulate the growth of flower parts in some plants.

Cytokinins

These can be naturally occurring or synthetic substances. Kinetin is the most common synthetic cytokinin; and zeatin,

which is found in sweetcorn, *Zea mays*, is the most common natural cytokinin. More than 200 cytokinins have been discovered and work continues to investigate the effects they have on plants.

Cytokinins are found at highest concentrations in the meristematic areas; they appear to stimulate cell division here. They are believed to be synthesised in the root and translocated up the xylem to the shoots.

Functions of cytokinins – Cytokinins stimulate cell division, stimulate the growth of lateral buds and stimulate leaf expansion.

Gibberellins

These are usually called gibberellic acids (GA_3) and given a number in the order they have been discovered. At present there are nearly 140 different gibberellic acids identified in plants and fungi.

Gibberellic acid is believed to be synthesised in young stems and seed tissues. Some growth regulators sold to 'dwarf' plants work by blocking the synthesis of gibberellic acid. This reduces the length of the internode and therefore the stem length and plant height.

Functions of gibberellic acids – Their functions vary depending on the type of gibberellic acid. They can stimulate stem elongation; stimulate flowering in response to long days; break seed dormancy in some plants; delay senescence in leaves and some fruits.

Ethylene

This is the only gaseous hormone known at present and is produced in all higher plants. It is associated with fruit ripening and appears to stimulate leaf abscission in conjunction with abscisic acid.

Functions of ethylene – Ethylene stimulates the breaking of dormancy; stimulates leaf and fruit abscission; promotes flowering in Bromeliads; stimulates flower opening; stimulates fruit ripening; and stimulates flower and leaf senescence.

It is likely that auxins of all types will be more widely used in the future for controlling plant growth, flowering and fruiting. They may lead to improved yields in some crops.

EUDICOTS

The term eudicot was introduced during the early 1990s and covers many plants that were formally classed as dicotyledons (having two seed leaves). The term has become more widely used since the millennium, as further research has led to a reclassification of the dicotyledons.

Recent research based on molecular phylogenetic evidence has indicated that dicotyledons are a natural or monophyletic group. At present there are approximately 320 families in the eudicots and these include 75% of the flowering plant species.

Phototropism – the action of plants growing towards the light is controlled by the auxin IAA. This is produced in the apical meristem and moves down the stem, it tends to move away from light and accumulates on the shaded side of stems. Indole-3-acetic acid encourages cell elongation and therefore the cells on the shaded side grow more than on the light side. This makes the stem grow towards the light.

Stem length – gibberellins increase the length of the internodes of stems. In good light conditions gibberellins seem to have little effect, but in poor light they encourage an increase in stem length so the plant grows taller and towards the light.

Dwarfing – there are a number of synthetic chemicals that can be used to reduce the stem length of plants. These are now widely used to produce short squat pot plants in commercial production. Plants including chrysanthemum, *Chrysanthemum* spp., and poinsettia, *Euphoria pulcherrima*, are often treated.

A product called Primomaxx is becoming widely used in the golf and sports-turf industry to control grass growth and produce dense short leaves on the grasses, producing a better surface for putting. It also appears to encourage root growth, which gives a better grass plant.

The eudicots are roughly divided up into the early diverging eudicots and core eudicots.

The early-diverging eudicots are very similar to the basal angiosperms in their floral structure. They have apetalous flowers with flower parts in twos or threes and a variable number of stamens and carpels. The early-diverging eudicots include the *Ranunculales* (buttercup order).

The core eudicots have flowers with parts in fours and fives and a clear differentiation in the sepals and petals. They have one or two stamen whorls and a gynoecium of three to five carpels. The core eudicots are the largest group of eudicots and include *Rosa, Aster, Brassica* and legumes.

One of the main morphological features of the eudicots is the presence of pollen grains having three grooved apertures. This has led to some authorities naming them tricolpates, which can be confusing.

The relationship between the eudicots and other flowering plants is still being researched. It is likely that there will be further changes in the classification of plants within the eudicot group in the future. See Chapter 1 for further details.

FLORAL FORMULA

Flowers are a key feature in helping to identify plants. To help ensure correct identification, the flower features need to be accurately described and where possible herbarium specimens and photographs used. It is important that everyone follows the same method of describing plants and includes the same details. To ensure this happens floral formulas and floral diagrams were developed.

The floral formula is a type of shorthand to describe the flower parts and their arrangement. A floral formula is often accompanied by a floral diagram, which is a schematic diagram of the flower arrangement; see Figure 2.40.

It is not a picture or drawing of the actual flower, but a diagram that shows the number and relationship of the flower parts in a formal way.

The floral formula tells you:

1. Whether the flower is **actinomorphic** or **zygomorphic**
2. Whether the flower has an inferior or superior ovary
3. The number of sepals, petals, stamens and carpels
4. Whether they are free of joined
5. Whether the parts of any whorl are united to another whorl.

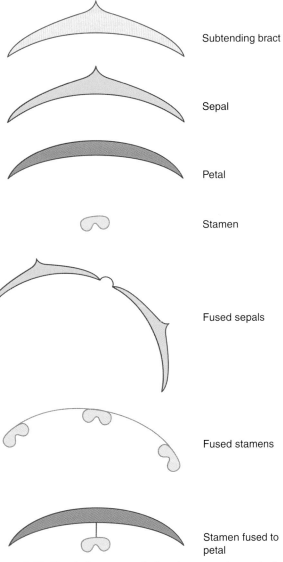

Figure 2.40 Floral diagram symbols – these are the symbols used to illustrate floral diagrams.

O – actinomorphic

⊙ – zygomorphic

K – calyx, sepals, the adjacent number indicates how many

C – corolla, petals, the adjacent number indicates how many

A – androecium, stamens, the adjacent number indicates how many

G – gynoecium, carpel, the adjacent number indicates how many

( ) – if the number is in brackets they are joined

(– if one whorl is joined to another

P – perianth, used if not a separate calyx and corolla but a perianth

♀ – female flower

♂ – male flower

X – hermaphrodite, bisexual, flower

∞ – many usually more than 12

G̲ – superior ovary

G̅ – inferior ovary

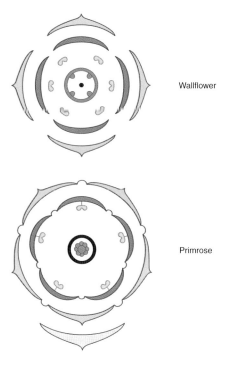

Wallflower

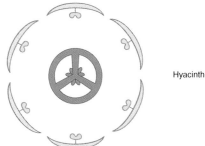

Primrose

Hyacinth

Figure 2.41 This shows the floral diagram for wallflower, *Erysimum cheiri*, primrose, *Primula vulgaris*, and hyacinth, *Hyacinthus orientalis*.

The recording of a floral formula is always in the same order, that being: (1) flower structure – actinomorphic or **zygomorphic**; (2) calyx; (3) corolla; (4) androecium (5) gynoecium.

Two examples of floral formulas are:

❀ Wallflower: O, K4, C4, A4+2, G̅(2)

❀ Tulip: O, P3+3, A3+3, G̲(3)

FLORAL DIAGRAM

This is a basic plan diagram of the flower: the flower parts are represented by recognised symbols. It is always drawn so that the bract (if present) is facing the artist and the flower axis is on the far side.

The floral parts are drawn using accepted symbols, see Figure 2.41.

These symbols include the axis of the inflorescence and any bract, which subtends the flower if present.

Figure 2.42 Photograph of the flower of a wallflower, *Erysimum cheiri*, for comparison to the floral diagram.

Chapter summary

Chapter 2 has covered the structure and workings of plants. You should have an understanding of the following key points.

✿ *Plant life cycles* – ephemerals are plants with a life cycle of less than a year; annuals have life cycles of approximately a year; biennials have life cycles of two years; and perennials have life cycles of over two years.

✿ *Geophytes* – are plants that have evolved part of their structure to enable them to overwinter or spread. Bulbs, corms and tubers are organs that are capable of storing food and surviving over winter.

✿ *Root system* – the function of the root system is to anchor the plant in the soil and absorb water and minerals. Some have the ability to store food and reproduce vegetatively. There are tap and fibrous root systems, which have lateral roots and root hairs. Many plants have modified roots that enable them to grow in difficult conditions. The internal structure of the root consists of epidermis, cortex, endodermis and stele. Within the stele are the phloem and xylem, which are the main transport systems.

✿ *Stem system* – the stem is the main transport link between the roots and leaves. The stems also hold the leaves in the best position for photosynthesis and the flowers for pollination. The stem consists of nodes and internodes with apical buds at the tips and axillary buds at the nodes. The internal structure changes as the stem ages owing to secondary thickening from the second year. Many plants have evolved modifications of the stem to enable the plant to grow in a range of environments. Internally the stem consists of the epidermis, cortex, phloem, cambium and xylem; other tissues develop as the plant ages.

✿ *Leaves* – the leaf is the powerhouse of the plant as this is where photosynthesis occurs. Transpiration is the other main function of the leaf. The size, shape and arrangement of the leaves vary between plants and are a useful means of identification. The leaf has an outer layer of cutin over the epidermis under which is the palisade layer where most of the photosynthesis takes place. The lower half of the leaf is made up of spongy mesophyll and lower epidermis with stomata present where gaseous exchange takes place.

✿ *Plant organs* – plants are made up of different types of cells, which come together to form tissues with a common function. The tissues are made up of parenchyma, collenchyma and sclerenchyma.

✿ *Plant processes* – photosynthesis is the process of making food using sunlight, carbon dioxide and water, oxygen is produced as a by product. Respiration is the burning of the food to enable the plant to grow. Transpiration is the movement of water and minerals from the root to the leaves.

✿ *Plant hormones* – are chemicals produced within the plant to control its growth and other functions such as flowering. The five main hormones are abscisic acid, auxins, cytokinins, gibberellins and ethylene.

✿ *Eudicots* – recent research has led to many plants being reclassified from the dicotyledon plant group to eudicots. Research work is continuing to complete this process.

✿ *Floral formulas and diagrams* – a shorthand method of describing and drawing flowers to help in the identification of plants.

Review questions

1. List and briefly **describe** the main plant life cycles.
2. Name and describe **four** types of geophytes.
3. List the **four** functions of roots and name **four** types of modified roots.
4. Draw a labelled diagram showing a **cross section** of a young dicotyledon stem.
5. Draw a **transverse** diagram of a dicotyledon leaf showing all the main parts.
6. Describe the **internal** structure of a dicotyledon root.
7. Name and describe the **two** primitive plants that have a symbiotic relationship with roots.
8. Draw and label a typical plant cell and state the **functions** of the main parts.
9. Briefly describe the **processes** of photosynthesis and transpiration.
10. Name **four** plant hormones and state the effects they have on plants.

FURTHER READING

Capon, B. (2005). *Botany for Gardeners*. Timber Press.
This book is a good introductory guide to botany and explains plant structure and function. It is related to what the gardener is required to know in order to grow and understand plants.

Cullen, J. (2006). *Practical Plant Identification: Including a Key to Native and Cultivated Flowering Plants in North Temperate Regions*. Cambridge: Cambridge University Press.
A publication that highlights practical ways to improve your skills in plant identification.

Ingram, D. S., Vince-Prue, D. & Gregory, P. J. (eds) **(2008)**. *Science and the Garden: the Scientific Basis of Horticultural Practice*, 2nd edn. RHS, Blackwell Publishing. ISBN 978-1-4051-6063-6.
This book covers more extensively the material discussed in the chapter.

Lack, A. J. & Evans, D. E. (2001). *Instant Notes: Plant Biology*. BIOS Scientific Publishers Limited.
Short notes on plant biology, each chapter covers a different topic briefly but gives the key points.

Mauseth, J. D. (1988). *Botany: an Introduction to Plant Biology*. Jones and Bartlett Publishers.
This is a well-illustrated book covering botany to a good depth yet easy to follow.

RHS level	Section heading	Page no.
2 2.2	Plant life cycles	26
2 2.3	Case study: Annual plants	26
2 2.4	Half-hardy perennials	27
	Tender perennials	27
	Perennials: woody plants	27
2 2.5	Perennials: woody plants	28
2 3.1	Plant organs	46
2 3.2	Root structure	41
	Stems	43
2 3.3	Vacuoles	46
2 4.1	Root system	29
2 4.2	Root system	29
2 4.3	Root modifications	30
2 4.4	Stem system	31
2 4.5	Stem modifications	32
2 4.6	Leaves	33
2 4.7	Leaf modifications	39
2 4.8	Internal structure of plants	40
2 7.1	Photosynthesis equation	47
2 7.2	Photosynthesis	46

CASE STUDY Plant responses to light

Acting as a source of information, light controls many plant responses that are important in the garden. A common example is the greening of potato (*Solanum tuberosum*) tubers where light induces the production of a toxic alkaloid. In this case, it is the exclusion of light that is important.

A less commonly recognised example is the effect of light to inhibit the development of adventitious roots on stem tissues. Keeping shoots in the dark is remarkably effective in increasing the formation of adventitious roots. The inhibitory effect of light is greatest in red light indicating that it is mediated by the pigment, phytochrome (see main text). This effect of light is used in the garden by inserting the shoot into the rooting medium to maintain the base in darkness. Similarly, the production of adventitious roots is increased by the practice of earthing up the base of plants that are to be used as a source of hardwood cuttings or clonal rootstocks. As with the greening of potatoes, it is the exclusion of light that is important here.

Light through trees.

Chapter 3
Light

Daphne Vince-Prue

INTRODUCTION

The 'feel good factor' of good light levels is well known to humans with the pleasing emotional response to sunshine and the longer days of summer being welcomed after a long, dark winter.

However, green plants (all those with leaves or stems that contain chlorophyll) are completely dependent on light for many aspects of their growth and development. In particular, light energy is required to fuel the process of photosynthesis, which results in the production of carbohydrates and, subsequently, leads to all of the other organic components of plants. Light is also an important source of information for plants, giving them the ability to sense their light environment and, in many cases, also their seasonal environment.

This chapter describes how plants use light under different environmental conditions to ensure growth is continued to successfully complete their life cycles and set seed to provide future generations. The chapter also considers how gardeners can manage light most efficiently in the context of the garden, especially as climate change is likely to lead to altered seasonal light levels. The interrelationship of light with temperature and water availability leads on to future management and the opportunities presented with the balance between light and shade. Shade perception and plant adaptations to shade are also introduced to provide a greater understanding of this common, but somewhat challenging, environment.

In order to understand the ways in which light produces these effects, it is important to know something about the physical properties of light itself, as a starting point, to enable the most successful outcome of plant growth in the widest range of garden environments.

Key concepts

- ✿ Light is essential for plant growth
- ✿ The definition of light and its measurement
- ✿ Leaf structure and light capture
- ✿ Light as an energy source: photosynthesis
- ✿ Light as a source of information

The Fundamentals of Horticulture: Theory and Practice, ed. C. Bird. Published by Cambridge University Press. © The Royal Horticultural Society 2014.

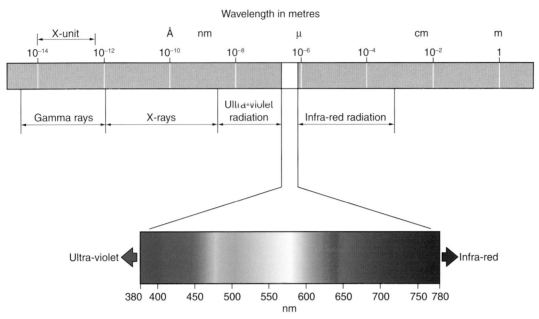

Figure 3.1 The electromagnetic spectrum showing regions most visible to the human eye, measured in nanometres (nm).

WHAT IS LIGHT?

Light is defined as the part of the electromagnetic spectrum that we can see, as shown in Figure 3.1.

For most people this occurs at wavelengths between about 350 and 750 nanometres (nm), although it varies somewhat between individuals. This band covers the spectrum of the rainbow from the shorter wavelengths of blue light, through green to the longer wavelengths of red light. Not all animals see the same as we do. Bees, for example, are sensitive at shorter wavelengths than 400 nm into the ultra-violet, which can be damaging to both plants and people. At wavelengths longer than 700 nm is the infra-red region, which primarily has a heating effect. However, the near infra-red region between 700 and 750 nm (now called **far-red**) is an important source of information for green plants (see below).

For most garden situations it is not necessary to measure light, as areas of relative shade and sun can be easily detected visually. On the other hand, it is important to measure light when supplementary artificial lighting is used under protected cultivation to stimulate growth in winter or to extend the natural day-length.

To be effective in any light-dependent process in plants, light must first be absorbed by a pigment. For photosynthesis (see below), the green chlorophylls and some associated yellow/orange carotenoids are the main light-absorbing pigments. Light absorption occurs throughout the visible spectrum but is least in the green region: hence the green colour of most leaves. A complication is that light is absorbed in the form of discrete units of energy called **quanta** (or **photons** for energy within the visible spectrum). The amount of energy in each quantum is inversely proportional to the wavelength of light so that for the same measured energy

LIGHT MEASUREMENT

Measuring the amount of light that is available to plants is not straightforward because the wavelengths of greatest importance to plants are not those that relate to human vision. Many simple measuring devices are calibrated in photometric units and measure light in terms of human vision. The measured amount of light is expressed as illuminance and the units are **lux** (an obsolete unit that is still sometimes encountered is the foot candle; 1 ft candle = 10.76 lux). The human eye is most sensitive to green light and is relatively insensitive to red and blue, which are the most important wavelengths for plants. Consequently, light measurements expressed in lux can be misleading when considered in relation to plants.

Light can also be measured in terms of radiant energy using radiometric detectors that measure all wavelengths equally. This can also be misleading because detectors of radiant energy are sensitive to wavelengths that are outside the range that is active in photosynthesis. One advantage of photometric detectors is that they measure only radiation that is within this range. Most simple detectors that are available to gardeners measure in photometric units (lux). These can be converted to radiometric units (watts per square metre; W m^{-2}) by using the conversion units given in Table 3.1.

(W m^{-2}) there will be fewer quanta available for photosynthesis in the shorter wavelengths of blue light at about 450 nm than in the longer wavelengths of red light at about 650 nm.

Recommendations for winter supplementary lighting are usually given as W m^{-2}. This can most easily be measured using a simple photometric detector and converting the lux values to W m^{-2} using Table 3.1.

Type of light	Conversion factor[a]
Natural sunlight	4.0
Tungsten filament lamp (199 W)	4.2
High pressure sodium lamp (SON)	2.4
Tubular fluorescent lamps:[b]	
Warm white	2.8
Deluxe warm white	1.6
Daylight	3.7

[a] Multiply by this factor to convert measurements in lux to $mW\ m^{-2}$.

[b] The conversion factors for the various colours of tubular fluorescent lamps may vary with the manufacturer but the values given are adequate in practice.

Table 3.1 Factors used for converting illuminance (lux) to irradiance (milliwatts per square metre, $mW\ m^{-2}$)

WHY GREEN PLANTS NEED LIGHT

When light is absorbed by green plants, the energy is used to convert water into its component parts of hydrogen (H) and oxygen (O_2). The overall process is known as the 'light' reaction of photosynthesis. Oxygen is released into the air and is the main source of the oxygen in the Earth's atmosphere. In several subsequent steps, hydrogen is used to reduce carbon dioxide (CO_2), which has been obtained from the air surrounding the leaf. These reactions do not need light and so are termed the 'dark' reactions of photosynthesis. The end products of many further reactions are carbohydrates, mainly sugars.

The overall process of photosynthesis is shown diagrammatically in Figure 3.2 and can be summarised as follows:

$$2H_2O + CO_2 + light\,energy \rightarrow O_2 + (CH_2O) + H_2O$$

(Water + carbon dioxide + light energy →
oxygen + carbohydrate + water)

Through photosynthesis plants on land and in the oceans and rivers maintain the atmosphere we breathe by consuming CO_2 and releasing O_2 and either directly or indirectly provide the food source for almost all living organisms.

THE LEAF

Although photosynthesis occurs in all green parts of the plant, the principal site is the leaf. The expanded external surfaces maximise the collection of light and the internal structure facilitates the movement of the gases CO_2 and O_2 between the air and the chloroplasts where photosynthesis takes place.

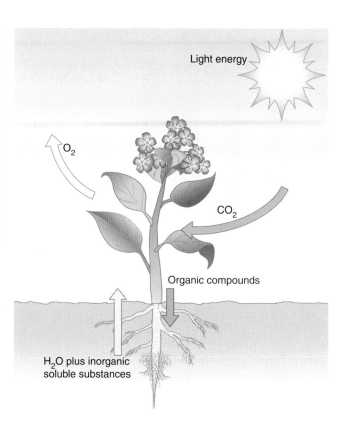

Figure 3.2 A diagrammatic representation of the basic reactions of photosynthesis.

Leaf structure

The leaf surface is covered with a waterproof layer, the cuticle, which provides an effective means of reducing water loss from the leaf. Below this layer are the epidermal cells. Special openings in the epidermis (the **stomata**) allow for the necessary gaseous exchange of water vapour, O_2 and CO_2 between the leaf and the external air. For how stomata control this exchange see Chapter 4.

The collection of light energy and most of the photosynthetic processes occur in the *palisade cells*. These consist of one or more layers of elongated cells containing numerous chloroplasts. Plants growing in the shade where light levels are low usually have only one layer of palisade cells, while plants growing in strong sunlight have two or even three layers in order to maximise the absorption of light energy. Below the palisade layer, the cells of the **spongy mesophyll** are irregularly shaped and contain few chloroplasts. There are large air spaces between the cells, especially around the stomatal pores where they form the **sub-stomatal cavities**. A simplified diagram of the overall structure of a typical dicotyledon leaf is shown in Figure 3.3.

The chloroplasts contain the chlorophylls (chlorophyll *a* and chlorophyll *b*) and accessory pigments (carotenoids) that are responsible for the absorption of light energy and its transfer to the reaction centres of photosynthesis.

Transport out of the leaf

The organic compounds synthesised in photosynthesis are transported out of the leaf via the veins. These also serve to bring in

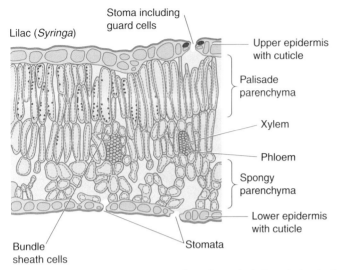

Lilac (*Syringa*)

Stoma including guard cells
Upper epidermis with cuticle
Palisade parenchyma
Xylem
Phloem
Spongy parenchyma
Lower epidermis with cuticle
Stomata
Bundle sheath cells

Figure 3.3 Diagrammatic cross-section of a typical dicotyledon leaf.

substances that are required for the functioning of the leaf itself. The main veins of most monocotyledons, such as grasses, are roughly parallel to one another along the main axis of the leaf, whereas the veins of dicotyledons form a complex network of branches. The leaf veins (or **vascular bundles**) consist of **xylem** tissues (vessels and fibres), **phloem** tissues (sieve tubes and companion cells) and **cambium**. They are surrounded by a single layer of cells called the **bundle-sheath**.

The phloem cells of the leaf transfer materials to the phloem tissues of the stem and from here they move both upwards to the shoot apex and young leaves, and downwards to the roots. The carbohydrate products of photosynthesis are usually converted to sucrose for transport in the phloem. Xylem vessels have no living contents and during development they lose the end cell walls and become linked end to end to form continuous open tubes. The xylem of the leaf veins is linked to that of the stem and thereby provides an unbroken cluster of channels from root to leaf. The major function of the xylem vessels is to transport water from the roots to the leaves as discussed in Chapter 4, but they also have a key role in transporting hormones synthesised in the root, such as abscisic acid, and minerals absorbed from the soil to the upper parts of the plant. Conifers and their relatives do not form vessels and instead possess less efficient water transporting cells called **tracheids**. Like vessels, they lose their living contents during development but they have closed ends and do not form unbroken tubes. Transport from cell to cell occurs through permeable areas of the impermeable walls, called pits.

The phloem cells of the leaf transfer materials to the phloem tissues of the stem and from here they move both upwards to the shoot apex and downwards to the root. The carbohydrate products of photosynthesis are usually converted to sucrose for transport in the living tissues of the phloem.

WHY LIGHT IS IMPORTANT IN THE GARDEN

Plant growth is directly (or indirectly when using storage materials) affected by the amount of photosynthesis. Consequently it is very important to understand what factors affect the amount of photosynthesis and which of these factors can be controlled or modified in the garden.

External factors that affect photosynthesis

A number of external factors influence the amount of photosynthesis that occurs. These include light (both intensity and daily duration), temperature, water and carbon dioxide concentration. As discussed above, photosynthesis is a complex process that involves reactions that are dependent on light as well as dark reactions that are dependent on CO_2 and temperature. Consequently the responses of plants to these environmental factors are also complex and any one of them can be limiting to the rate at which photosynthesis is occurring and ultimately lead to the cessation of growth.

Light quantity – When a single leaf is exposed to different amounts of light at normal atmospheric concentrations of CO_2, the rate at which photosynthesis occurs increases with increasing light levels until, above a certain value, there is no further increase as shown in Figure 3.4. This value is known as the **light-saturation point** and it varies with both species and the past light history of the plants. Plants with C-4 type photosynthesis (see below) usually have higher light saturation values than plants with normal C-3 photosynthesis.

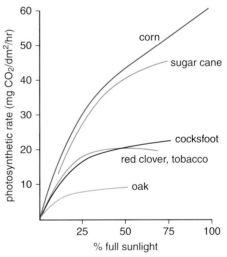

Figure 3.4 The relationship between light intensity and the rate of photosynthesis.

Because less light can penetrate to the lower leaves of the plant, the rate of photosynthesis of the whole plant usually increases with the amount of light being received to a much higher value than for single leaves. Consequently growth and subsequently yield are both influenced by the total amount of light received. For example, for tomatoes, *Lycopersicon esculentum*, growing under glass, it was found that a 1% reduction in light resulted in a 1% reduction in yield. It should also be noted that the amount of photosynthesis (and the yield) depends on the daily duration of light as well as on its intensity at any one time, that is on the total amount of light received, which is known as the **light integral**.

There are only a few ways in which gardeners can influence the amount of light received by the plant. One obvious way is the choice of shady or sunny locations within the garden and plants that are suited to these conditions. Some examples of plants that are suitable for growing in sunny or shady conditions are listed in Table 3.2.

Sunny	Shady (dry)	Shady (damp)
Broom, *Genista*	Ivy, *Hedera helix*	Blueberry, *Vaccinium corymbosum*
Catmint, *Nepeta*	Periwinkle, *Vinca*	Bog myrtle, *Myrica gale*
Lavender, *Lavandula*	Portuguese laurel, *Prunus lusitanica*	Cardinal flower, *Lobelia cardinalis*
Marjorum, *Origanum*	Spotted laurel, *Aucuba japonica*	Dog's tooth violet, *Erythronium*
Rhizomatous iris, *Iris*	Sweet box, *Sarcococca*	Lady fern, *Athyrium filix-femina*
Rosemary, *Rosmarinus*		Wood lily, *Trillium*
Spanish broom, *Spartium junceum*		
Stonecrop (*Sedum*)		
Sun rose		
Thyme (*Thymus*)		
Treasure flower (*Gazania*)		
Wormwood ((*Artemesia*)		
Yucca (*Yucca*)		

Table 3.2 Examples of plants that are suitable for growing in sunny or shady locations

A less obvious way is by the choice of spacing between the plants. More light can penetrate to the lower leaves when plants are spaced more widely apart and, in low light conditions (for example under protected cultivation during winter), wider spacing may achieve better results.

In the UK, the amount of available light is particularly important under glass during the winter months, when both the intensity and the daily duration of light is dramatically decreased compared with the summer months as shown in Figure 3.5. In addition, both plastics and glass reduce the amount of light reaching the plants compared with the open air. Clean glass transmits about 86% of the incoming light while dirty glass significantly reduces this and can cause more than 30% loss of available light. Clearly, clean glass is of great importance, especially in the winter when light is low.

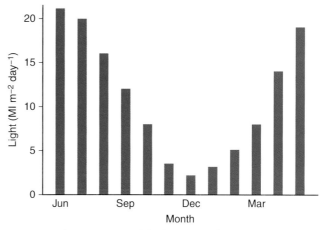

Figure 3.5 The average total daily amount of light received throughout the year at a site on the south coast of the UK.

Nearby shading structures also contribute to the overall reduction in available light and, where possible, it is important to avoid these when initially considering where to locate the glasshouse. A rule of thumb is that a greenhouse should be no closer than four times the height of a nearby shading structure, although this is often not possible in small gardens.

Too much light – It is also important to remember that too much light can also cause problems for plants. For example, plants previously growing in low light conditions can be damaged when they are transferred to bright light. This phenomenon (called **solarisation**) is due to the inhibition of photosynthesis by bright light followed by bleaching of the leaves. It is not commonly encountered in the UK. Most other forms of damage by strong sunlight are due to high leaf temperatures and consequent water stress. These are considered in Chapter 4.

The daily duration of light – The rate at which photosynthesis occurs depends on the level of light at any one time, but the daily total amount of photosynthesis is a product of both the duration of light (the **day-length**) and its overall intensity during the day, that is on the light integral. However, day-length also acts as a source of information (see below) and frequently influences patterns of growth and development such as, for example, increasing leaf size, which in turn may increase the amount of photosynthesis taking place.

Carbon dioxide – When light is not limiting, the rate of photosynthesis increases with increasing external concentrations of CO_2 until a certain level is reached. As with light, therefore, the amount of CO_2 available to the plant can affect the rate at which photosynthesis occurs. The level at which CO_2 becomes a 'limiting factor' depends on both the amount of light and the type of photosynthesis. In normal C-3 type photosynthesis the first product is a molecule with three carbon atoms. Plants with this type of photosynthesis have a strong response to CO_2 and, when light is not limiting, the rate increases until quite high levels of CO_2 are reached. In contrast, photosynthesis in C-4 type plants is generally saturated at about 100 ppm, which is only just above the normal content of CO_2 in the air. For more details about C-4 photosynthesis see Chapter 4.

As with light, there is little that gardeners can do about CO_2 levels. However, this is generally not a problem, except under protected cultivation where the CO_2 level can fall well below that needed for rapid photosynthesis, especially in bright light with inadequate ventilation and a full crop. Gardeners can improve the situation by ensuring that ventilators are open sufficiently early in the day. This increases the exchange between the CO_2 depleted air inside and the normal air outside. Ventilation also increases air movement and so can increase photosynthesis by displacing the CO_2 depleted air immediately around the leaf. The use of fans can augment the air movement. Some commercial growers add CO_2 to the glasshouse atmosphere (see Chapter 11).

Water – The basic features of water uptake and its loss to the surrounding air are discussed in Chapter 4.

If water is lost from the leaf at a faster rate than it is taken up from the soil, the plant wilts and the stomata close. Consequently photosynthesis decreases or may cease altogether because CO_2 is no longer available from the atmosphere. Preventing water stress is, therefore, of major importance for maintaining photosynthesis and healthy plant growth.

Temperature – As with other metabolic reactions, the 'dark' reactions of photosynthesis increase with increasing temperature until the enzymes begin to become denatured (photochemical reactions such as the 'light' reactions of photosynthesis are not influenced by temperature). However, the reactions of 'dark' respiration, which break down carbohydrate for the processes of growth also increase with temperature. Because of this the net amount of photosynthesis (i.e. CO_2 fixed minus CO_2 lost to respiration) is not promoted by an increase in temperature as much as might be expected. Plants with C-4 type photosynthesis have a stronger response to temperature.

The simplest way of reducing the leaf temperature is by shading the plant but this will also reduce the rate of photosynthesis by decreasing the amount of available light. Shading and other means of reducing leaf temperature are discussed further in Chapter 4.

Interaction of factors

Photosynthesis depends on a number of external factors including light intensity, day-length, CO_2, temperature and water. These factors interact in a complex way and any one of them can affect the response to one or other of the others. For example, an increase in the available CO_2 will have little effect if light is limiting.

Internal factors that affect photosynthesis – A number of factors relating to the leaf itself can also affect the rate at which photosynthesis occurs. These include the light-collecting area of the leaf and the amount of light-harvesting chlorophyll, both of which can be affected by light. The amount of chlorophyll is also decreased by the lack of availability of some nutrient elements. Magnesium is a constituent of the chlorophyll molecule and iron is required for its synthesis; a deficiency of either of these therefore leads to a reduction in chlorophyll synthesis, paler leaves and a lower rate of photosynthesis. The white parts of variegated leaves also have substantially reduced rates of photosynthesis and it is important to remove any part of the plant that has reverted to green leaves, as these have a faster rate of photosynthesis and can rapidly swamp the growth of the variegated shoots.

LIGHT ALSO GIVES INFORMATION TO THE PLANT

Green plants are entirely dependent on light for photosynthesis and growth but they are stationary and unable to relocate to better light conditions. They have, however, evolved mechanisms to perceive the surrounding light environment and, in many cases, make adjustments to it. A summary of these mechanisms is shown in Figure 3.6.

LEVEL 3 BOX

PHOTORESPIRATION

In addition to normal dark respiration, C-3 plants carry out a process known as photorespiration in the light. Like dark respiration, the rate of loss of carbon dioxide by photorespiration increases with increasing temperature so that the net amount of carbon fixed by photosynthesis is less than it would be if photorespiration were not occurring. C-4 plants do not carry out photorespiration so this source of carbon loss is absent and net photosynthesis in C-4 plants such as maize, *Zea mays*, increases with temperature to a higher value than in C-3 plants such as oak, *Quercus robur*.

signal	detection	response
light or dark	light vs dark	germination seedling morphogenesis
shade	spectral discrimination or intensity perception	adaptation avoidance
direction	direction perception	orientation
day-length	light vs dark plus timer	seasonal adaptations

Figure 3.6 The information signals that enable plants to make adjustments to their environment.

Gardeners know that plants grown in the dark look different from those grown in the light. These **etiolated** plants lack chlorophyll, fail to expand their leaves and are taller than those growing in the light as shown in Figure 3.7. In nature, plants normally only grow in the dark as seedlings germinating under the soil. Rapid elongation and small leaves enable them to reach the light quickly and without damage. Once in the light, leaves expand and green up and stems become shorter and sturdier.

Figure 3.7 The effects of (a) light and (b) darkness (etiolation) on the growth of mung bean, *Vigna radiata*, seedlings.

When plants are growing through the soil there is little need for the production of fibrous materials to support the stem and gardeners make use of this response to darkness in the production of a number of crops.

For example, early forced rhubarb, *Rheum × hybridum*, is grown commercially in dark sheds in order to produce tender stems with little fibre. The only light they receive is the minimum necessary for harvesting the crop. In the garden, terracotta cloches are used to keep out light for crops such as chicory, *Cichorium intybus*, and rhubarb.

For over 200 years rhubarb, *Rheum × hybridum*, has been 'forced' to meet early seasonal demands for the tenderest stems, which are, in fact, the leaf petioles.

Within a relatively small area of Yorkshire, between Morley, Rothwell and Wakefield, often called 'the Rhubarb Triangle', a very successful commercial industry developed producing over 200 tonnes of forced rhubarb annually.

Production peaked in 1939 when the area increased to around 78 km^2, stretching between Bradford, Leeds and Wakefield.

Traditionally lit by candlelight, to provide the minimum light levels, this crop is just one example showing the value of a sound understanding of the science behind the growing techniques employed.

Now holding, since February 2010, protected Designation of Origin (DPO) status, Yorkshire Forced Rhubarb is legally protected by the European Commission's Protected Food Name Scheme and joins Champagne, Stilton Cheese and Parma Ham.

How do plants obtain information about their light environment?

A light-sensitive pigment is necessary in order to detect light. Two light-sensitive pigments are known to be involved in giving information to the plant about its light environment. The most important of these is a red-light sensitive pigment called **phytochrome**; the other pigment is sensitive only to blue light and is called **cryptochrome**.

The phytochrome system – The discovery of phytochrome stemmed from an observation in 1935 that the germination of some varieties of lettuce seed was suppressed by light of wavelengths in the near infra-red region of the spectrum between 700 and 750 nm, now called **far-red light**. It was later found that the germination of lettuce seeds was stimulated by giving only a few minutes of red light (between 600 and 700 nm) and that this was prevented when the red light was followed immediately by a brief exposure to light in the far-red region. Exposure to an alternating series of red and far-red light revealed that the response was determined by the final exposure. When the series ended in red, germination was promoted, but when the series ended in far-red, it was inhibited as shown in Table 3.3.

It was concluded that the light-sensitive pigment involved in bringing about these responses must exist in two forms and that these could be converted from one to the other by exposure to the appropriate wavelength of light. This pigment was called phytochrome (a word that simply means plant pigment) and has the following reactions:

Pr — — — — — — — — ▶ Pfr (in red light)

Pr ◀ — — — — — — — — — Pfr (in far-red light)

Light treatment	Percentage germination
Dark	9
1 minute of red light	98
1 minute of red, 4 minutes of far-red light	54
1 minute red, 4 minutes far-red, 1 minute red light	100
1 minute red, 4 minutes far-red, 1 minute red, 4 minutes far-red light	43

Sequences ending in red light promote germination while sequences ending in far-red light depress it.

Table 3.3 The effect of red and far-red light on the germination of lettuce (*Lactuca sativa*) seeds

Thus the Pr form of phytochrome is converted to the Pfr form in red light, while the Pfr form is converted back to Pr in far-red light. Because the germination of lettuce seeds required exposure to red light, it was concluded that the Pfr formed by red light is the biologically active form and that, in darkness, the pigment must be synthesised in the inactive form, Pr.

Phytochrome was later found to be a protein associated with a light-absorbing region, called a **chromophore**. More recently it has been found that plants contain more than one kind of phytochrome and that these different molecules may have different functions.

At first sight, it is difficult to understand why the phytochrome system evolved in the first place since plants are not exposed to sequences of red and far-red light under natural conditions. However, red light is strongly absorbed by leaves while far-red light passes through them as shown in Figure 3.8.

Because far-red light strongly inhibits germination as shown in the lettuce seed experiments, germination is prevented in the far-red enriched light under the leaf canopy close to the parent plant where the seedling would have little chance of survival. Moreover, the fact that only the inactive form of phytochrome is present in darkness means that buried seeds normally fail to germinate until they are exposed to light by cultivation or the activities of animals.

Shade perception and adaptations to shade
Because of their need for light in photosynthesis, green plants have evolved a number of responses that enable them to modify their growth in relation to the prevailing light conditions. Plants in the shade may be able to grow taller and, in this way, outgrow their neighbours to reach better light conditions. This would be of no use to plants of the woodland floor and there are species (the **shade tolerators**) that do not show the elongation response but have other responses that enable them to flourish in the shade. For example, in the shade their leaves are often larger and thinner and

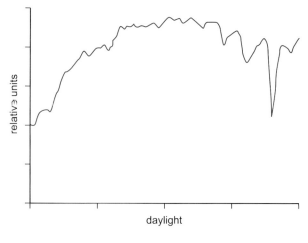

daylight

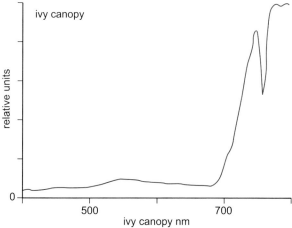

ivy canopy

Figure 3.8 The transmission and reflection of red and far-red light by leaves.

have more light-harvesting chlorophyll; both of these responses increase their ability to collect light.

The **shade avoiders**, on the other hand, have the ability to redirect their growth into stems at the first signs of shading. Sunlight contains roughly equal amounts of red and far-red light. Due to the absorption of light by chlorophyll the effect of a leaf canopy is to decrease the amount of red light while far-red light is readily transmitted as shown in Figure 3.8. This means that the ratio of red to far-red light is considerably lower under a leaf canopy than it is in sunlight. Shade avoiders respond to canopy shade by detecting changes in the ratio of red to far-red light through the red/far-red reversible reaction of phytochrome. We have seen that red light produces the biologically active form Pfr, while far-red light converts this back to the inactive form, Pr. The amount of Pfr present is therefore dependent on the relative amounts of red to far-red light (the R/FR ratio). With increasing canopy shade, the proportion of red light decreases and so the amount of Pfr also decreases. Since Pfr (the form that actually controls the response) *inhibits* stem elongation, the result is that stem length increases as the amount of shade increases.

The effect of a changing ratio of red to far-red light on elongation in a shade-avoiding plant is illustrated in Figure 3.9. Plants growing in a light environment with a low ratio of red to far-red light, as under a leaf canopy, are considerably taller than those grown in a light environment with a high ratio of red to far-red light. This type of light environment occurs under fluorescent lamps, many of which emit very little far-red light. Consequently plants growing under fluorescent lamps (which are often used for supplementary lighting under glass) are very compact in their growth habit. Plants growing under light with an approximately equal amount of red and far-red light, as in sunlight, are intermediate in height.

Figure 3.9 The effect of the ratio of red (R) to far-red (FR) light on the growth of potato, *Solanum tuberosum*, plants.

How does this response relate to the garden? One obvious point is that shade tolerators (mainly woodland species) are the plants of choice for shaded areas and some examples are given in Table 3.2.

CASE STUDY   **Plant selection for habitat**

Over the last few years the whole concept of selecting plants for their proven growing environment has come to have a much higher profile with specialist nurseries, such as MacGregors Plants for Shade, Romsey, Hampshire, improved plant labelling and the re-arrangement of some garden centres from alphabetical parades into grouping plants in suitable light levels – a much more practical approach to those gardeners who like the plants but are less familiar with their botanical names. Look out for developments such as these in your local area or request, in discussion with your local plant supplier, a higher emphasis on the light level, of lack of it, required by the plants they provide. Indeed, this can be extended to other environmental considerations such as water requirement, which is dealt with in much more detail in the following chapter, with the whole concept also discussed in a number of later chapters.

Less obvious is the question of spacing. The far-red rich light passing through leaves is also reflected to neighbouring plants thus decreasing the R/FR ratio of the light that they receive. This has two effects on shade-avoiding species; stem elongation is increased and axillary growth is suppressed leading to the development of 'leggy' plants. Close spacing may, therefore, be undesirable but the ideal spacing will depend on both the species and the type of growth required. For example, the close spacing of young trees can lead to a suppression of the development of axillary branches, which may be a desirable feature, especially for some forest trees.

Germination – The original discovery of phytochrome arose from studies of the effects of light on the germination of lettuce seeds. Many other seeds also require light to promote germination (including most leaf grasses) and show similar responses to those of lettuce seeds, being promoted by red light and inhibited by far-red light. In a study of 142 species, 75% were promoted by light, 22% were indifferent to light and only 2% were actually inhibited by light. However, few modern cultivars have a light requirement for germination as this has been lost through breeding.

As a generalisation it can be said that most species with small seeds are promoted by light while those with large seeds will germinate equally well in the dark. Small seeds have relatively small food reserves, which would run out before reaching the light if the seeds germinated when deeply buried in the soil. In contrast, the greater food reserves in large seeds enables them to reach the light and begin photosynthesis before the seed reserves are exhausted. Most soils are very opaque and only transmit enough light to promote germination to a depth of a few millimetres (mm). For example, 80% of white mustard, *Sinapis alba*, seeds germinated when sown at the surface and only 50% when the seeds were buried at a depth of 3.5 mm. At 9 mm depth, none of the seeds germinated. A light requirement for germination explains the appearance of many seedlings following disturbance of the soil by cultivation, especially where the soil has not previously been disturbed for some time. As can be seen in Figure 3.10, cress, *Lepidum sativum*, seeds show a similar

Figure 3.10 The effect of sowing depth on the germination of light-requiring cress, *Lepidum sativum*, seeds. (a) Sown on surface; (b) partially buried; (c) buried deeply.

response. Seeds germinated well when sown at the surface or where the surface was disturbed, whereas few seeds germinated when they were covered with soil.

When sowing seeds, the possibility that they require light to promote germination needs to be borne in mind. Unless they are known to be inhibited by light, seeds should be sown at the soil surface and covered with vermiculite to a depth of a few millimetres in order to admit sufficient light for germination.

There are a few plants in which germination is sensitive mainly or only to blue light. For example, germination in *Phacelia isignis* is inhibited by light and this appears to be a blue light response.

RESPONSES TO BLUE LIGHT

The phytochrome system controls many aspects of plant growth, including the germination of most light-sensitive seeds, growth responses to canopy shading and seasonal responses to the length of day. There are, however, some important responses that are not sensitive to red and far-red light. These are controlled by a group of pigments known collectively as **cryptochrome** and are sensitive only to blue light.

Stomatal movements – Stomata close in darkness and open in the light. Dark closure conserves water during the night while light-dependent opening allows CO_2 to enter as soon as light energy becomes available. The opening response to light is sensitive only to blue light and red wavelengths are ineffective. The mechanisms for the opening and closing responses are discussed in Chapter 4.

Phototropism – The fact that plants grow towards the light is observed by the many amateur gardeners who raise seedlings on their window sills. This bending is known as **phototropism** and is controlled by the blue-light-sensitive pigment. It appears that when irradiated unilaterally, light causes the migration of the hormone, **auxin** (see Chapter 8) towards the shaded side of the shoot tip and, consequently, more auxin reaches the elongating zone on the shaded side. Since auxin is required for elongation growth, the shaded side grows more rapidly than the illuminated side and the plant bends towards the light: the classic result experienced by growing on window sills.

This window-sill problem can be alleviated by turning the seedlings around each day. Alternatively, a reflective surface such as aluminium foil can be placed on the shady side in order to equalise the light more efficiently. Little can be done in the garden but growth often becomes more upright later, unless the plants are close to a wall and very heavily shaded from one side.

Responses to shade – The shade-avoidance response under a leaf canopy is dependent on the R/FR ratio of the light received by the plant, as discussed above. However, unlike shade from trees, shade from neighbouring buildings does not alter the R/FR ratio. Nevertheless, many plants show typical shade responses, probably detecting the light conditions through the blue-light-sensitive system.

Day-length perception

By coupling the response to a **circadian clock**, the phytochrome system also allows plants to measure the duration of light and/or darkness each day. This enables them to make seasonal adjustments to their growth patterns and is also important in their ability to grow well at different latitudes. For details of day-length responses see Chapter 5.

Summary

All green plants are completely dependent on light for growth and also for many aspects of their development. In particular, light energy fuels the process of photosynthesis, which results in the production of carbohydrates and subsequently leads to the formation of all other organic components of the plants. Light is also an important source of information for plants, giving them the ability to sense their light environment and, in many cases, also their seasonal environment. Understanding how light acts is, therefore, of great importance in the production of healthy plants and high-yielding crops.

Light is what we see and, for most people, this occurs at wavelengths between about 350 nm in blue light and 750 nm in far-red light with the most effective region being in the green waveband. Plants, in contrast, are least sensitive to green light. Measuring light using detectors that are designed for human vision gives misleading information for plant growth and correction factors have to be used. Light measurement is most important under glass, especially when artificial light sources are being used.

The structure of leaves is designed to maximise the collection of light energy by developing an expanded, flat surface. The internal structure facilitates the movement of the carbon dioxide gas to the chloroplast sites where photosynthesis takes place. Pores in the leaf surface allow the entry of carbon dioxide. The organic products of photosynthesis are transported out of the leaves to the stem via the network of veins and from there to all other parts of the plant. The veins also serve to bring in substances that are required for the functioning of the leaf itself.

Plant growth is directly (or indirectly when utilising storage materials) dependent on photosynthesis. The main factors affecting the rate of photosynthesis are the quantity of light reaching the plants, the amount of carbon dioxide in the surrounding atmosphere and the leaf temperature. Water supply can also drastically limit the rate of photosynthesis. If leaves lose water faster that it can be absorbed, the gas exchange ports in the leaf (the stomata) close, carbon dioxide cannot enter and photosynthesis decreases or stops completely.

Green plants are dependent on receiving sufficient energy from sunlight for the manufacture of organic compounds and, consequently, they have evolved mechanisms that enable them to adapt to the light environment in which they are growing. Light is here acting as a source of information. Plants mainly detect their light environment through a red-light-sensitive pigment called phytochrome. The phytochrome system controls many aspects of plant development including shade perception, germination, stem elongation and seasonal responses. Some responses are, however, sensitive only to blue light; these include bending towards the light (phototropism) and the opening and closing of the stomata.

Review questions

1. Discuss light as a source of **information**.
2. Explain light as an energy source within the process of **photosynthesis**.
3. Draw and label a sketch showing **leaf structure** and state its role in **light capture**.
4. State a definition of light and describe **two** methods of its **measurement**.
5. Light is essential for plant growth. List **four** methods of ensuring **quality** light is obtained for plant use.

FURTHER READING

Fish, M. (2004). *Gardening in the Shade*. Betterway Books.
A practical book looking at the problems and opportunities presented by reduced light levels.

Ingram, D. S., Vince-Prue, D. & Gregory, P. J. (eds) **(2008)**. *Science and the Garden: the Scientific Basis of Horticultural Practice*, 2nd edn. RHS by Blackwell Publishing. ISBN-13: 978-1-4051-6063-6.
This book covers more extensively the material discussed in the chapter.

Orr, S. (2011). *Tomorrow's Garden: Design and Inspiration for a New Age of Sustainable Gardening*. Rodale Books.

This gives a broad discussion of using plants in a sustainable way, selecting suitable examples for their correct habitats.

Salisbury, J. B. & Ross, C. W. (1992). *Plant Physiology*, 4th edn. Wadsworth Publishing.
This advanced textbook of plant physiology is a useful source of reference and gives detailed information on the main topics discussed in the chapter.

Smith, H. (1994). Sensing the light environment; the functions of the phytochrome family. In **R. E. Kendrick & G. H. M. Kronenberg** (eds), *Photomorphogenesis in Plants*, 2nd edn. Kluwer Academic Publishers.
This is an advanced discussion but would be of interest to anyone wanting to understand more about how phytochrome functions to perceive shade light.

RHS level	Section heading	Page no.
2 2.2	Case study: Plant selection for habitat	65
2 3.7	Case study: Plant selection for habitat	65
2 3.9	Shade perception and adaptations to shade	64
3 1.4	Germination	65
3 4.1	Why green plants need light	59
3 4.2	Transport out of the leaf	59
3 5.1	Shade perception and adaptations to shade	64
3 5.2	Light quantity	60
3 6.1	Phototropism	66
3 6.3	Phototropism	66

CASE STUDY Abscisic acid and vine cultivation

Experiments have shown that when the soil begins to dry out around them, roots send a hormone signal (**abscisic acid**) to the leaves, which causes the stomata to partially close and so reduce water loss. This observation has led to field studies in which plants have been exposed to drying conditions for part of the time and to part of the root system.

A practical application has been found in the cultivation of irrigated vines where water is applied to alternate sides of the row in turn, so that first one side and then the other is drying out at any one time. The roots in the drying side then send the abscisic acid signal to the leaves. This results in better quality fruit by reducing excessive leaf growth and also conserves water, an important factor if we consider the probable effect of climate change to decrease the water supply, especially in summer. This technique has been very successfully used in Australian vineyards.

Baskets and containers require careful moisture control.

Chapter 4

Water and its importance for plants

Daphne Vince-Prue

INTRODUCTION

Water is an essential component of all living cells. In addition, for plants water is needed to maintain the rigidity or **turgor** of cells and as a substrate for photosynthesis (see also Chapter 3). However, the amount of water in a plant at any one time is small compared with the amount that passes through on a bright day.

The problem for land plants is that a supply of CO_2 from the surrounding air is needed for photosynthesis. This necessitates pores (the stomata) through which the gas can enter into the cells of the leaf. The plant must therefore be able to maintain an adequate supply of CO_2 to the photosynthesising cells while preventing a net loss of water vapour to the atmosphere. If water is lost from the leaves at a faster rate than it can be supplied from the soil, turgor is lost and the leaves become flaccid and wilt.

This chapter describes how plants have evolved to prevent a net loss of water under different environmental conditions. The chapter also considers how gardeners can manage water most efficiently in the context of the garden, especially as climate change is likely to pose increasing problems of water management in the future, together with the issues of cost, sourcing, storage and quality as well as quantity, including practical opportunities such as utilising 'grey water'.

Instances of excess water, such as flooding, are also discussed highlighting plants' reactions to these situations.

Key concepts

✿ The problem for plants

✿ Water loss from the leaves in relation to water stress

✿ Strategies that have evolved to minimise water stress

✿ Problems associated with excess water

✿ Climate change in relation to water management

WATER POTENTIAL AND TURGOR PRESSURE

The **vacuoles** of plant cells are contained within a membrane known as the **tonoplast**. This **semi-permeable membrane** blocks the movement of many dissolved solutes but allows the passage of water molecules. Water moves into and out of the vacuoles by simple diffusion from a region with a low concentration of solute molecules (i.e. one of a high concentration of water molecules or high **water potential**) to a region with a high concentration of solute molecules (i.e. one of low water concentration or low water potential). This process, known as **osmosis**, will continue until the concentrations are the same on both sides of the membrane.

In plant cells, water moves into the vacuole until the actual or **hydrostatic pressure** exerted on the rigid cell wall prevents any further movement of water. The pressure on the cell wall that is exerted by a fully rigid cell is called the **turgor** pressure, while the containing pressure exerted by the cell wall on the cell contents is known as the **wall pressure**. When turgor pressure is lost, the cells become flaccid and the leaf wilts.

How plants lose water to the atmosphere

Plants lose water to the atmosphere by **evapo-transpiration** from the wet cell walls of the cavity below the stomata, known as the **sub-stomatal cavity**. The rate of loss is regulated both by the opening and closing of the guard cells and by the **vapour-pressure gradient** between the water vapour content of the atmosphere and the vapour pressure within the sub-stomatal cavity. In order to understand how water is lost from the leaf to the atmosphere, it is necessary to understand both of these processes.

The role of the stomata – Each of the stomata is surrounded by two kidney-shaped guard cells. As shown in Figure 4.1 the walls of these cells are thickened in such a way that when they lose water and become flaccid, the surfaces surrounding the stomata move together and close the pores. This reduces the loss of water vapour from the leaf and improves the water balance as long as there is available water present in the soil and water continues to

THE STOMATA OF MOST PLANTS OPEN IN THE LIGHT AND CLOSE IN DARKNESS

Exposure to light increases the concentration of solutes (mainly potassium ions) in the guard cells causing water to move into the cells by osmosis thus increasing their turgor and causing the stomata to open. The guard cells also respond to the concentration of CO_2 in the atmosphere surrounding them. When this decreases due to photosynthesis by the leaf cells, the concentration of potassium ions increases in the guard cells, causing water to enter them and the stomata to open as discussed above. Thus the opening of stomata in the light is partly due to a direct response to light and partly due to a reduction in the CO_2 concentration because of photosynthesis. Darkness and an increase in CO_2 concentration when photosynthesis ceases has the opposite effect and causes water to enter the guard cells and the stomata to close.

move into the root cells. The guard cells then re-absorb water and the stomata open once again.

Water loss from the leaf – Leaves lose water to the air in the form of water vapour diffusing through the stomata. The absolute amount of water vapour present in the atmosphere at any one time is called the **vapour pressure** of the air, and like any other gas water vapour moves by diffusion from a region of high vapour pressure to one of lower vapour pressure. The **relative humidity** of the atmosphere is a measure of the amount of water vapour present in the air compared with the amount that would be present if the air were saturated with water vapour. When the air is saturated any additional water vapour would condense out as liquid water. The difference between the amount of water vapour present in the atmosphere at any one time and the amount that would be present at saturation is known as the **vapour-pressure deficit**.

The air within the stomatal cavities is saturated with water vapour because it is in contact with the wet cell walls that line the

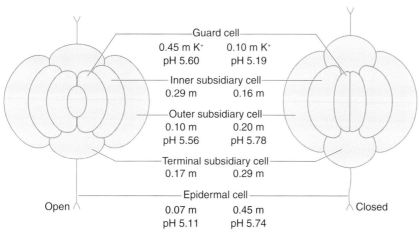

Figure 4.1 Simple diagram showing open and closed stomata noting thickeness of guard cells.

cavity. As the air surrounding the leaf is normally less than saturated, water vapour diffuses along the concentration gradient between the higher vapour pressure of air within the cavity and the lower vapour pressure of the unsaturated atmosphere surrounding the leaf, the process known as **transpiration**. Warmer air holds more water vapour than cooler air when both are saturated. This means that, even when the atmosphere surrounding the leaf is saturated with water vapour (i.e. at 100% relative humidity), water will still be lost provided that the temperature of the air inside the leaf is higher than that in the surrounding atmosphere. Leaves in sunlight are often warmer than the air surrounding them and so plants can still lose water even when the atmosphere is saturated.

Permanent and temporary wilting – When water is lost from the leaves faster than it can be supplied from the soil as, for example, on a sunny day when plants are in a hot dry atmosphere and the stomata are fully open, the leaf cells lose water, turgidity is lost and the plants wilt. Recovery will normally occur at night when the temperature falls and the stomata close in darkness, provided that water is available to the roots. This type of behaviour is known as **temporary wilting**. In contrast, **permanent wilting** occurs when the amount of water in the soil is insufficient to restore the water balance of the cells. In this case, recovery from wilting occurs only when water is added to the soil.

Wilting depresses photosynthesis and slows the rate of growth as shown in Figure 4.2. Moreover, recovery of the growth rate can take some considerable time, especially if wilting is prolonged. It should, therefore, be avoided where possible.

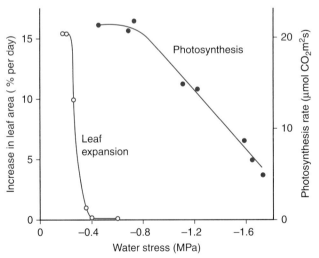

Figure 4.2 The effect of water stress on leaf growth and photosynthesis.

Water uptake by the roots – The roots absorb both water and nutrients from the soil and transport them to the shoot. Water can be taken up through the whole length of the root provided that the roots and soil are in close contact and water is available. Water moves to the shoot through the xylem vessels, which consist of elongated cells that have lost their living contents. They also lose

their cross walls and become joined from end to end to provide a continuous channel of open tubes between the roots and the leaves. The pH of the xylem is slightly acid, which helps to prevent precipitation of dissolved nutrients. The **tracheids** of conifers and their relatives also transport water but less efficiently than the xylem vessels.

Water moves upwards from the roots to the shoots because the evaporation of water from the leaves results in a tension in the xylem columns. Because water molecules are strongly bonded together, this tension results in a pulling force that is transmitted through the water columns in the xylem vessels and causes water to be drawn upwards through the plant and into the vessels in leaf veins, from where it moves into the living cells of the leaf. Since movement through living cells is relatively slow, the distance that water must travel in this way is minimised by the extensive network of fine veins that permeate the leaf.

Leaf structure in relation to water loss – The most obvious feature of the leaf in the majority of flowering plants is the large, thin leaf blade (the **lamina**), which is designed to maximise the collection of light for photosynthesis. There are, however, many modifications of this basic shape, especially in plants adapted for life in the arid and semi-arid zones of the world as will be discussed later in the chapter.

A surface with a large area relative to its volume will inevitably lose water very rapidly by evaporation. This is prevented in leaves by the presence of a waterproof layer (the cuticle) that covers the entire surface of the leaf. A single layer of cells called the epidermis lies under the cuticle. These cells do not contain chloroplasts with the exception of the kidney-shaped guard cells that surround the stomata. An important feature of both the epidermis and cuticle is that the cells contain chemicals that filter out damaging ultra-violet radiation. The presence of the stomata allows the leaf to take up CO_2 for photosynthesis despite the

THE SPACING OF THE STOMATA ON THE LEAF OPTIMISES THEIR FUNCTION

The spacing of the stomata on the leaf surface has been found to be approximately ten times the maximum diameter of the stomatal pores. There are good reasons for this. The diffusion of any gas occurs in all directions and a spacing of ten times the diameter of the pores allows CO_2 to diffuse into the leaf almost as readily as if the cuticle were not present. The spacing is less optimal for the diffusion of water vapour out of the leaf, which occurs only at a rate proportional to the area of the pores themselves. This means that, when the stomata begin to close, the loss of water vapour is restricted far more than the entry of CO_2. Consequently water can be conserved by partly closing the stomata without reducing the supply of CO_2 for photosynthesis. This is important when considering the function of the hormone abscisic acid, as discussed later.

impermeable properties of the cuticle. In most of the higher land plants, the stomata are located on the undersurface of the leaf where there is relatively more protection from the drying effects of wind and sun.

The main photosynthetic cells lie just beneath the epidermis and consist of one or more layers of elongated cells (the palisade) containing numerous chloroplasts. The moist walls of these cells have a large surface area exposed to the internal air spaces of the leaf, allowing for rapid diffusion of CO_2 and water vapour between these cells and the external atmosphere. The large cells of the **spongy mesophyll** that lie beneath the palisade layer have few chloroplasts. There are large air spaces between these cells and especially around the stomata where they form the sub-stomatal cavities.

The geometry of the leaf is designed to maximise the entry of CO_2 and its movement to the palisade cells where photosynthesis occurs, but it also allows water to be lost rapidly from the wet cell walls into the sub-stomatal cavities and then through the stomatal pores into the surrounding atmosphere.

FACTORS AFFECTING THE RATE OF WATER LOSS FROM THE LEAF

A number of factors affect the rate at which water is lost from plants, some of which are under the control of the gardener.

Leaf temperature
When it is saturated, cool air hold less vapour than warmer air. Cooling the leaves therefore reduces the water vapour content of the saturated air inside the sub-stomatal cavities. In turn, this reduces the steepness of the concentration gradient of water vapour between the leaves and the surrounding air and, consequently, slows down the rate of water loss. The easiest way to cool a leaf and so reduce water loss is to shade it from direct sunlight, and plants in the shade will clearly lose water at a slower rate than those in the sun. However, shading will also reduce the rate of photosynthesis and slow the rate of growth, although much less than if the plants were allowed to wilt.

In practice, it may be advisable to give some form of temporary shading to cool the leaves and reduce water loss, especially for plants that have yet to become established and are exposed to hot sunlight.

The most effective way of cooling a leaf, however, is by spraying it with water. The subsequent evaporation of water, which consumes energy in the form of heat, results in cooling of the leaf surface. Under glass, a number of systems such as intermittent mist are used to maintain the leaf surface wet and, therefore, cooler (see also Chapter 8). In the open, however, spraying in direct sunlight is inadvisable as it can cause the leaves to become scorched.

Wind
A boundary layer of saturated or partly saturated air builds up around the leaf when the air is still. Wind blows some of this saturated air away and reduces the thickness of the boundary layer. This action of wind results in cooling the leaves because the transfer of heat by convection from the leaf surface to the air is faster when the boundary layer is thin than when the boundary layer is thick.

Although wind helps to reduce water loss by reducing the leaf temperature in this way, the overall effect of high wind velocity is usually to increase the rate of loss, especially when the heat load on the leaf is low. This is obviously a major factor in exposed sites. The drying effects of wind are particularly evident where the cuticle is thin and does not offer a significant barrier to water loss, and so plants with thick cuticles, such as Portuguese laurel, *Prunus lusitanica*, shown in Figure 4.3, are better able to withstand the effects of wind on exposed sites. Some form of shelter from drying winds is therefore necessary where the site is exposed. This may take the form of a windbreak of resistant plants such as Portuguese laurel or willows or some kind of artificial boundary such as a netted structure that reduces the wind speed. Solid boundaries such as brick and wooden fences are less desirable because the wind can travel over them with little decrease in its velocity.

Figure 4.3 Portuguese laurel, *Prunus lusitanica*. This plant has a thick cuticle and can withstand the drying effect of wind. It can also be used as an effective windbreak.

The best windbreaks are 60% permeable as this type of structure is most effective in reducing damaging wind eddies. Windbreaks are effective over a distance of about ten times their height.

Wind is particularly damaging to young developing leaves before the protective cuticle has fully developed and wind scorch on young leaves is a common event on exposed sites. Particular care is needed at this time to prevent damage and some sort of temporary shelter may be desirable.

Humidity
The rate of water loss from the leaves is lower when they are in air with a high relative humidity. The higher content of water vapour in the surrounding atmosphere reduces the concentration gradient between the air surrounding the leaf and the saturated

air in the sub-stomatal cavities and, consequently, slows the rate of diffusion of water vapour out of the leaf. However, as discussed above, some water continues to be lost into an atmosphere of 100% relative humidity, if the leaf temperature is higher than the temperature of the air surrounding it.

Increasing the relative humidity of the air is only possible under protected cultivation by, for example, repeated **damping down**. Commercial growers also use devices such as **fogging** to increase the water vapour content of the atmosphere surrounding the plants.

WATER STRESS

Natural habitats often suffer from a shortage of water. This may occur in the form of a seasonal drought as in Mediterranean regions where the annual rainfall occurs mainly in winter. Water is more permanently scarce in arid and semi-arid regions. Even in the UK, periods of severe water shortage occur from time to time and these may become more frequent with climate change.

Strategies that minimise water stress
Several strategies, both biochemical and structural, have evolved that allow plants to survive periods of water stress. Some knowledge of these different types of adaptation to water stress is helpful in the choice of plants for dry or relatively dry conditions.

The role of abscisic acid – Plants are subjected to water stress when there is an insufficient supply of water to the roots, or too rapid a loss of water from the leaves. It has been known for some time that when plants are subjected to water stress the stomata close and the hormone **abscisic acid** (ABA) begins to increase in the leaves. The stomata also close in response to an application of ABA directly to the leaves, and it was originally thought that ABA was produced in them when they experience water stress. However, partial closure of the stomata and a reduction in leaf expansion begins to occur before there is any measurable water stress in the leaves. Experiments have shown that when the root tips begin to encounter dryness in the soil they generate a signal that is transported to the leaves. This signal has been found to be ABA, and its arrival in the leaf results in a partial closing of the stomata and a consequent improvement in the water economy of the plant. Thus the ABA signal from the roots in drying soils has a regulatory role in improving the water economy of the leaves before wilting takes place. This conclusion is supported by the discovery of 'wilty mutants' that are deficient in ABA and are always partially wilted.

As described in the case study at the beginning of the chapter, understanding the function of the ABA signal from drying soils has found a practical use in the cultivation of irrigated vines. If only half of the root system is allowed to dry out while the remaining half is supplied with water, ABA from the drying roots causes the stomata to partly close. This both conserves irrigation water and results in better quality fruit by reducing excessive leaf growth. The possibility of applying this approach to other irrigated crops is being investigated.

Once wilting has occurred many cell functions are depressed leading to an often severe reduction in growth and yield. Since recovery can take some time, even when water is supplied to the roots and the water balance is restored, the prevention of wilting is of great importance for healthy plant growth.

C-4 plants – Two biochemical strategies that conserve water are modifications to the normal process of photosynthesis. These are **crassulacean acid metabolism (CAM)** and **C-4 photosynthesis**, both of which have evolved in response to the stress imposed by low levels of soil water combined with high temperatures. Although partial closure of the stomata in response to the ABA signal is effective in reducing the loss of water vapour from the leaves without significantly reducing the uptake of CO_2, further closure begins to limit the availability of CO_2 and decrease the rate of photosynthesis. Complete closure of the stomata leads to a severe reduction in the internal concentration of CO_2 and will ultimately prevent photosynthesis.

In normal C-3 photosynthesis, the initial product is a molecule with three atoms of carbon. In contrast, C-4 plants produce organic acids that contain four carbon atoms. These organic acids are produced in the chloroplast-containing cells of the palisade tissue and are subsequently transferred to specially modified bundle-sheath cells, which contain chloroplasts, unlike the bundle-sheath cells of C-3 plants, which do not. The organic acids are broken down in the modified bundle-sheath cells to release CO_2, which diffuses to the palisade cells where it is utilised in normal C-3 photosynthesis. Plants possessing such modified bundle-sheath cells are said to exhibit **Krantz anatomy**.

Plants with C-4 photosynthesis are adapted to growing in regions of high light intensity with periods of water stress. The incorporation of CO_2 into organic acids occurs rapidly when the stomata are open and the light intensity and temperature are high. When water stress occurs and the stomata close, the release of CO_2 from the organic acids enables plants to continue to carry out normal C-3 photosynthesis in the light.

Although the C-4 pathway is less efficient and uses more light energy to fix CO_2, C-4 plants often exceed C-3 plants in their productivity, especially when light intensity and temperature are high and water is limiting. At lower light intensities and with water freely available C-3 plants are usually more efficient.

Many of the North American prairie grasses that have become very popular for some types of landscape planting are C-4 plants. They include, for example, Indian grass, *Sorghastrum nutans*, and little bluestem, *Schizachyrium scoparium*. These plants are drought tolerant and grow best in high summer. Many other tropical grasses, including the important crop plant maize, *Zea mays*, shown in Figure 4.4, are C-4 plants and require high light intensities and warm temperatures for maximum growth.

Crassulacean acid metabolism (CAM) – Many drought tolerant species, such as the *Crassula* plant illustrated in Figure 4.5, have thick, succulent leaves with a low surface area to volume

Figure 4.4 Maize, *Zea mays*, is a C-4 plant and requires high light intensities and high temperatures. By utilising stored carbon dioxide it can continue to carry out photosynthesis when water becomes limiting and the stomata close.

Figure 4.5 Drought-tolerant plants have adaptations that enable them to exist in regions where water is in extremely short supply. Many cacti, such as the *Echinocereus* (on the left) have leaves that are modified into spines. *Crassula* plants (on the right) have crassulacean acid metabolism (CAM) and so are able to carry out photosynthesis when the stomata are closed during the day-time. Both types of plant are succulent and able to store water that can be utilised in times of shortage.

ratio. This together with a thick cuticle restricts water loss when the stomata are closed. However, if the stomata remained closed during the day-time, plants would not be able to obtain CO_2. Consequently, many succulent plants have evolved a biochemical pathway that enables them to open their stomata only at night and still continue to carry out photosynthesis. This pathway was first observed in members of the family *Crassulaceae* and is therefore called crassulacean acid metabolism or CAM.

The CAM pathway is in many ways similar to C-4 photosynthesis. In CAM plants, CO_2 is taken up when the stomata are open during the night and is stored in the vacuoles of the leaf cells in the form of an organic acid (malic acid) that contains four carbon atoms. The increases the concentration of solute in the vacuoles and allows them to absorb and store water, leading to the succulence shown in Figure 4.5. During daylight, the malic acid is broken down to release CO_2, which diffuses to the chloroplasts where normal C-3 photosynthesis occurs. Although there is little water loss during the day when the stomata are closed, the absence of cooling by transpiration can lead to high cell temperatures, which must be tolerated by the plants. Moreover, the capacity of CAM plants to store malic acid overnight is limited, and the rate of growth is consequently much slower than in C-4 plants where the 4-carbon acids can be metabolised to release their CO_2 immediately.

Figure 4.6 Short-lived desert annuals can survive drought conditions as dry seeds.

The CAM pathway was first observed in members of the *Crassulaceae* but it also occurs in many other *Angiosperm* families and even in some ferns. CAM plants normally inhabit places where water is scarce or is difficult to obtain, such as in arid or semi-arid regions or in salt marshes where the high content of salt reduces the amount of water that can be taken in by root cells. Some epiphytic orchids that obtain their water from the surrounding air also exhibit CAM.

The CAM pathway is restricted to certain plants such as species of *Crassula* and *Kalanchoe* and is determined by the plant's genes. However, many CAM plants can switch to normal C-3 photosynthesis under favourable conditions, when the stomata remain open for longer periods during the day.

Avoidance and tolerance – Some plants survive periods of water shortage by simply escaping them. Many short-lived desert annuals, such as those shown in Figure 4.6, avoid dry periods in the form of seeds. These only germinate when there has been enough water for them to complete their life cycle.

Other plants avoid dry periods as underground storage organs. These lose only small amounts of water into the surrounding soil, which is usually moister than the atmosphere. Many spring and autumn-flowering garden plants that originate in Mediterranean regions pass the hot, dry summers as underground bulbs or other types of storage organ. Common examples are tulips and crocuses. **Geophytes** of this kind are usually started into growth by late autumn and winter rainfall.

Succulent plants that store water in their vacuoles, such as those shown in Figure 4.5, survive periods of severe water shortage by utilising their stored water, which is replenished when conditions allow.

Several drought-tolerant plants such as alfalfa, *Medicago sativa*, are able to survive periods without rain because they have deep roots that are able to reach the permanent water table.

MORPHOLOGICAL AND STRUCTURAL ADAPTATIONS

As discussed earlier in the chapter, two important ways of reducing the rate of water loss from the leaf are to increase the length of the diffusion pathway between the saturated air in the sub-stomatal cavities and the surrounding atmosphere and/or to reduce the heat load on the leaf. Both are employed in different ways by drought-tolerant plants.

Changes that increase the length of the diffusion pathway

When water vapour diffuses from the leaf into still air, a **boundary layer** forms around the leaf. This boundary layer has a higher content of water vapour than the surrounding atmosphere and, consequently, decreases the rate of water loss from the leaf by increasing the length of the diffusion path between the saturated air in the sub-stomatal cavities and the external atmosphere. Morphological adaptations that help to maintain the boundary layer and increase the length of the diffusion pathway therefore help to decrease the rate of water loss from the leaf.

The presence of hairs on the leaf surface helps to maintain a layer of moister air around the leaf and also reduces the drying effect of wind, which tends to blow away the protective boundary layer. Examples include drought-tolerant plants, such as *Senecio cineraria* shown in Figure 4.7, which has soft leaves and hairs on both surfaces, and the rigid leaves of the holm oak, *Quercus ilex*, which has dense hairs on the undersurface where most of the stomata are found.

Another way of increasing the length of the diffusion pathway found in many drought-tolerant plants is to have stomata that are sunk below the surface of the leaves as, for example, in oleander, *Nerium oleander*. A different strategy is to have leaves that roll under in response to drought since most of the stomata are on the undersurface.

Figure 4.7 Plants with hairy leaves such as *Senecio cineraria*, shown here, reduce the loss of water from the leaves because moist air is trapped around them. This increases the length of the diffusion pathway between the wet cell surfaces of the leaf and the outside air and so slows down the rate of diffusion of water vapour from the leaf to the atmosphere.

Strategies that reduce the heat load on the leaf

Leaf surfaces that reflect at least some of the incoming radiation help to reduce the heat load on the leaf, and many drought-tolerant plants have shiny leaves, e.g. *Ilex* spp. and *Ceanothus* spp., or grey foliage, as in the desert plant, *Atriplex*, shown in Figure 4.8.

Figure 4.8 Plants of the salt bush, *Atriplex* spp., are native to the arid regions of Australia. They have grey leaves that reflect light and so help to maintain a lower leaf temperature. They also have small leaves, which have a thin layer of moist air around the leaf and so lose heat rapidly to the air, especially when this is hot and dry. Reducing the heat load on the leaf slows down the rate of water loss because the cooler saturated air inside the leaf contains less water vapour than would warmer air. This reduces the vapour pressure gradient between the leaf and the warmer air and slows down the rate of transpiration.

The presence of reflective hairs on leaves in plants such as *Stachys byzantina* and *Lavandula lanata* also helps to lower the temperature. Leaf orientation can also reduce the heat load and many of the *Eucalyptus* species that are characteristic of the hot, dry areas of Australia have leaves that are angled downwards so that their surfaces are not at right angles to the mid day sun

Leaf area

The loss of heat from leaves by convection is greatest when the boundary layer is thin. In general the boundary layer is thinnest for small leaves, especially when the wind velocity is relatively high. Small leaves are therefore a factor in drought tolerance as exemplified by the characteristic small leaves of many shrubs that are native to Mediterranean areas, such as *Cytisus*, *Genista* and *Spartium junceum*, Spanish broom. Where the reduction in leaf area is carried to extremes by modification of leaves into spines, as in the desert cactus shown in Figure 4.5, the stems are often green and able to carry out photosynthesis.

Perhaps the most obvious way of reducing leaf area and so minimising the loss of water from the plant is to shed all or some of the leaves in times of water shortage. Drought-deciduous plants have evolved both in arid zones and in regions with seasonal dry periods. For example *Lotus scoparius*, a shrub native to California, sheds its leaves in summer and remains dormant until the arrival of the winter rains. This is an extreme example and many plants shed only some of their leaves, especially the older ones, when they are subjected to water stress. This can result in unsightly plants with bare basal stems and only a few leaves remaining on the uppermost part of the shoot. Such plants are undesirable in the context of the garden and emphasise the need to avoid water stress where possible.

The usefulness of the various morphological adaptations for reducing water loss depends on the environment in which the plant is growing. When water is short, it is clearly advantageous for the plant to conserve water by slowing down the rate of water loss by transpiration, and a number of adaptations have evolved to achieve this. When water is plentiful, however, high rates of transpiration cool the leaves by evaporation, which in turn reduces the heat load on the leaf and so reduces the rate at which water is lost to the atmosphere without impeding the uptake of CO_2 into the leaf.

CONSERVING WATER IN THE GARDEN

With even the most conservative of climate change models it is clear that water is likely to become increasingly scarce in the south eastern part of the UK. Even in areas in which water continues to be plentiful, abnormal weather systems may occur more frequently and disrupt supplies. Hence conserving water is becoming an increasingly important topic for horticulture.

Gardeners can do a number of things to reduce the amount of water used in the garden. One of the most important is to consider the choice of plants and, where possible, to use those that are

Dry sunny areas	Dry shady areas
Agave, *Agave* spp.	Box, *Buxus sempervirens*
Anthemis, *Anthemis tinctoria*	Butcher's broom, *Ruscus aculeatus*
Broom, *Cytisus* and *Genista*	Brunnera, *Brunnera* spp.
Catmint, *Nepeta* spp.	Bugle, *Ajuga reptans*
Chinese privet, *Ligustrum lucidum*	Cotoneaster, *Cotoneaster*
Eryngium, *Eryngium* spp.	Cyclamen, *Cyclamen coum*
Fig, *Ficus carica*	Dead nettle, *Lamium*
Globe thistle, *Echinops ritro*	Dutchman's breeches, *Dicentra cucullaria*
Hyssop, *Hyssopus officinalis*	Euonymus, *Euonymus fortunei*
Marjorum, *Origanum vulgare*	Elaeagnus, *Elaeagnus × ebbingei*
New Zealand flax, *Phormium tenax*	Foxglove, *Digitalis purpurea*
Lamb's ears, *Stachys byzantina*	Garrya, *Garrya elliptica*
Lavender, *Lavandula angustifolia*	Hart's tongue fern, *Asplenium scolopendrium*
Olive, *Olea europaea*	Hellebore, *Helleborus foetidus*
Phlomis, *Phlomis* spp.	Herb robert, *Geranium robertianum*
Rosemary, *Rosmarinus officinalis*	Periwinkle, *Vinca* spp.
Shrubby potentilla, *Potentilla fruticosa*	Portuguese laurel, *Prunus lusitanica*
Spanish broom, *Spartium junceum*	Skimmia, *Skimmia japonica*
Stonecrop, *Sedum* spp.	Spotted laurel, *Aucuba japonica*
Sun rose, *Cistus* spp.	St John's wort, *Hypericum × inodorum*
Yucca, *Yucca* spp.	Sweet box, *Sarcococca* spp.
Many spring and autumn flowering bulbs	Wood spurge, *Euphorbia amygdaloides*

Table 4.1 Examples of plants that tolerate dry conditions

tolerant of dry conditions. Several examples of drought-tolerant plants are mentioned in the preceding sections of this chapter and a brief further list is given in Table 4.1. At the present time, choosing such plants is largely confined to the ornamental garden and, even here, other less tolerant plants may be needed. Moreover, when gardening for food, there are relatively few drought-tolerant plants available. However, the pressing need to conserve water means that drought tolerance is becoming an increasingly important objective for plant breeders and new more resistant cultivars are continuing to be produced such as, for example, the new drought-tolerant lawn grasses.

CASE STUDY   Drought-tolerant grasses

Increased study over recent years has revolved around the selection of grass species and cultivars showing a greater tolerance to seasonal drought, e.g. chewings fescue, *Festuca rubra commutata*, slender creeping red fescue, *F. trichophylla*, and smooth-stalked meadow grass, *Poa pratensis*.

For longer swards, of a minimum of 150 mm, tall fescue, *F. arundinacea*, is recommended and has proven to be the most drought-tolerant of all the fescues.

Also, selective breeding and genetic modification have been brought to bear in the search for increased tolerance to more extreme conditions without compromising 'normal' UK growing conditions and the traditional look of British lawns.

In addition to choosing appropriate plants, the gardener can reduce water use by incorporating a number of garden practices that either conserve water, or use it more efficiently.

Grey water

The water used for personal washing and cleaning and cooking vegetables is termed **light-grey water** and can be collected and used in the garden. In contrast, laundry water usually contains pollutants, and if such **dark-grey water** is to be of use in the garden, eco-friendly washing products must be used.

Grey water should not be kept for more than 24 hours in order to minimise bacterial growth. It is best applied with a watering can as it is likely to contain grease, which can clog irrigation systems. In general there should be no problems with using grey water to tide plants over during a period of summer drought but long-term, extensive use is not advisable.

Grey water should preferably be used only for ornamentals and woody plants, and should not be used for leafy or root vegetables due to the risk of contamination.

Softened tap water and dishwater water contain salts that can damage soil structure, especially if the soil is rich in clay. However, short-term use may be worth considering as an emergency measure and should not cause serious damage.

Rainwater harvesting

Rainwater run-off from roofs can be collected in water butts or other containers for use in the garden. The containers should be covered to exclude organic matter, which may break down to produce unpleasant smelling gases. At least annual emptying and cleaning is recommended. Rainwater is particularly useful in hard water areas for **ericaceous** plants that are being grown in containers.

Unfortunately, a typical rainwater butt holds only about 200 litres, which is just enough to keep one container supplied throughout the summer. An alternative solution, albeit very expensive at about £5,000, that may be worth considering, especially where a large area of food crops is grown, is to install a large under-surface storage tank.

Reducing evaporation

In the open garden water retention in the soil can be increased by mulching with coarse organic matter. Gravel can also be used to develop a **gravel garden**, which can be particularly useful in dry regions (Figure 4.9). To be effective the mulch needs to be at least 5 to 10 mm thick. Pierced, plastic soil covers can also be used, provided that they are biodegradable.

Water loss from the soil can also be decreased by incorporating large quantities of organic matter, such as compost. This is a relatively short-term solution, however, as the incorporated organic matter will break down fairly rapidly. Peat should not be used either as **a mulch** or as a soil conditioner because of the environmental damage to peat beds that occurs with peat extraction.

CASE STUDY   Beth Chatto's gravel garden

Over 25 years ago Beth Chatto took the momentous decision to redevelop the car park area to her garden nursery into a new gravel garden. This area has subsequently become a model for other similar developments, including at the nearby RHS Gardens of Hyde Hall.

For a more detailed look at the process, the reasoning behind the design and the recommendations made, read *The Dry Garden* by Beth Chatto, originally published in 1978 by J. M. Dent & Sons Ltd.

Irrigation

Where irrigation is necessary, water economy can be achieved by employing devices that deliver water directly to the plant, such as drip feeds and perforated hoses or irrigation systems that supply accurately measured amounts of water at predetermined times. Sprinklers are very wasteful of water and their use is not recommended. If they are used any run-off water should be collected and re-applied.

Watering

The golden rule for gardeners is to 'water well, not often', because small amounts of water stay in the soil close to the surface and can

Figure 4.9 Beth Chatto's now well established Gravel Garden in Essex, UK.

be easily lost by evaporation without ever reaching the roots. Moreover, if water is only available from the top few inches (7–10 cm) of the soil, deep rooting is discouraged. Where possible, water should be applied directly to the soil around the base of the plants so that it reaches the water-absorbing roots quickly and is not lost by evaporation from bare soil. The best time of day for watering is either evening or early morning when there will be minimal loss of water by evaporation.

Enough water should be given at any one time to wet the soil to the root depth. This amount will vary depending on the depth of rooting and on the type of soil, but is likely to be more than most gardeners would expect. When the soil water has been completely depleted, the amount needed to replenish this varies from about 18 to 26 litres per square metre in different types of loamy soil. How long the water lasts depends on the rates of water loss by evaporation from the soil and by transpiration by plants. These rates are likely to increase due to climate change with higher summer temperatures. Although the rate of water loss varies from day to day with changes in the weather, a reasonable average for summer in the UK is 3.5 mm depth of soil per day. Thus a plant with a root depth of 500 mm has enough water for 31 days before permanent wilting occurs. However, as discussed earlier, water is needed long before permanent wilting occurs since this results in a drastic restriction in growth. Although this will vary with both the type of soil and with the plant, water is needed about every three to seven days to prevent water shortage when there is no rainfall.

Planting and other horticultural practices

It is important when planting to ensure that the newly planted roots are in good contact with the soil, as air spaces between the root and the soil water slows down the movement of water to the root surfaces. The soil should therefore be firmed around the roots and water applied to the plants. This is necessary even in moist soil in order to ensure good contact between the roots and

water in the soil. Water should be applied to the base of the plant as wet leaves in hot sun can be scorched.

Planting should preferably be carried out in the early morning or evening when the heat load on the plant is lowest and transpiration is consequently reduced. The best times of year for planting are spring and autumn when the soils are moist and not drying out.

Weeds deplete soil water through their transpiration, and weeding should therefore be undertaken as regularly as possible.

In the open garden some degree of shading reduces the heat load and is particularly helpful after planting when the root system has not yet become established. The various ways of reducing water loss in protected cultivation (i.e. misting, fogging, damping down and fan-and-pad cooling) are discussed in Chapter 11.

EXCESS WATER

It seems likely that flooding will happen more often in the future if, as predicted by most climate change models, the occurrence of heavy storms increases. Excess water can pose a number of problems for plants. Hairy leaves may become waterlogged in heavy rain and may rot, while most plants cannot tolerate waterlogged soils due to their lack of oxygen.

Roots normally obtain their oxygen from air spaces in the surrounding soil. When the soil is well drained, air can diffuse through these spaces to a depth of several metres. In contrast, water fills the spaces between the particles of soil when it becomes flooded and oxygen must diffuse through liquid rather than through air. Diffusion through liquid occurs at a much slower rate than through air, and oxygen depletion begins to occur. This happens only slowly when the temperature is low and the plants are dormant. Consequently temporary flooding under these conditions in winter may be relatively harmless. However, when the temperature is above 20°C, oxygen consumption by the roots and by micro-organisms in the soil increases and can rapidly deplete oxygen from the bulk of the soil. Under these conditions total depletion of the oxygen stores in the soil can take place in as little as 24 hours. The absence of oxygen severely depresses aerobic respiration and growth in most plants and the subsequent yield can be markedly reduced. In peas, *Pisum sativum*, for example, yield may be halved if the soil is flooded for only 24 hours.

Flood tolerance

Specialised plants such as marsh and bog species, e.g. the American skunk cabbage, *Lysichiton americana*, shown in Figure 4.10, can survive flooding for long periods.

As with drought tolerance, several different strategies have evolved to make this possible, although most of these involve some means of getting oxygen to the roots, rather than the ability of the root cells to tolerate low oxygen levels. In many wetland species, e.g. the salt marsh bullrush, *Scirpus maritimus*, oxygen movement to the roots is aided by the development of **aerenchyma**, a tissue consisting of long files of interconnected air-filled

Figure 4.10 Bog and swamp plants such as the American skunk cabbage, *Lysichiton americana*, are able to tolerate flooded soils and are suitable for planting at the margins of ponds.

spaces that allow gaseous oxygen from the air to diffuse rapidly through the plant to the cells of the root. In other plants, such as the bald or swamp cypress, *Taxodium distichum*, and black mangroves, *Avicennia germinans*, portions of the root (known as 'breathing knees' in *Taxodium* or more formally as pneumatophores) remain above the water or soil surface and allow oxygen to diffuse down to the submerged parts of the root. In addition, the roots of many wetland plants have an impermeable corky layer around them, which prevents the loss of oxygen from the root cells into the surrounding soil.

In flood-tolerant plants such as willow, *Salix* spp., there is little aerenchyma present when plants are growing in well-aerated soils. Aerenchyma only develops when the surrounding oxygen level is low and this process involves the gaseous hormone, **ethylene**. At low oxygen concentrations, a precursor of ethylene is produced in the roots. From here it is transported to the aerial parts of the plant where it is converted to ethylene and results in the consequent development of aerenchyma.

Unfortunately, the production of ethylene in the shoots of flooded plants has effects other than to induce the formation of aerenchyma. Ethylene gas is the 'ripening hormone' for many fruits and one of its many effects is to causes **senescence**. Thus although ethylene is a key hormone in the ability of some species to tolerate flooding by inducing the development of aerenchyma, prolonged exposure to increased levels of ethylene leads to yellowing of the leaves and ultimately death in species that are not tolerant of water-logged soils.

Causes of plant death due to flooding – The death of plants in flooded soils is mainly caused by the lack of oxygen for normal aerobic respiration by the roots. However, other factors are also important. As discussed above, the production of ethylene by the plant results in senescence and death in many species. Aerobic micro-organisms also become replaced by anaerobic ones that produce toxic compounds such as hydrogen sulphide, which is a respiratory poison. The production of hydrogen sulphide and other toxic gases also accounts for the unpleasant smell that often develops in flooded soils.

Ponds and aquatic plants

Well-managed ponds that have a thriving population of oxygenating plants have more available oxygen than in flooded soils. However, the diffusion of CO_2 through water occurs more slowly than through air so that submerged leaves have a less readily available source of this essential gas. There are a number of adaptations that increase the ability of plants to survive in an aquatic environment. They include, for example, long petioles that enable leaves to be in an aerial environment while their roots remain in the soil at the bottom of the pond. Floating leaves, such as those of water lilies, *Nymphaea* spp., have most of their stomata on the upper surface of the leaf, unlike those of the majority of land plants where the stomata are mainly located on the under-surface.

Submerged leaves are often finely divided with a high surface to volume ratio that allows a more rapid diffusion of CO_2 into the photosynthesising cells of the leaf. In some aquatic plants such as the water crowfoot, *Ranunculus aquaticus*, the submerged leaves are finely divided while those that emerge into the air are undivided. As with adaptations for boggy conditions, the presence of aerenchyma in many aquatic plants allows gases to diffuse rapidly through the plant.

MANAGING EXCESS WATER

The problems arising from too much water can be tackled in two ways: by choosing plants that are tolerant of boggy conditions and flooding, or by improving the drainage of the soil.

Plants that are tolerant of periods of flooding or more persistent boggy conditions include a number of plants that are native to the UK and their non-native relatives. Among these are willows, *Salix* spp., dogwoods, *Cornus* spp., mountain ash, *Sorbus aucuparia*, and summer snowflake, *Leucojum aestivum*. Several ferns are also highly tolerant of wet soils.

Amelioration of soil drainage can be achieved by the use of various types of drainage systems, such as **mole drains** (tunnels made in the soil by dragging an implement known as a mole plough through it), permanent clay or plastic drainage pipes, or open drainage ditches. All of these solutions are demanding of labour and relatively expensive. The cheapest solution is to use mole drains but these have several disadvantages. They are only suitable for heavy clay soils because smearing of the clay particles as the mole plough is dragged through the soils holds the drains open. Although there is a benefit in cost compared with installing permanent drainage pipes, mole drains have a limited life and are likely to collapse after a period of time.

The drainage of wet lawns can be improved by spiking them all over with hollow-tined forks. Other ways of improving their drainage are to add material to the soils that will improve their structure and increase the movement of water through them. In order to be effective, this requires adding large quantities of coarse organic material to the soil. This is not a permanent solution, however, as the organic material is quickly broken down. For further details see Chapter 6.

One of the problems of excess water, especially from heavy rainfall and storm conditions, is the run-off that occurs from paved and other types of impermeable surface. When front gardens are paved to accommodate cars, water cannot penetrate to the soil and run-off is increased. The use of water-permeable surfaces such as gravel for these and other areas of the garden such as patios, is now strongly recommended and should be used instead of impermeable paving wherever possible. Where paving is essential, particular attention should be given to the provision of permeable areas to accommodate the run-off of water after heavy rain storms.

CLIMATE CHANGE

Most climate change models predict that there will be a change in the weather patterns of the UK. In particular, a rise in temperature will increase the probability of more severe summer droughts, especially in the South East. An increase in the frequency of severe storms is also predicted, leading to temporary flooding. Both of these occurrences can lead to problems in the garden.

Summer droughts

Solutions to summer droughts include the use of plants that are relatively resistant to dry soils. Many of those that are currently available to gardeners have their origin in Mediterranean regions where the winters are milder and relatively frost free. Consequently, many of these plants are not sufficiently frost tolerant to be grown in the UK, except in the milder parts of the country. This may change as the climate warms. In addition, plant breeders are studying the possibility of improving drought resistance by **target breeding**. Lawns require large amounts of water to maintain their greenness and, although drying out does not necessarily kill the grasses, brown lawns are unsightly in the ornamental garden. As a result of breeding, some cultivars of perennial rye grass, *Lolium* spp., and fescue, *Festuca* spp., have increased drought tolerance through the incorporation of a **green gene**.

The problems of conserving water by changing watering practices are also being investigated. For example, it has been found that applying water only during critical periods of growth did not affect the flowering performance of poinsettias, *Euphorbia pulcherrima*, growing in protected cultivation. By using this **regulated water deficit** treatment, up to 50% water saving could be achieved without compromising yield. Changes in the watering of irrigated vines leading to significant savings in the amount of water used have already been discussed. Both approaches are being investigated for use with other crops.

Flooding

Plants can also suffer from the periods of flooding that may follow the predicted increase in the frequency of heavy storms. Although plants from regions with Mediterranean climates are more tolerant of dry periods, they are likely to be more susceptible to flooding. Experiments have shown that there are differences between Mediterranean plants in their tolerance of flooded soils. For example, lavender, *Lavandula angustifolia*, and sage, *Salvia officinalis*, are less tolerant of temporary flooding than is *Stachys byzantina*. Plants are more likely to be harmed by flooding in the spring and summer than in the winter, probably because the higher soil temperatures increase the respiration rates of the roots and soil micro-organisms, resulting in a more rapid depletion of oxygen than in the lower temperatures of winter as discussed above.

In conclusion, gardeners need to be aware of the potential effects of climate change in the UK and be prepared to adjust their own horticultural practices appropriately.

Summary

In this chapter we have seen that managing their water economy is a major problem for plants. Green plants are completely dependent on obtaining a supply of carbon dioxide from the air in order to carry out the essential process of photosynthesis. This necessitates the provision of pores in the leaf so that carbon dioxide can enter. However, these pores (stomata) also allow the rapid loss of water vapour from the wet cell surfaces of the sub-stomatal cavities into the surrounding air. If water is lost faster than it can be supplied from the soil, the leaf cells lose water and the leaves wilt, the stomata close, and photosynthesis ceases, leading to a rapid slow down of growth.

In this chapter, the basic physiology of water uptake from the soil, movement through the plant and loss from the leaves is discussed. This is followed by a consideration of ways in which plants are able to minimise their water loss. These include biochemical and physiological methods, such as the action of abscisic acid to regulate the closing of stomata and modifications that allow photosynthesis to continue in the light even when the stomata are closed. Plants have also evolved a diversity of structural and metabolic adaptations that enable them to succeed under different climatic conditions including both a shortage of available water and an excess supply. Understanding these can help the gardener to plan with confidence planting schemes that are appropriate for the prevailing environmental conditions, and those that are predicted to result from any future climate change.

Practical ways of managing water in the garden are also discussed. These include methods of conserving water, such as rain-water harvesting and using waste water from the house, as well as optimising management practices such as irrigation, drainage and watering techniques.

In conclusion, there are many ways to optimise water management in the garden. These include the choice of appropriate plants for any particular situation or habitat and the adoption of practical methods to conserve water and use it economically.

FURTHER READING

Barron, P. (1999). *Create a Mediterranean Garden.* Lorenz Books.
Plant selection and development of this popular garden style.

Gildemeister, H. (2002). *Mediterranean Gardening: a Water-wise Approach.* Editorial Moll.
Provides even further detail of reduced water requirement and use, in a practical way.

Ingram, D. S., Vince-Prue, D. & Gregory, P. J. (eds) **(2008).** *Science and the Garden: the Scientific Basis of Horticultural Practice*, 2nd edn. RHS by Blackwell Publishing. ISBN 978-1-4051-6063-6.
A complementary publication providing a broad overview of related topics.

Simmons, J. (2008). *Managing the Wet Garden: Plants that Flourish in Problem Places.* Portland, Oregon: Timber Press.
A helpful book that highlights the opportunities presented by difficult habitats.

Weinstein, G. (1999). *Xeriscape Handbook: a How-to Guide to Natural Resource-wise Gardening.* Xeriscape.
This publication provides discussion on gardening with reduced water availability.

Revision questions

1. List **five** forms of water saving that could be used in a **garden** situation.
2. Explain **four** problems associated with excess water and outline possible **solutions**.
3. State the method of water loss from the leaves and describe **four** natural methods used by plants in relation to **water stress**.
4. Define the term **aerenchyma** and name **two** examples.
5. Discuss **two** climatic changes that have been highlighted as possible future outcomes, including adjustments to **plant selections** made.

CASE STUDY   All-year-round flowering in chrysanthemums

One of the outstanding successes of applying plant physiology to commercial horticulture has been the development of all-year-round (AYR) flowering in chrysanthemums. In the 1950s, growers had considerable difficulty in producing flowers for the valuable Christmas market and only one or two varieties were suitable. The problem was that buds developed too early in the year and flowering could only be somewhat delayed by substantially lowering the temperature.

The discovery that the florists' chrysanthemum is a short-day plant and develops flower buds when the natural days begin to shorten sufficiently enabled growers to overcome the problem. By extending the day-length artificially in the autumn, flower bud formation could be delayed and flowers could be produced for the Christmas market without sacrificing quality and with a much greater range of varieties. Subsequent development has led to the marketing of flowers all year round by 'shortening' the days in summer by covering plants with an opaque black cloth and 'lengthening' the days in winter with an artificial light source. To do this successfully necessitates a detailed knowledge of the day-length responses of the cultivar being grown.

Commercial all-year-round chrysanthemum crop in RHS Wisley display.

Chapter 5

Climate, weather and seasonal effects

Daphne Vince-Prue

INTRODUCTION

Ideally, a garden should be attractive throughout the year. To achieve this requires an understanding of the factors that change with the changing seasons and how plants react to these changes. It is also important to recognise that the climate varies considerably throughout the UK, not only from north to south but also from the warmer, wetter west influenced by the Gulf Stream to the drier, more continental climate of the east. Finally, our climate is changing due to the influence of global warming and this too will affect the plants that can be grown successfully.

Seasonal interest in the ornamental garden can be maintained by choosing plants that are attractive at different times in the year. A succession of early spring bulbs (*Galanthus, Crocus* and *Narcissus*) is followed by spring-, summer- and autumn-flowering annuals, perennials, shrubs and trees. In winter, interest is maintained with evergreens, often with colourful berries: *Pyracantha, Cotoneaster* and *Sorbus* spp.; coloured bark: *Prunus serrula* and *Acer griseum*; variegated leaves, e.g. *Ilex* and *Euonymus*, or flowers, such as *Hamamelis mollis*. Autumn colour is another important item in the seasonally attractive garden.

Season and climate are also of great importance in the production of outdoor fruit and vegetables. Many crops are responsive to winter low temperatures, e.g. biennial vegetables, carrots, *Daucus carota*; or day-length, e.g. lettuce varieties, *Lactuca sativa*; while others are responsive to both low temperatures and day-length, such as celery, *Apium graveolens*. Understanding how crops respond to the changing seasons can also help the gardener to extend the harvest for as long as possible, ideally for the whole year.

This chapter describes how plants have evolved to prevent damage under different environmental conditions. The chapter also considers how gardeners can manage the environment most efficiently in the context of the garden, especially as climate change is likely to pose increasing problems of plant management in the future.

Key concepts

✿ Seasonal environmental changes are important in regulating the timing of plant development

✿ How changes in day-length during the year control the timing of plant processes

✿ Low temperatures during winter are essential for some processes

✿ Below freezing temperatures are important factors in the ability of plants to survive

✿ Understand the ways in which environmental factors control development

The Fundamentals of Horticulture: Theory and Practice, ed. C. Bird. Published by Cambridge University Press. © The Royal Horticultural Society 2014.

Favoured by short days	Type of storage organ	Favoured by long days
Apios tuberosus (peanut)	Root tubers	
Begonia socotrana	Aerial stem tubers	
Begonia tuberhybrida (tuberous begonia)	Underground stem tubers	
Dahlia	Root tubers	
Helianthus tuberosus (Jerusalem artichoke)	Underground stem tubers	
Solanum tuberosum (potato)	Underground stem tubers	
	Bulbs	*Allium cepa* (shallot and onion)
	Underground and aerial bulbs	*Allium sativum* (garlic)
	Corms	*Tritelia laxa*

Table 5.3 Examples of plants with different day-length requirements for the formation of storage organs

CASE STUDY **Know your onions: practical implications of the control of storage-organ formation by day-length**

Understanding the role of day-length in the formation of storage organs is important in practice. Long days are necessary for the formation of bulbs in onions. Many cultivars normally grown in the UK do not form bulbs at lower latitudes, such as those that occur in Egypt where onions are an important crop, and the so-called 'short-day bulbing' types are grown. However, these are also LDP but have a shorter critical day-length. For example, the variety 'Sweet Spanish' has a critical day-length of between 10 and 12 hours and so will form bulbs in regions where the day-length is longer than this, although a day-length of 14 hours is required if most of the plants are to form bulbs. In contrast, the cultivar 'Yellow Zittau' will only achieve maximum bulbing when the day-length is 16 hours or longer.

Many European and North American cultivars of potato form tubers in both long and short days, but do so earlier in the latter. However, the earlier onset of tuber formation in short days results in a lower yield than in the longer day-lengths at more northerly latitudes. In the USA, soya bean, *Glycine max*, cultivars with a wide range of critical day-lengths have been developed for use at different latitudes in order to maximise the yield in each location. These examples demonstrate the importance of choosing cultivars that are adapted to the day-length conditions that are experienced in any particular geographical location.

TEMPERATURE

The temperatures experienced by a plant can have beneficial effects or may damage it. The most damaging effects of temperature occur at below freezing temperatures, although some tropical plants such as African violet, *Saintpaulia* cultivars, can be damaged at higher temperatures. Plants can also be damaged by exposure to temperatures that are too high. Many plants have evolved mechanisms that enable them to avoid or survive under such conditions. This may simply be avoidance, by the shedding of leaves or die back of the shoots and over-wintering by underground organs. In other cases changes that increase the resistance to damage take place within the cells.

DAMAGE BY BELOW-FREEZING TEMPERATURES

When exposed to below-freezing temperatures, ice crystals begin to form in the cell walls of plants where the concentration of **solutes** is lower than inside the cells and consequently the freezing point is higher. Water then begins to move out of the cells into the cell walls, thus lowering the freezing point of the cell contents even further and preventing the damaging effect of ice formation within the living contents of the cell itself. However, the loss of water increasingly dehydrates the cells, leading to possible damage by dehydration. Indeed, in many cases, resistance to frost damage is based on the ability to tolerate severe dehydration and the cellular changes that occur are similar to those that are found in drought-resistant plants. Many of the symptoms of frost damage (for example, yellowing and scorching of the leaves) are similar to those that occur in dehydrated plants.

When thawing occurs, the ice crystals in the cell walls melt and water re-enters the cells. Damage is therefore more likely to occur when thawing is rapid and ice crystals melt before cells have regained their ability to absorb water quickly enough. For this reason direct exposure to early sunlight, such as against an east-facing wall, is more likely to cause damage than in situations where thawing occurs more slowly.

The ability to tolerate low temperatures is not necessarily constant. Plants are able to undergo a process known as **frost-hardening**. In woody plants, frost tolerance is partly induced by exposure to autumn short days but is also dependent on a later exposure to low temperatures. This process was interrupted by the rapid change of seasons in late autumn 2010, with resulting stem tip damage seen the following spring. In hardy herbaceous plants, exposure to a period of relatively low temperatures above freezing increases their ability to withstand the damaging effects of low temperatures in the process known as **hardening-off**.

Hardening-off

When hardy plants are raised in the glasshouse they have to be acclimatised (hardened-off) to lower temperatures for about three weeks before planting out in the garden. The usual procedure is to cover plants with two layers of horticultural fleece during the day for the first week, reducing that to a single layer in the second week. The plants need to be returned to the glasshouse during the night. Depending on the **ambient** temperature at the time, no covering is needed during the day by the end of the second week and the plants can remain in the open during the night. Cold frames are ideal for the hardening-off process, gradually increasing the ventilation over the first three weeks. Frost-tender plants should not be planted out until the danger of freezing temperatures is past. This is normally after the third week of May in the south of England and later in the north.

Frost hardiness

Plants have different abilities to withstand below-freezing temperatures. Hardy plants, for example many *Euphorbias*, are frost tolerant although even here plants vary widely in the degree of freezing that they can survive. Resistance to freezing also varies during the year. For example, plants that have tolerated temperatures well below freezing during their winter dormancy often show damage to the emerging young leaves in spring after dormancy has been broken. Non-hardy or tender plants have evolved in warmer climates and many are unable to tolerate exposure to low temperature, even though this may be above freezing, e.g. *Saintpaulia* cvs. There are now a number of hardy cultivars that have been bred for hardiness in the UK in species that are normally not frost tolerant, for example the new 'Everlast' cultivars of *Gerbera jamesonii*.

Figure 5.8 Many cultivars of *Euphorbia* are frost hardy and can withstand temperatures below freezing in winter.

LEAF FALL AND DORMANCY

Dormancy or rest is usually defined as failure to grow under conditions that are normally favourable. Consequently, only by exposing plants to favourable growing conditions can it be determined whether or not they are dormant.

Winter dormancy

In most cases the entry into winter dormancy or rest in trees and shrubs of high latitudes is a response to the decreasing day-length during autumn. The response to autumnal short days is of value to the plants as it precedes the onset of low temperatures and enables the plant to make cellular and **morphological** adjustments that will enable it to survive the ensuing winter. Thus the time at which they become dormant and frost hardy may be crucial for the survival of plants in any particular geographical location.

The onset of winter dormancy

The photoperiodic mechanism for the control of the onset of winter dormancy appears to be similar to that controlling flowering. It is the leaves that are sensitive to day-length and, consequently, some signal or signals must be exported from the leaves to the responding tissues, namely the apical and lateral buds. Plants also have a critical day-length and do not enter into dormancy unless the days are shorter than a critical value, which varies widely between species and cultivars. As with the photoperiodic control of flowering, the dormancy response involves the *FT* family genes and probably also one or more of the gibberellins.

The entry into dormancy is often accompanied by the formation of specialised structures that protect the internal tissues of the bud examples are the sticky bud scales of horse chestnut, *Aesculus hippocastanum*, and the black bud scales of ash, *Fraxinus excelsior*. In deciduous trees, one obvious morphological change is the shedding of leaves that precedes the entry into dormancy. Although in some cases as, for example, in the tulip tree, *Liriodendron tulipifera*, leaf fall is directly under the control of short days, in others, as in the false acacia, *Robinia pseudoacacia*, leaves are retained while the temperatures remain high irrespective of the day-length.

The effect of latitude on winter dormancy

Trees have been found to be strongly adapted to the location at which they evolved. For example, when seeds were collected from trees growing at different latitudes it was found that, irrespective of species, plants originating from the same latitude had approximately the same critical day-length for the maintenance of stem growth and prevention of entry into dormancy.

The day-length response for the entry into dormancy is critical for the survival of trees in any particular latitude. For example,

Figure 5.10 Protective bud scales develop in horse chestnut, *Aesculus hippocastanum*, during the onset of dormancy in the autumn.

plants from the far north at latitude 70° N have a critical day-length of about 22 hours for the maintenance of growth and prevention of entry into dormancy. This may never be achieved when they are grown further south. This means that, following the breaking of winter dormancy by low temperature as discussed below, the shoots cease to elongate and become dormant again after making only a small amount of elongation growth resulting in stunted trees such as those that have been observed experimentally in birch, *Betula pendula*. The shorter critical day-lengths of more southerly species means that they will continue to grow for longer if they are moved further north. For example, the onset of dormancy and development of frost hardiness in a willow, *Salix pentandra*, from 60° N (with a critical day-length of 14 hours) was delayed when they were grown out of doors in the much longer days of Tromsø at latitude 70° N and they were severely damaged by frost. Similarly, frost damage has sometimes been observed on trees growing close to street lamps where the exposure to long days may delay the onset of dormancy.

The altitude at which the trees have evolved has also been found to affect their critical day-length. Trees from high altitudes have a longer critical day-length than those from a lower altitude and behave like trees originating in lowland sites further north.

These effects of latitude on seedling behaviour mean that it is important to know the provenance of seeds in relation to their suitability for being grown in the local area.

Summer dormancy

Winter cold is not the only seasonal stress that can be avoided by a response to day-length. Some plants from Mediterranean climates where the summers are usually hot and dry often shed their leaves during the summer months, thus allowing the plants to survive when water is in short supply. In some cases, this has been shown to be a response to long days as, for example, in *Anemone coronaria*.

BREAKING WINTER DORMANCY

Once plants have become fully dormant in winter, growth will not be resumed until dormancy has been broken by exposure to a sufficiently long period at low temperature. The time of leafing out then depends on the temperatures in spring. However, once rest has been partially broken by cold, exposure to long days may also accelerate bud break and the resumption of shoot growth as, for example, in beech, *Fagus sylvatica.*

The effective temperatures for the breaking of winter dormancy are similar to those for vernalisation, ranging from just above freezing to about 10°C. However, the duration of winter cold required to break dormancy varies widely between species and cultivars and the failure to break dormancy can be a major problem in regions with mild winters. For example, in temperate-zone fruit trees such as apple, *Malus domestica*, insufficient winter cold can result in erratic or delayed leafing out in spring. This affects the timing of crop spraying programmes and, in the worst cases, can reduce leaf expansion to such a degree that growth is affected. This problem could become increasingly important in

Figure 5.11 The onset of dormancy in beech, *Fagus sylvatica*, is accompanied by leaf fall. However, leaves remain on the lower branches because these retain their juvenility. The fact that juvenile leaves are retained during winter is a feature of beech, *Fagus sylvatica*, hedges, which are kept in their juvenile state by repeated cutting back.

some areas, and perhaps even in the UK, if winters become warmer due to climate change.

Cultivars differ considerably in their cold requirements for the breaking of winter dormancy, and consequently it is important to know both the special requirements of any cultivar and the likely winter temperature regime in the locality before selecting the plants to be grown there.

Various chemical sprays (including mineral oils, often in combination with dinitro-ortho-cresol) have been found to promote dormancy breaking and are used commercially in some regions with insufficient winter cold. However, where possible, the use of cultivars with shorter cold requirements is the better option.

AUTUMN COLOUR

An important seasonal display in the garden is that of autumn colour, which precedes leaf fall in many deciduous woody plants. Depending on the species and cultivar, red, yellow, orange or purple colours may develop as chlorophylls begin to break down revealing the colours of the other pigments in the leaf.

The accumulation of anthocyanins and other flavonoid pigments leads to the development of yellow, bronze and red tones. These pigments accumulate because the movement of sugars out of the leaf begins to decrease as the temperatures fall, and the excess sugars are converted into various flavonoid pigments. The particular ones that are formed depend on species and cultivar leading to the differing colours of autumn leaves in different plants.

Figure 5.12 The leaves of *Acer palmatum* 'Osakazuki' turn a brilliant orange red in autumn.

The intensity of the display varies widely from year to year, with the autumn of 2010 being exceptional. Bright autumn days increase photosynthesis and thus increase the amount of sugars and, consequently, colour that develops, while plants growing in the shade often colour poorly. Low temperatures, particularly at night, decrease the export of sugars from the leaves and so result in a more colourful display. Nutrition is also a factor and high levels of nitrogen in the soil may decrease the amount of colour because sugars in the leaf can combine with nitrogen to form proteins rather than coloured pigments.

Despite the influence of environmental factors such as light and temperature, much of the variation in autumn leaf colour between plants is genetic. This emphasises the importance of selecting strains for good autumn colour, which is usually most effective when strains with good autumn colour are propagated vegetatively.

Although autumn colour is influenced by factors such as temperature and light that are beyond the control of the gardener, the intensity of the display can be increased by selecting the best cultivars, growing plants where the light is good and not over-feeding with nitrogen.

LIGHT AND WATER

The daily amount of light as well as its duration and the availability of water also vary seasonally. Both have significant effects on plant growth and are considered elsewhere. Adaptations to water stress and its effects on growth are discussed in Chapter 4, especially in relation to the selection of plants suitable for growing in either dry or excessively wet locations, while the **amelioration** of water stress by shedding leaves and/or the development of summer dormancy have been discussed in this chapter. The importance of the daily amount of light is discussed in Chapter 11 with relation to the problems associated over low winter light when growing plants in protected cultivation.

Summary

This chapter is concerned with seasonal gardening and the ways in which the changing seasons affect the growth and development of ornamental and crop plants. The effects of climate, especially as modified by latitude, are also considered.

The first part of the chapter considers the ways in which aspects of the environment change with the changing season. For both crop and ornamental plants the most important seasonal events are changes in temperature and day-length.

Day-length controls the time of flowering in many plants. Some plants require exposure to short days in order to form flowers; others require long days, while some require a sequential exposure to both. A response to long days enables plants to time their flowering and seed set to coincide with good weather conditions in summer, while a response to short days enables woodland plants to flower before a tree canopy reduces the available light for photosynthesis. The day-length signal is perceived by the leaves but the changes that lead to flowering take place at the shoot apices. Consequently a transmissible signal must be involved. The nature of this signal is discussed.

Day-length also controls the entry to winter dormancy in many woody plants. The cessation of growth is a response to the short days of autumn and is accompanied by an increase in frost hardiness that enables the plant to survive the below-freezing temperatures of winter.

Day-length controls the formation of storage organs in many plants. In most cases this is a short-day response and storage organs form during the decreasing day-lengths of autumn. Onion species, *Allium cepa*, are an exception and require long days for the formation of bulbs. These responses are discussed in relation to the problems of growing plants at different latitudes.

The temperature also affects many aspects of seasonal behaviour. Once dormancy has been induced by autumnal short days, a period of exposure to low temperatures during winter is needed to break dormancy and bring about the resumption of shoot growth. Inadequate winter cold leads to delayed leafing out in spring and consequent problems in the timing of crop sprays.

Some plants need particular temperature regimes for their development. These are discussed with particular reference to the effect of exposing plants to chilling temperatures early in their life, on their later flowering in plants requiring vernalisation, and the direct effect of temperature on flowering in other plants, especially in relation to bulb forcing.

Damage may result from exposure to extremes of temperature. In the UK, temperatures below freezing are a particular concern for some crops and for some localities. The causes of frost damage are outlined and the importance of frost hardiness is discussed.

Overall the importance of seasonal effects is discussed with respect to the timing of events such as flowering, the development of storage organs and frost hardiness. Climate and latitude are also discussed in relation to the ability of plants to survive and reproduce in any particular locality. Understanding the ways in which seasonal and climatic signals affect plant development can help gardeners to maintain interest in the ornamental garden throughout the year and also to obtain a succession of harvestable crops for as long as possible.

Revision questions

1. How do plants record **time** and the changing seasons?
2. Discuss the use of **hardening-off** as a horticultural technique.
3. What **environmental** conditions improve the onset and longevity of autumn colour?
4. Define the term **long-day plants**, including **four** named examples.
5. Discuss the term **dormancy** and the situations where the plants benefit from this state.

REFERENCE

Smithers, R. & Sparks, S. (2010) A review of spring 2010. www.naturescalendar.org.uk [Accessed 23 August 2013].

FURTHER READING

Bisgrove, R. & Hadley, P. (2002). Gardening in the global greenhouse: the impacts of climate change on gardens in the UK. Technical report. UKCIP, Oxford.
The UK Climate Impacts Programme (UKCIP) is based at the University of Oxford and funded by DEFRA to co-ordinate an assessment of how climate change will affect the UK and help organisations assess how they might be affected. The research behind this report was carried out by the University of Reading School of Plant Sciences.

Dehgan, B. (1998). *Landscape Plants for Subtropical Climates.*University Press of Florida.
The reasons for, selection and use of plants for this habitat are discussed.

Duffield, M. R. & Jones, W. (2001). *Plants for Dry Climates: How to Select, Grow and Enjoy*, revised edn. Da Capo Press.
General advice covering a broad range of plant types, modified for dry conditions.

Ingram, D. S., Vince-Prue, D. & Gregory, P. J. (eds) (2008). *Science and the Garden: the Scientific Basis of Horticultural Practice*, 2nd edn. RHS by Blackwell Publishing. ISBN 978-1-4051-6063-6.
A complementary publication providing a broad overview of related topics.

Springer Ogden, L. & Protor, B. (2000). *Passionate Gardening: Good Advice for Challenging Climates*. Fulcrum Publishing.
General and broad coverage of the area, including reasons, progressing to deal with a range of extremes, in a practical way.

RHS level	Section heading	Page no.
2 3.1	Day-length and the formation of storage organs	93
2 3.3	Day-length and the formation of storage organs	93
2 5.2	Damage by below-freezing temperatures	94
2 6.2	The critical day-length	90
3 1.1	Frost hardiness	95
	Damage by high temperature	96
3 1.2	De-vernalisation	97
3 1.3	De-vernalisation	97
3 1.4	De-vernalisation	97
3 4.2	Which environmental factors change throughout the year?	88
3 5.2	Hardening-off	95
3 7.1	Effective temperatures for vernalisation	97
3 7.2	Plant hormones and flowering	91
3 7.3	What controls the time of flowering?	88

CASE STUDY PlantWorks Limited, Sittingbourne, Kent, UK

PlantWorks Limited are one of the main companies that have, since 2000, brought the use and development of beneficial fungi, as an additive at planting time, into a garden product now available at any garden centre. As we shall see in this chapter, the use of mycorrhizal products, where fungi form a relationship with their host plants, has important economic and environmental benefits as they are used for the improved establishment and subsequent growth of plants, especially adopted for roses and plants in extreme growing conditions. To give an indication of the scope of their relationships it is worth noting that the only common plant families not using mycorrhizae are *Amaranthaceae* and *Brassicaceae*.

An introduction to the different types is given in Chapter 2, with their current impact given in this chapter and possible future developments are discussed in Chapter 18.

Lifting potatoes, *Solanum tuberosum*.

Chapter 6

Soils and plant nutrients

Chris Bird

INTRODUCTION

Using the soil and other products to grow and sustain plant material has been practised since the earliest times, when the volcanic dust settled and the spores and seeds began to grow. An understanding of the type of soil that we are operating on, its capabilities, how to improve, cultivate and manage it, has been the main topic of conversation when and wherever growers of plants meet. A number of different alternatives, e.g. peat free, have become available, but which is the best route to follow? Or is a combination of recommendations the one to produce the most balanced, productive and sustainable media?

The area of cultivation or non-cultivation is another topic of discussion, with guidance as to the terms employed and the order of use. Unfortunately, short cuts have been used with the aid of modern machinery that can do untold damage, which may take years to correct. Equally, what are the benefits and limitations presented by the no-dig policy and is it successful in all situations?

Once germinated, planted or established, the nutrient requirements of a wide range of plants, their situation and the timing of use, may depend on the origin or source: be it artificial or organic. This is discussed, with linkage to nutrient cycling, organic matter breakdown and pH, together with the identification of different plant growth stages and the nutrient requirements that these bring. How to prevent damage or delay in the harvest or aesthetic quality of whichever plant species you are growing is also discussed, including how plants uptake nutrients and adapt themselves to the media present.

This then leads us on to the use of artificial soils and the wide range of alternative products now available, and those that are emerging as future contenders, including soil-less growth systems, such as nutrient film technique (NFT).

Key concepts

- ✿ Introducing the physical, chemical and biological properties of soil
- ✿ Outlining the active relationship between soil and roots
- ✿ Explaining the opportunities for cultivating and managing soils
- ✿ Discussing the impact of humus, nutrients, fertilisers and pH levels
- ✿ Describing other growing media and their management

The Fundamentals of Horticulture: Theory and Practice, ed. C. Bird. Published by Cambridge University Press. © The Royal Horticultural Society 2014.

FACTORS INFLUENCING SOIL TYPE IN THE GARDEN

Time

One of the things often misunderstood is the duration of the various processes that result in soil formation. Yes, we talk of the ending of the last Ice Age, some 10,000 years ago, but forget that soil is still being formed, and also destroyed, today. The other reminder is that soil is a precious and reducing resource, with vast areas being lost to development, e.g. roads and buildings, together with erosion, but this rearranges the soil rather than losing it altogether. This includes some of the best, highest quality types, especially light clay loam, which has the best balance of moisture and nutrient retention together with the ability of rapid drainage of any excess water. This results in pulling in and through the soil, of gases to supply and refresh the needs of plant roots and the wider soil flora and fauna. At its most basic the main part of soil is finely crushed rock, with decomposed organic matter and water holding it all together.

Climate

With most temperate regions of the world the temperature range of −20 to +40°C allows relatively rapid weathering to occur by a successful combination of physical, chemical and biological methods. Rainfall will both physically, by water flow and the freeze-and-thaw action, and chemically cause change, especially with chalk or high limestone content soils, due to carbonic acid being present, resulting in a naturally acid rain, with a pH of 5.5. This should not be confused with the often-discussed acid rain, when the type of acid is sulphuric, usually from industrial processes, resulting in plant damage or death.

Other physical methods of weathering

These include dust particles, especially in hot dry situations, where they are taken up by the wind and strike against rock surfaces, causing some considerable damage over time. This is also a problem on historical buildings made of sand or limestone.

Parent material

This is the phrase that is used to describe the mineral matter from which the soil develops. The types of rocks that are the starting point relate strongly to the resulting soil produced: its physical and chemical properties, together with the range, or lack of it, of flora and fauna it can successfully sustain. Even to the underlying stability and its ability to withstand erosion. The oldest group of rocks, igneous (think about igniting or ignition as a memory aid), result from the forming of the Earth's crust, when the magma and larva cooled, examples include: granite, basalt and grist. These are very hard wearing and slow to weather. However, the resulting soil, when it is formed, is clay based and very high in nutrients and also has the ability to store and exchange them with plant material. Sand, which is based on silica, has the opposite characteristics: very free draining with no water or nutrient-holding capacity and therefore very poorly regarded for horticultural use, unless you are growing early crops, where the speed of soil temperature increase in the spring is seen as a major advantage.

The next type of rock to consider is sedimentary, as the name indicates, these rocks are the result of older rocks weathering and being washed or blown down to resettle in the valleys, rivers, streams and out to sea. They then build up as sediment and as the pressure increases they re-cement together to make the new rocks, e.g. sandstone, limestone and chalk. In the case of chalk, this is laid down in a slightly different way: it is formed from the remains of sea shells, which have fallen to the sea bed and again under pressure and over time have formed the soft multilayered material we call chalk. As a range of organisms and different materials were involved, this results in the multilayered look seen where chalk slopes are exposed.

Figure 6.1 Twyford cutting, Winchester, showing chalk layers and *Buddleja davidii*.

The third of the major rock types is metamorphic, meaning change, and these can be derived from both igneous and sedimentary rocks, which have been subsequently re-heated and sometimes also pressured to form a new material, e.g. shale to slate, and limestone to marble.

The other major group of soil formers are the organisms, which include vegetation (both living and dead), animals and micro-organisms. The pioneers of this group are the lichens and mosses that colonised the bare rock surfaces, starting to trap dust particles that laid the foundations for other plants to obtain a foothold in the cracks and crevices in and between the major rocks. This resulted in stabilising areas like scree beds to allow a progressive colonisation of the areas, by the higher plants, e.g. shrubs and trees.

Figure 6.2 Mountain side colonisation: bare rock to plant establishment.

Another factor in the formation of soils is the topography, that is, the shape of the ground surface, resulting in gravity-assisted settlement of material into the valley bottoms being washed or blown off the exposed hilltops, causing more highly fertile productive soils in these areas. These are also normally much deeper in the formation of the topsoil layer, but may also result in poorly developed drainage, commonly seen in silt soils, which are well known for being unstable and can easily be eroded by wind and water.

TRANSLOCATION WITHIN THE PROFILE

This term relates to the movement of soil material, in solution or suspension, from one depth of soil level to another. These levels or layers are called horizons and together make up the soil profile.

Eluvial horizon, or E horizon, is the term given to the level losing material: where the maximum area of leaching or eluviation takes place.

However, lower down the soil profile another horizon, often the subsoil or B horizon, gains the material and this is called the illuvial horizon: the level of maximum accumulation.

A quick way to remember the direction of movement is to think of 'E' for exit and 'I' for in.

The prevailing wind also has an effect on the loss and deposition of material in different places, usually for the majority of the UK this is south westerly in direction, but changes at regular intervals depending on the season.

PROPERTIES OF GARDEN SOILS

Physical properties

Soil texture is one of the most over-used phrases, usually related by most people to the 'feel' of the soil. However, a technical definition of 'the relative proportions of sand, silt and clay, given as a percentage' is the correct definition. The other term that is often confused with it is 'soil structure', which is defined as 'the arrangement of the sand, silt and clay particles into aggregates', i.e. into crumbs or clods, depending on their size.

Therefore texture is more fixed, and can only be changed by the addition of more of one of the constituent components. In contrast, a soil's structure can be adjusted, for good or ill, by our actions and reactions. The classic destroyer of soil structure is cultivation in wet weather: the phrase 'if it sticks to your boots, keep off it' is as true today as it ever was. The classic producer of soil structure is the use of **green manures** (or adding organic matter in some form), but more of these later in this chapter; their use is also further discussed in Chapters 12 and 18. A soil with a well-developed structure is in fact half air spaces, of different sizes, called pores. The size is important as their function changes with it. **Macropores** are the largest, greater than 50 μm in size, and provide the exchange of air for a soil together with the ability for rapid drainage. Mesopores, being between 0.5 and 50 μm, provide the water retention but that is still readily available to plants. This size of pore is also used by plant roots and earthworms as a pilot hole for spreading throughout the soil. The final pore type, commonly discussed, is that of micropores, being anything in size below 0.5 μm and having the role of keeping the soil together, acting as a glue, together with **humus**, forming the resultant gaps, therefore producing the other pore types. See Table 6.1.

These different pore types all need to be present to obtain the maximum performance from any soil, and this can be visually checked and confirmed by looking for the presence of interconnecting cracks down the soil profile.

Equally important, but much less understood, are the chemical properties of soils. These include components derived from igneous rocks: clay minerals, alumino-silicate materials, oxides of iron and aluminium.

Humus is formed by the breakdown of organic materials and is a black, non-sticky, jelly-like material, which is one of the 'Holy Grails' of gardening. This is because it holds together the soil particles, causing the development of aggregates, has a massive (for its size) water-holding capacity and nutrient-holding and exchange capacity. In short: it's good stuff. Eventually it will itself breakdown into inorganic materials and nutrients, which in turn will be held on other particles of humus or be taken up by the roots for plant use.

Pore title	Alternative water title	Size range	Limitation percentages	Main uses	Produced and maintained by
Micropores: sometimes called residual pores	Hygroscopic water: held too tightly for the plant roots to extract	<0.5 μm	If <20% of the soil volume then mechanical difficulties may be seen	Holding the particles of soil together to form the structure	A good level of humus present within the soil providing stability of fine clay particles
Mesopores: sometimes called storage pores	Capillary water: held in these pores by surface tension	0.5 up to 50 μm	If <15% of the soil volume then available water may be restricted	Retaining the major volume of usable water accessed by plant roots	A good level of organic matter providing good water-holding capacity
Macropores: sometimes called transitional pores	Gravitational water	>50 μm	If <10% of the soil volume then drainage may be a problem	Water passing through the profile draws in fresh air for root use and due to their size ensures rapid drainage	Well-cultivated soil with an air-filled porosity (AFP) of between 20 and 30%
Developed from Ingram *et al.*, *Science and the Garden* (2008) based on Rowell (1994)					

Table 6.1 Soil pore sizes and roles.

The pH is a term that indicates the balance between acidity and alkalinity on a 14-point scale, that is from 0 to 14; values from 0 to 7 being called acid, and from 7 to 14 alkaline. The natural soils in the UK have a pH from between 4 to 10; however, if pollution has occurred extreme acidity may be experienced, down to pH 3 in some cases. pH 7 is called the 'neutral' point, but most plants prefer a pH of 6.5. Most people will have come across the term pH, but less understood is the fact that the scale is **logarithmic**, that is, that the scale is multiplied by a factor to make the scale easier to use. With this one it is to the power of ten, so the difference in alkalinity between points of the scale is ten times, e.g. for one point difference, say, between pH 6 and 7 is ten, for two points, e.g. pH 6 and 8 it is one hundred. This is important, especially between 6 and 8 on the scale, as the range of plants that may be selected in light of a pH test will not thrive if the result is incorrect or given in general terms, e.g. 'it's about pH 7'.

Biological properties

A soil in good condition is teeming with micro-organisms, which include algae, fungi and bacteria. Each of these groups have their role to play and this is an emerging area of understanding that has been long overdue in being recognised as vital to balanced and successful growing. With the recent, and developing, interest in growing organically a welcome change of emphasis is shown with natural methods of low or no environmental impact being championed. The over-arching theme is that if your soil is well prepared and in balance then any crop can be successfully grown with minimum damage being caused by pests and diseases.

The breakdown of organic matter: (regular supplies of fresh organic matter are vital to keep the living systems of the soil in balance) each stage of breakdown relies on the next and all are interconnected. If things do go wrong, then systems are also available to undo years of good work, so keeping things in balance is very important. Too much change, such as the addition of new material, to an established soil will still have a disrupting effect. Earthworms and their importance – Within the UK we have more than 25 native species of earthworms, ranging from the largest terrestrial land or lob worm, *Lumbricus terrestris*, reaching the impressive length of 35 cm (14 inches) when fully grown, down to the little tree worm, *Satchellis mammalis*, barely 3 cm (1¼ inches) long, which lives within leaf litter. Earthworms keep the soil open and also pull down fresh organic matter, especially in the autumn period. They supply numerous bacteria through their worm casts. They provide a good seed-sowing thermometer for spring (because of the link between worms casting and the soil temperature). Worms form casts when the soil temperature is between 10 and 12°C, which is the ideal temperature for seed germination for a very wide range of plants.

This is a much better system for use on allotment sites and in public places than the method of using the germination and growth of ephemeral or annual weeds as a gauge for when your soil has obtained a suitable germination temperature. Please select your choice of method but spare a thought for those around you trying to keeping a weed-free plot.

Nematodes – Unseen by the naked eye, nematodes have come into common use as an example of biological control for slugs and vine weevil, where a different species controls each, but they also play a vital role within the soil and growing media areas.

CASE STUDY	Which nematode to use?	
Pest controlled	**Nematode species**	**Application notes**
Slugs	*Phasmarhabditis hermaphrodita*	Moist media and temperatures between 5 and 20°C are required for successful establishment.
Vine weevil	*Steinemema kraussei*	Use in autumn when the media is warm (5 to 20°C) and the new generation of weevil larvae has not yet affected the crop roots.

Protozoa – These are a group of microscopic organisms, found in three forms: amoebae, ciliates and flagellates, closely related to bacteria, but larger in size. They play a role in the breakdown of organic matter from fresh deposits, transferring them to semi-broken down, humified, material to be finished off by fungi for the woody hard material and bacteria for the quicker breakdown of soft green material. They are single-celled containing a nucleus and are included within the domain *Eukarya*. Moisture is very important to their movement through soil and they also form symbiotic relationships, usually with bacteria. A major link in the soil food web: both eating and being eaten in turn and so promoting the quicker transfer of fresh organic matter into usable materials.

Molluscs – This group includes the all too familiar slugs and snails, which play an important role in clearing up any rotting organic green, non-woody, material and provide a vital link between the soft raw primary materials and the bacterial level of the media in the complex food webs that develop in healthy, active fertile soils. This is especially true of the great black slug, *Arion ater*, which can grow up to 15 cm (6 inches), and confusingly is also found in grey, brown or even orange forms; they prefer the natural rotting soft green material rather than any precious plants that may available. On the other hand, one of the most troublesome species is the black keeled slug, *Milax* spp., which is commonly found in potato, *Solanum tuberosum*, tubers having burrowed into the tuber making a chamber. Therefore do not over tidy your plot; leave some dead leaves under shrubs for your mollusc population to work on, together with your earthworms, *Lumbricus terrestris*.

Insects – A group that can be recognised by the presence of six legs and a body with three parts: head, thorax and abdomen. A wide range of species are found in a healthy growing media, these include some that can be identified as pest species, such as: wasps, *Vespa vulgaris*, black ants, *Lasius niger*, and hornets, *Vespa crabro*. However, they all have a place in a well-balanced healthy media, including the consummation of smaller insect problems, e.g. aphid species, *Aphis faba*.

Mammals – Common mammals include moles, *Talpa europaeus*, rabbits, *Oryctolagus cuniculus* and badgers, *Meles meles*. They mix the soil when they dig and also provide aeration with their network of burrows. These actions throw up new soil to be weathered and also the lower levels can be more nutrient rich than the well-weathered surface levels, thus making these new nutrients available to the developing system of plant roots.

WHAT DO GARDEN PLANTS NEED FROM THE SOIL?

A range of factors is required from the selected growing media for successful plant establishment and growth.

Anchorage
All media are required to provide anchorage for plant roots, to hold the plant upright in the most extreme conditions. Plants anchored in the soil range from large trees, e.g. common oak, *Quercus robur*, to very small annuals, e.g. basil, *Ocimum basilicum*. However, not all media, including soils, provide the same level of anchorage, with clay providing most support and silt being quite unstable and resulting in landslips or soil creep on sloping land.

The scope or range of a plant's root system also impacts on its ability to find and extract water.

Water
Without the selected growing media being able to retain water and provide it, as required, to the plant root system then growth will be absent or reduced. The perfect media would be able to absorb, retain and release the required moisture with minimum effort being made by the plant roots. The ideal moisture level should be between 20 and 30%, allowing a wide range of crops to be successfully grown. However, those media based on clays, i.e. soils that contain more than 20% clay particles, react like clay, meaning that although they retain a high level of water it cannot all be released to the plants growing in it. This is due to the size of the particles and surface tension of the water molecules. The clay holds on to the water more strongly than the root system can extract it. The ideal water point for maximum growth is called the field capacity, which can be defined as 'the maximum water held after free drainage'. The volume of water held and made available to plants is different for each media but the physical point is the same, i.e. free (non-pumped) drainage stops. It is worth noting that wilting point is when the green material of the plant cannot support itself, i.e. the cells of the plant do not remain turgid.

However, if water, either rain or irrigation, is supplied then the plant can recover. This method is sometimes used to force a crop into a survival response, e.g. the onset of flowering.

At the opposite end of the spectrum is the term **permanent wilting point**, which is the point where a plant has gone beyond the point of no return, i.e. if it is given water it will not recover, as the cells themselves are damaged. Water will still be in the soil; however, this is not available to the plant, it is holding the media together. If, subsequently, this water is also lost, then a dust bowl develops and it blows away, which has occurred within the UK, in East Anglia on the Fens, where the soil is an organic one based on low-moor sedge peat. The management of organic soils is discussed in Chapter 18. These areas can also be called basin histosols, where high levels of peat build up in shallow depressions. However, because water-containing nutrients flow into the area a wider range of plant material can establish, dominated by soft rush, *Juncus effusus*, and sedges, *Carex* spp.

Air

In balance with the available water, should be the volume of air present within the media, the ideal being between 20 and 30%. Again, some soils, e.g. sands, have a very high air content, with many large pore spaces (macropores); this is good for fast drainage but is at the expense of water-holding ability. This area of media air content, called **air-filled porosity (AFP)**, has been one of the most problematical items in plant growing for some years. This has been shown in discussion around the areas of growing plants in regularly cultivated soils and the move towards the 'no-dig' method of production, with the perceived risk of developing compaction, if mismanaged, through to the outdoor overwintering drainage problems of growing in container media, especially seen in wet winters. With the result of an increased emphasis on the production and use of very open media, including the addition of chipped bark and other coarse, high-fibre organic products.

Mineral nutrients (macro- and micro-nutrients)

The soil also has the potential to provide a range of both macro-nutrients (i.e. those required in large amounts and supplied in terms of percentages, e.g. nitrogen, phosphorus and potassium) and micro-nutrients (i.e. those required in very small amounts and supplied in terms of mg/l, or parts per million, ppm). A well-formulated soil, such as light clay loam, should have the total range of 18 mineral nutrients required for successful growth. This should not be assumed by the colour or texture of the soil, but confirmed by analysis or estimated by the range and quality of growth of any plants present in the area. This also can be an indicator of general soil quality, such as drainage, aeration balance and pH.

A temperature and pH buffer

These are two functions that the soil can provide. They are often overlooked but are vitally important to soil flora and fauna, and

Figure 6.3 The contrast between (a) good, and (b) poor, soil structure.

the successful cropping of cultivated plants. The temperature changes occur within the soil at a very gradual and even rate, which allows a wider range of living organisms to be successful within this habitat than if these changes were rapid and constant.

Organism group	Active outline pH range	Active outline temperature range	Light levels	Inter relationship
Actinomycetes	4–7	−10 to 50°C	Low levels	Soil detoxification
Algae	4.5–10	−20 to 50°C	Not required	Water pollution
Arthropoda	5.5–10	−5 to 40°C	Required	Organic matter recycling
Bacteria	6–12	4 to 86°C	Not required	Nitrogen fixing
Fungi	4–7	2 to 75°C	Low levels	Mycorrhizae
Insects	3–12	−10 to 40°C	Low levels	Pollination of plants
Mammals	4–10	−10 to 35°C	Required	Organic matter processing
Molluscs	5–10	2 to 30°C	Not required	Organic matter recycling
Nematodes	4–10	5 to 35°C	Required	Organic pest controls
Protozoa	6–12	2 to 35°C	Not required	Soil detoxification

Table 6.2 The growing media as a temperature and pH buffer.

Also provided by many soils is a very effective buffering capacity related to the changes in pH, or the balance between acid and alkaline conditions. A number of species, including crop plants, can be unsettled in their rapid establishment and growth if the pH changes rapidly or repeatedly. The most suitable pH in which to grow the maximum range of plants is pH 6.5, just on the acid side of the neutral point of pH 7, which also provides a balancing point between bacterial and fungal colonies. More acid and fungi become dominant; more alkaline and bacteria gain the upper hand.

MANAGING SOIL PROPERTIES IN THE GARDEN

Texture
Remember that this relates to the size of the soil particles and the amount of them in percentage terms, e.g. a light clay loam contains 20% clay, 40% sand and 40% silt and is regarded as the best all-round soil to have. This is due to the balance of properties that it shows: it is water retaining but also well drained. The only way to change it is to change the volume of any of the particle sizes, e.g. by adding extra clay to a cricket wicket; this is a traditional way to increase the bounce of the ball.

Composition
Also called soil structure, this relates to the arrangement of the soil particles provided by your textural class. Look at the cracks in your soil, do they inter-connect to allow for water drainage and air movement? Are there any hard compacted layers, either on the surface or at the normal digging depth of a spade, known as a 'spit'? These things need to be noted and provide valuable information as to the physical state of your soil, together with the species range and quality of any plant material present.

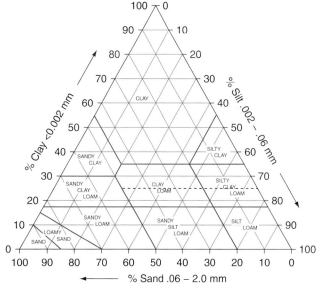

Figure 6.4 Soil textural classes triangle showing the relationship between particles of clay, silt and sand.

Nutritional status
By noting the range and quality of plant materials present this also can give clues as to the current nutritional status of your soil. Different dominant species can indicate an increased level of a particular nutrient, e.g. stinging nettles, *Urtica dioica*, indicate an elevated level of phosphates and rosebay willow herb, *Chamerion angustifolium*, an elevated level of potassium. Equally, remember that problems can also be encountered through elevated levels of some nutrients, e.g. phosphate levels can be high in some soil types but in a form that plants cannot easily obtain, so adding more via the use of a compound fertiliser will increase the potential for problems to develop.

Aeration

Maintaining minimum compaction is the key point to remember: the old adage, as stated above, of 'if the soil sticks to your boots keep off it', is as true today, as it ever was, if not more so, as the pressure on high-quality soil and land in general increases. One of the main causes of lack of rapid plant establishment and development is poor soil preparation. This is especially so if the area is new to cultivation, such as making or expanding a vegetable area from a lawn. As a general rule, the deeper this is cultivated the better, especially to break up any hard layers, called pans, that build up due to regular mowing over the years.

For those who wish to start a no-dig policy, usually the first instruction is to double-dig your plot and then don't dig it again or stand on it; see the section on raised beds in Chapter 12 for further details and construction instructions.

Once good aeration is established then organic matter can be added to the soil surface and this will be incorporated by earthworms over a period of time. This is the best way to cultivate your soil if you have a bad back or limited mobility.

pH

It is very important to confirm your current pH level before attempting to alter it, do remember that normal rainfall is acid, around 5.5, so the soil will become naturally more acid over time. Also some chemical weedkillers, e.g. glyphosate (Roundup®), and fertilisers, e.g. sulphate of ammonia (21% nitrogen), have an acid reaction with soil, so again their repeated use will build up. Conversely, some other fertilisers, e.g. Growmore® or blood, fish and bonemeal, will increase the pH level, i.e. make it more alkaline by adding some form of calcium. Calcium-containing products and their neutralising values are discussed later in the chapter.

However, plants have adjusted their needs to respond to the natural pH of your soil. It is more usual these days, to select a group of plants that will grow successfully on a naturally unimproved soil, rather that replacing it for an improved type of media or going for expensive and extensive changes for short-term gain.

ROOTS: THEIR GROWTH AND FUNCTIONING IN GARDEN SOILS

Structure and growth

Roots are the main link between the plant and the soil and take a wide range of forms for different functions: tap roots for storage of overwintering carbohydrates, and fibrous for transfer of water and nutrients, with the actual transfer being undertaken by the total root size. Modifications such as nodules and 'breathing knees' or pneumatophores are discussed in more detail later and show the range and diversity of the role of roots.

Uptake of minerals and water and their movement in the root

Roots are selective in their uptake of plant nutrients and this is controlled by the surface of the root hairs, the semi-permeable membrane of the cell wall and a layer of cells called the Casparian strip sited on the outside of the stele (as illustrated in Chapter 2, Figure 2.33).

All the minerals and nutrients that are taken up by the plant are dissolved within water to allow them to be mobile and more easily transported around the plant system: root hair, across the cortex, through the Casparian strip, into the stele and up the xylem tissue to the stem and beyond. Not all of them are collected and transported at an even rate, with some such as calcium being well-known poor travellers, resulting in disorders developing, such as, bitter pit in apples, *Malus domestica* and blossom end rot in tomatoes, *Lycopersicon esculentum*.

Green manures

This is a term used to indicate the selection and growth of a crop of living plants grown deliberately to be incorporated into the soil or chopped up on the surface to assist in improving the quality of the growing media. They can help in a number of ways:

- ✿ they retain nutrients, especially excess nitrogen at the end of the growing season, holding them until the plants forming the green manure break down themselves releasing the nutrients to be used by subsequent crops
- ✿ they reduce the damage caused to the soil surface by extreme weather conditions
- ✿ they also reduce the disturbance to soil fauna when growing media is left fallow, without a new crop being grown
- ✿ they reduce water lost when a longer term crop, e.g. early potatoes, *Solanum tuberosum*, finish earlier than planned and the time period is too short to use a short-term productive crop
- ✿ they provide surface coverage for beneficial organisms, offering continuity of control numbers over the seasons
- ✿ some green manures, e.g. red clover, *Trifolium pratensis*, also fix nitrogen through friendly bacteria from the *Rhizobia* group, who grow root nodules on the plant's root system
- ✿ other examples, e.g. *Tagetes minuta*, can provide a soil-cleaning service through their deposition of natural chemicals via the root system. Golden potato cyst eelworm, *Heterodera rostochiensis*, can be eradicated through this method and the numbers of keeled slugs, *Milax* spp., reduced.

Further discussion on green manures can be found in Chapter 18, including a table of plant examples and their contributing properties.

Mineral deficiencies

As mentioned above, these develop when minerals and nutrients are reduced in supply, either on a temporary or longer term basis. This could be because the element is not present or is present in low amounts in the media, or it could be that there are instances of nutrient antagonism.

See Tables A 1.1, A 1.2 and A 1.3 in Appendix 1, which show the nutrient groups and their outline roles within the plant.

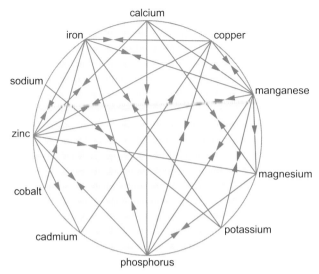

Figure 6.5 Nutrient antagonisms shown related to each other in a diagrammatical format.

Root modifications encountered in garden plants

The most commonly known of these are usually nodules, which are naturally occurring on the roots of plants within the family *Papilionaceae* and are formed by bacteria, e.g. the *Rhizobia*, and 'breathing knees' or pneumatophores found on established *Taxodium distichum*, the bald or swamp cypress. The latter allow plants to grow in soils with low oxygen levels, such as in mangrove swamps, by taking a section of the root structure permanently above ground level to enable the plant to undertake gaseous exchange, that is to release the potentially toxic or harmful gases, e.g. carbon dioxide, and replace them with fresh air, which contains the oxygen needed for new root growth.

Another modification seen on plants such as sweet corn, *Zea mays*, is that of buttress or stilt roots: growing out of the base of the stem they provide a method of improved stability to the plant, which has a

Figure 6.6 'Breathing knees' or pneumatophores found on established *Taxodium distichum*, the bald or swamp cypress.

narrow upright **morphology** that can be blown over and damaged. They look just like the guy-ropes on a tent and act in the same way.

Mycorrhizae and their significance for garden plants

The term 'mycorrhizae' translates loosely from the Greek as 'fungus root' and describes a group of fungi that have over millions of years developed a close relationship with the root systems of selected plant species, in a mutually rewarding way, called symbiosis.

A mycorrhizal fungus grows out into the soil around the plant host's root system forming a dense network of long, tiny hair-like threads known as hyphae: estimates indicate that 20 to 30 metres of these strands can exist in the space of a sugar cube. This secondary root system helps its plant partner to absorb essential nutrients, especially phosphates, which can be held in the soil in a form that is unavailable to the plant alone. Also provided, long after the host has given up, is water, where due to their size the hyphae can access the finer pore spaces precluded from the roots. However, the host initially does provide the fungus with carbon, in the form of carbohydrates, as a 'starter pack' to enable the growth of the hyphae.

There are both economic and environmental benefits to be gained by their use.

- ✿ Plants with mycorrhizal associations can have higher survival rates compared with those without, under low-nutrient conditions.
- ✿ Mycorrhizal fungi can help restore complex natural plant communities more quickly than if fertilisers are used.
- ✿ Mycorrhizal fungi can increase flowering of plants compared with non-mycorrhizal plants in nutrient-limited conditions.
- ✿ The growth resulting from the early production of mycorrhizal plants can lead to the healthy plant effectively crowding out weeds, which are frequently non-mycorrhizal.
- ✿ Mycorrhizal-treated plants have been used successfully in restoring sites, such as old coal and mineral mines. Mycorrhizal trees and legumes, e.g. white clover, *Trifolium repens*, planted at such sites have been shown to have much better survival and growth rates.
- ✿ Mycorrhizal plants can also help to contain environmental pollution of a site by effectively 'fixing' heavy metal pollutants, e.g. cadmium, in the network of hyphae, thus reducing or preventing leaching away of those contaminants into underground aquifers and hence into water sources.

Mycorrhizal fungi can be used with most garden plants, the notable exceptions being plants within *Amaranthaceae*, e.g. beetroot, *Beta vulgaris* var. *vulgaris*, and *Brassicaceae*, e.g. radish, *Raphanus sativa*. Young saplings, whips and shrubs treated with mycorrhizal fungi would be expected to have a higher survival and improved growth rates in most reduced-nutrient environments, compared with non-treated plants.

Mycorrhizal fungi are a sustainable alternative to repeated inputs of fast-release fertilisers. Mycorrhizal products can be used within a sustainable garden strategy allowing plants to co-exist naturally using optimal levels of more appropriate fertiliser additions. They have now become readily available in garden centres and nurseries, and can be attuned for specific plant ranges, e.g. roses, *Rosa* spp., or available as cosmopolitan mixes for use in a specified habitat, e.g. woodland.

The types of mycorrhizal fungi are introduced in Chapter 2 and further detail is given in Chapter 18.

REASONS FOR CULTIVATION

Think about the things that could be achieved as a result of undertaking cultivation:

- ✿ incorporation of trash/rubbish from the last crop
- ✿ exposing soil pests and diseases to winter weather
- ✿ providing drainage
- ✿ providing aeration
- ✿ allowing incorporation of introduced organic matter
- ✿ stimulating soil flora and fauna growth
- ✿ breaking up of pans at depth.

Seed beds

The best and most careful cultivation needs to be undertaken for these sites, as a seed crop failure would be most costly in terms of both time and materials. Sites for this activity to take place are best to be in a well-lit site with some protection from excessive rain and very high winds. Crops that are traditionally grown using this method are: cabbages and cauliflowers, *Brassica* spp., and leeks, *Allium porrum*; with the same skills and techniques being employed for a lawn from seed.

Figure 6.7 The front cover from *Fungal Friends: an Introduction to Mycorrhizal Fungi*, the supporting leaflet produced by Sparsholt College for the RHS Chelsea Flower Show in 2001 to highlight their value to a wider audience.

Figure 6.8 Seed bed preparation stages: (a) forking, (b) rough raking, (c) consolidation and (d) final raking.

Rectifying defects and damage

If you are left with an uneven or unsuitable area of the garden or site after the creation of a new feature then this must be thoroughly and deeply cultivated to a depth of 45 cm (18 inches) to relieve any compaction of hard layers of soil, called pans, which do not allow the passage of air or water. It is also important not to mix the lower levels of soil with the surface levels of material as this results in subsoil being brought to the surface, with a lower nutrient level and a less well-weathered structure, resulting in poor or delayed plant establishment after the new plant display or crop is planted.

DRAINAGE

In most situations the use of very expensive and time-consuming pipe drainage is a thing of the past. Only when vast volumes of water need to be collected and redirected do they come into their own. The more modern, holistic, solution is to adjust your planting list to suit a moister, even wet, site. More cost-effective methods for selective use include the following.

Soakaways and French drains

Soakaways consist of a cubic metre of hard-core or stone rejects, 75 mm in size topped with a landscape membrane, to prevent silt build-up, and above this 15 cm (6 inches) of topsoil and turf, or whatever is the surface material, e.g. chipped bark. This is needed to let the surface water through and the soakaway is sited at the lowest point of the slope or site. The excess water runs into it and then it gradually soaks away, either to be taken up by plants or drained into the groundwater.

French drains are a long narrow soakaway: a trench is taken out across the slope, usually at the lowest point, to a depth of 45 cm (18 inches) and backfilled with clean rubble or stone rejects 75 mm in size, which can be left exposed and may run along a path or road. French drains can also be used at the top of a slope to intercept water from higher ground and prevent it flowing over your ground. On sports fields they are mainly used to intercept water and prevent it from getting onto the pitches.

Figure 6.8 (cont.)

Figure 6.9 Slit drainage being installed in a sports field situation.

Type of drainage system	Soil texture	Diameter of pipe	Silt traps required?
Grid	Clay	100 mm	Yes at each intersection of the mains, not laterals like below
	Silty clay	75 mm	As above
	Silty loam	50 mm	As above
Herringbone	Clay	120 mm	Only for 90° turns and an adjustment of level across a slope
	Silty clay	100 mm	As above
	Silty loam	75 mm	As above
Mole	Clay only	75 mm (smeared clay)	Not usually required
French	Clay	Not applicable	Not usually required but can be used at the base of slopes on unstable silty soils

N.B. Hooghoundt's theory provides the basis for the calculation of lateral spacing on individual sites.

Table 6.3 Pipe sizes related to soil types.

Equally, it can be covered over as a soakaway, depending on the situation. A development of these types of basic drainage is called slit drainage, where minimum soil and land disturbance is caused as the required trench is only up to 10 cm (4 inches) wide, and then filled with graded sand. This method is especially helpful for fine turf situations or where large-scale disturbance is a problem, e.g. in heritage gardens.

Pipe drainage: for maximum coverage of an area, with minimum use of materials, a traditional herring-bone arrangement of pipes is used. If, however, a more even drainage result is required (e.g. on sports fields) then a grid system is recommended. The spacing of the pipe runs (for both methods) would depend on the soil texture (e.g. in clay soils they will be closer together) and also the soil structure to allow a good drainage link from the surface to the pipe system. Hooghoundt's theory provides a basis for the calculation of pipe spacings.

The depth of land drainage systems should be set at a minimum of 45 cm (18 inches), as this is below the normal cultivation level. For turf areas 30 cm (12 inches) is more normal, as the maximum depth of hollow tining is 20 cm (8 inches). The final depth should be set in relation to the soil porosity, which can be established using laboratory equipment and would normally be undertaken by a drainage consultant.

Clay tiles: these are the traditional drainage pipes made of porous clay, resulting in a good transfer of water through the tile when they are butted together, but not allowing the passage of soil particles, such as silt, in to block the pipe. However, over the last few years continuous plastic pipes with ribs and slits have replaced the clay tiles, these being easier and cheaper to install. The extra item that is required is that the whole system has to be covered by a polyester 'sock' of fleece-like material, to stop the silt and clay particles washing through the slits in the bottoms of the ribs and preventing the silting up of the whole system over a period of time.

Silt traps are a method of reducing the impact of any fine soil particles that do get through, with either clay or plastic pipes, and are also good for changing the direction or depth of the drain. A chamber is used with the incoming pipe at the top and the outgoing one at a lower level, but also incorporating a drop of at least 30 cm (12 inches) and a place for the fine particles to drop out of suspension and not 'silt-up' the next section of pipework.

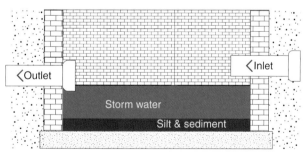

Figure 6.10 A silt trap fitted into a piped drainage system to reduce the risk of silting up.

Headwalls, usually made of concrete, are the final outcome of a drainage system into a river or stream. These should be fitted with an anti-rodent mesh cover to prevent them nesting in the pipe and be well above the normal water level to prevent any risk of backflow of water returning to the pipe system.

INCORPORATING ORGANIC MATTER

Another, much cheaper, method of improving soil drainage is to use the regular application of well-rotted organic matter to maintain or improve the soil's structure, i.e. the arrangement of soil particles into peds or crumbs, and seen in the soil profile as the interconnecting cracks that should be seen in a well-structured soil from the surface level down the profile into the subsoil. Benefits of using organic matter include:

✿ improved drainage
✿ improved aeration
✿ improved water retention
✿ increased micro-flora and micro-fauna activity
✿ supply of major plant nutrients
✿ supply of minor plant nutrients.

Seasonal application to the soil surface of items such as leaf mould keeps the organic matter levels renewed for those areas under deciduous woodland, but for productive areas then regular applications should be made. The timing of these will depend on your soil type and the crops previously grown or those intended to be subsequently grown. Well-rotted manures can still be incorporated at the normal cultivation depth of 30 cm (12 inches) as with traditional single digging.

However, in more modern times, the practice of laying organic matter to a depth of between 5 and 10 cm (2 to 4 inches) onto the soil surface, both in the autumn and spring periods, has gained use, especially with organic gardeners. The benefits of this include:

✿ less digging; in fact, no digging
✿ protection of the surface soil structure
✿ encouragement of the soil macro-fauna
✿ reduction of nutrient leaching (if spring application used)
✿ application can take place in a wider range of weather conditions
✿ absorption of large amounts of carbon dioxide.

MANURES AND FERTILISERS

Manures are usually the waste products of animals, but the broad group can include garden compost, which is derived from plant materials and normally has a lower level of available nutrients.

The best all-round type is bullock manure, if this is obtainable, as with the higher nutrient feeds given to these animals, the higher the waste nutrient levels. Second in the league is horse manure, which is more commonly obtainable, but the quality for horticultural use depends on the bedding material: straw being good, rather than wood chip, sawdust or waste paper. These others require an elevated level of nitrogen to break them down into usable products, so well-rotted samples are required. Many people do ask 'When is manure or garden compost well rotted?', as a general rule, if the component materials cannot be identified then it is ready for use. For garden compost the lack of scent is also an important clue; high-quality compost has little or no smell.

Figure 6.11 Digging stages: (a) trench opening, (b) forking the base, (c) application of organic matter and (d) trench closure.

With straw and wood products then it is when the edges have become ragged due to the breakdown processes.

Cow manure is usually more commonly available but lower in nutrients; pig manure is less recommended as this has a high smell and is lower in available bacteria; poultry manure has higher levels of nitrogen and a higher pH level, due to the extra calcium provided in their feed for quality egg shells; and spent mushroom and allow the plants to offset any pest attack they may be subject to during the growing season.

The complete order of secondary cultivation is:

- ✿ harrowing or forking
- ✿ rolling or consolidation (penguin walk)
- ✿ application of a balanced base fertiliser
- ✿ chain harrowing or raking
- ✿ further consolidation, if on a light sandy soil
- ✿ final raking.

Rotary cultivator – This is a very successful machine for final cultivations, but unfortunately it has become over-used. They produce, depending on the type and number of blades fitted, a very fine finish (tilth); sometimes too fine. This can result in a **pan** developing at the cultivation depth, especially when used repeatedly at the same depth, usually only a maximum of 15 cm (6 inches) is obtainable. Also if heavy rain follows its use the water can get trapped above the pan, causing a perched water table, this reduces the depth of available soil and results in a growth check for developing crops or delayed and patchy germination. If prolonged, an increased incidence of diseases, especially fungi, can develop, e.g. foot rot, *Fusarium* or *Verticillium* spp.

Another limitation of repeated use or incorrect use, such as on soils that are too moist for the benefits to be fully experienced, is that they shatter or smear soil structure and disrupt earthworm activity.

No-dig methods – This is a developing area or method, which as the name suggests does not involve the stages outlined above, but rather these stages are undertake by others not you. However, the catch is that this method works superbly on high-quality, well-structured and well-cultivated soils. The method consists of heavy applications of well-rotted organic matter in the spring period to the soil surface, which is then incorporated by earthworms and broken down by bacteria and fungi. To protect the soil surface during the winter months and to prevent the leaching of nutrients, especially nitrogen, green manures should be grown during the autumn period and allowed to be frosted and rot or be hoed off and chopped up as fresh organic matter.

The other vitally important area that this method of cultivation contributes to is that of providing a **carbon sink** to absorb an elevated volume of carbon dioxide to offset climate change. The reason for this is that the network of soil flora and fauna that is built up over the growing season is not disturbed or destroyed by heavy cultivations over the autumn period, rather it continues the organic matter breakdown process providing a higher volume and increased range of nutrients available in a state usable to the next range of crops grown.

Managing soil nutrients – One of the most concerning subjects discussed as part of the responsible growing of plants or crops is that of nutrient supply, management, recycling and waste. All plants require nutrients to grow and prosper but providing them in a sustainable and non-damaging way, which is also cost effective, is a challenge that all of us face irrespective of the size of our area or the range of plants we intend to grow.

Nutrient requirements – As with all living things plants require nutrients to survive, but unlike the rest, green plants do take things to a higher level by using the biochemical process called photosynthesis to transfer light energy into chemical energy. However, as a plant grows and develops the need for the range and quantity of other nutrients increases. Also nutrients are not required on an even basis or on an even timescale but each one, although needed only in very small amounts, is vital for successful germination, establishment, growth, storage, flowering and setting of seeds. Each has a role or roles to play and like all activities involving a number of players, some get along with everyone and others with very few. Also in some cases they will block the uptake or effectiveness of each other or are required in ratios or in combinations to be the most effective and in the correct state to be absorbed by the plants.

Types of nutrients (macro- and micro-)

The standard number of nutrients that a plant requires to successfully go through its life cycle for many years was stated to be 16; but according to further research by the American space agency NASA the number has increased. These are divided up depending on the volume required by the plants.

Major (macro-) nutrients – These are required by the plants in large amounts and are usually supplied expressed as percentages. See Appendix 1, Table A 1.1 for examples.

Minor (micro-) nutrients – These nutrients are required in much smaller amounts and are usually expressed in milligrams per litre, they are also known as trace elements. Normally they are required for a small but vital role, which may not be activated until later in the plant's life and could include flowering or seed setting. Germination of the next generation could also be affected if the nutrients in this group are in short supply or totally missing. They can also control other processes vital to the long-term survival of the species. See Appendix 1, Table A 1.2 for examples.

Functions in the plant

Horticultural research is still, to this day, revealing the interrelationship of the roles that the nutrients across all the groups play in plant growth and development. A different balance of nutrients is required for each stage of a plant's life but also for those features that impact on the next generation: seed setting, viability and germination. See Appendix 1, Table A 1.3 for examples.

Symptoms of deficiency in garden plants

As an increased return to soil care is seen and we champion a holistic approach to the growing of plants, thought and care should be paid to the prevention of deficiencies across the range of plant nutrients. Consideration for the media as a vital living, fertile resource, where we could successfully grow any plant range we wished, is very much the modern viewpoint taken.

Rather than selecting, and limiting, the nutrients supplied a broader view is recommended: providing a whole range for a longer time period and allowing a balance of supply, use and re-use to be achieved, especially within the area of the trace elements.

Also within this area of nutrient deficiencies, refinements in analytical techniques and the current level of sophistication have resulted in more accurate recognition and diagnoses being achieved.

Figure 6.12 Blossom end rot: a plant disorder that develops due to a reduced level of available calcium.

Sources of major nutrients in soils

Originally the sources of all nutrients were from the weathering and breakdown of rocks in the presence of air and water. The range and amount of major nutrients found depends on the type of rocks that were present and how these rock fragments were moved and then re-settled. Soils with a high clay content have a very high content of major (and minor) nutrients, as the particle size is very small and the particles have a very large surface area for their size. These broken edges allow nutrients to be attracted to negative clay plates, building up a system of nutrient exchange called **cation exchange capacity**, which is usually seen in its abbreviated form as **CEC**.

Compost and composting

In recent years the creation of high-quality compost has become an art form, with much written and talked about it. Different recipes and methods abound, but here the basic guidelines or instructions will get you started and then you can adjust them to suit your site, situation and available components, in both type and volume.

Figure 6.13 Nutrient antagonisms, e.g. tomato, *Lycopersicon esculentum*, showing the impact of calcium and potassium imbalance.

Also in sharp focus are the methods of undertaking home composting using all your waste plant materials.

The basic guidelines to follow are:

✿ Set up your compost bins in a sunny site, or at least a site that is exposed to some hours of sun (minimum two hours) per day.

✿ One of the best configurations of bins is known as 'New Zealand bins', which consist of a pair of bins, each with a volume of 1 m³ (19 sq.ft.) mounted alongside each other. The idea was that one would be filled totally and then be decomposing while the other was being filled, and you could then use the contents of the first, once the second was filled. However, some problems were found with this system, as in practice, depending on the size of your plot, the time of year and the type of plant material used, it could take a longer period of time to fully break down the first bin and less time to completely fill the second.

✿ So, if this is the case with you, the modified New Zealand system is used, that is, adding an extra bin alongside the second and also turning the compost once a bin becomes full to allow aeration to occur more effectively, giving an hydroponic system. This last display will require the normal care, level of irrigation and nutrient supply through fertilisers, either with base dressings within the compost or by liquid feeding as part of the regular watering regime.

INTRODUCTION TO AEROPONICS

These growth systems are a development of hydroponics, and are where additional oxygen is used, together with a very high humidity level, but again in a controlled environment in a protected situation. Crops are provided with nutrients in the form of regular mists. Exceptional vegetative growth rates are the result of careful management and monitoring of the artificial balanced environment produced in these compact systems. Reduced incidences of media-based diseases are seen as there is no media of any kind involved: the plants are held in position via support collars providing no resistance to the plant's root system and so can be grown at any angle. A quick return is shown and the system has favoured the production of high-value salad crops, e.g. rocket, *Eruca vesicaria*.

MANAGING NUTRITION WITH 'ARTIFICIAL SOILS'

Regular checking of both the nutrient solution and the look of the plants themselves is very important for the successful growth of crops under these media selections.

Avoid 'boom and bust' problems with the stabilisation of the available nutrients and test regularly, using a sap or leaf test, for a build-up or imbalance of any nutrients being supplied.

Another main area of concern is the maintenance of a viable media structure. The main problem, which is experienced over time, is slumping of the media; this is a physical collapse of the pore spaces causing compaction and a rapid reduction in the air-filled porosity. The regular application of semi-decomposed organic matter is the recommended method of providing stabilisation over the longer term. A balanced media flora and fauna is also required for the recycling of nutrients and production of humus to successfully hold the media together.

Summary

Throughout time different media have been used for growing plant material, with a mixture of species making the best of whatever was available and close at hand. In more difficult and challenging sites, the dominant plants successfully developed. Either on their own or working with bacteria and fungi to provide a link with water or nutrient availability, humans have gradually developed highly skilled manipulation of whatever situation has been presented, including these species selections, seasonal changes or opportunities that are noted in relation to the different media identified.

One of the main foundations of success has been the correct identification of the underlying geology and soils present, the opportunities this gives and the range of plant species that can be successfully grown, together with the cultivation opportunities or limitations, both primary and secondary, that your type of media presents.

Add to this an understanding of the physical structure of the media, its drainage potential and methods to adjust this, either way, then increased cropping can be achieved. The choices have been outlined, be it ditch, pipe or soak away. Price, skill level, disturbance and the volume of water to be moved have to be taken into consideration, together with regular or intermittent use. Overlay this with the living part of the media: the microbes and organisms, both micro- and macro-, and a balanced holistic growth media can be established.

Or take another pathway and control your own destiny following a total replacement system with the use of artificial media or just nutrient-enriched water, namely the nutrient film technique (NFT), one of the leading growing methods in the world of hydroponics.

If a more traditional route is selected then the skills of cultivation, when and what to use, come into play. Be clear that over-cultivation, whatever the scale, can be very damaging and upset not only the plants but the whole finely balanced ecosystem of your growing media. Therefore in times when labour is one of the greatest costs and time is short the no-dig policy may not only be good for your media ecosystem but also contribute to your work–life balance.

Review questions

1. Define the **terms** 'soil texture' and 'soil structure'
2. Relate the pH scale to soil **profile** types
3. State **two** benefits and **two** limitations of **cultivating** the soil
4. Explain the terms: base, top and liquid feeding, in relation to the **supply** of plant nutrients
5. Outline the **role** of organic matter in the formation and maintenance of soil structure
6. Define the term mycorrhizae and their **role** in soil reclamation

FURTHER READING

Ashman, M. R. & Puri, G. (2002). *Essential Soil Science: a Clear and Concise Introduction to Soil Science*. Blackwell Publishing. ISBN 978-0-632-04885-4.
One of the best balanced books on this area, which is very readable, with excellent examples to underline the major points discussed.

Cooper, A. (2002). *The ABC of NFT*. Grower Books.
A clear, well-illustrated, publication outlining the value, design and methods used for this specialist growing system.

Kilham, K. (1994). *Soil Ecology*. Cambridge University Press.
An excellent development for obtaining further understanding in this vital area of media management.

Lowenfels, J. & Lewis, W. (2006). *Teaming with Microbes*. Timber Press.
This comprehensive book makes you reconsider the value of soil as a living media and how to obtain the best results from it.

White, R. E. (1997). *Principles and Practice of Soil Science: the Soil as a Natural Resource,* 3rd edn. Blackwell Science.
A more in-depth publication, especially good in the soil chemistry and pollution areas.

RHS level	Section heading	Page no.
2 2.1	Properties of garden soils	107
	Types of cultivation	119
2 2.2	Properties of garden soils	107
2 2.3	Properties of garden soils	107
2 2.5	No-dig methods	120
2 3.1 to 2 3.7	Properties of garden soils	107
2 5.1 to 2 5.8	Mammals	109
2 6.1 to 2 6.5	pH	112
	Managing soil pH in the garden	123
2 6.6	pH	112
	Reducing the pH level	123
2 7.3 to 2 7.5	Fertilisers	122
3 1 1	Properties of garden soils	107
3 3.1	Mammals	109
	Soakaways and French drains	115
3 3.4	Mycorrhizae and their significance for garden plants	113
3 4.2	Increasing the pH level	123
3 5.1	Minor (micro-) nutrients	120
	Water conservation	124
3 5.3	Fertilisers	122
3 6.2	No-dig methods	120

CASE STUDY Long-term seed storage

The end result of flowers and fruits are seeds; these are required for the perpetuation of the species. If biodiversity in the world is to be maintained it is important that existing species are encouraged and not lost. As well as maintaining plants in their natural habitat another method of encouraging biodiversity is to save and store seeds for the future. During recent years we have started to do this by using seed banks.

Bedding begonias.

There are throughout the world a number of seed banks to store seed for the long term. The purpose of these banks is to ensure the survival of the plant species, even if there is a catastrophe in their natural environment. There are also banks to ensure the survival of heritage and unusual cultivars.

The main seed bank in the UK is run by Royal Botanic Gardens, Kew at Wakehurst Place in West Sussex. It is also called the Millennium Seed Bank. At the end of 2009, it held 10% of all known species in the world.

Garden Organic, based at Ryton in Warwickshire, runs a heritage seed library with the aim to maintain old cultivars that are no longer produced by the seed firms.

A global seed vault has been constructed on the island of Spitsbergen in Arctic Norway. This is a secure vault cut into the sandstone rock under the permafrost. The seeds are packaged in four-ply packets and heat sealed to exclude moisture and gases. They are maintained at a temperature of −18°C (−0.4°F). In 2008 there were over 400,000 seed samples stored in the vault, including 90,000 food crops.

Chapter 7

Flowers, fruits and seeds

Kelvin Mason

INTRODUCTION

Why is it important for the reader to have a knowledge and understanding of flowers, fruits and seeds? First, the vast majority of the food we eat is produced from plants grown from seeds: including large-scale farming and vegetable production as well as 'grow your own' crops.

Having an understanding of flower parts and the types of inflorescence can help you to identify a plant. This chapter covers flower parts and structure, including the various types of inflorescence.

The function of flowers is to produce fruit and seed to perpetuate the species. The flower facilitates the pollination and fertilisation of the ovules that ensures the production of seed. Fertilisation results in the production of fruit, which ensures the survival and spread of the seed.

With the recent loss of pollinating insects, including bees, it is important to have a good understanding of pollination and fertilisation to ensure a good crop of fruit is produced. Factors affecting pollination and fertilisation and how the grower can improve these are discussed in this chapter.

Seeds also have the ability to survive very difficult climatic conditions that would kill many plants, thus the plant can survive over winter and grow again in the spring. A good understanding of seed germination requirements, overcoming dormancy and ensuring the correct conditions for growth helps to ensure a good crop. The germination process is explained along with the types of dormancy and how these control when the seed will germinate. Methods of breaking seed dormancy are also explained.

Finally, how to propagate plants from seed both indoors and outdoors and the various sowing techniques are covered.

Key concepts

Once you have read this chapter you should be able to:

✿ Describe and name the flower parts and their functions

✿ Describe and name the various types of inflorescence

✿ Describe the pollination and fertilisation of flowers

✿ Describe the types of fruits produced and their methods of dispersal

✿ Describe seed structure, dormancy and germination

✿ Explain the types of seed available and sowing techniques used

The Fundamentals of Horticulture: Theory and Practice, ed. C. Bird. Published by CAMBRIDGE UNIVERSITY PRESS. © The Royal Horticultural Society 2014.

FLOWER STRUCTURE

A flower is the end of a stem that has become specialised and has developed flower parts instead of leaves. The flower contains the plant's reproductive organs. Once a stem has changed from vegetative growth to flower production it cannot change back.

Flower size and shape vary considerably between the various genera as is shown in the range of flower sizes in the case study. The shape of flowers has evolved and adapted over centuries to encourage outside agencies to assist in their pollination.

> **CASE STUDY** **Range of flower sizes**
>
> The largest flower in the world is produced by a parasitic plant called *Raffiesia* – it is up to 1 metre (3 ft 3 inches) in diameter and weighs up to 11 kg (24 lb 4 oz). The smallest flower in the world is the *Wolffia* duckweed, which is a water plant native to Australia and the flower is only 0.6 mm (1/32 inch) in diameter. Both of the above flowers carry out the same function but have evolved in different ways to suit the environment and circumstances where they grow.

Structure of flower parts

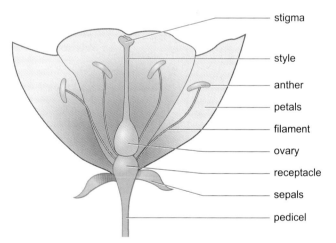

Figure 7.1 This shows a simple flower and structure. The flower parts are labelled to show their position and arrangement.

Dicotyledons – The basic flower is made up of five main parts.

Sepals – these form the outer whorl (ring) of a simple flower. Their function is to protect the bud and flower as it grows. They are usually green although they can be coloured in some species, e.g. *Anemone*.

Collectively the sepals are called the calyx, and the number of sepals will vary depending on the species of plant.

Petals – these are the next whorl on the inside of the sepals and are usually highly coloured. Their main function is to attract insects to the flower for pollination. Many have markings on them that are nectar guides to guide the pollinating insect to the nectaries at the base of the petals. The nectaries produce nectar, a sweet sugary substance that insects use for food and energy.

Collectively the petals are called the corolla and if the petals and sepals are referred to together they are called the perianth.

In some plant families (e.g. *Liliaceae*) the petals and sepals look the same and form a colourful part of the flower and are called tepals.

Stamens – consist of the anther containing the pollen grains for pollination. Anthers are supported by a stalk called the filament, which holds them in the best position to release the pollen for pollination. Stamens are the male part of the flower and are called the androecium.

Carpels – also called the pistil, which contain the female parts of the flower; carpels consist of stigma, style and ovary. The stigma is at the top of the carpel and its function is to receive the pollen. The style is a section between the stigma and ovary down which the pollen tube grows during fertilisation. The ovary is at the base and this contains the ovules (female gametes), which will form the seed once fertilised. The number of ovules in the ovary will depend on the species of the flower. These female parts are referred to as the gynoecium.

Receptacle – this is the base of the flower to which all the above parts are connected and is called the floral axis. It is held on one or two types of stalks a pedicel or peduncle.

Pedicel – this is the simple stalk of a single flower and connects the receptacle to the peduncle or stem.

Peduncle – this is the main stalk if the inflorescence consists of more than one flower e.g. foxglove, *Digitalis purpurea*.

The simple flower described and illustrated above has a number of variations depending on whether male, female or both parts are present. The three types are:

Hermaphrodite flowers

Hermaphrodite flowers have both male and female reproductive organs in the same flower. This is the most common plant form and is also called the perfect flower.

Monoecious flowers

Plants that have separate male and female flowers on the same plant but in different flowers are called monoecious; examples include oak, *Quercus robur*, hazel, *Corylus avellana* and sweet corn, *Zea mays*. Figure 7.2a and b show the male and female flowers of *Zea mays*.

Monoecious flowers are most common in plants that are wind pollinated. The male tassels or catkins release the pollen and it drops or drifts on the wind to the female silks or catkins.

Dioecious flowers

If the plants have separate male and female flowers on different plants of the same species they are called dioecious, examples include holly, *Ilex* spp. and poplar, *Populus* spp. There is not a wide range of dioecious plants.

Figure 7.2 (a) Picture of *Zea mays* male flowers called tassels, the anthers containing the pollen can be clearly seen hanging from the flower. (b) Picture of *Zea mays* female flowers, called silks, on which the pollen lands to fertilise the flowers.

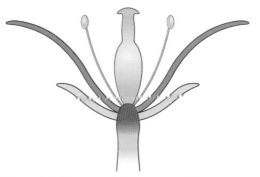

Figure 7.3 Diagram of a hypogynous flower.

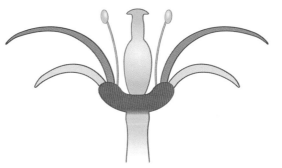

Figure 7.4 Diagram of a perigynous flower.

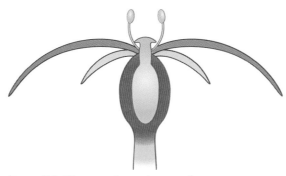

Figure 7.5 Diagram of an epigynous flower.

Dioecious flowers help to ensure cross-pollination, but to ensure pollination both male and female plants are required. Two common plants used in the garden that are dioecious are holly, *Ilex* spp., and *Skimmia* spp., so to get berries you need one male and one or more female plants.

The position of the ovary in relation to the other flower parts is important as it can help with the identification of the plant. There are three main positions:

1. Superior ovary – **hypogynous** flower

   In these flowers the ovary is borne above all the other flower parts. The sepals, petals and stamens are connected to the receptacle *below* the ovary as can be seen in Figure 7.3. Examples of flowers with this type of structure include butter-cup, *Ranunculus* spp., and bloodroot, *Sanguinaria canadensis*.

2. Superior ovary – **perigynous** flower

   Perigynous flowers have the floral parts connected along the rim of a cup-like structure around the top of the ovary as shown in Figure 7.4. This is not a common structure but is seen in some cherries, *Prunus* spp.

3. Inferior ovary – **epigynous** flower

   Epigynous flowers have the other floral parts above the ovary. Examples of this can be seen in *Narcissi* and *Fuchsia* (shown in Figure 7.5).

Monocotyledons – Flowers of monocotyledonous plants have some similarities to dicotyledons, but a number of differences. Using grasses as an example, the flower parts are:

Rachilla – a very short stem that connects the flower to the pedicel.

Glumes, lemma and palea – these form the outer rings of the flower structure. These protect the male and female reproductive parts. They serve a similar function to the sepals.

Lodicules – these are two small rounded structures at the base of the palea. When the flower is ready to open these fill up with sap and expand. This forces the palea and lemma open allowing the stamens and stigma to emerge.

Stamens – as in the dicotyledons these consist of filament and anther and serve the same function. There are three stamens in grasses that hold vast amounts of pollen.

Pistil – this serves the same function as in the dicotyledonous plants. The main difference is that there are two feathery stigmas, and they are considerably larger and hang outside of the flower. This is so that they can 'catch' windblown pollen.

The various parts can be seen in Figure 7.6, which shows a simple monocotyledon flower typical of a grass.

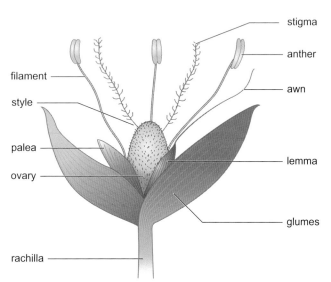

Figure 7.6 Diagram of monocot flower.

INFLORESCENCE

Flowers may be borne as a single flower – these are called solitary flowers; or they can be in groups on a branched system – and are called multiple flowers.

Solitary flowers
These have one flower on each peduncle (stem) and include *Anemone nemorosa* and some tulips such as the Darwin and Triumph types, *Tulipa* spp.

Multiple flowers
The branched flower groups are called inflorescences and the arrangement of the flowers on the stem (peduncle) is important in helping to identify the plant.

Examples of these are bluebell, *Hyacinthoides non-scripta*, foxglove, *Digitalis purpurea*, and sweet pea, *Lathyrus odoratus*.

Figure 7.7 shows the peduncle, pedicel and bracts that form the stems that support the flowers.

The various inflorescences are named depending on the arrangement of the flowers. Some of the common inflorescences are set out below with examples of plants with that arrangement.

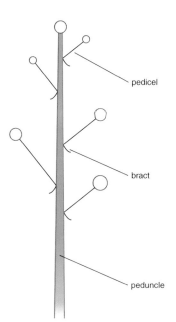

Figure 7.7 This diagram shows the position of the peduncle, pedicel and bract of a flower.

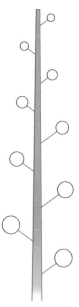

Figure 7.8 This diagram shows the layout of a raceme.

Raceme – Each flower is held on a short stem (pedicel). The flower at the bottom opens first and then flower opening progressively moves up the stem. Examples include foxglove, *Digitalis purpurea*, and lupin, *Lupinus* spp.

Spike – This looks similar to the raceme but each flower is attached directly to the main stem (peduncle). There are no pedicels. Examples include the lawn weed plantain, *Plantago* spp., and gladioli, *Gladiolus* spp.

Figure 7.9 This diagram shows the layout of a spike.

Figure 7.11 This diagram shows the layout of a verticillaster.

Panicle – Panicles have a main peduncle (stem) with branch peduncles spaced along its length. At the top of each peduncle are pedicels with the flowers at the top.

Examples include *Yucca filamentosa*.

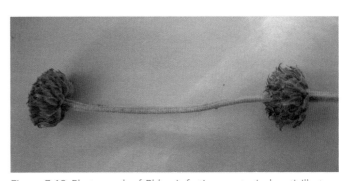

Figure 7.12 Photograph of *Phlomis fruticosa*: a typical verticillaster flower.

Verticillaster – A verticillaster is a stem with whorled inflorescences. The flowers are borne in rings at intervals up the stem. The tip of the stem continues to grow producing more whorls. Examples of this type of flower are the dead nettle, *Lamium purpureum*, and *Phlomis fruticosa* shown in Figure 7.12.

Corymb – These have a peduncle with pedicels branching off. The lower pedicels are longer than the upper ones, giving the flower arrangement a rounded top. Each pedicel ends with a flower. The flowering starts from the outside moving inwards.

Examples include candytuft, *Iberis sempervirens*.

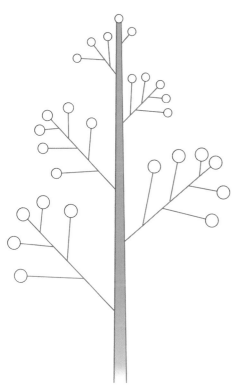

Figure 7.10 This diagram shows the layout of a panicle.

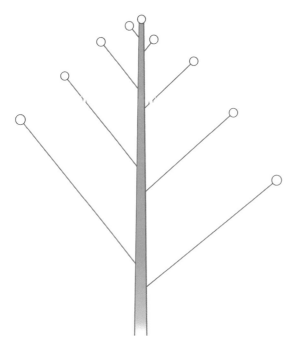

Figure 7.13 This shows the layout of a corymb.

The compound umbel has a main peduncle that branches at the top end. At the end of these branches are pedicels with the flower attached. These could be considered an umbel of umbels! This arrangement produces a large flat-topped inflorescence. Examples include parsley, *Petroselinum crispum*, carrot, *Daucus carota* (see Figure 7.15), and hemlock, *Conium maculatum*.

Figure 7.15 Photograph of carrot, *Daucus carota*, flowers: a typical umbel flower.

Umbel – The simple umbel has a number of pedicels at the top of the peduncle. These pedicels vary slightly in length giving the inflorescence a flat-topped arrangement of flowers.

Capitulum – The capitulum is a round flattened disc with a large number of small florets on top. The outer ring of florets, which look like petals, are called ray florets. The inner florets, which are usually yellow as in daisies, are small disc florets.

Examples include dandelion, *Taraxacum officinale*, daisy, *Bellis perennis*, *Dahlia* spp. and many other compositae (now members of *Asteraceae*) flowers.

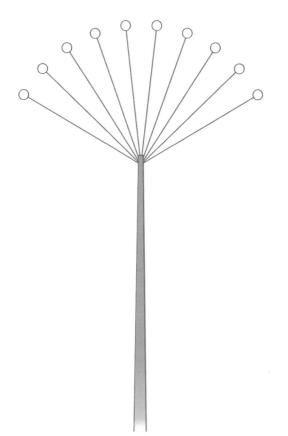

Figure 7.14 This shows the layout of a simple umbel.

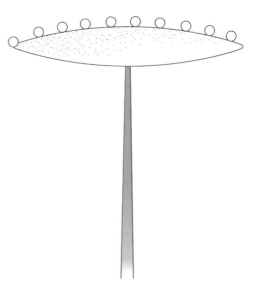

Figure 7.16 This shows the layout of a capitulum flower.

Cyme – Figure 7.17 shows the layout of a cyme. There are different types of cyme inflorescence arrangements. They usually look flat topped or slightly rounded. The terminal (top or centre) flower opens first working outwards or downwards. Cymes can be divided into two main groups, these being monochasium and dichasium.

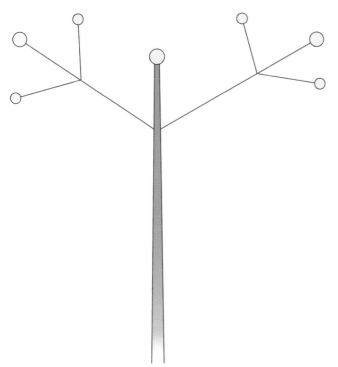

Figure 7.17 Diagram of cyme.

Monochasium means the flower has single branching pedicels and the dichasium has two branching pedicels.

Figure 7.18 shows the layout of monochasium and dichasium flowers.

Examples of cymes are *Iris* spp. and *Hypericum* spp.

POLLINATION AND FERTILISATION

The terms pollination and fertilisation are often used loosely and inaccurately as if they were interchangeable. These are two separate parts of the process in the plant's sexual reproductive phase.

Pollination – this is when compatible pollen lands on a receptive stigma. The pollen grain (male) has been transferred to the stigma (female) of a flower of the same species.

Fertilisation – this occurs when the pollen tube grows from the pollen grain down through the style and enters the ovule. This should result in seed production.

There are two main types of pollination:

Self-pollination – this is the transfer of pollen from the anthers to the stigma of the same flower or a flower on the same plant. This should result in the fertilisation of an ovule in the same flower or plant.

Cross-pollination – this is the transfer of pollen from the anthers on one plant to the stigma of a flower on another plant of the same species.

METHODS OF POLLINATION

For cross-pollination to take place pollen has to be transferred from one plant to another, there are four ways this can be done:

1. Insect
2. Wind
3. Birds/animals
4. Water

Insect pollination
Many insects carry out pollination including:

bees – honey, bumble and solitary
wasps
butterflies/moths

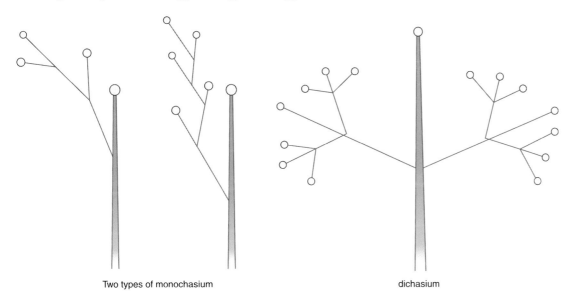

Two types of monochasium　　　dichasium

Figure 7.18 Diagram of mono/dichasium flowers showing the arrangement of the flowers on the stems.

hoverflies

flies, e.g. bluebottles and gnats

lacewings.

These insects carry out pollination in return for nectar and some may feed on the pollen as well.

In order for the flower to attract insects for pollination they have some or all of the following characteristics:

brightly coloured flowers to attract the insects

many are scented

pollen is large and sticky

reproductive parts are inside the flower

nectaries to feed the insect

markings on the petals to guide the insect.

Wind pollination

Pollination is carried out by the pollen being carried on the wind/ breeze. As these flowers do not rely on insects they do not require many of the above characteristics. The flowers therefore tend to be:

small and insignificant

often green in colour

produce large amounts of pollen

have feathery stigmas to catch the pollen and these often hang outside the flower.

scentless

no nectaries

pollen is small, light and not sticky

not particularly attractive.

Examples of plants pollinated by wind include:

grasses

many conifers

trees such as hazel, *Corylus avellana*, willow, *Salix* spp., and oak, *Quercus* spp.

Bird/animal pollination

Even if pollinated by small birds the flowers are usually large compared to insect-pollinated flowers. They have:

solid structure

large nectaries

sticky pollen.

Water pollination

Plants pollinated by water are mainly aquatic or bog plants where the pollen is carried by the flow of water from one plant downstream to another.

FERTILISATION

This is the fusion of the male and female gametes in the ovule.

Definition for some specialist terms are:

Gamete – a haploid sex cell, which when mated with the opposite sex will produce the zygote (diploid) cell. The pollen

Figure 7.19 his shows the flowers of *Helliconia* spp., these flowers are pollinated by humming birds.

grain contains the male gamete; the ovule contains the female gamete.

Zygote – a diploid cell, which is produced by the fusion of the two haploid gametes. A fertilised ovule is a zygote. The zygote develops into the seed embryo.

The pollen grain containing the male gamete lands on a receptive stigma and is then nurtured by the stigma. The pollen grain then germinates and a tube grows down the inside of the style to the ovary. See Figure 7.20.

At the ovary the tube enters an ovule usually through the micropyle. When the tube reaches the embryo sac in the ovule the tip of the tube breaks down. This allows the two generative nuclei inside the tube to enter the embryo sac.

One nucleus will fuse with the ovum fertilising it and this produces the embryo.

The second nucleus fuses with the endosperm nucleus and this forms the endosperm of the seed. When the two nuclei have fused it is called fusion.

FACTORS AFFECTING POLLINATION/ FERTILISATION

Weather conditions

Wind – if too windy it can blow the pollen away from receptive plants, a lot of pollen can be wasted in high winds which results in no or very low pollination.

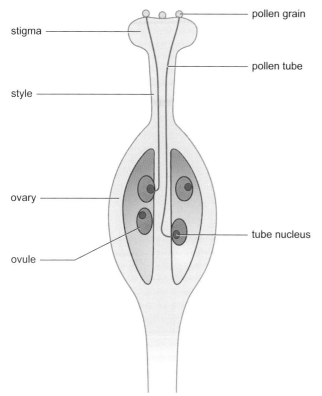

Figure 7.20 This shows the process of fertilisation of a flower. The pollen grains have landed on the stigma and have grown down the style to the ovary where fertilisation of the female gamete has taken place.

Compatibility – pollen can land on any surface, but for it to grow it must be compatible with the plant species stigma it lands on. Compatible pollen is capable of fertilising the female gamete. If it is not compatible the pollen will die and fertilisation will not occur.

Incompatibility – if the pollen grain does not grow or does not have the ability to penetrate the ovule then it is incompatible. One of the reasons for incompatibility is to prevent inbreeding. Inbreeding can result in the deterioration of the gene pool in the species and the species dying out in extreme cases. Inbreeding (selfing) can result in plants that are of poor vigour and prone to various problems.

Sterility – many hybrid plants are sterile, that is they do not have the ability to produce seed. This may be because the plant produces double flowers and lacks male and/or female parts. Double flowers have extra sets of petals but this is at the expense of other flower parts.

Self-sterile – some plants are self-sterile to prevent self-pollination and inbreeding. Examples of these are apples and plums. A different cultivar is required to ensure good pollination. This is the reason orchards are planted with two cultivars, the main crop cultivar and a pollinator cultivar.

LEVEL 3 BOX

TRIPLOIDS

A triploid plant is one that has three sets of **chromosomes**. Most plants are diploid – with two sets of chromosomes. Triploid plants can have advantages such as heavier cropping, better flowering or more vigorous growth.

Apples, *Malus domestica*, have some triploid cultivars including 'Bramley's Seedling'. The normal diploid apple has 34 chromosomes, when these divide during **meiosis** this produces the haploid number of 17. At fertilisation in the flower the two haploid sets of chromosomes recombine to return to the diploid number of 34. (17 + 17 = 34)

Triploid cultivars have an extra set of chromosomes thus having a number of 51 (34 + 17 = 51).

Because of this their pollen is incompatible with other cultivars, so triploids are not effective pollinators.

Therefore orchards containing the cultivar 'Bramley's Seedling' or other triploids will require two pollinators. These must pollinate the 'Bramley's Seedling' as well as pollinating each other. All three trees must flower at the same time to ensure good pollination and good crops of apples.

OTHER BREEDING TERMS

Parthenocarpy – the production of *fruit* without pollination, fertilisation or seed development. It can happen naturally in some plants such as cucumber, *Cucumis sativus*, or it can be induced by spraying the plant with high levels of auxin. The fruit develops without the production of any seed. Parthenocarpic fruit are seedless, like some grapes.

CASE STUDY **Pollination aids**

Many insects do not fly in windy conditions so this will reduce pollination in plants pollinated by insects. This is one reason why orchards are protected on one or two sides by windbreaks to provide sheltered, warmer conditions to encourage insects to fly. At present there is a shortage of bees and other pollinating insects, which could lead to reduced yields in fruit crops. Therefore it is important to do everything possible to encourage bees and other pollinating insects; this can include purchasing red masonry bees to assist with pollination.

Many fruit growers have hives of bees or ask beekeepers to put hives in the orchards when the fruit trees are in flower. Lack of pollinators can reduce the crop size to an uneconomic level. Most orchards have windbreaks on the windward side to reduce the wind to encourage the bees to fly even on windy days.

Crops that are wind pollinated such as sweet corn, *Zea mays*, and hazel, *Corylus avellana*, should be planted in blocks and not rows to assist with pollination.

Temperature – bees and many other insects require a reasonable temperature or they get chilled. Insects will not fly in frosty or cold weather.

Heavy rain – deters many insects from flying.

Apomixes – the production of a *viable seed* without pollination or fertilisation. There is no fusion of the gametes in the ovule. Apomixes occurs when certain cells in the ovule (not the gametes) develop into a viable embryo and seed is produced. This seed will only have the genetic characteristics of the one parent (the plant that has produced the seed). This can happen in some *Citrus* species, also the two common weeds that have apomixes are annual meadow grass, *Poa annua*, and dandelion, *Taraxacum officinale*. This is one reason why they are such prolific weeds.

F¹ HYBRIDS

These are the result of a controlled crossing of two distinct types of plant. F¹ is an abbreviation of first filial generation. That is they are the first generation of the controlled cross made by the breeder from known parents.

F¹ hybrids are usually:

- superior to ordinary cultivars
- show hybrid vigour
- may have pest and/or disease resistance
- be higher cropping or bigger flowers
- have uniform cropping for machine harvesting
- uniform growth
- better quality

Breeding F¹ hybrids

F¹ hybrids are usually produced from plants that cross-pollinate (outbreeders).

The first stage in breeding F¹ hybrids is to select two strains of the plant that have the desirable characteristics. This may be high cropping and a certain disease resistance. These two strains are grown on in separate areas well away from other similar plants, often in compartments in a glasshouse.

The two strains are self-pollinated for a number of generations to produce a stable plant with the desirable characteristics. The resultant plants are often poor growing after all the inbreeding. These are known as inbred lines.

When ready the two strains are crossed under controlled conditions to prevent any other unwanted pollination occurring. The seed resulting from this crossing is the F¹ hybrid seed. This will have the two desirable characteristics and also hybrid vigour.

It is necessary to purchase new F¹ hybrid seed each year, seed saved from any F¹ hybrid plants will not come true to type. The seed saved from the F¹ plants will be the F² generation.

See Chapter 9 for further details on plant breeding.

FRUITS

The fruit is formed from the ovary after fertilisation. Its function is to protect the seed and often helps in seed dispersal. Fruits, like seeds, vary considerable in size and shape: they can be large and soft like a pear or smaller and very hard like a nut.

Fruit may contain one seed like an acorn, *Quercus* spp., plum or cherry, both *Prunus* spp., or contain many seeds such as a tomato, *Lycopersicon esculentum*, or poppy, *Papaver rhoeas*, capsule.

Fruits are important in helping to disperse the seed away from the parent plant to reduce the competition between the parent plant and seedlings, and also to spread the species onto new areas of land.

Knowing the type of fruit a plant produces can help in the identification of the plant and also when the fruit or seeds are ready to harvest. Fruits can be divided into two main groups and then subdivided into other groups.

The first characteristic of fruit to identify is whether it is dry or succulent (fleshy).

If dry these subdivide into two main groups:

dehiscent
indehiscent.

A dehiscent (means split) fruit is one that can open and expel the seed (dehisce the seed) on its own.

An indehiscent fruit uses other means to disperse the seed such as animals or wind.

Dry dehiscent	Dry indehiscent	Fleshy
Follicles: monkshood, aquilegia, delphinium	**Nuts:** acorn, walnut, sweet chestnut, hazel	**Drupe:** plum, cherry, peach, sloe
Siliques: wallflower, honesty, *Brassicaceae*	**Achene:** sunflower, strawberry, calendula, *Zinnia*	**Pome:** apple, pear, quince, hawthorn
Capsules: poppy, snapdragon, horse chestnut, primrose	**Samara:** maple, elm, ash	**Pepo:** cucumber, marrow
Legume: pea, beans, sweet pea	**Caryopsis:** sweet corn, grasses, wheat	**Berry:** tomato, grape, blackcurrants

Table 7.1 Chart to show the classification of some fruits and whether they are dry dehiscent, dry indehiscent or fleshy.

Dry dehiscent fruit

These are more complex than indehiscent fruit in that they open and mechanically spread the seed. Some split open with force when dry and throw the seed out so that it lands a reasonable distance from the parent plant. Plants that can do this include pea, *Pisum sativum* (see Figure 7.21), broad bean, *Vicia faba*, and busy lizzie, *Impatiens* spp. The seed pods (fruits) spring open and eject the seed with force.

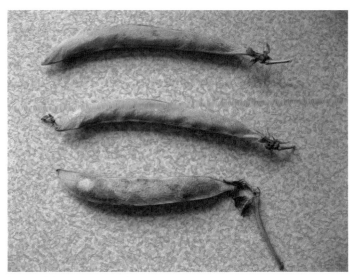

Figure 7.21 Picture of a legume (pea pod).

Figure 7.23 Picture of a wallflower, *Erysimum cheiri*, seed head showing a silique fruit.

Figure 7.22 Picture of poppy seed head: a dry dehiscent fruit.

Dry indehiscent fruit

These are simple fruits and the seed is usually dispersed by wind or animals. The seed or fruit will often have an attachment that helps in the dispersal, such as a hook to catch onto fur or clothing, an example being achenes.

Other dry indehiscent fruit, such as nuts, rely on animals to disperse them: for example, squirrels burying nuts for use during the winter; some are lost and will germinate and grow.

Examples of dry indehiscent fruit types:

achenes – *Calendula officinalis*, buttercup, *Ranunculus* spp., and *Zinnia* spp.

nut – acorn, *Quercus* spp., sweet chestnut, *Castanea sativa*, hazel, *Corylus avellana*

samara – maple, *Acer* spp., ash, *Fraxinus* spp., and elm, *Ulmus* spp.

caryopsis – sweet corn, *Zea mays* (see Figure 7.24), grasses including wheat, *Triticum aestivum*, oats, *Avena sativa*, and barley, *Hordeum vulgare*.

Other dry dehiscent fruit rely on the wind moving the seed head (fruit) and the seeds being thrown out of holes in the top of the fruit. The poppy, *Papaver rhoeas*, is a good example of this type of fruit shown in Figure 7.22.

Examples of dry dehiscent fruit types:

legume – pea, *Pisum sativum* (Figure 7.21), beans, sweet pea, *Lathyrus odoratus*, Judas tree, *Cercis siliquastrum*

follicle – aquilegia, delphinium

silique – wallflower, *Erysimum* spp. (see Figure 7.23), radish, *Raphanus sativus*, honesty, *Lunaria annua*

capsule – violet, *Viola* spp., poppy, *Papaver rhoeas* (see Figure 7.22), primrose, *Primula vulgaris*, horse chestnut, *Aesculus hippocastanum*.

Figure 7.24 Photograph of sweet corn cob, *Zea mays*, a caryopsis fruit. It also shows the silks still attached.

Testa – This is the outer coat, which protects the seed and embryo. It is derived from the outer layers of the ovule; it can vary in thickness and be very hard. Some are coated in a layer of suberin (a wax-like substance); others are made up of hard cells. The suberin/hard cells have to be broken down before the seed will germinate.

Plumule – The embryonic stem, which consists of a stem tip and a few leaflets. This is all compressed into a very small space, even in the pea it will only be a couple of millimetres in size.

Radicle – This is the primary (first) root. It consists of the root tip and first root hairs.

Cotyledons – These are the seed leaves. There may be one, as in grasses, and these are called monocotyledons. Other plants have two, as in many flowering plants, and are called dicotyledons.

Hilum – This is the scar from where the seed was attached to the fruit.

Micropyle – This is a minute hole at one end of the scar. This is where the male gamete entered the ovule to fertilise it.

Hypocotyl – The term given to the short length of stem *below* the cotyledon.

Epicotyl – The term given to the short length of stem *above* the cotyledon.

Monocotyledon seed structure

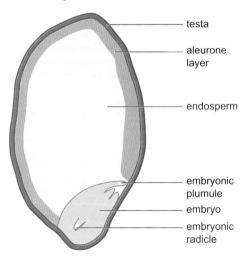

Figure 7.29 This shows the structure of a monocotyledon seed showing the main parts of the seed.

The monocotyledons, such as grasses, have a slightly different structure than dicotyledons. The main difference is that they only have one cotyledon, hence their name. Their food is stored in the endosperm. The embryo is where the seed will start to grow as this is where the radicle and plumule are situated.

Germination – The seed is a dormant shell containing the miniscule plant parts to form a new plant. Once in the right environmental conditions, and if the seed is viable, it will germinate and grow. The seed consists of an embryo (immature young plant)

with a food supply that is contained in a protective coat (testa). Seeds are a living part of the plant's life cycle.

The right environmental conditions for germination are:

sufficient water
oxygen
suitable temperature
suitable growing media
light – for light-sensitive species
freedom from pest, diseases and poisons.

GERMINATION PROCESS

This is set out below as it occurs in nature.

Imbibition

This is the start of the process; the seed imbibes (takes in) water via the micropyle. The water is used by enzymes to break down the stored food in the seed. This changes the food from starches to sugars, which are more soluble and are used for respiration.

The seed swells (caused by the absorbed water) and the testa splits. This allows the seed to take up water quicker and allows better access for air (oxygen).

Cell multiplication

This will only happen if the environmental conditions are suitable. The seed needs to be at the correct temperature for the species and have a good supply of oxygen. Also all dormancy mechanisms must be overcome before germination will start.

Respiration and enzyme processes take place in the seed at a rapid rate. The first outward sign of cell multiplication is the emergence of the radicle (young root) from the testa. The radicle usually emerges near the micropyle and will rupture the testa.

Digestion and translocation

This is the breaking down of stored starches, fats and proteins in the endosperm or cotyledons into simpler substances such as sugars. The sugars are used for respiration and for the growth of the embryo.

Water is starting to be taken in by the young root at this stage.

Seedling growth

This is the very rapid cell division and cell elongation, which enables the radicle and plumule to expand and grow. Respiration will increase and the radicle will grow in a downwards direction owing to geotropism and the plumule will grow towards the light owing to phototropism.

The young seed leaves will appear above the soil and the gardener will know that successful germination has occurred. The growing plant can now start to photosynthesise and produce

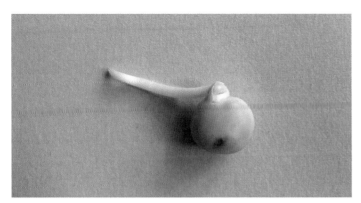

Figure 7.30(a) Photograph of a pea seed showing the emergence of the radicle.

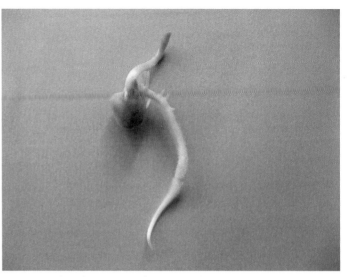

Figure 7.30(b) Photograph of pea seed showing both the radicle and plumule.

Figure 7.30(c) Photograph of pea seedling at the point of emerging from the soil.

its own food. It is no longer reliant on the stored food in the seed, which will wither away.

Once the radicle has emerged it is capable of absorbing water and nutrients.

Figure 7.30 show the sequence of germination of a pea, *Pisum sativum*.

TYPES OF GERMINATION

The way the seedling emerges from the soil is used to describe the type of germination. It can be either:

    epigeal
    or hypogeal.

Epigeal germination

'Epi' means above – and refers to the cotyledons coming above the soil level.

The cotyledons come up and out of the soil. As the seed germinates and starts to grow, the hypocotyl elongates and this pushes the cotyledons above ground. When above the soil level the epicotyl will elongate and the first leaves will appear.

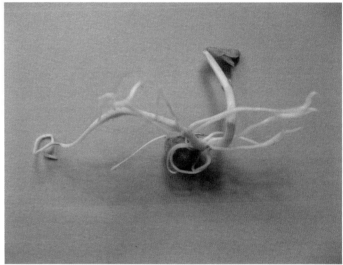

Figure 7.30(d) Photograph of a young pea seedling making both root and shoot growth.

Examples of plants with epigeal germination include: marrow, *Cucurbita pepo*, cucumber, *Cucumis sativus*, and French bean, *Phaseolus vulgaris*. See Figure 7.31.

Hypogeal germination

'Hypo' means below – and refers to the cotyledons remaining below the soil surface. The hypocotyl will remain below the soil surface and the epicotyl will elongate to come above the surface. As a useful mnemonic to help remember the difference between hypogeal and epigeal germination think of the 'h' in hypogeal as 'h' for hidden – as the cotyledons are hidden below the soil surface.

The plumule will emerge with the tip bent over like a hook. This protects the very young delicate leaves and growing point as they are pushed through the soil. Once the epicotyl has broken through the surface it will straighten out and grow towards the light.

Examples of plants with hypogeal germination include: pea, *Pisum sativum*, and broad bean, *Vicia faba*.

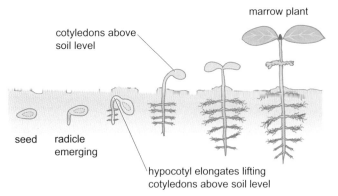

Figure 7.31 Diagram of epigeal germination sequence showing the cotyledons coming above the soil surface.

REQUIREMENTS FOR GERMINATION

Water

This is important to swell the seed and to start the chemical process of germination that initiates growth.

There should be sufficient water in the growing media for the seed to imbibe. If the soil is too dry the seed will stay dormant. If the soil is too wet there may be insufficient oxygen for respiration and the seed will again stay dormant or may even rot. The seed needs to be in contact with moist soil or growing media.

Oxygen (air)

The seed requires oxygen for respiration, particularly once germination has started as the rate of respiration increases rapidly. If the oxygen levels are low, germination levels will be low.

Figure 7.32(a) Photograph of sweet corn germinating.

Figure 7.32(b) Photograph of sweet corn showing the radicle and plumule growing.

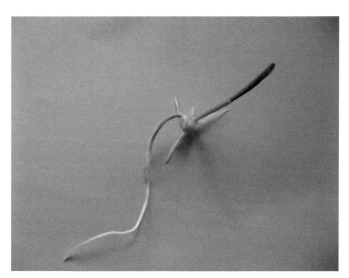

Figure 7.32(c) Photograph of sweet corn at the point of emerging from the soil.

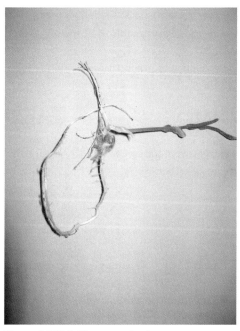

Figure 7.32(d) Photograph of a young sweet corn seedling.

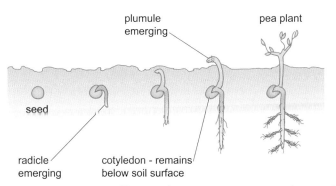

Figure 7.33 Diagram of hypogeal germination sequence showing the cotyledons remaining below the soil surface.

Once the seed has started to germinate it is important that any carbon dioxide produced by the seedling is dispersed quickly and can escape to the atmosphere, because carbon dioxide can inhibit germination.

Other gases that can affect germination include hydrogen, ethylene and nitrogen.

Temperature

The ideal temperature for germination will vary depending on the plant species. Plants native to the arctic tundra will germinate at much lower temperatures than plants native to the tropical rain-forests. Therefore the temperature required will be fairly specific to the species being grown. The highest germination percentage will be at the optimal temperature for that species.

Temperature is also an important factor in the speed of germination. The higher the temperature, within reason, the quicker the seed germinates. Seeds sown into cold soils will tend to stay dormant or germinate very slowly. This makes them prone to rotting or pest and disease attack.

Some seeds need alternating temperatures to germinate, this helps to break the dormancy.

Plants that require a warm temperature to germinate include tomato, *Lycopersicon esculentum*, cucumber, *Cucumis sativus*, pepper, *Capsicum annuus*, and busy lizzies, *Impatiens walleriana*. These need a temperature of 20°C to 25°C.

Plants native to the UK and other temperate regions, such as carrot, *Daucus carota*, cabbage, *Brassica oleracea* (Capitata Group), and wallflowers, *Erysimum* spp. will germinate at much lower temperatures, e.g. 7°C to 12°C.

Light

Most seeds require dark conditions to germinate and will not germinate if sown in light conditions. However, some seeds require light to germinate. Many annual weed seeds will not germinate until exposed to light, the seed can then detect that it is not buried too deep to germinate and will grow. Cultivating the soil and bringing weed seeds to the surface encourages them to germinate.

Small seeds such as *Begonia semperflorens* and *Impatiens walleriana* are often sensitive to light.

Most seeds of cultivated plants do not have a light/dark sensitivity and do not react to light so should be covered when sown.

VIABILITY

A seed will only germinate if it is viable – that is the embryo is still alive. Not all seeds are viable even just after they have been dispersed by the parent plant.

The viability of seeds depends on:

the species of plant
fruit and seed health
harvesting conditions
storage conditions.

Species

The viability of seed varies enormously between different species. Plants bred and cultivated by man tend to have a high viability as this has been bred into them. This includes most vegetables and annual flowers. But species that have had little work carried out on them or are recent discoveries can have very low and erratic viability, and this can result in slow and poor germination.

Fruit and seed health

The fruit and seed must be from a healthy and well-fed plant, growing in reasonable conditions to ensure good viable seed is produced. Poor-quality, starved plants will only produce poor-quality seed.

The seed should be harvested at the correct stage for the species, cleaned and dried properly.

Storage conditions

The seed needs to be kept alive, but not respiring too much or it will use up the food reserves required for germination. The seed should be stored in the following conditions.

Dry – This is to prevent the seed absorbing moisture and the germination process starting. Dry conditions include keeping the air humidity very low.

Cool/cold – Keeping the respiration rate as low as possible will increase the length of time the seed can be stored. Therefore to keep seed for long periods they are stored in cold and dry conditions. This keeps the respiration rate very low. The seed must be dry when put in the store to prevent damage by freezing.

Low oxygen

Again low levels of oxygen slow down respiration, and so increase the length of time the seed can be stored.

Many seed firms sell their seed in hermetically sealed packets nowadays. This is to keep them dry, in low humidity and a low oxygen level. The packets should be stored unopened and in cool conditions.

DORMANCY

A seed is said to be dormant if the environmental conditions are suitable for germination and the seed is viable yet does not germinate.

Dormancy is a method of controlling the germination of the seed until the conditions are ideal for both germination and plant growth. For example, a seed dispersed in the autumn will remain dormant over winter until the conditions improve in the spring and then will germinate.

The types of dormancy can be divided into two main groups:

physical
physiological.

Physical dormancy

This is usually caused by the seed coat (testa), which can be very hard and impervious to water and gases. The embryo cannot get access to water and oxygen so the germination process cannot start. The testa needs breaking down to allow water and oxygen through. These seed coats are normally broken down in the soil by bacteria and fungi, and this will take a number of months.

Some seeds have a waxy or suberised coating, which prevents water access. This again needs to be broken down.

Plants with hard seed coats include: sweet pea, *Lathyrus odorata*, *Lupinus* spp. and *Robinia pseudoacacia*.

Physiological dormancy

This is an internal dormancy where the embryo needs to undergo some changes before it can germinate. Some of these seeds need to go through a period of after ripening.

One form of after ripening requires an inhibitor to be broken down or washed out of the seed. The inhibitor can be in the form of abscisic acid. This can be washed out of the seed by rain or broken down as a result of cold temperatures.

Breaking dormancy

If growing crops of trees, shrubs or other plants that have a dormancy problem, the grower needs to break the dormancy to get quick and even germination. If the grower had to wait up to two years for a good germination it would be an expensive and slow process. So to ensure quick and even germination various methods are used to break the dormancy.

TYPES OF DORMANCY

Physiological dormancy occurs in three types: deep, intermediate and non-deep.

Deep – this is the longest type of dormancy, which could be a period of months, up to 18 months through two winters and a summer.

Intermediate – this is a shorter period of a few weeks up to a couple of months.

Non-deep – these seeds require a few days up to a couple of weeks.

Morphological dormancy

This is where germination is prevented owing to morphological characteristics of the embryo. For example, the embryo is made up of cells but these have not yet differentiated into their various types. In other seeds the cells have differentiated but have not fully grown when the seed is dispersed. The cells mature during dormancy and the seed will germinate. Morphological dormancy occurs in some plants in the families: *Apiaceae*, *Liliaceae* and *Ranunculaceae*.

Secondary dormancy

This occurs after the seed has dispersed and would germinate, but is exposed to conditions that are not favourable for germination. One cause of this is high temperature. This occurs in lettuce, *Lactuca sativa*, that are sown in temperatures of 25°C or higher. The seed will not germinate until the temperature has cooled and often germination is lower.

Stratification – The seed is mixed with damp sand, peat, coir or vermiculite and put into a container like a large pot or deep wooden box. The container is then put in a cool place over winter, usually on the north side of a wall or in a cold store.

It is possible to speed the process up by putting the seed and mixture into a polythene bag and placing it into a refrigerator set at 1.5°C or lower for some species. This mimics the winter period. The length of time the seeds need treating for varies depending on the species. Seeds that are stratified include *Sorbus* spp., *Crataegus* spp. and some *Rosa* spp.

Scarification – This involves wearing away or chipping part of the testa. On a small scale this can be done by filing or rubbing the testa with a file or sandpaper. Commercially it is done by putting the seed through emery paper rollers until the testa has worn very thin. Scarification is used on cherry, *Prunus* spp, seed.

Chipping – Also called nicking. Using a sharp knife the testa is chipped to make a hole that will allow water and oxygen to enter. Care must be taken not to cut your fingers! Chipping is used for some cultivars of sweet peas, *Lathyrus odoratus*, *Ipomoea* spp., and *Canna indica*.

Soaking – Acid – commercially seeds can be soaked in a strong acid solution to break down the testa. Great care needs to be taken not to damage the embryo. The length of time the seed is in the solution is critical and varies depending on species. Seed that can be treated with acid to break dormancy include *Hamamelis* spp*., Tilia* spp., and some *Cotoneaster* spp. Acid soaking is used less now owing to the health and safety implications of using concentrated acid.

Hot/warm water – some seeds can be soaked in hot/warm water for a period of time. This weakens the testa and allows water and oxygen to enter.

This method is for *Acacia* spp., *Gleditsia triacanthos* and *Cercis siliquastrum*.

Smoke treatment

Seeds from certain environments will only germinate after a bush fire. It is believed the flames help to scarify the seed coat and chemicals in the smoke break the dormancy in the seed. This can be simulated by covering the seed with dry leaves 60 mm to 100 mm deep, burning them and then watering the compost to wash in the chemicals produced by the burned leaves. Smoke water or paper can be purchased to have the same effect.

Smoke treatment can be used for *Eucalyptus* spp. and *Dendromecon rigida*.

SEED COLLECTION

Before collecting seed the following points should be considered.

Seed from cultivars/hybrids will not breed true to type.

Seed must be collected when ripe but before it disperses. Some seed is collected just before it is ripe as this can prevent it going into a dormancy phase if sown quickly.

The seed **provenance** can be important, so try to choose a local provenance if possible. The provenance is the area in which the plant is growing such as the south of France or the north of Scotland. This can over time produce small variations in the species, which may have adapted to grow in certain temperature ranges.

Ensure the plant is correctly identified and that it will breed true to type. Collect seed only from healthy plants. Ensure all packets are correctly labelled.

Collecting seed

Seed can be collected by machine or hand. Large-scale production is collected by combine harvester-type machines. Small-scale or tree seeds are usually collected by hand.

Dry seeds – seeds from dry dehiscent and indehiscent fruits can be collected by hand and placed into paper or cloth bags. Dehiscent fruits may need a paper bag putting over the fruit to catch the seed unless collected before it is fully ripe and dehisces.

Once collected the seed is cleaned usually by sieving through very fine sieves and gently blowing to remove any light chaff. Once cleaned it is dried thoroughly to a very low moisture level, bagged, labelled and stored.

Fleshy fruits – these are usually berries or similar fleshy fruits and are collected when ripe, slightly earlier if birds usually eat them. Some fruits can be stratified and then sown once dormancy is broken; other fruits require the flesh to be removed before the seed is stored. Tomato, *Lycopersicon esculentum*, and similar fruits need the flesh removing, which is done by macerating (pulping) the fruit in a container to retain it all. The flesh is then allowed to ferment for two to three days, giving it an occasional stir to help the seed settle out. Most of the flesh and skins float to the surface and the viable seeds sink to the bottom. The flesh and other waste is skimmed off and disposed of. The remaining water is passed through a sieve to collect the seeds. These are then washed to remove any remaining flesh and thoroughly dried ready for storing.

CASE STUDY   **Sources of seed**

Collection from the wild – this is how many plants were brought to the UK from various parts of the world. Permission is now required before seeds can be collected to help prevent the destruction of native habitats.

Collection from cultivated plants – if the plants are natural species the seed if viable should breed true to type. If the plant is a cultivar (bred by man) the seedlings produced will be variable as they will have inherited characteristics of both pollen parents.

Purchase from seed companies – there are a number of large-scale producers in the UK who produce seed catalogues each year. They sell a wide range of common plants and cultivars. It is worth getting catalogues from a couple of companies to see the range available. There are also a number of small specialist suppliers who specialise in a limited range of plants. These could be a particular species of plant, e.g. sweet peas, *Lathyrus odoratus*, or exhibition-size vegetables.

Seed swaps – are carried out on a large scale between botanic gardens such as Royal Botanic Garden, Kew or the various Royal Horticultural Society gardens, where the aim is to keep species from becoming extinct. The more places the plant is grown, the less chance of a catastrophe happening and the plant being lost. Also rare or unusual plants are exchanged between gardens throughout the world for research and educational purposes. There are also specialist growers who specialise in a particular species, such as the national plant collections, and will swap seeds with like-minded people. A recent development are seed swaps being organised by various societies or local groups where you can swap your surplus seed with people who have seed they wish to exchange.

The viability of the seed will be checked by sowing a small batch of seed onto tissue paper in a petri dish and noting the speed and number of seeds that germinate.

Seed storage

To keep the seed for a reasonable period and still ensure good germination the aim is to slow down the metabolic processes in the seed. That is to slow down the respiration rate to prevent the seed using up its food reserves. Kept in ideal conditions most seed will keep for three to five years. Viability will decrease with age, therefore so will the germination percentage.

The seed should be stored in:

low temperature 0°C to 2°C
low humidity 4% to 6%
low oxygen level/high carbon dioxide level (up to 21%).

Plants that are stored in dry, cool conditions include most of the vegetables and annual flowering plants.

Plants that require storing in cool moist conditions are some of the trees, shrubs and alpine plants. Species include *Quercus robur*, *Castanea sativa*, *Fagus sylvatica*, *Corylus avellana* and *Acer palmatum*.

Advantages of propagation by seed

Large quantities of plants can be produced quickly.
Low cost of seed and low-cost methods of producing a crop.
Simple techniques used to propagate seed.
Very wide range of species and cultivars available.
Seed can be stored in good conditions for a number of years.
Very few pests, diseases or virus are carried by seed.
New types of plants can be produced.
Seed is easy to store and transport.

Disadvantages of propagation by seed

Some plants do not come true to type from seed.
Some plants are sterile and will not produce seed.
Slow-growing plants can take a long time to mature and fruit.
Difficulty in germinating and growing some species.
Some plants do not produce viable seed in sufficient quantities.

TYPES OF SEED AVAILABLE

Seed is sold in a number of forms: this can be the basic raw untreated seed or treatments are carried out on the seed to enhance its germination prospects.

Raw seed – this is seed straight from the parent plant. It will have been cleaned, stored and put in packets for sale.

Pelleted seed – the seed has been coated in an inert material to make it larger and easier to handle. These are useful for machine sowing or space sowing.

Primed seed – the seeds are partly germinated in ideal conditions and often treated with chemicals (e.g. polyethylene glycol), which improves the germination when sown. Once the germination process has started the seed is dried off and then put in packets. This gives the seed faster and more uniform germination.

Clipped seed – seeds like marigolds, *Tagetes* spp., and seeds with feathery tails, e.g. *Clematis* spp., are clipped to remove the unwanted part of the seed. This makes the seed easier to sow and also reduces disease risks.

Coated seed – the seed is covered in a bright water-soluble dye that makes it easier to see when sowing by machine.

Pesticide treatments – seeds are coated in an insecticide or fungicide to help prevent pest or disease attack.

Seed tapes – the seed is embedded onto a water-soluble tape at a set spacing. The tape is placed onto the compost/soil, lightly covered and watered. The seeds germinate ready spaced out and do not require pricking out or thinning.

SOWING

Indoors

Hygiene – this is critical when sowing seeds indoors as the environment is ideal for the growth and spread of pests and diseases. The compost, containers, work benches and propagating area must be scrupulously clean. Dirty work areas and containers can soon lead to fungal disease attack and whole trays of seedlings being killed.

All compost should be purchased new each year and kept in sealed polythene bags or covered if brought loose in bulk.

If using homemade compost it should be sterilised to kill any weeds, pests or diseases.

New containers should be purchased each year if in commercial production. Amateurs should wash any containers in hot water with some bleach or detergent added.

Work areas should be washed down annually with a detergent solution and any plant remains removed and disposed of. The propagating areas including glasshouses, polytunnels, growing rooms and propagators should again be thoroughly cleaned.

Propagating areas

If sowing at home, seedlings can be raised on a window ledge or in small plastic propagators. Propagators and small frames with heating cables in the base are very useful to create the right temperature and humidity for propagating plants.

Larger facilities for propagation include glasshouses, polytunnels and growing rooms. These types of areas are used commercially and are expensive to heat owing to the high energy costs. These structures are only used for large-scale production.

The very large-scale seedling producers use growing rooms. These are like insulated sheds with no windows, but have high levels of artificial lighting. They are kept humid and at a high temperature. As they are insulated the heating costs are lower.

The seed is sown into seed or plug trays and put onto shelving under fluorescent lights. These provide the lighting when required and also some of the heat. These conditions are ideal for quick germination and growth of young seedlings. A very large number of seedlings can be produced in a small area very quickly.

Containers

There is a wide range of containers used for seed raising, including seed trays, modular trays (see Figure 7.34), pots and pans. The container needs to hold sufficient compost, allow good drainage, and be clean and not too expensive. Most seedlings are not in the container very long so they can be quite shallow (40 mm to 50 mm).

Containers can be made of plastic, polystyrene, wood, etc. Amateur growers can use recycled containers such as margarine tubs, meat trays and other food packaging.

Figure 7.34 Photograph of a range of seed propagation containers.

Composts

John Innes seed compost was widely used prior to multipurpose compost being developed. Recently peat-free composts have come onto the market and are being widely promoted to reduce the use of peat.

Multipurpose composts are a mixture of peat and coarse sand, the ratio of each will depend on the product purchased. They also have some fertiliser added for the seedlings to use once germinated and growing.

Peat-free composts are manufactured from a number of recycled materials. What they contain will depend on which product is purchased. Materials used include bark, wood waste, coir and composted municipal waste. Trials of the compost need to be carried out so the grower can adapt their growing techniques for the compost used.

Homemade peat-free compost can be made from worm compost, leaf-mould and sand.

Sowing techniques

Hand – If the compost has been stored outside, bring it inside to warm it up before using, otherwise if cold it could slow or reduce germination. Fill the container with compost and tap gently on the work surface to settle the compost. Scrape off any surplus compost level with the top of the container, and then firm it with a presser board.

Either water the compost with a fine rose on a watering can or put the container and compost in water for a short period. Always wet the compost before sowing as watering after sowing can result in the seeds being washed off the compost or all washed to one end of the container. It can also result in the surface of the compost capping, which can affect germination.

Sow the seed either from the palm of your hand or directly from the packet. Hold the packet or your hand over the compost and gently tap so that the seeds drop evenly onto the compost. Sowing the seed in two directions should result in even spacing. Large seeds can be space sown at an even distance. Very large seeds are usually sown individually into modular trays at one seed per module or directly into 75 mm pots.

Very small seed can be mixed with dry silver sand to make even sowing easier, and so that you can see where you have sown.

Cover the seed to approximately its own depth with compost or vermiculite. Very small seeds will not require covering or this will result in poor or no germination.

Seeds that take a long time to germinate (such as tree, shrub or alpines) are often covered with fine grit (6 mm). This protects the compost from capping and helps to prevent moisture loss.

Machine sowing – There are now a number of machines available that sow a wide range of seed directly into seed or modular trays. They can either space sow or broadcast very evenly.

Commercially seed can be sown directly into modular trays at one seed per module and given a light covering of vermiculite.

Management of seedlings

Once sown the compost should not need re-wetting until the seed has germinated if it has been kept in the correct conditions. The trays should be kept in:

a humid environment
an even temperature suitable for the species
with a good oxygen supply.

Trays are usually covered with a sheet of clean glass or polythene to keep the humidity levels high.

Propagation facilities have thermostatically controlled heating, usually bottom heat to maintain an even temperature.

The trays should be checked daily for signs of germination. Once germinated the seedlings should be moved into good light, but not direct sunlight.

Aftercare

Watch for any sign of damping off disease; if this appears treat the seed tray with a suitable fungicide.

Once the seedlings are large enough to handle they should be pricked out or potted up.

Pricking out – this is the careful removal of the seedling from its original container, which is then replanted into fresh compost in a new container. The aim is to give the plant sufficient space and compost to develop into a size ready to sell or plant out into the garden. Prick out before the seedlings become leggy or too large as they become difficult to handle.

Water the seedlings before pricking out, this reduces the stress and makes it easier to remove them from the compost. Ensure sufficient compost and containers are available and the work area is tidy and ready for work.

Tap the box gently on the bench to help loosen the seedlings. Ease the seedlings out of the compost using a plant label, spatula or pencil. Hold the seedling by the seed leaves, never the stem, which if damaged will result in the death of the seedling.

Make a hole in the compost in the new tray/module with a dibber. Place the seedling into the hole so that the seed leaves are just above the compost and gently firm into place. Keep the seedlings in straight lines and evenly spaced.

Prick out one seed per module or if using seed trays 35 or 40 a tray (7 rows × 5 rows, or 8 rows × 5 rows).

Put the modules or trays into the glasshouse or polytunnel at a suitable temperature, and water them to settle them into the compost. Use a watering can or hose with a fine rose on the end.

Potting up – large seedling such as runner beans, marrows and squashes are usually potted up into individual containers of 75 mm diameter (if not sown directly in them). Half fill the pot. The seedling is then placed into the centre of the container and compost placed around the roots up to the top of the pot. Remove any surplus compost and tap the pot on the work surface to settle the compost around the roots.

Place the pots in the growing area and water in with a fine rose on a watering can or hose.

Plug plants – There has been a trend in recent years to the growing and selling of plug plants. These are available to both amateur and professional growers. Plug trays are plastic trays with a number of small individual compartments set in them. The number varies depending on the size of the plugs. Small plugs are approximately 10 mm diameter and larger plugs are 40 mm to 50 mm in diameter.

The seeds are sown directly into compost in each plug by machine. The trays are put into growing rooms to germinate. The seedlings germinate very evenly and are grown to a saleable size quickly in ideal conditions. Producers of plugs have a very high throughput of plants and the operations are highly mechanised to reduce costs.

Any gaps owing to seed not germinating are filled by machine so that when the plug tray is dispatched it has the correct number of seedlings.

When delivered the plug trays should be unpacked if in parcels and watered if dry. The plugs should be potted up into pots or modules as soon as possible.

Growing on

Once seedlings have been pricked out or potted up they are usually placed in a glasshouse, polytunnel or cold frame. The choice of structure will depend on the temperature required to grow the plant on and the facilities the grower has available. The plants should be grown in good light to prevent them becoming leggy and at a suitable temperature for the plant being grown.

If grown in good compost it should not be necessary to start feeding the plant for the first six weeks, as there should be sufficient food in the compost.

Check regularly for signs of pests and diseases and take the appropriate action. (See Chapter 17.)

Once the plants are of a suitable size the process of hardening off can start. This is carried out to acclimatise the plant to the cooler and more stressful conditions outside. Years ago the containers and plants were moved into cold frames and covered with the lights (the frame top). Over a period of time the lights were opened further and for longer periods until the plants were hardened off. Nowadays with the cost of moving the plants the process has changed. The plants remain in the growing areas and the temperature is slowly lowered, followed by the doors and ventilators being left open for longer periods. When ready the plants are then planted out or dispatched for sale at garden centres.

OUTDOORS

Seed bed preparation

If the area is covered in weeds these will need removing by chemical or physical methods.

The soil should then be cultivated by digging or ploughing. This will bury any weed/crop remains and incorporate organic matter if required. The soil can be left over winter to weather if cultivated in the autumn. The next phase is to produce a tilth for sowing. This is done using a fork, cultivator or, on a large scale, disc harrows and rollers. Any large stones and debris should be removed and soil lumps should be broken down. If required a fertiliser can be added and then raked to produce a fine tilth.

The soil should not be too cold or wet as this will reduce or prevent germination.

Protection

To produce early crops it is necessary to cover the soil to warm it up in the spring. This can be done using:

cloches of glass, plastic or polythene
fleece or mesh sheets
germination sheets.

These should be put over the soil 10 to 14 days before sowing to warm it up and then put back into place after sowing until the crop has germinated and is established. Germination sheets are often used by grounds staff when they carry out repairs to cricket wickets and football pitches to get a quick even germination.

SOWING METHODS

There are two main methods of sowing: broadcast or drills.

Broadcast

This is the method used to sow a lawn, wildflower meadow or large area. It is quick and gives a reasonably even coverage if carried out carefully. The seed is either sown through a fertiliser distributor or by hand.

The fertiliser distributor has a spinning disc at the base of the hopper that holds the seed. As the seed drops onto the spinning disc it is thrown out by centrifugal force and gives a reasonable coverage of the ground.

The soil is prepared to a tilth as above; the seed is sown and then raked into the surface of the soil.

To sow by hand take a handful of seed and throw it forward across your body and as you release the seed flick you wrist to propel the seed out of your hand with an even spread.

Drills

Sowing in rows makes it easier to maintain the plants when weed control is required and to get access to them. It is also easier to identify the weeds as all the crop plants should be in a straight line.

The soil should be prepared to a tilth then a line should be put down to ensure the drill is straight. Use a stick, draw hoe or corner of the rake to draw out a drill to the correct depth. This will vary depending on the seed being sown. If the soil is dry water the drill to wet the bottom so the seed can germinate.

The spacing of the drills will again vary depending on the seed being sown. Once an even depth drill is formed the seed should be sown thinly along it. The soil is then raked back over the seed and it can be lightly firmed to ensure the seed is in contact with the soil.

Station sowing – Seed is sown at 'stations'. This is done to save thinning and seed as the plants will germinate and grow the correct distance apart. Station sowing can be done by hand or machine.

Machine sowing – On a large scale seed can be sown using machinery. The drills are set the correct width apart and at the correct depth. The spacing of the seed in the drill is controlled by belts or wheels that are pre-set. As the seed drill is pulled along by the tractor it pulls out a drill, drops the seed into the drill and then covers it with soil, which is firmed by a small roller at the back. Large acreages can quickly be sown using this method.

Fluid drilling – This method was developed at what is now the University of Warwick (Wellsbourne Horticultural Research International). Seed is chitted (partly germinated) in aerated tanks or on tissue paper. It is then mixed with a gel or wallpaper paste to act as a carrier. The gel is then sown by machine or hand into drills in the soil. It can be sown as a continuous stream of gel or blobs of gel spaced out at a set distance.

Fluid drilling gives a quicker, more even germination and is useful for crops such as parsnip, *Pastinaca sativa*, which often have uneven and poor germination.

See also Chapter 12.

CASE STUDY **Dunemann seed beds**

Research has shown that conifers, and some deciduous trees and shrubs will germinate better and grow quicker if their roots are inoculated with mycorrhizal fungi. Dunemann seed beds were developed by Alfred Dunemann as a means of ensuring the seedlings came into contact with the mycorrhizal fungi.

The seed beds have been modified over the years but the basic principle is the same. They are approximately 35 cm high with wooden board edging. The base of the seed bed consists of a free-draining material such as pea gravel to a depth of 10 cm to15 cm. This is topped up to a depth of 25 cm with leaf litter from a woodland or forest. The litter contains the fungi. The seed is then broadcast over this and covered with grit; this stops the surface panning down during watering or heavy rain.

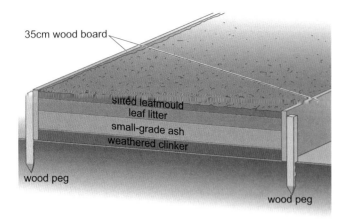

35cm wood board

sifted leafmould
leaf litter
small-grade ash
weathered clinker

wood peg

wood peg

Figure 7.35 Diagram of a Dunemann seed bed.

Timing

The timing of sowing will depend on the crop being sown. The main sowing period is the spring (March to May). This gives the longest growing period before the following winter. Most crops will be harvested or will have flowered by then. If a perennial crop, it gives a long growing season for the plant to become established.

Temperature – the timing of the sowing often depends on the temperature of the soil. It must be within the range for the crop being sown. Pea, *Pisum sativum*, will germinate and grow at 4°C but sweet corn, *Zea mays*, needs at least 10°C. Therefore pea, *Pisum sativum*, can usually be sown in early March in southern UK, but the soil will not have warmed up to 10°C until at least May.

Successional sowing

This is where the grower wants to achieve a continuous supply of a crop and avoid a glut. By sowing at 10- to 14-day intervals the crops will mature at a regular rate. When the first sowing has germinated and is showing the first seed leaves, the second crop is sown; this is then repeated until the last crop is sown. See also Chapter 12.

Aftercare

Once the seedlings have germinated they should be inspected regularly for signs of pest and disease attack, particularly slugs and snails. Preventative action should be taken promptly. If the weather is dry the seedlings may need irrigation.

It is important that weeds are controlled to prevent them smothering the crop and reducing the yield.

When the seedlings are a suitable size they should be thinned to the intended spacing.

Summary

Chapter 7 has covered flowers, fruits and seed. It has included the parts of a flower and the types of inflorescences; and the methods of pollination and fertilisation of flowers and production of fruit and seed. Moving on to how the seed is dispersed and the storage and germination of seeds, including the sowing and growing of the seedlings both indoors and outside. The following are the key points covered, which you should understand and be able to define or describe.

Flower parts – Flowers are made up of the following parts: sepals (calyx), petals (corolla), stamens (androecium), carpels (gynoecium), receptacle, pedicel and peduncle. You should be able to name the parts and know their function. There are three types of ovaries in flowers, these are superior ovary (hypogynous and perigynous) and inferior ovary (epigynous).

Flower sex – Flowers on plants come in three sexes. Plants with monoecious flowers have different male and female flowers on the same plant. Plants with dioecious flowers have the male and female flowers on separate plants. Hermaphrodite flowers have both male and female parts in the same flower.

Flower inflorescence – The arrangement of the flower inflorescence varies between species. The following are some of the common arrangements: solitary, multiple, raceme, spike, panicle, corymbs, umbel, capitulum and cyme. The flower layout helps in the identification of the plant.

Pollination – This is the transfer of pollen from the male (stamens) to the female (stigma). There are a number of ways this is carried out, which include insects, wind, birds/animals and water. The structure of the flower will be different depending on the method of pollination. There are two methods of pollination: self-pollination and cross-pollination. You should be aware of the factors affecting pollination.

Fertilisation – This is the fusion of the male and female gametes in the ovary following the male gamete growing down the style.

Seed structure – The outer coat of the seed is the testa with the hilum and micropyle. Inside can be found the plumule, radicle and cotyledons.

Fruits – Fruits are described as being dehiscent, indehiscent or fleshy. These are then grouped into the following types: legume, follicle, silique, capsule, achenes, nut, samara, caryopsis, pome, drupe, berry, pepo or aggregate. The fruit is produced as a means of dispersing the seed. The main methods of dispersal are wind, water, animals and mechanical.

Seed – There are a number of sources of seed including purchase from seed companies or collecting and saving your own. Seed should be stored in the correct conditions if it is to remain viable with good germination. There are a number of reasons seed may not germinate including dormancy, which can be physical or physiological. If seed is to germinate the dormancy needs breaking.

Germination – The requirements for germination are water, oxygen, temperature and in some cases light. The germination process is imbibition, followed by cell multiplication, and the radicle and plumule growing. There are two main types of germination: epigeal and hypogeal.

Review questions

1. Draw and label the basic parts of a flower.
2. What is the function of the following parts of a flower? Sepal, petal, stamen, carpel.
3. Define the term superior ovary and inferior ovary.
4. State the difference between pollination and fertilisation.
5. State the differences between insect-pollinated flowers and wind-pollinated flowers.
6. Draw and label the main parts of a seed.
7. State the differences between a dehiscent and indehiscent fruit.
8. Describe the process of germination.
9. State the conditions necessary for the storage of seed.
10. What methods can be used to overcome seed dormancy?

FURTHER READING

Brown, L. V. (2008). *Applied Principles of Horticultural Science*. Butterworth-Heinemann.
A book that goes into more detail on some of the topics covered in this chapter. It is suitable for students who need a greater depth of detail than is possible in this chapter.

Capon, B. (2005). *Botany for Gardeners*. Timber Press.
This book is a good introductory guide to botany that explains the flower structure and function. It is related to what the gardener requires to know to grow and understand plants.

McMillan-Browse, P. D. A. (1979). *Hardy Woody Plants from Seed*. Grower Books.
This book gives comprehensive detail of how to grow a wide range of woody plants. It covers a number of techniques and gives practical information on how to propagate the plants from seed.

Titchmarsh, A. (1981). *Gardening Techniques*. Mitchell Beazley.
This well-illustrated book covers a wide range of techniques that the reader can carry out in their garden. It gives easy-to-follow details of how to carry out many of the techniques required in flower growing.

Wheeler, C. & Wheeler, V (2003). *Success with Seeds*. Guild of Master Craftsman Publications.
A book giving practical advice on how to grow a range of plants from seed. It has plenty of photographs and gives easy-to-follow advice on propagation from seed.

RHS level	Section heading	Page no.
2 1.2	Advantages of propagation by seed	152
2 1.3	Disadvantages of propagation by seed	152
2 2.1	Collecting seed	151
2 2.2	Seed storage	152
2 2.3	Seed storage	152
2 2.4	Storage conditions	149
2 2.7	Machine sowing	153
2 3.1	Requirements for germination	148
2 3.2	Dormancy	150
2 3.3	Breaking dormancy	150
2 3.4	Germination process	146
2 3.5	Containers	153
2 3.6	Seed bed preparation	154
2 3.7	Sowing techniques	153
	Sowing methods	155
2 3.8	Management of seedlings	154
2 3.9	Aftercare	154
2 5.1 to 2 5.2	Dicotyledons	134
2 5.3	Hermaphrodite flowers	134
	Monoecious flowers	134
	Dioecious flowers	134
2 5.4	Pollination and fertilisation	139
2 5.6	Methods of pollination	139

PART 2
The adjustments

INTRODUCTION

Within this section the direction is provided to start influencing the growing or display conditions for plants, that provide the best or quickest results. Leading on from the foundations, the opportunities are outlined and discussed for adjusting our thinking of how to obtain the maximum benefit from our present knowledge level and build on this to meet our objectives, whatever they may be.

One of the most pleasing aspects of gardening is the reward of new plants that are free apart from a little time and effort. Propagating plants vegetatively provides a plethora of methods, with the background reasons for the selected methods being discussed, together with the alternatives on offer. Technical language is fully explained and images are provided to enhance understanding of the practical aspects.

The whole subject of designing plants leads on from home propagational techniques into a world filled with a new language and some fears about the unknown, and worries as to the pathway of genetic modification being a one-way street or dead end to some species.

Shaping plants, as a topic, is always high on the agenda when plant people meet or undertake garden visits. The discussion runs from the 'only prune to shape, once in a lifetime' through to the 'cut it to the ground, once a week, if not more often' view, and everything in between. However, often the reasons and the methods that can be used are unknown or have been forgotten over time. Whereas, newer methods or techniques, such as the Chelsea chop, are being adjusted or renamed for modern times.

Also the developing and expanding area of protected cultivation has been transformed, using different materials for both structures and cladding, together with energy use and sources, with an eye always on the resulting environmental impact and the development of sustainable practices when and wherever possible. Size is not an issue with materials and systems being now available for all growing situations, crops and aspects.

The coming together of all these adjustments to plant propagation, growing, display and use, has meant that everyone can now take the opportunity presented, even in a very limited space, to produce the plants or display of their dreams, added with the labour-saving methods and materials at our disposal, which are also continually emerging; think of the impact that products such as swell gel, horticultural fleece and F^1 hybrid cultivars have and continue to have.

However, do remember two main phrases related to pruning: 'growth follows the knife' and 'the quality of the pruning does not depend on the size of the bonfire'.

CASE STUDY Plant hormones

The role of plant hormones in the regeneration of roots and shoots was first discovered by plant physiologists who were trying to understand how initially similar plant cells were able to differentiate and form new tissues and organs. When pieces of tobacco pith, *Nicotiana* spp., were cultured aseptically on a medium containing an auxin together with essential plant nutrients and a source of carbon, it was found that the cells continued to divide and develop into a mass of undifferentiated cells called a callus. At the time many substances ranging from yeast cells to tomato, *Lycopersicon esculentum*, juice were tested (with little success) in an effort to get the callus cells to develop into normal tissues. In the 1950s, it was observed that the addition of coconut, *Cocos nucifera*, milk to the culture medium strongly promoted the division and proliferation of undifferentiated cells but they still did not develop into organised tissue(s). Much later it was discovered that coconut milk is a good source of a natural cytokinin, called *zeatin* and experiments with the synthetic cytokinin, kinetin had already shown that a main function of cytokinins is to stimulate cell division.

The breakthrough came when Folke Skoog and his colleagues (Skoog *et al.*, 1965) found that by manipulating the relative levels of auxin and cytokinin in the culture medium the undifferentiated callus cells could be induced to develop into new tissues and organs. When equal amounts of cytokinin and auxin were present in the nutrient medium, the callus tissue continued to produce undifferentiated cells, but when the ratio of the two hormones was changed, the cells began to differentiate into new tissues. As described later in the chapter, a high ratio of cytokinin to auxin favoured the development of shoots, while a high ratio of auxin to cytokinin favoured the development of roots.

The commercial development of micropropagation as a method of multiplying plants has developed from these early studies and the manipulation of the auxin/cytokinin ratios is still an important part of the techniques used. Initially a high ratio of cytokinin to auxin stimulates the regeneration of shoots while the level of auxin increases later to promote the initiation of roots.

Rare and endangered rhododendrons being conserved by micropropagation.

Chapter 8

Propagating plants vegetatively

Daphne Vince-Prue and Rosie Yeomans

INTRODUCTION

In the wild, the majority of higher plants reproduce by means of seeds. However, in the garden this is often not desirable, or may not be possible and other means must be used to obtain new plants. Therefore vegetative methods have been developed where a section of the parent plant is selected and roots are established so it can live independently.

A variety of different methods can be used to propagate plants vegetatively. These include softwood and hardwood cuttings, root cuttings, leaf cuttings, division, layering and a number of specialised techniques, such as grafting, budding and reproduction from structures such as bulbs and corms. The first section (Basic principles) is concerned with the physiological principles underlying the regeneration of new plants by vegetative propagation, with the second section (Practical techniques) focusing on the practicalities of the various procedures. Some techniques are very simple and success is guaranteed, while others require patience, skill, equipment and perfect timing. Micropropagation has become an important method for the regeneration of large numbers of plants by vegetative means and more details are given in the text below. Other methods are outlined with the time of year or age of the material highlighted. This allows a more informed view for the propagator to select a more suitable, potentially easier, method to be chosen with the aim of maximum success as the result. The environmental options available are also discussed, together with the required equipment, for all technical levels.

The propagation of plants is enormously satisfying, the concept of removing the shoot of a plant and rooting it to form a new plant is awe inspiring for the novice and once accomplished will encourage the grower to try ever more challenging species and techniques. This together with the satisfaction of saving money and contributing to a sustainable gardening ethos makes propagation a horticultural highlight.

Key concepts

- ✿ The reasons for propagating plants vegetatively

- ✿ The regeneration of new organs and the organisation of root and shoot apical meristems

- ✿ Etiolation increases the ability of plants to form roots

- ✿ The role of plant hormones in cell differentiation and micropropagation

- ✿ Equipment used to propagate plants and tips for success over a range of methods

- ✿ Techniques for different plant material and aftercare to promote successful rooting

The Fundamentals of Horticulture: Theory and Practice, ed. C. Bird. Published by Cambridge University Press. © The Royal Horticultural Society 2014.

BASIC PRINCIPLES

THE REASONS FOR PROPAGATING PLANTS BY VEGETATIVE MEANS

Many garden plants, including many fruit trees, are complex hybrids that do not come true from seed or may be infertile. Seeds from F^1 hybrid plants are also not true to the parent F^1 type. Even when plants produce seed that is true to type, they may not be self-fertile and so need a pollinator close by. In a small garden, there is often only room for one plant of a particular type and so seeds will not be produced or may not be fertile. There are many other reasons for using vegetative propagation. Juvenile plants do not flower and when the juvenile phase is of long duration (as in many trees) it may be desirable to propagate from adult plants. However, some plants do not root easily from cuttings especially where adult material is used and other techniques such as grafting may be necessary.

GRAFTING

Grafting is a specialised technique in which two different plants are joined together to produce a new plant with different characteristics. A graft is formed when a piece of shoot (the **scion**) is placed in contact with a root system for another plant (the **rootstock**) and the two pieces grow together as a single plant.

There are many reasons for grafting. For example, the use of adult scions (see juvenility below) bypasses the juvenile phase during which flowering and fruiting do not occur so that the grafted plant comes into flowering much more rapidly than one raised from seed. Grafting can also be used for cultivars that are difficult to propagate by other means. These include most fruit tree cultivars.

There are other special reasons for grafting. For ornamentals, it is possible to produce plants with enhanced decorative features such as the combination of *Prunus × subhirtella* 'Autumnalis', which has autumn flowers grafted onto *Prunus serrula* with coloured bark. Different fruit cultivars can be grafted onto one plant to produce a family tree to ensure pollination in small gardens where there is no room for several trees.

One of the most important reasons for grafting is the effect of the rootstock on the growth of the scion. For some plants, rootstocks are available that can tolerate unfavourable soil conditions. For example *Rhododendron* 'Cunningham's White' will tolerate a neutral soil and can be used as a rootstock for cultivars that require acid conditions. For some plants rootstocks can be used to obtain freedom from a disease. For example the cucumber, *Cucumis sativus*, can be grafted to the wilt-resistant species *Cucumis ficifolia*.

A further important reason for grafting is to reduce the size of the tree and rootstocks are available that cause the grafted tree to become partially or extremely dwarfed. Among these are the extremely dwarfing apple, *Malus domestica*, rootstocks M9, M27 and EMLA M9, the resultant trees being about one third the size of plants on the vigorous rootstocks. The dwarfing rootstocks result in early fruiting and the smaller trees have many advantages for orchard management as they are easier both to spray and to prune.

CASE STUDY — **Dwarf cherry rootstocks**

Over the last ten years great propagational developments have become available to the home gardener including a range of dwarfing rootstocks that enable tree fruits, which formerly have taken up a considerable area, to grow in a much more modest space.

One of the most welcome examples of these is Gisela 5, which enables fruiting cherries, *Prunus avium*, to be grown on a tree with a height of 2 metres (6 ft 6 inches). This means they can be planted in a wider range of gardens, including urban situations, and makes netting the fruit a practical possibility again.

Successful grafting depends initially on placing the cambium of stock and scion together, as this is where the new cells arise that will subsequently form the graft union. Ultimately, the success of the graft depends on the formation of a union that can transport organic compounds between rootstock and scion, since many organic compounds cannot cross by simple diffusion. A strong union between stock and scion is also necessary for stability between the two partners of the graft. Some practical methods of grafting are given below.

Incompatibility

Grafts are more likely to be successful when the two partners are closely related botanically. For example, successful grafts between different genera within the same family are rare, and grafts between plants in different families is usually considered to be impossible; whereas grafts between two cultivars within the same species are almost always successful. When two plants cannot produce a satisfactory union they are said to be *incompatible*. The graft union may fail completely or appear to be successful initially but fail later, sometimes after several years.

THE REGENERATION OF NEW ORGANS FROM PLANT PROPAGULES

When plants are produced from stem cuttings, they must regenerate roots. Root cuttings must be able to form shoots, leaf cuttings have to produce both shoots and roots, while budded or grafted plants must develop new tissue in order to form a union between the two grafted plants. Development into different types of cell is under genetic control and most plant cells contain all the genetic information that is needed to produce a new individual. They are, therefore, said to be **totipotent**.

Pre-formed root initials are present in the stems (usually at the leaf nodes) of some easily rooted plants and remain dormant until the cutting is made, when they emerge as **adventitious** roots.

Where there are no preformed initials, new roots may be induced to form by wounding.

Shoot and root meristems

Irrespective of the method used, propagation by vegetative means depends on the stimulation of meristematic activity (i.e. cell division) to form new organs. The most important meristems in the plant are those at the growing points of the root and shoot, and the cambium, which divides to form the vascular tissues. Except for micropropagation from shoot tips and embryos (see below), vegetative propagation depends on the development of new cell divisions at the cut surfaces. Cutting alone stimulates meristematic activity and the movement of hormones to the cut surfaces probably causes most of the increase in cell division.

In flowering plants (*Angiosperms*) the shoot apical meristems usually consist of an outer layer called the tunica (from one to several cell layers thick), which encloses an inner mass of cells called the corpus, shown in Figure 8.1. The cells of the tunica give rise to the outer layers of the stem and leaves, while cell divisions in the corpus give rise to the inner cells of the stem and leaves.

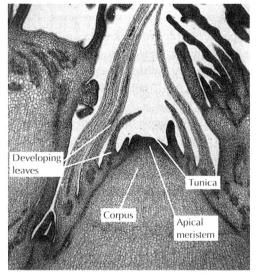

Figure 8.1 Diagrammatic section through a shoot apex showing layered structure.

Chimaeras

Several examples of variegation in leaves and flowers depend on the fact that the growing point of the stem is organised in layers. This type of variation (called a **chimaera**) occurs when one or more of the outer layers of one partner in a graft grow over the inner layers of the other, producing a plant consisting of cell layers of two different genetic types. A well-known chimaera of this type is +*Laburnocytisus* 'Adamii', which is the result of grafting purple broom, *Cytisus purpurea*, and laburnum, *Laburnum* spp.

Chimaeras can also result when a mutation occurs in the growing point of the shoot causing part or all of one layer of cells to be genetically different from the other layers in the same plant. When the mutation affects the production of chlorophyll, plants with attractively variegated leaves can be produced as in *Euonymus*

fortunei 'Emerald Gaiety', for example. It is clearly necessary to maintain the variegation by taking cuttings since such chimaeras do not breed true from seed. However, they cannot be propagated by taking root cuttings, which would disrupt the stable layered structure of the original growing point (see below).

In a number of cases, the mutation is not stable and reversion may lead to the development of some all-green shoots. These have more chlorophyll than the variegated ones, and therefore photosynthesise more rapidly. Consequently they grow more strongly and should be pruned out.

Figure 8.2 *Euonymus fortunei* 'Emerald Gaiety' – a chimaera in which the outer cell layers are unable to make chlorophyll and are colourless. The inner cell layers are able to produce chlorophyll and so are green.

The root apex

Unlike the shoot apex, the root tip (Figure 8.3) does not have a clearly defined layered structure. In the root the dividing cells of the meristem give rise to the cells that form the epidermis (outer cell layers), cortex (central tissues) and the transporting tissues (xylem and phloem cells). As noted above, this difference is important when considering the propagation of chimaeras that consist of differently mutated cell layers. In the thornless blackberry, *Rubus fruticosus* 'Oregon Thornless', for example, only the outer layer contains the gene for thornlessness and plants grown from root cuttings revert to the normal thorny genotype because the layered structure is disrupted and the new shoots arise only from the inner tissues.

CLONES

Many garden plants have arisen by selecting them for desirable characteristics, such as dwarfness or flower colour, and propagating them vegetatively. Such plants constitute a **clone** and all contain the same genes. Even so, they do not necessarily all have the same appearance, or **phenotype**, as this results from the interaction of a plant's genes (its **genotype**) with its environment, also see Chapter 10 for further details.

Although the life of a clone is theoretically unlimited since it can be rejuvenated by propagating it vegetatively, in practice

Auxins can be applied to the base of a cutting to promote root formation in subjects that are otherwise difficult to root. Indole-3-acetic acid is not used because it is readily converted to an inactive form in the plant, especially on exposure to air. The chemicals most commonly used are the synthetic auxin, NAA (*α naphthalene acetic acid*) and the natural auxin IBA (*indole butyric acid*). The latter is closely related to IAA and has been found in several plants. Indole butyric acid is the safer option because, like IAA, it is gradually broken down and so results in a more optimal concentration for root elongation once the root initials have developed. Synthetic auxins, such as NAA, are not readily degraded by the plant cells and care must be taken in their use. However, some plants are difficult to root even when rooting hormones are used, suggesting that other factors than auxins may be involved. In some case rooting potential has been found to depend on the time of year, as in some lilac cultivars, *Syringa vulgaris*, such as 'Mme. Lemoine', although the reasons for this are not understood.

CYTOKININS

Although they are not directly involved in root initiation, cytokinins play an indirect role in the successful regeneration of plants from various types of cutting. Cytokinins are present in coconut milk, which was found to increase cell division in plant tissues cultured on an artificial growth medium. The active ingredients were later identified as *zeatin* and a related compound, *zeatin riboside*. These and other similar compounds are known as cytokinins. Most are not available commercially with the exception of BA (*benzyladenine*, which occurs naturally but is not commonly found) and the synthetic cytokinin, kinetin. Cytokinins are not used in the propagation of cuttings because a high cytokinin to auxin ratio is inhibitory to root formation. They are used, however, in micropropagation to stimulate the development of shoots (see under 'Micropropagation' above).

As well as promoting cell division and shoot formation, cytokinins also play an important part in preventing senescence. For example, when leaves are removed from the plant, they rapidly age and become yellow. However, this is delayed if adventitious roots form on the leaf. This effect of the roots can partially be replaced by applying cytokinins, indicating that cytokinins arriving from the roots have an important function to delay leaf senescence. Therefore the successful regeneration of adventitious roots on leafy cuttings is important not only for water uptake but for the supply of essential hormones such as cytokinins and abscisic acid, see Chapter 2 for further information.

PRACTICAL TECHNIQUES

BASIC EQUIPMENT

Most important
Most types of vegetative propagation require a horticultural knife, a pair of secateurs, a small hand mister and a polythene bag.

To root the **propagule** a low-nutrient compost is used such as loam-based John Innes No. 1 or a multipurpose peat-free compost. Other options for successful propagation mixes include: 50/50 peat (or coir)/grit mixtures or the use of perlite or vermiculite, which are discussed in Chapter 6.

For techniques such as division, an old knife or a pair of forks are used depending on the root type. Labels written with a pencil will retain the information in any conditions.

Useful
A ceramic tile can be used safely as a cutting board for those less proficient at knife work, it is also quickly and easily sterilised with methylated spirits to prevent cross-contamination.

A rooting hormone compound is used on species that are difficult to root, or out of season, but it must be fresh. Specialist grafting and budding knives are useful but the technique can still be achieved with a sharpened horticultural knife. More specialist equipment is needed for grafting; this is discussed under that heading later in this chapter.

Propagation facilities
A range of facilities can be used to propagate plants, from the simplest nursery bed to sophisticated fogging systems. The crucial point about these facilities is that the propagule is protected and water stress is minimised. Also Chapter 4 discusses water stress and the plant's reaction to it in more detail.

Outdoor facilities
Nursery beds – It is possible to root cuttings, layer and grow on divisions in the open ground. The site must be sheltered, protected from vermin and ameliorated with washed grit to ensure the soil is free draining and therefore aerated.

Frames – Frames are the simplest way of protecting an area for propagation. Cuttings can be rooted slowly in this facility and protected from scorching sun or drying winds. Covers for frames are traditionally glass but these are heavy to handle and need to be

Figure 8.5(a) Cold frame.

Figure 8.5(b) Sun frame.

shaded in hot sun so the use of polycarbonate sheet offers the advantage of protection, diffused light transmission and ease of handling.

The terms cold frame and sun frame are used to describe the use and site of the frame. Cold frames are normally glass covered, north facing and used for slow-rooting cuttings, hardening off plants and stratifying seed for germination. Sun frames have a milky polythene cover supported by hoops and are used to root spring and summer cuttings.

Basal heat can be provided using soil heating cables to speed rooting of cuttings but this requires daily management of air flow to prevent fungal rotting in the enclosed frame. Porous pipe irrigation can also be used.

Facilities under glass or polythene tunnels

Closed case – Polythene-covered benches or beds with soil-warming cables to give a basal heat of 18 to 20°C (64.4 to 68.0°F) make an excellent rooting facility for cuttings. Milky polythene is best and reduces direct light to avoid scorching. Careful management is required to prevent fungal rotting. Cuttings should be aired regularly but never allowed to dry out.

Mist propagation – Mist units were hailed as the modern solution to problems of water stress in the rooting of cuttings in the mid 20th century. The continuous supply of a water film on the surface of the cutting serves to cool the leaf surface and increase humidity, which in turn reduces transpiration. This lack of water stress together with basal heat of 18 to 20°C (64.4 to 68°F) increases the success and speed of rooting for cuttings.

There are, however, issues of water management as too much water will encourage fungal rotting in these humid, warm conditions. Slow-to-root species may also suffer from nutrient leaching (Thompson, 2005) and require some nutrient misting to ensure success.

The frequency of misting can be controlled in several ways.

* *Electronic timer*: the best timing system is to use two timers, one turning the whole system off at night and on during the day, with the other operating the frequency during the day. The frequency is adjustable and is operated by the propagator according to season, cutting type and temperature.
* *Electronic leaf*: this control mechanism consists of a plastic surface containing two terminals placed among the cuttings. The surface, like the leaves, should be continuously wet and when it dries and no longer connects the circuit between two terminals, a valve is activated and applies the mist. There is a similar system that uses water weight on a stainless steel surface to regulate mist application.
* *Photoelectric control*: light transmission to the cutting surface will affect the transpiration rate by heating the leaf as discussed in Chapter 4. This system uses that principle to monitor light intensity and regulates the mist frequency. A photoelectric cell is placed above the cuttings, and the higher the light intensity the more frequently the mist is applied.
* *Thermostatic control*: using the same principle as the photoelectric cell, this system uses a thermostat placed among the cuttings to increase mist application as the temperature rises.

Maintenance of the mist unit is essential. An even supply of mist to the bench must be maintained. Mist nozzles are easily blocked by algal growth and limescale deposits and should be cleaned regularly. Sufficient water pressure to operate the nozzles is also crucial and a pump may be necessary to overcome intermittent pressure problems. The efficiency of soil-warming cables should also be checked as **heat banding** can be a problem and cause hot and cold spots on a bench.

The advantage of mist units is their efficiency in rooting cuttings, soft cuttings can be inserted earlier and cuttings rooted without a cover in maximum light conditions will acclimatise quickly to normal light levels for growth.

The disadvantages are predominantly the expense of setting the system up and specialised management required. It should also be noted that some species such as succulent plants or those with hairy foliage that require drier growing conditions, see Chapter 4, are not suitable for this system.

Figure 8.6 Cuttings on a mist bench.

Figure 8.7 Cuttings in a fogging system.

Fogging system – Fogging is a very efficient way of rooting cuttings, working on a similar principle to mist units but using less moisture to maintain 100% air humidity around the cutting rather than leave a film of moisture on the leaf surface. This means that an entire structure such as a glasshouse or polytunnel provides the fogging environment, unless an enclosed structure within a house can be used. Basal heat is provided to speed rooting.

Fogging systems vary in pressure supply, with low pressure being the easiest to install and high pressure being the most expensive but efficient to use. Portable fogging units are useful and can make the use of protected structures for propagation very flexible.

In a similar way to mist systems, fogging requires a reliable power and water supply, and a control method.

The fogging unit is controlled with the use of an **aspirated** humidity sensor or rod-type **humidistat**.

Fogging systems have the same advantages and disadvantages as mist units but are not suitable for species that require high light levels as the continuous fog reduces light transmission to the leaf surface.

There is also a health and safety issue with high humidity in an enclosed environment using recirculated water, with a risk of Legionnaire's disease.

CASE STUDY Legionnaire's disease

With the advent of more sophisticated propagation and water-recycling systems an increased risk of bacterial diseases is seen, for humans, if the water is heated over a threshold temperature. This can give rise to an acute pneumonia caused by species of bacterium belonging to the genus *Legionella*. Recommendations to prevent this are set out by the Health & Safety Executive (HSE) in Approved Code of Practice & Guidance: L8.

SUCCESSFUL PROPAGATION

There are some general aspects of propagation that must be considered to ensure success.

Choosing the right stock material

Plants used for the provision of material for propagation are known as **stock plants**. The success of cuttings or other vegetative propagation methods is dependent on the health and quality of the stock plants.

Stock plants should be healthy, vigorously growing, correctly named and true to type.

Young vigorous juvenile growth, discussed above, is the best source of material and is promoted by **hedging** or **stooling** the stock plant in order to maximise the number of new lateral shoots for making cuttings.

Cuttings are best collected from laterals in the middle area of the plant where auxin concentrations in the shoot tips are at their highest (Dirr & Heuser, 1987). Lateral shoots should be chosen because material from the top of the plant tends to produce leggy plants, while material from the base may be slow to root and reluctant to make a **leader**. Growth on the side of the stock plant that gets the most light will also contain the most reserves of carbohydrates to provide energy for rooting.

Clump-forming stock plants will provide the best juvenile growth for propagation by division on the outer edges of the plants so should be grown with maximum space for development and access.

All stock plants should be fed with a compound fertiliser such as blood, fish and bone in the spring, watered, and checked regularly for pest and disease infestation.

Minimising water stress

Water stress is a major factor at all stages of propagation.

When collecting material for propagation, shoots should be taken during the coolest periods of the day and covered or put immediately into a polythene bag. If material is to be kept for some days before being made into cuttings the temperature must be lowered to reduce the transpiration rate, see Chapter 4 for reasons why leaf temperature affects the rate of water loss. The cold box of a refrigerator or a cold store is ideal with material moist and kept covered in polythene to avoid desiccation. During the activity of making the cuttings a hand mister can be used to keep the material turgid and cool.

Large-leaved plants can have the leaves on the cuttings trimmed by 50% to reduce the surface area and potential for water loss.

The propagation facility (see previous section on facilities) must provide humid conditions for rooting and growth.

Once rooted, water management is crucial to success as the new plant must be **weaned** from this generous water regime and acclimatised to the erratic water supply of a natural growing environment.

Hygiene

Hygiene at all stages of the propagation process is important to maximise success. Tools should be kept clean and working surfaces free of dust contaminants. The basal cut surface of a cutting must be clean.

Pests and diseases are less likely to be a problem where hygiene is given priority. See Chapter 17 for a full account of plant pests and diseases. Fungal diseases such as *Botrytis cinerea* (sexual stage, *Botryotinia fuckeliana,*) and soil-borne *Pythium* and *Rhizoctonia*

spp. are minimised by using clean containers for striking cuttings, sterilised compost and the regular cleaning of fixtures and fittings in the protected structure.

Sciarid flies favour the warm, humid conditions of a propagation unit. The larvae will feed on the young developing roots of the cuttings. They can be controlled using a biological control of either predatory mites or a nematode. Further detail is given in Chapter 17.

Algal growth must be removed and glass structures should be regularly cleaned to maximise light transmission.

During the rooting process, any fallen leaves or failed cuttings must be removed daily to prevent rotting.

Rooting media

The medium into which a propagule is put to make root or develop shoot growth must be chosen carefully. The range of composts is covered in more detail in Chapter 6.

In an outdoor nursery bed, soil will be the main constituent. An open, well-drained soil is best, so soils dominated by clay, for example, can be improved with the addition of washed grit and organic matter. The soil used to layer plants *in situ* should also be treated in this way.

All other facilities will use horticultural compost. Some specialist plants such as succulents will need mineral-rich gritty compost but most are best rooted in compost where the bulk ingredient is organic.

Low-nutrient compost is generally recommended but the speed of rooting will affect this. The longer a cutting takes to root, the more it needs nutrients to maintain the cell tissue during the rooting process. Cuttings rooted straight into a modular container will also be better in compost with nutrient as it will remain in this container during the stage of formative shoot growth.

Whichever compost material is chosen, a good air to water relationship is essential (Dirr & Heuser, 1987) with an air-filled porosity (AFP) of 25 to 40%. The stability of the cutting once it is struck and firmed into the compost is also a consideration when choosing bulk ingredients for propagation composts.

MAKING CUTTINGS

The following are general tips for making, inserting and rooting cuttings of all the types described in this section.

- ✿ Cutting material is removed from the stock plant using secateurs and should be sturdy non-flowering shoots of the current season's growth.
- ✿ Keep knives and secateurs sharp so that clean cuts are made with no **snags**, which are vulnerable to die-back and rots.
- ✿ The lower third of leafy shoots are trimmed so that the cutting can be inserted into compost without risk of soft leaf material rotting. When trimming the lower leaves from the base of the cutting, turn the cutting upside down in your hand and support the leaf stalk with your thumb under it before cutting through it. This will ensure a clean cut with no tearing into the main stem.

Figure 8.8 Cutting prepared in the hand.

- ✿ The size of cutting is variable depending on the species being propagated but it should be physically and visually balanced.
- ✿ The longer the cutting takes to root, the more leaves are needed to sustain the cell tissue while it roots.
- ✿ The efficiency of a rooting hormone compound is often in question, as discussed above. The principle is to make root-promoting hormone available at the base of the cutting quickly to aid rooting. However, the chemical is unstable and goes off quickly so should be as fresh as possible.
- ✿ Cuttings made from difficult-to-root species using mature wood are **wounded** to improve rooting. MacDonald (2006) explains that this is where the base of the cutting is gently wounded with a knife to reveal more cambial tissue either taking a 1 to 2 cm (0.5 to 1 inch) long slither of bark or making a slit up 1 to 2 cm (0.5 to 1 inch) starting from the base of the cutting. The resulting damage stimulates rapid growth of callus material followed by rooting. The exposed tissue is treated with a rooting hormone compound at the time of preparation.

Figure 8.9 Cutting prepared with rooting hormone compound at the time of preparation.

- ✿ Cuttings must be inserted with enough space between them to ensure that wet leaves do not touch each other. It is useful to grade cuttings into sizes and insert each grade together with spacing judged for each group. Apart from being practical, this uniformity also helps the propagator when making visual inspections of the cuttings during the rooting process.

The broad categories of cuttings made from stems relate to the maturity of the plant material used.

Softwood cuttings come from the soft early season growth. These are shoots with tips and are fast and easy to root. Many deciduous shrubs and herbaceous plants can be propagated in this way. Water stress is a major factor for this soft material, which makes mist and fogging the best facilities to root softwood cuttings. Success can also be achieved with the careful management of a closed case facility or sun frame, or even in a simple polythene bag. Basal heat should be 18 to 20°C (64.4 to 68.0°F), with the air temperature 2 to 3°C lower. The higher basal temperature speeds up cell division as the roots develop and the lower air temperature slows down the transpiration rate from the leaf and discourages fungal disease.

A rooting hormone compound of 0.2% IBA can be used to aid rooting, although softwood cuttings are the easiest group and will root reliably without it.

Abelia	Fuchsia	Perovskia
Abutilon	Halesia	Potentilla
Acer	Hebe	Prunus
Amelanchier	Helenium	Rosmarinus
Anchusa	Humulus	Ruta
Aster	Kolkwitzia	Salvia
Callistemon	Lamium	Santolina
Campanula	Lavandula	Sarcococca
Caryopteris	Lavatera	Scrophularia
Cistus	Ligustrum	Sedum
Convolvulus	Linaria	Skimmia
Cornus	Lippia	Spiraea
Corokia	Lonicera	Syringa
Cotoneaster	Lupinus	Trachelospermum
Deutzia	Magnolia	Viburnum
Erica	Nepeta	Vinca
Fabiana	Penstemon	Weigela

Table 8.1 Examples of plant genera suitable for propagation by softwood cuttings.

Semi-ripe cuttings are made in summer from wood that is soft at the tip but ripening in the stem. Semi-ripe material is found from late June onwards. These cuttings are made with tips, heels, mallets and leaf buds, described later in this section. As the cuttings consist of hardened wood and leaves with well-developed cuticles, water stress is less immediate so the rooting facility used can be slower with the use of a closed case or frame. Mist and fog are also very successful for this type of cutting. The rooting hormone IBA at 0.5% is used to aid rooting.

Actinidia	*Euonymus*	*Passiflora*
Akebia	*Fatshedera*	*Pernettya*
Ballotta	*Forsythia*	*Philodendron*
Brachyglottis	*Genista*	*Phlomis*
Bupleurum	*Grevillea*	*Pieris*
Buxus	*Griselinia*	*Pittosporum*
Camellia	*Hebe*	*Pyracantha*
Ceanothus	*Hoheria*	*Rhododendron*
Chaenomeles	*Hydrangea*	*Rhus*
Chamaecyparis	*Hypericum*	*Ribes*
Clethra	*Ilex*	*Sambucus*
Coronilla	*Indigofera*	*Sarcococca*
Cotoneaster	*Kerria*	*Skimmia*
Cytisus	*Myrtus*	*Symphoricarpos*
Daphne	*Nandina*	*Teucrium*
Elaeagnus	*Osmanthus*	*Ulex*
Escallonia	*Osmarea*	*Viburnum*

Table 8.2 Examples of plant genera suitable for propagation by semi-ripe cuttings.

Hardwood cuttings are made from wood hardened for winter frost protection. These are taken from the current season's wood from leaf fall onwards. Deciduous subjects are leafless. Cuttings can be nodal tip, straight or with a heel at the base. These are described later in this section.

Hardwood cuttings are the slowest to root as there is an element of dormancy in the shoots taken. They can be rooted in frames, outdoor nursery beds or closed cases. Water management is crucial to avoid rotting of this woody plant material. Basal heat can be used to accelerate the callusing of the cut surface, a stage that must take place for rooting to occur. With warmth, cuttings can be rooted and potted off by the following spring but those in outdoor nursery beds will be spaced or lined out to grow on outside in the following growing season as in leaf they cannot be lifted until the following autumn. Cuttings inserted in containers and placed in a frame can solve this problem. The rooting hormone compound, IBA at 0.8% aids rooting.

Examples of plants propagated by hardwood cuttings are shown in Table 8.3.

Acer	*Garrya*	*Populus*
Arbutus	*Grevillea*	*Potentilla*
Berberis	*Griselinia*	*Prunus*
Buddleja	*Ilex*	*Rosa*
Buxus	*Itea*	*Rubus*
Cornus	*Kalmia*	*Salix*
Cotinus	*Laurus*	*Sambucus*
Cotoneaster	*Morus*	*Spiraea*
Elaeagnus	*Philadelphus*	*Stephanandra*
Escallonia	*Photinia*	*Tamarix*
Euonymus	*Physocarpus*	*Viburnum*
Fremontedendron	*Platanus*	*Weigela*

Table 8.3 Examples of plant genera suitable for propagation by hardwood cuttings.

STEM CUTTINGS

Basal cuttings

These softwood cuttings are made from the newly emerged spring shoots found at the base of herbaceous perennials or young biennial shrubs, such as *Euphorbia*. These shoots root easily, occasionally have roots emerging already and are known as 'Irishman's cuttings' (Bird, 1993) They are an excellent way to bulk up herbaceous plants quickly. They can be grown on and planted into the border later that season.

Stock plants can be manipulated to produce early shoots by lifting and potting in February and then brought into a cool greenhouse for forcing. Alternatively take cuttings from plants grown outside when the shoots are about 5 cm (2 inches) long.

Cuttings are removed from the stock plant and cut close to the point where they emerge and must have some firm tissue at the base. The lower leaves are trimmed and the upper leaf surfaces can be trimmed if they are particularly big, but care must be taken to leave the growing tip in place.

Dip the base of the cuttings in rooting hormone compound 0.2% IBA for softwood material.

Large cuttings such as *Dahlia, Delphinium* and *Euphorbia* (Thompson, 2005) are inserted into individual 7 cm (3 inch) pots, which have the depth to make them stable. Small cuttings such as *Helenium, Phlox* and *Penstemon* are struck into trays, modules or several to a pot.

The cuttings must be watered thoroughly and placed in a closed case with basal heat of 18°C (64.4°F), mist bench or frame. Those placed in a frame are less prone to fungal problems but take longer to root.

Rooting will take place in three to six weeks depending on the species. Rooted cuttings must have the tip pinched out to encourage new basal flowering shoots and be potted off into larger pots. They will rapidly grow into plants strong enough to be planted out later in the same growing season.

Nodal tip cuttings

Nodal tips are the most commonly made cuttings and can be made from softwood, semi-ripe or hardwood material.

Cutting material is removed from the stock plant and should be sturdy shoots of the current season's growth.

Figure 8.10 Nodal tip cutting: leaf surface reduced by trimming.

The cutting has a node or bud at the base of the cutting, the basal leaves are trimmed and the tip is left intact. A good rule of thumb is to leave at least two sets of leaves at the top of the cutting. Softwood and semi-ripe cuttings can have the leaf surface reduced by trimming if appropriate.

Treat the base of the cutting with an appropriate rooting hormone compound for the maturity of the stem used (see above).

When these cuttings are rooting there is a danger that flowering shoots are generated. These flower shoots must be pinched out when they appear as the energy used by the cutting to support the new flower shoot will jeopardise rooting.

When rooted, the tips are pinched out or cuttings trimmed to encourage lateral shoots to develop and form a bushy plant. Cuttings are then potted off and weaned from their propagation environment.

Heel cuttings

Heel cuttings are made from semi-ripe and hardwood plant material.

The material used for these cuttings needs to be short laterals of the current season's growth, which are torn from the main stem to leave a small slither of mature wood at the base of the cutting. This rough slither must be trimmed and the lower leaves removed to minimise fungal rotting once the cutting is inserted into the compost. A rooting hormone compound is applied to the trimmed heel if required.

Cuttings taken as heels are more successful than nodal tips made from the same batch of material (Thompson, 2005).

Heel cuttings are typically made from species such as *Thuja, Chamaecyparis, Buxus, Osmarea* and *Ilex.*

Rooted cuttings of conifers should not be trimmed at the tips as the apical dominance exhibited by these plants creates the desired shape. All other evergreens can be trimmed to encourage laterals to make a bushy plant, unless required to grow for use as standards or topiary.

Mallet cuttings

Mallet cuttings are made using semi-ripe and hardwood plant material.

The prepared cutting looks like a mallet with the youngest, current season's shoot being the long handle and the short 4 cm (1.5 inch) piece of second-year wood making the mallet. It could also be described as an 'L' shape. This type of cutting is prepared using secateurs.

The cutting will vary in size according to the species being propagated, but as guide one would expect a mallet cutting to be 10 cm (4 inches) long. The soft tip is removed to avoid fungal rotting.

The mallet end is treated with rooting hormone compound and inserted into the compost.

Mallet cuttings are typically made for species where the current season's growth is particularly spindly, with the second-year wood at the base providing starch (carbohydrate) reserves to support the cutting during the long rooting process. Examples include *Berberis, Syringa* and *Vitis.*

Straight or hardwood cuttings

Straight cuttings are made from deciduous hardwood plant material. They are made during the dormant period and so are leafless, and use one-year-old wood from the previous growing season.

The cutting is prepared using secateurs and must have a bud at the top of the cutting and one at the base. Any soft material at the tip is removed. The number of buds between depends on the **internode** length; cuttings of *Buddleja* would have just three buds in total whereas *Spiraea*, for example, may have seven or more.

These cuttings are 15 to 20 cm (6 to 8 inches) long and as a guide should be of pencil thickness in order to have enough stored starch (carbohydrates) to support the rooting process.

A rooting hormone compound is used at the base of the cutting, and they are inserted to at least two thirds of their length or they desiccate easily. Where propagation space is at a premium, these cuttings can be inserted as bundles or **Dutch rolls** and separated once rooting has taken place. The newly rooted cuttings require cutting back to a strong bud to encourage lateral shoot development.

Plant examples for this method include *Buddleja*, *Weigela*, *Ribes*, *Cornus* and *Spiraea*.

Leaf-bud cutting

Leaf-bud cuttings are made from semi-ripe and hardwood plant material.

The cutting is prepared using secateurs or a knife depending on the maturity of the plant material. A shoot of matured growth from the current season is selected and small sections are prepared consisting of a short piece of stem with a leaf and axillary bud. There is no need to have a bud at the base of the stem material, but the axillary bud must be intact. This method greatly maximises the number of propagules that can be made from the stock material. Leaf-bud cuttings are 3 to 5 cm (1.5 to 2 inches) long depending on the material used. Species such as *Mahonia* with large compound leaves have the leaves cut back by 50% or to 10 cm (4 inches) long.

Rooting hormone compound is applied to the stem base only and the cutting inserted, with the bud lightly covered, into the compost.

The newly rooted cutting will shoot from the axillary bud at which stage the deteriorated leaf from the original cutting is removed.

Plant examples propagated by leaf-bud cuttings include *Mahonia*, *Camellia*, *Rhododendron (Azalea)*, *Clematis* and *Hedera*.

LEAF CUTTINGS

Some plants have the ability to generate both roots and shoots from the leaf. Plants that do this tend to be clump- or rosette-forming plants that do not readily produce stem material. All leaf cuttings are softwood plant material.

Leaf petiole cuttings

These are simple cuttings prepared with a knife, leaving the entire leaf attached to the leaf stalk. The leaf stalk or petiole is trimmed to 3 to 4 cm (1 to 1.5 inches) for stability.

A rooting hormone compound is not used on the base of the leaf petiole, and the cutting is inserted into the compost with the leaf lamina left above the surface (McMillan Browse, 1992). The base of the cutting will first produce roots within three to four weeks followed by new shoots emerging after another three to four weeks. At this stage the original leaf can be removed and the new plant potted off when it is large enough to handle.

The rooting facility must be chosen carefully, as a mist or warm closed-case system can cause rotting of hairy-leaved or succulent specimens, many of which are propagated in this way. Basal heat is required for these tender plants. Smooth-leaved plant examples such as *Peperomia* can be put under mist but others such as *Saintpaulia* need careful management of humidity in a closed case to avoid fungal rotting.

Heel cutting   Straight cutting   Mallet cutting

Figure 8.11 Drawing of heel, straight and mallet cuttings.

Examples of plants propagated by leaf petiole cuttings include *Peperomia, Saintpaulia, Begonia, Echeveria* and *Aeonium*.

Leaf lamina cuttings

These cuttings are made from sections of the leaf blade or lamina, and then roots are formed from the tissue around the damaged leaf veins. The same sequence of growth as for leaf petioles takes place, with roots formed within a few weeks followed by shoots forming the new plant some weeks later.

The simplest cuttings are 4 to 5 cm (1 to 1.5 inch) sections of the leaf lamina cut across the midrib. Orientation is important, so the correct way up is indicated by the leaf veins making an arrowhead, which should point downwards, showing the lower cut surface and this is inserted into the compost. Where the main veins are cut, rooting will take place followed by the generation of a new plant.

Alternatively the whole leaf is cut longitudinally down the length of the midrib, removing the midrib and inserting the long cut surface into the compost. Where the main veins running into the midrib have been damaged, roots are formed followed by new plants.

Entire leaves can be used by slashing the main veins in several places with a knife on the underside of the leaf. The whole leaf is then pinned down to the compost surface. Roots, then young plants are formed at the damaged points. Where propagation material is scarce, the leaf can be cut with a knife into small 2 to 3 cm (1 to 1.5 inch) squares. Each square 'postage stamp' is inserted vertically by two thirds into the compost and will produce roots; another reason for using this method is when the leaf is naturally undulating. If the leaf is variegated, all the colours should be present on each propagule. However, in the case of *Sansevieria trifasciata* 'Laurentii' most of the resulting plants will be green, so division of the whole plant would guarantee variegated leaves.

The propagation facility has to be carefully managed. These cuttings require basal heat and high humidity as they are particularly vulnerable to drying out. However, many suitable species are hairy leaved and succulent so careful management of air and humidity in the closed case is required. Plants propagated in this way include *Sansevieria, Begonia, Streptocarpus* and *Lachenalia*.

Figure 8.12 Leaf lamina cuttings: (a) sections, (b) rooted lateral vein, (c) slashed and pinned, (d) rooted squares.

ROOT CUTTINGS

Root cuttings are used to propagate a range of herbaceous perennials and woody shrubs. Many of our garden weeds such as field bindweed, *Convolvulus arvensis*, and dandelion, *Taraxacum officinale*, naturally exhibit the capability to grow back from small sections of root left in the ground. Making root cuttings of ornamental plants uses the plant's natural ability to do exactly the same as those garden weeds.

Some plants such as *Dahlia* develop root tubers as storage organs to ensure survival over the dormant period. These can be lifted and separated from the main plant but must be grown on as whole tubers in the spring.

Cuttings are taken at the end of the winter and in early spring (Bird, 1993). The plant is lifted or removed from the pot. The biggest, fleshy roots are cut away from the crown and the stock plant repotted or planted back into the ground. The root material is cut into sections of the appropriate size for the specimen. *Acanthus* is 5 cm (2 inches) long and 1 cm (0.5 inch) in diameter, *Primula* 2.5 cm (1 inch) long and 0.2 cm (1/8 inch) in diameter. Fibrous roots must be removed and the tip of the cutting notched or cut diagonally to guide the propagator when inserting the cutting to orientate it the correct way.

Cuttings are inserted so that the top is just below the surface. Some cuttings are laid horizontally on the surface and covered with compost. Rooting hormone compound is not needed for these cuttings as it is the shoot that is needed for success followed by more root development.

A cold frame can be used for this method of propagation but some basal heat will speed up the process of shooting in the spring.

This method of propagation is fast with new plants ready to be potted off in early summer.

Examples of plants propagated in this way include: *Primula, Acanthus, Rhus, Papaver, Anchusa, Aralia, Romneya, Catalpa* and *Campsis*.

LAYERING

Layering is a propagation technique that requires the propagule to remain attached to the stock plant for the duration of root development. Layering uses the plant's natural ability to produce root when a stem comes into contact with the surface of the soil.

The success rate for layering is very high because the new plant is being supported by the mother plant during the whole process: therefore little aftercare is required from the propagator.

It is a useful technique for the garden but is no longer relied on commercially as it requires a lot of space and stock material for relatively few new plants.

The timing is also less crucial than when taking cuttings, but better results are experienced when the layers are put down in spring or autumn as conditions are mild. Shoots to be layered need to be young and vigorous, which may present a problem on older plants with few new shoots at the base. To overcome this, low lateral shoots can be cut hard back during the dormant season, with the resulting new growth used for layering the following season.

The soil where the layer is to be pinned down should be improved with a mixture of grit and organic compost; the new roots will need well-aerated and drained soil for successful development.

There are five main layering techniques as described below.

Simple layering

Take a low shoot and pull it down to the soil surface. The point at which the shoot comes into contact with the soil should be of the current year's growth and not too soft.

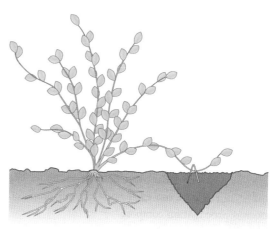

simple layering: shoot pegged down into prepared ground

Figure 8.13 Line drawing of a simple layer.

Examples of plants propagated using this technique are: *Magnolia, Cotinus, Viburnum, Rhododendron* and *Cotoneaster* (McMillan Browse, 1992).

Remove the leaves from the stem where it will be pinned beneath the soil surface. Using a knife, take a slither of stem from just below a bud or leaf node. Apply a rooting hormone compound to the wound. Use a wire staple or peg to pin the shoot down into the newly prepared soil. Alternatively, twist the stem at the point you require rooting, this method prevents an open wound being made, thereby reducing the incidence of infection. Soak the ground around the new layer to bed it in.

This will take a growing season to root, and once growth is visible the layer can be removed from the parent plant and potted or planted out into the garden.

Tip layering can also be done using a similar technique for species such as *Rubus*, which have the natural ability to produce roots when the tips come into contact with the ground.

French layering

French layering is a technique that was developed to maximise the number of plants from one stock plant using a reliable technique difficult to propagate in other ways. Plants like tomato, *Lycopersicon esculentum*, and cucumber, *Cucumis sativus*, with soft tissue can also be grafted. Shoots can also be grafted at the base of the stem onto root tissue; this is known as root grafting.

The two portions of plant are known as the rootstock, which is normally a one- or two-year-old seedling with roots, and the scion, which looks like a cutting, is of the current season's growth and is from the plant to be propagated. The cuts are made just under the surface of the bark so that the two pieces fit together securely but they also need to be tied and sealed. The resulting propagule has the appearance of a young plant wearing a plaster.

This requires extra equipment: grafting knife, grafting ties and heated wax with a brush.

Factors involved in achieving a successful graft union are as follows.

- ✿ The stock and scion must be genetically compatible, disease free and both at the right stage of growth.
- ✿ Knives must be sharp and clean to keep cut surfaces free of dust and infection.
- ✿ Cut surfaces must be physically matched to achieve cambial contact and secured with a tie. The tie either degrades or is easily removed.
- ✿ The union must be sealed to prevent the tissues drying out, wax is normally used.
- ✿ Aftercare must be conducive to callus formation, the correct temperature maintained and monitored for fungal disease.

Field grafting
Field grafting takes place in the spring, when the cambial tissue is just active. Fruit trees and ornamental trees are typically propagated in this way.

Rootstocks are lined out in the field or potted but kept uncovered in frames outside. Scion material is collected on the day of grafting and kept cool and moist.

The graft union is made, secured with a polythene tie and sealed with wax. The actual cuts made depend on the species being grafted, see examples later in this section.

Bench grafting
Many species are bench grafted; this can be done over the dormant season and into the spring with rootstocks being brought inside to initiate cambial activity prior to grafting. Bench grafting allows the propagator to manipulate timing and have complete control over the aftercare of the propagules.

The graft union is made, tied with rubber ties, waxed string or raffia and sealed with wax.

Once grafted the plants are kept warm and moist while the union seals. Commercial growers have developed a facility called 'hot pipe callusing' where the grafts are leant against hot water pipes to increase the temperature immediately around the union. This has greatly increased the success rates for bench grafting for many growers.

GRAFTING TECHNIQUES

There are many specialised grafting techniques for a range of purposes; the following four techniques are the most commonly used.

The whip or splice graft
This technique is used mainly for deciduous trees and shrubs, the stock and scion must be the same diameter. Some species are root grafted with a splice graft.

Examples of plants grafted using this technique: *Acer, Aralia, Fraxinus, Hamamelis, Magnolia, Clematis, Syringa, Platanus, Malus* and *Prunus*.

The stock is cut back to the point at which the graft is to be made and a shallow 4 cm (1.5 inch) sloping cut made to the top of the rootstock. An identical cut is made at the base of the scion; the cut should end just below a bud. The scion is trimmed at the top.

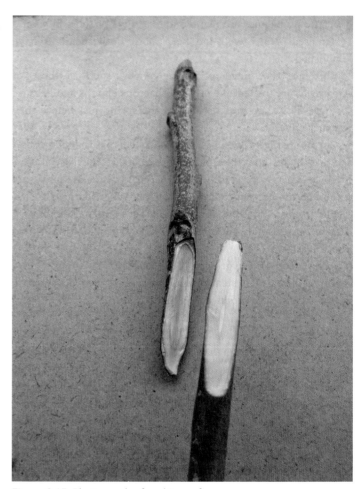

Figure 8.15 Photograph of a whip graft.

Whip and tongue graft
This technique is mainly used for field grafting, where the tongue exposes more surface area of cambial tissue and strengthens the union. Stocks and scions of different diameters can be grafted in this way.

Examples of plants grafted using this technique: *Aesculus, Crataegus, Larix, Prunus, Malus, Sorbus, Tilia* and *Ulmus*.

The stock is cut back to the point at which the graft will be made, a 4 cm (1.5 inch) sloping cut is made at the top of the rootstock. An identical sloping cut is made at the base of the scion wood with a bud behind it. A single slice is then made into the rootstock, downwards at the opposite angle to the exposed cut surface. An identical cut is made upwards to the base of the scion. The two cuts will then interlock when the surfaces are joined.

If the scion is smaller than the rootstock, slide the scion across to one side in order to ensure the cambiums match on one side (Garner, 1988). If it is put centrally the graft will fail. The graft is tied and waxed.

Apical wedge graft

This technique is one of the easiest and is used to bench graft shoots low down at root level. *Wisteria, Gypsophila, Daphne, Hibiscus* and *Syringa* are grafted this way.

The rootstock is trimmed with a horizontal cut to the root tissue. A single cut is made vertically 3 cm (1 inch) down into the centre of the rootstock.

The scion is prepared by making two 4 cm (1.5 inches) cuts at the base to form a wedge shape. The scion wedge is inserted into the cut in the centre of the rootstock, offset to match the cambial tissue if they are not the same diameter. The exposed tissue of the scion at the top of the graft will promote extra callus material and strengthen the union.

The graft is tied and waxed.

Side veneer graft

This technique is used for bench grafting deciduous trees and shrubs, and conifers. The cuts are modified for some specimens where the rootstock is left intact as a sap drawer. The rootstocks are often bigger than the scion material.

Rootstocks are potted and brought into the glasshouse four to six weeks before grafting (MacDonald, 2006) with a temperature of 10 to 12°C (50 to 53.6°F) to encourage sap flow. They are kept dry so that the union will not be flooded.

The rootstock is cut at the point of grafting, 3 to 4 cm (1 to 1.5 inches) from soil level with a short sloping cut made into the rootstock. A long slither of wood is then removed from 3 cm (1.5 inches) above and joining the first cut. If the rootstock is headed back, these cuts are made on a 4 cm (1.5 inch) stump, where the rootstock is left intact; the cuts are made on the side of the stock where any foliage must have been cleared.

Two cuts are made on the base of the scion to match the rootstock, a short one at the base and a sloping shallow cut made on the opposite side.

The two cuts are matched, the union tied and waxed before being put into a warm moist propagation facility. Hot pipe callusing is ideal for this technique of grafting.

After eight to ten weeks, the scion begins to shoot at which point the rootstock if left intact is headed back in two stages at six-week intervals until the scion growth is strong enough to take over.

Examples of side grafted species include: *Acer palmatum, Cedrus* spp., *Betula* spp. and *Juniperus* spp.

BUDDING

The technique of budding works on the same principles as grafting, with the difference being in the size of the scion material used. A small slither of wood supporting a bud is cut from the scion material and inserted behind the lifted bark of the rootstock.

A specialist budding knife can be used, which has a small wedge at the top of the blade for handling the bud and lifting the bark of the rootstock.

Budding is a very economical technique as it is fast and uses less scion wood. Budding is used on deciduous trees and shrubs, and particularly roses. This technique is carried out in the summer when there is a high sap flow.

The bud selection on the scion wood is important. The softest tips of the scion wood are discarded and buds used are taken from close to the maturing wood of the current season's growth.

There are two main techniques: chip budding, and T budding or shield budding.

Chip budding

This is favoured by many for mainly fruit trees and some ornamentals, and uses a small slither of scion wood with a bud, applied to a matching cut surface on the rootstock at 20 to 30 cm (8 to 12 inches) above soil level. Both cuts have a lip at the base to support the union. This is secured with a tie or rubber patch and growth is expected from the bud during the growing season. The rootstock is cut back in the autumn and the new scion growth supported with a cane or bud clip to achieve a straight-stemmed specimen.

Figure 8.16 Photograph of chip budding.

T budding or shield budding

This technique is used mainly for roses and relies on the bark being easily lifted, a phenomenon known as the 'slip factor'.

A shallow T cut is made in the rootstock close to soil level on the **hypocotyl**, which facilitates the gentle lifting of the bark so that the bud can be slipped in behind it to make contact with the cambial layers beneath but be supported by the bark closing back to it. The bud and trimmed petiole is left exposed in the centre. This is then tied and growth from the bud will occur during the growing season. The rootstock is headed back the following winter.

Summary

This chapter looks at both the theory and practice of vegetative propagation. It is important that an understanding of the physiological processes described early in the chapter informs the horticultural practice outlined later in the chapter; each is complementary to the other, and repeated successful plant propagation the result of developing a balance of both. It is possible to learn how to carry out a propagation method but the opportunities to vary the propagation material, manipulate the rooting facilities and maximise the chances of success come with a deeper understanding of the whole process.

The regeneration of new organs and the organisation of root and shoot apical meristems underpins the practical choices. In addition the role of plant hormones in cell differentiation and their use in micropropagation are outlined. Also, as a result of plant hormone activity, etiolation increases the ability of plants to form roots.

Many plants can be propagated in a wide range of ways and the final choice of method will depend on factors such as: the time of year; the time available; the numbers of propagules required; the delay in use or sale of the resulting plants; the skill level available; and the volume, age and quality of stock material.

The critical factors for any method are choosing well-grown, healthy, juvenile plant material for propagation, getting the timing right, minimising water stress and keeping the propagation facility disease free.

Sometimes pretreatments are required, especially for more mature plants where juvenile material is just not available. Therefore, hard pruning during the previous growing season is required to simulate this quality and volume of material.

The care taken during material collection including clear labelling (vital when dealing with a range of cultivars of the same species) can not be understated, together with the often missed area of mother plant care after the material has been taken.

Do not overlook the importance of aftercare of the newly rooted plants, as this can be the most stressful time, having used up their available carbohydrates in the act of rooting and subsequently not yet able to amass any new reserves.

Review questions

1. Explain the role played by growth hormones in the **successful** rooting of a cutting.
2. Describe how water loss can be **minimised** in the process of preparing and rooting a cutting.
3. Discuss the **benefits** of grafting plants.
4. Explain why the methods of division and layering have been overtaken **commercially** by micropropagation.
5. Describe the cell activity that gives a **named** plant the ability to propagate from leaf cuttings.

REFERENCES

Bird, R. (1993). *The Propagation of Hardy Perennials*. London: Batsford Ltd.

Dirr, M. A. & Heuser, C. W. (1987). *The Reference Manual of Woody Plant Propagation from Seed to Tissue Culture*. Georgia: Varsity Press.

Garner, R. J. (1988). *The Grafters Handbook*. Cassel.

MacDonald, B. (2006). *Practical Woody Plant Propagation for Nursery Growers*. Timber Press.

McMillan Browse, P. (1992). *Plant Propagation*. Mitchell Beasley.

Skoog, F., Strong, F. M. & Miller, C. O. (1965). Cytokinins. *Science*, **148**, 532–3.

Thompson, P. (2005). *Creative Propagation*, 2nd edn. Cambridge: Timber Press.

FURTHER READING

Gardiner, J. (1997). *Propagation from Cuttings. Wisley Handbook*. Wisley: Royal Horticultural Society.
Excellent, well-illustrated round-up of propagational methods.

Hartmann, H., Koster, D. E., Davies Jr. F.T. & Geneve, T.T. (eds.) **(1997)**. *Plant Propagation. Principles and Practices*, 6th edn. Prentice Hall International Inc.
A more in-depth technical publication covering the methods and the reasons behind their development.

Ingram, D. S. Vince-Prue, D. & Gregory, P. J. (2008) *Science and the Garden*, 2nd edn. Blackwell Publishing.
This publication provides detailed information related to the science behind propagational methods and how the plant can be manipulated.

Skoog, F. & Armstrong, D. J. (1979). Cytokinins. *Ann Rev Plant Physiol*, **21**, 359–84.
An in-depth scientific review of these plant hormones and their use in micropropagation.

Toogood, A. (1999). *Propagating Plants.* London: Dorling Kindersley.
An excellent book dealing with the practical aspects of the subject in a step-by-step way.

RHS level	Section heading	Page no.
2 1	Making cuttings	171
2 1.5	Making cuttings	171
2 1.6	Making cuttings	171
2 1.8	Making cuttings	171
2 2.1	Juvenility	166
2 3.2	Shoot and root meristems	165
2 3.3	Shoot and root meristems	165
2 4	Making cuttings	171
2 5	Making cuttings	171
2 5.1 to 2 5.11	Successful propagation	170
2 6	Making cuttings	171
2 7	Making cuttings	171
2 8	Making cuttings	171
2 9.4	The role of plant hormones	167
3 2.1	Successful propagation	170
	Hygiene	171
3 2.2	Successful propagation	170
3 2.3	Successful propagation	170
3 3.2	Practical grafting	179
3 3.1	Basic equipment	168
3 7.1	Micropropagation	167

CASE STUDY   The lectin gene in snowdrops

There is a gene in the snowdrop, *Galanthus nivalis*, that produces a compound (lectin) that acts as a natural insecticide. Insects that try and eat the snowdrop are killed by ingesting the lectin. This lectin gene has been taken out of the snowdrop and inserted into potato plants, *Solanum tuberosum*. These potato plants are now able to produce lectin and insects feeding off them will die. The benefit of this is that less chemical insecticide will be applied onto the potato crop.

A detriment of this technology has been that non-target insects may be affected (e.g. predators of the insects that have fed off the potato such as ladybirds). This landmark study in 1995 caused controversy two years later when it was suggested that rats fed on the modified potatoes produced thicker stomach mucosa than rats fed with unmodified potatoes. The scientist in charge then went on to claim that the rats with the thicker mucosa must have suffered nutritional deficiencies – a claim that was received with derision from the scientific community. It is plain that the scientific and moral question of genetically modified organisms are disputed not just between non-scientists and scientists; but also between scientists themselves.

Snowdrops, *Galanthus nivalis*, are a welcome sight to herald in the spring season; produces lectin that acts as a natural insecticide.

Chapter 9

Gardening for science I

Aaron Mills

INTRODUCTION

Horticulturists throughout the ages have striven for plants that suit their needs: be they flower colour, odour, fruit size, shape, uniform harvest or resistance to pests and diseases. This chapter explains the biological processes that produce these characteristics (genetics and the inheritance of characteristics) and goes someway to explore the methods used by horticulturists in achieving a 'designed' plant. The techniques and morals of genetic engineering are explored in some detail, as this is an emotive and relatively modern technique of designing plants that is often cited in the media, commonly with some bias. It is important that you, as a student of horticulture, are informed of this technique and are able to make decisions of the validity of its use, for yourself, based upon a sound technical knowledge and understanding of the processes involved.

Genetics is, among other things, the study of the inheritance of characteristics. Characteristics may be observable, such as flower colour, shape or odour, or they may be 'hidden', such as the plant's abilities of disease resistance or resilience to dry conditions, either naturally or in times of stress.

The way these characteristics are passed from one generation to another is by means of genes (it is, however, important to point out that the environment also has an effect on a plant's characteristics although we will focus mainly on the genes' role). Genes are lengths of DNA (deoxyribonucleic acid), a long string-like chemical that chromosomes are made from, and are located within the nucleus of cells.

Genes are passed from parent to offspring within the pollen and eggs of the anthers and ovaries within the flower (and within sperm and eggs of animals). Genes hold the 'key' for each characteristic in a similar way to Morse code holding the 'key' to each of the words in the English language.

Key concepts

✿ Introducing genetics and the inheritance of characteristics

✿ Explaining DNA and its function in breeding methods

✿ Implementing practical uses for plant adjustment

✿ Outlining limitations and risks to implementation

✿ Discussing the safeguarding of biodiversity

The Fundamentals of Horticulture: Theory and Practice, ed. C. Bird. Published by Cambridge University Press. © The Royal Horticultural Society 2014.

GENETIC SELECTION

Over the past 60 years, scientists have been able to understand and decode the information that genes hold. It is possible to write down on paper the code for many of the characteristics that horticulturists are interested in (e.g. the gene that forms purple flowers in pea plants, *Pisum sativum*).

You need to be aware that all organisms (bacteria and fungi included) have genes and DNA just like plants and animals, and as the structure of DNA is the same whatever organism it is in (it is just the code that is different – much like the Morse code for each of the letters that make up this sentence is different) it is possible to introduce genes from, say, a jellyfish, *Aurelia aurita*, and introduce that gene into, say, a tomato plant. The tomato plant, *Lycopersicon esculentum*, is oblivious to the gene being foreign, all it 'knows' is that the gene must be read and the characteristic must be produced.

For many centuries, people have been selecting for 'desirable' characteristics using the process of artificial selection – e.g. apples, *Malus domestica*, have been bred to produce large quantities of fruit and to be uniform in size and shape. Here the apple grower crosses pollen from a desirable tree with the flower of another desirable tree. The result being an offspring that contains the desirable genes, thus characteristics, of both parents.

Before we go on to look at some examples of designing garden plants, we need to first understand the structure and function of the chemical that carries the information from one individual to another – DNA.

Structure of DNA

DNA is a long, linear (string-like) molecule that is found in the nuclei of plants and animals. Each DNA molecule is wrapped around protein molecules and together these structures are called chromosomes (see Figure 9.1).

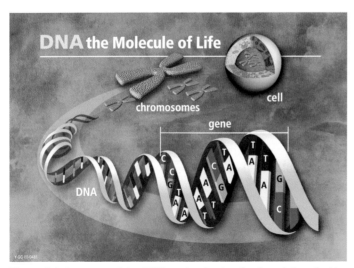

Figure 9.1 Structure of the DNA molecule showing how it is coiled in order to fit into the nucleus of the cell.

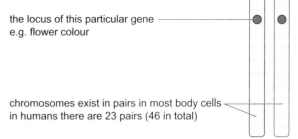

the locus of this particular gene e.g. flower colour

chromosomes exist in pairs in most body cells in humans there are 23 pairs (46 in total)

Figure 9.2 A gene is a length of DNA that codes for a characteristic, for example flower colour.

How many chromosomes are there in each cell? – You and I have 46 chromosomes in each of our cells (except our gametes – sperm and eggs that have 23). The chromosomes are incredibly long and thin. It is estimated that if all of the DNA in just one of our cells were strung together it would be 2 metres (6 feet 6 inches) long. That is enough DNA in you to stretch from here to the moon and back several times! Luckily the DNA is so thin that it can be coiled into a space the size of each cell's nucleus.

Plant chromosomes have exactly the same structure as animal chromosomes and work in the same way. Different species of plants have different numbers of chromosomes: potato, *Solanum tuberosum*, has 48; pineapple, *Ananas comosus*, has 50; radish, *Raphanus sativa*, has 18; and rice, *Oryza sativa*, has 24. You can see that there is great variation in the number of chromosomes of each species; however, it is important to remember that the chromosome structure of each species (whether animal, fungus, bacteria, virus or plant) is the same.

How do chromosomes that are the same structure for each species produce such different organisms? – Even though the structure of each species' chromosomes is the same, the information that the chromosomes contain are very different. This can be said to be true of the Morse code example stated above. Each letter of the alphabet has its own Morse code combination of dots and/or dashes (e.g. the letter 'S' is dot-dot-dot; the letter 'O' is dash-dash-dash). By knowing the Morse codes for each of the 26 letters in the alphabet, it is possible to produce an infinite number of sentences, each having a slightly different meaning. Well, DNA works using a similar principle but instead of dots and dashes it uses the different sequences of the four chemicals that make up the 'rungs' of the DNA molecule: A, G, T and C. Imagine that in order for a pea plant, *Pisum sativum*, to have red flowers, it will have a DNA sequence of ATTCCGTC. A different plant with white flowers will have the sequence TTTAGCCA. This method of using DNA codes is true of all the characteristics present in plants such as pest resistance, height, leaf shape, etc. In reality, of course, it is much more complex than the example used above, but hopefully you understand the principle.

What is a gene? – A gene is a length of DNA that makes up part of the chromosome. The gene holds the code for a particular characteristic, e.g. flower colour, disease resistance, height, etc.

Inheritance of characteristics from parent to offspring

Now we know the way DNA holds its information for each characteristic, we need to look at how the information is passed on from parent to offspring.

You need to be aware that the chromosomes within each cell exist in pairs. Humans have 23 pairs of chromosomes (thus 46 in total); potato, *Solanum tuberosum*, has 24 pairs (48 in total); pineapple, *Ananas comosus*, has 25 pairs (50 in total); radish, *Raphanus sativa*, has 9 pairs (18 in total); and rice, *Oryza sativa*, has 12 pairs (24 in total).

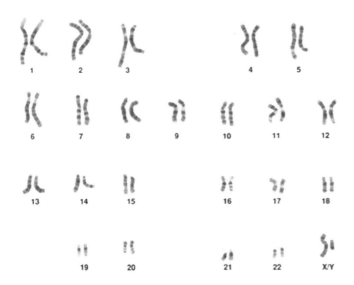

Figure 9.3 Microscope image of the 23 pairs of a human male.

You will be aware that there is a wide variety of characteristics in living organisms. Let us take eye colour in humans as an example. You may have blue, brown, green or hazel eye colour. Whatever eye colour you have, there is an 'eye colour locus' on chromosome number 15 that codes your body to have that particular colour. The locus for eye colour is at exactly the same position for everyone, on exactly the same chromosome number 15. These different genes (blue, brown, green or hazel) for the same characteristic (eye colour) results in the variety of eye colour for each of us.

The important thing to remember is that there is variety in the genes for eye colour in humans and that those genes are located on the same chromosome at the same locus for every person.

The same can be said of plant genes too, each pea plant, *Pisum sativum*, will have variation in genes for height (e.g. tall or dwarf) that are located on the same chromosome at the same locus.

In the pea plant height example, there are two variations for height: tall or dwarf. Tall is said to be dominant to dwarf; that is, if a tall gene is present the plant will always be tall, and dwarf is said to be recessive to tall; that is, a plant will only be dwarf if no tall gene is present.

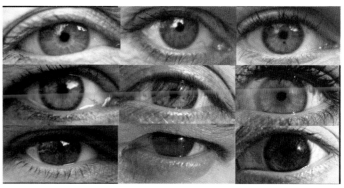

Figure 9.4 Eye colour in humans is determined by genes that code for particular colours – blue, brown and hazel.

Any one plant may have one of a combination of genes as shown in Figures 9.5a, b and c.

So, we now know that chromosomes exist in pairs and that genes for characteristics are found on the chromosomes at

T = allele for tall,
t = allele for dwarf

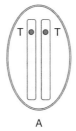

A

Figure 9.5(a) This plant has the tall gene on both chromosomes on this pair of chromosomes and will be tall.

T = allele for tall,
t = allele for dwarf

B

Figure 9.5(b) This plant has one tall and one dwarf gene in this pair of chromosomes and will be tall.

T = allele for tall,
t = allele for dwarf

C

Figure 9.5(c) This plant has two dwarf genes on this pair of chromosomes and will be dwarf.

particular loci. We also know that there are variations of genes that can give variety in the offspring.

How are the genes passed on to the offspring? – You will remember that earlier in the chapter gametes (sex cells – pollen and egg) were mentioned as having half the number of chromosomes than the rest of the body's cells.

During the process of gamete production (pollen in anthers and eggs in ovaries) only one of each pair of chromosomes goes into each gamete. This division is random – there is no way of knowing which of the pair will go into which gamete – thus each gamete will carry one of the genes of the parent to the resulting offspring, see Figure 9.6.

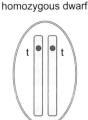

homozygous tall   homozygous dwarf

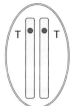

T = allele for tall
t = allele for dwarf
tall is dominant to dwarf
(dwarf is recessive to tall)

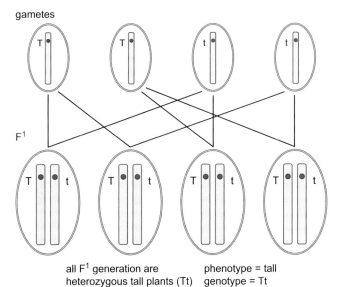

all F¹ generation are   phenotype = tall
heterozygous tall plants (Tt)   genotype = Tt

Figure 9.8 All of the offspring from this cross is called the F¹ generation (first generation from these two parents).

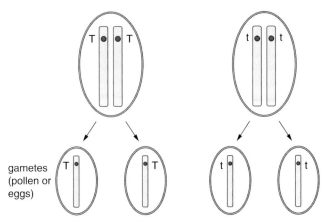

Each gamete contains half the number of chromosomes as the parent cell. It is random as to which chromosome ends up in each gamete.

Figure 9.6 Genetic diagram showing the segregation of chromosomes of the adult cells into the gamete cells.

Each gamete is now 'free' to fertilise any of the other gender's gametes in order to form a new offspring.

In the above example, any of the pollen grains could fertilise any of the eggs, which would result in an offspring that has genes as shown in Figure 9.7.

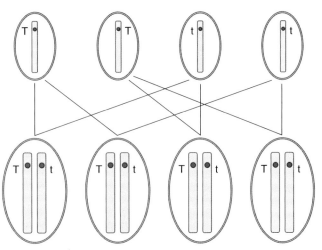

all F¹ generation are heterozygous tall plants (Tt)

Figure 9.7 Any of the pollen grains could fertilise any of the eggs.

They all have a combination of the genes T (tall) and t (dwarf), as tall is dominant to dwarf, all of the F¹'s will be tall.

What would happen if these F¹ plants were able to fertilise each other?

The resulting F² generation (the second generation from the original cross) would have a mix of three tall to one dwarf as shown in Figure 9.9.

Hopefully you can see that genes being on a chromosome, and that because the chromosomes exist in pairs and are segregated into gametes that have only one of the pairs, can lead to the variety of plants that exist today. Mutations can also lead to variation in offspring.

Formation of mutations

The reshuffling of genes in sexual reproduction is not the only way offspring can vary from their parents. Sometimes errors occur in the sequences of A, G, T and C that form a gene. These alterations of a gene usually lead to the gamete not being fertile thus not being passed into the next generation; however, occasionally the mutation can be carried on the chromosome, just as a normal gene is, and become a new gene that may or may not be useful to the offspring (e.g. it is more resistant to dry conditions or has larger fruits than its parents had).

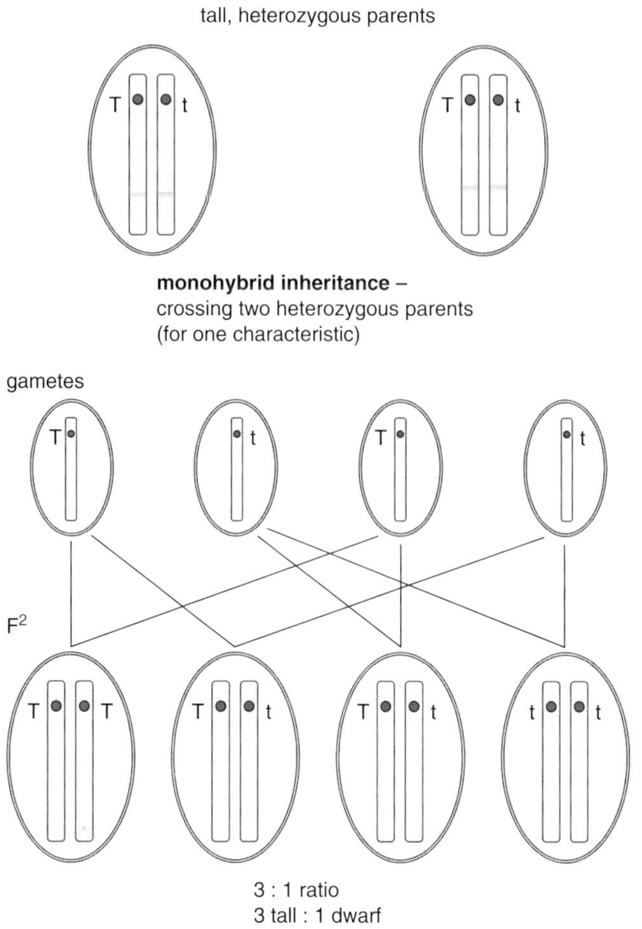

tall, heterozygous parents

T ● t   T ● t

monohybrid inheritance –
crossing two heterozygous parents
(for one characteristic)

gametes

T ●   ● t   T ●   ● t

F²

T ● T   T ● t   T ● t   t ● t

3 : 1 ratio
3 tall : 1 dwarf

Figure 9.9 The resulting F² generation (the second generation from the original cross) would have a mix of three tall to one dwarf.

Mutations occur spontaneously and at random, and although there are many 'proofreading' mechanisms within each cell in order to decrease the number of mutations, they will occur. In fact, everyone contains mutations within them that do us no harm and may, or may not, be passed on to any children we might have. Over the course of millions of years and millions of generations, these tiny mutations add up and give rise to new species over the course of time.

In horticulture, plant breeders actively seek these variations and artificially select which individuals are to be crossed. They artificially select plants with a desirable characteristic (e.g. pleasing flower colour) and transfer its pollen onto the stigma of another plant that has a different desirable characteristic (e.g. pest resistance), in the hope that at least some of the offspring will have a combination of both of the desirable characteristics (i.e. desirable colour and pest resistance).

This method of artificial selection has been the mainstay of horticulture for many centuries and has given rise to many of the beautiful cultivars of ornamental flowers and high-yielding vegetables that we see today.

It should be said, however, that this is a time-, space- and labour-intensive method of producing new plants as hundreds

of undesirable plants are produced as a by product and the process can take many years to complete.

Figure 9.10 In horticulture, plant breeders actively seek these variations and artificially select which individuals are to be crossed.

CASE STUDY   Inducing mutations

Recent technology has allowed horticulturists to increase the frequency of mutations by chemical and technical means. Chemicals such as ethyl methanesulphonate have been used to produce 'ice cube' lettuce, *Lactuca sativa*, cultivars and exposure of pollen grains to electromagnetic radiation (e.g. X-rays) has been used to produce field bean, *Vicia faba*, cultivars. From the many thousands of offspring plants that are produced using these methods, a few will have the desired characteristics required and can be marketed as a new cultivar.

Genetic modification (genetic engineering)

You will certainly be aware that another method of producing new cultivars is to use the relatively new technique of genetic modification (also known as genetic engineering). This technique has, probably more than any other in history, led to questions of the morality and possible safety implications of using such a plant (and animal) breeding technique. These moral and possible safety implications will be reviewed after we have first looked at the theoretical principles of genetic engineering.

It was seen initially that the advantages of genetically engineering a plant could potentially improve its quality and productivity by inserting an advantageous gene, or genes, without the need for labour- and time-intensive traditional techniques of back-crossing or hybridisation.

The principles of genetic modification – Remember from earlier in the chapter where it was stated that all organisms (plants, animals and bacteria alike) had chromosomes that were made of DNA and that the only difference between each organism was the sequences of A, G, T, and C 'rungs' in their chromosomes? It is this and are vegetatively propagated, commonly inbreed or are crossed with very close relatives. These processes lead to cultivar populations that have limited genetic diversity that leaves them open to pathogen attack. More recently, micropropagation techniques have been employed in order to increase the numbers of individuals of a desired variety. This technique, along with vegetative propagation, results in individuals that not only look identical, but have identical genetic diversity.

It is thought that introducing pathogen-resistant genes directly into the population by genetic engineering could combat some of these problems without the need for time- and labour-intensive outcrossing trials. Another benefit of inserting single genes is that none of the desired traits of the crop are lost when outcrossing with other individuals.

Recently there has been much interest in the conservation of genes from heritage varieties. Many of the crops that we are familiar with have been consistently bred for traits such as large fruit, high yield, uniform harvest time and stem length. As a result, the genetic diversity of modern cultivars has diminished. Heritage varieties (traditional varieties that are not commercially grown but are ancestors to our modern varieties) still carry many of the genes lost through subsequent inbreeding. There are many schemes currently running whose primary aim is to conserve these heritage varieties (thus the genetic diversity of that crop). This is one way amateur gardeners can play a vitally important role in horticultural conservation. Also see Chapter 1 for a discussion on the values of the work of the UK charity Plant Heritage.

Methods of decreasing the chance of contaminating non-GM crops and wild plants

There are several methods employed in order to decrease the chances of contaminating non-GM plants with GM pollen; some of which are outlined below.

Terminator genes can be added to the GM crop. In this instance the seed that is produced is sterile, thus cannot grow into an adult plant and pass its GM genes on to other plants. This does, however, mean a farmer cannot keep back any seed from that year's crop in order to grow in the second year. Regular purchase of that particular seed is required from the biotechnology company that supplies it.

Similarly, GM can result in the production of sterile pollen that works in a similar fashion.

Isolation zones are a physical distance barrier between GM crops and non-GM crops that decrease the chance of pollen drift or insect transport of the GM pollen.

C-4/C-3 SWITCH

You will be aware that plants photosynthesise in order to make organic molecules (e.g. sugars such as glucose) and produce oxygen as a waste product from the raw materials, carbon dioxide and water.

You may not be aware that there are different types of photosynthesis, called C-3 and C-4, and that different plant species use different pathways. Examples of C-3 plants are wheat, *Triticum* spp., rye, *Secale cereale*, field bean, *Vicia faba*, and tobacco, *Nicotiana tabacum*. Examples of C-4 plants are maize, *Zea mays*, sugarcane, *Saccharum officinale*, and tomato, *Lycopersicon esculentum*.

A general rule of thumb is that C-3 plants are temperate plants whose photosynthetic rate is more efficient at lower temperatures and light levels than C-4 plants.

C-4 plants are, generally, tropical plants whose photosynthetic rate is more efficient at higher temperatures and light levels.

The result of this is that it is difficult to grow C-3 plants, e.g. wheat, *Triticum* spp., in tropical regions, as is growing C-4 plants, e.g. sugarcane, *Saccharum officinale*, in temperate regions. As these photosynthetic pathways are under genetic control, it may be possible to alter the pathway of a C-3 plant to become a C-4 plant, and successfully grow it in tropical third world countries for food. Theoretically the same C-4 to C-3 switch is possible for crops such as sugarcane, *Saccharum officinale*, and maize, *Zea mays*. The advantages of employing these C-3/C-4 switches would be an increase in the geographical range in which food crops could be grown; potentially an important factor in the production of global calories, and in controlled environments, e.g. glasshouses, a C4 plant such as tomato, *Lycopersicon esculentum*, that has been genetically modified to exhibit C-3 photosynthesis, may increase its production output when supplied with the correct environmental conditions.

Please also see Chapter 4 for further information.

Strips of non-GM plants can be planted around the GM crop to create a 'buffer crop' that acts as insect and wind-blown 'stop-shops' that catch GM pollen and dilute the GM pollen volume by producing its own non-GM pollen before it contaminates any other plant.

Of course, it is important to note that these methods can decrease the chances of GM gene transfer, but they cannot be considered 100% failsafe – much in the same way as when spraying with pesticide it cannot be considered 100% safe that the spray will not carry on to non-crop plants.

Summary

Designing plants is a complex issue dependent upon the rules of genetics and an organism's interaction with its environment. People have been designing plants for millennia: initially by selecting and crossing plants that showed desirable characteristics and, more recently, by selection of specific genes that may come from unrelated organisms and their insertion into the plant genome of choice.

This high-technology method of designing plants has led to debate about the safety and moral implications of using such techniques. This debate still goes on.

A plant's characteristics are carried by genes (lengths of DNA) and can be passed on to future generations following the rules of Mendelian genetics. The DNA molecule is able to hold coded information of the physiological and morphological aspects of the organism. This code is held in the sequence of nucleotide bases (A, T, G and C) of the genes. That is not to say that organisms are slaves to their genes as the environment also exhibits a contributing effect on the developing embryo (be it plant or animal). All organisms from the simplest bacterium to the most complex mammal (would it be too arrogant to suggest humans, *Homo sapiens*?) share this information-containing molecule. This being the case; it is now possible to remove a gene from one organism and insert it into a completely unrelated organism enabling it to produce the phenotype desired by industry.

Desirable characteristics can be observable, e.g. flower colour or odour, or hidden, e.g. pest resistance or resilience to dry conditions.

Desirable characteristics focused upon by genetic modification are primarily herbicide and insect resistance. Other desirable characteristics being researched are increasing nutritional value of foodstuffs, flower colour, odour, C-3/C-4 shifts, resilience to dry conditions, anthocyanin production, improving fruit colour or size and uniform size, and harvest time of vegetables.

It is up to you, scientists, policy makers and society in general to decide whether genetically modifying a plant is worth the risks involved. Society is facing ever-increasing environmental pressures, such as climate change, and the insatiable need for energy to drive our technologically led lifestyles. Education about these issues of adults and our children will enable us all to make informed decisions of how we design plants in the future.

Revision questions

1. Name the molecule that holds the genetic information of an organism.
2. Explain how the structure of the molecule you have named in Question 1 is able to store that information.
3. Compare and contrast natural and artificial selection.
4. Explain how mutations may be advantageous or disadvantageous to organisms.
5. Write a list with the headings 'advantages of GMOs' and 'disadvantages of GMOs'.

FURTHER READING

Capon, B. (2010). *Botany for Gardeners*, 3rd edn. Timber Press.
An excellent introductory book for newcomers to botany. While it lacks the detail of the rest of the books below, it serves as a gateway to some of the technical concepts and terminology peculiar to science subjects. Capon also applies botanic principles to horticultural situations in an easily digestible form.

Curtis, H. & Barnes, S. (1989) *Biology*. New York: Worth Publishers.
An excellent all-round biological text book that is well designed and illustrated. As the title suggests, this is a general biological text book whose animal topics may put off some hardy botanist readers.

Green, N., Stout, G. & Taylor, D. (1997). *Biological Sciences 1 & 2*. Worcester. Cambridge University Press.

A particular favourite of mine. May seem a little dated now although the basic concepts of biological sciences are well addressed and complemented with fine black-and-white line drawings.

Raven, P., Evert, R. & Eichhorn, S. (2004). *Biology of Plants*, 7th edn. New York: W.H. Freeman and Company.

A must for all serious botanists. Extensively illustrated and written in a way that effortlessly makes some of the more complex topics seem understandable. There is also an interactive website that accompanies this text with student-friendly information and activities.

www.youtube.com/user/UCBerkeley/videos?query=angiosperms [Accessed 29 July 2013].

The University of California, Berkley College (UC Berkley) has uploaded a series of excellent botany lectures on YouTube that takes the viewer from the evolution of plants, through the plant's tissues and organs, to the diversity of plant phyla.

RHS level	Section heading	Page no.
3 4.1	Level 3 box: C-4/C-3 switch	196
3 5.7	Level 3 box: C-4/C-3 switch	196

CASE STUDY Little Somborne Wood, Stockbridge, Hampshire

Coppicing of hazel, *Corylus avellana*, planted in the reign of Henry VIII and still productive today, producing materials for hurdle making and thatching spars used by craftsmen within the local area. After some years of neglect the wood is now thriving again but is under pressure from the building deer numbers, who cause major damage by removing the vigorously growing tips of coppiced stools. This reduces the number of high-quality straight regrowths required of the same thickness and length for high-quality hurdles, together with increasing wastage.

Espalier apple, *Malus domestica*.

Chapter 10

Shaping plants

Daphne Vince-Prue and Chris Bird

INTRODUCTION

The shapes and forms of plants are the structural backbone of any well-designed garden. They continue to provide interest throughout the year, especially in winter when the plants that provide colour through the changing seasons are much less in evidence. Size is also important for a number of reasons. Over-large plants have no place in small gardens, supermarkets increasingly require produce of a uniform size as well as shape, while saleable pot plants need to be compact and in proportion to their containers.

There is a great natural variation in the shapes and sizes of plants. The branching habit determines the architecture of the plant and shapes vary from upright forms such as *Prunus* 'Spire', where branching is largely suppressed, to those with spreading canopies such as oak, *Quercus robur*, and *Prunus × blireana*. Even small plants vary in habit from low spreading types, *Thymus serpyllum*, to those consisting of compact rosettes such as *Saxifraga cochlearis* 'Minor'. Although shape is ultimately regulated by their genetic constitution, plants also have a remarkable degree of plasticity, which enables them to be modelled into a variety of different forms by the gardener. One only has to consider the difference between a step-over apple tree and a fully grown standard to realise the extent to which size and shape can be modified even in the same cultivar. This chapter considers the many factors that can modify shape and size, emphasising those that are under the control of the gardener.

Also dealt with is the selection and correct use of a range of pruning tools, together with techniques involving chemicals and non-pruning physical methods, e.g. shaking and stroking. Technical language and techniques are fully explained, with the background reasons for their development and use being included, with the most suitable timing or timescale required.

Key concepts

- ✿ Manipulating the shape, size and natural form of plants for decorative and other purposes and where this variation can be used to create interest in the garden

- ✿ Using genetic and chemical means; genetic constraints ultimately determine shape and size

- ✿ Endogenous hormones have major effects on shape, and apical dominance is a major factor in determining branching patterns

- ✿ Environmental factors also affect shape and size

- ✿ A review of the tools and equipment available for practical pruning

- ✿ An explanation of the practical methods and techniques available

The Fundamentals of Horticulture: Theory and Practice, ed. C. Bird. Published by Cambridge University Press. © The Royal Horticultural Society 2014.

rootstocks and more compact forms of many garden plants root pruning is now normally only undertaken where a specimen plant is outgrowing its space. Repeated cutting back is, however, routinely carried out for some species (see under 'Stooling' below).

Pruning must also be considered in relation to the flowering habit of the plant in question. For plants that flower early in the year on the previous season's wood, pruning normally takes place after flowering has occurred. In contrast, plants that flower late on the current season's wood are usually cut back early in the year to allow maximum growth before flowering. Good examples of varying the time of pruning to ensure maximum flowering can be found in the different cultivars of *Clematis* whose pruning is described below.

PRUNING METHODS

Often as a starting point, described as the '3Ds', the material to first identify and remove is any **d**ead, **d**amaged or **d**iseased sections, followed then by a 'C', which means **c**rossing branches. Do remember on badly overgrown or neglected plants, by the time you have completed the 3Ds and C that may be all that is required in the first year, to prevent uncontrolled new growth, which then has to be removed at a subsequent time.

Established trees and shrubs

This often requires the removal of sizeable branches and the recommendations for doing this have changed in recent years. Originally cuts were made flush with the trunk in order not to leave a stump that might die back and become the entry port for pathogenic organisms. However, cutting flush with the trunk tends to leave large wounds with dead areas that are not covered by wound callus. The current recommendation is to carry out target pruning, which cuts through that part of the branch where cell division in the reaction zones immediately begins to form a defensive callus that covers the entire surface of the smaller cut.

In order to carry out target pruning it is first necessary to identify both the bark collar, which is a swollen area where the underside of the branch meets the trunk, and the bark ridge, which is a conspicuous line of folded bark at the upper point of the union between the branch and the trunk. A sloping cut is then made from just outside the bark ridge to a point where the branch meets the collar, without penetrating either area. The practice of applying tree paint to seal the wound against possible infection often prevents the cut from healing and is not now recommended.

Stooling and pollarding

Stooling is the name given to the practice of cutting shrubs back almost to ground level in order to encourage the development of new shoots. The stems are usually shortened to about 1 to 2 cm (½ inch) above the bases of the old shoots in early spring. This is often done for seasonal interest as, for example, in the production of highly coloured young shoots in shrubs such as *Salix alba* subsp. v*itellina* and *Cornus alba* 'Sibirica'. Shrubs in which the

Figure 10.1 The bark collar and ridge shown on a mature branch.

flowers are produced at the end of the current year's shoots, e.g. *Hydrangea arborescens* 'Annabelle', are treated in the same way. Stooling is also practised to obtain wood for stock plants that are to be grafted, see Chapter 8 for further details.

Pollarding is the practice of pruning the plant back to a stump, usually about 60 cm (2 feet) high in early spring. It is often used with the cider gum, *Eucalyptus gunnii*, to encourage the production of the bluish juvenile foliage that is considered to be more attractive than the dark green adult leaves, see also 'Juvenility' in Chapter 8.

Commonly also seen for other tree species, e.g. *Populus* × *canadensis*, alongside roads, where a height of 3.1 metres (10 feet) is used. Historically this was due to the height that was reached by the pruner standing in a haywain, a type of agricultural cart.

Coppicing (a form of stooling) and pollarding were once widely used in woodland management to encourage the growth of long straight branches on trees such as hazel, *Corylus avellana*, and oak, *Quercus* spp., for use in building and furniture making. These procedures are still used to provide wood for making items such as hurdles for garden use and some specialist furniture. Coppicing may also be practised in some woodland areas, to let in light in order to encourage flowering in species such as bluebell, *Hyacinthoides non-scripta*, although branch thinning is the more usual option, depending on the tree species.

Figure 10.2 Hazel, *Corylus avellana*, newly coppiced after seven years' growth showing strong regrowth.

LEVEL 3 BOX

ROOT PRUNING

Because of the relative consistency of the root to shoot ratio, removal of part of the root system at the same time as removing some of the shoots is a reliable method of restricting growth. In the open ground root pruning should be carried out over two winter seasons when the plant is dormant, removing half of the root in each year. A 30 to 40 cm (12 to 14 inch) deep trench is dug out about 120 cm (4 feet) from, and half way round, the trunk during the winter and the large roots are cut through. The soil is then replaced and the tree should be mulched in the following spring to conserve moisture. Part of the canopy should be removed at the same time as the root system is cut back in order to maintain the root to shoot ratio.

Root pruning is also used for container grown plants that have become pot-bound and where transfer to a larger pot is not desirable; here, the outer roots should be cut back by approximately two thirds and the plants repotted. About one third of the top growth should be removed at the same time in order to maintain the balance of growth.

The ultimate reduction in tree size that can be accomplished by restriction of the root system is in the production of **bonsai**, which depends for its success on root pruning and restriction of the root system in very small containers, together with bending shoots from the vertical by twisting them round wires. Careful shoot pruning is then used to achieve the desired form.

PRUNING TOOLS

We now turn to discuss the expanding range and cost of tools and equipment that could be used for pruning of plant material, without damaging the plant, bystanders or the operator.

Selection of tools for the task in hand

As with any practical task, over time specialist tools and equipment have been developed, each to cover their own task and, in some cases, be complementary to others.

General safety

Tool selection – always select the correct tool for the job in hand, if the task changes then change the tool. This is undertaken for a number of reasons: the main one being to prevent damage to you, your plant or the tool itself. The number and array of pruning equipment that is available seems to increase with every gardening magazine read or visit to the local garden centre or national garden show. Remember the improvements in ergonomics with design of both hand and powered tools, together with the use of lighter metals, such as aluminium, which is still very strong and robust.

Correct use – before obtaining pruning equipment assess the tasks that will need to be undertaken and on what timescale. Can you cope with a reduced number of tools for the early years of your grand plan? It may be, however, that major vegetation clearance is required in the first instance? Clearance tools, such as brush cutters or bill hooks, can be very dangerous to use, especially if in a group of people. Make room for each person to be able to work: with a minimum distance of 2 metres (6 feet 6 inches), between them when slashing or for clearance of overgrown vegetation.

Personal protective equipment (PPE) – the minimum for clearance tasks should be stout boots or preferably those with steel toe caps. The investment is well worth it; especially as these days they come in a wide range of sizes, including ladies, and are much lighter: the quality of the boot is not dependent on the weight. With the use of materials such as Kevlar™, which is also used for bullet-proof vests, they have become much more comfortable. Also, if any machinery is to be used then steel-toe-capped boots are a must for protection against cutting or piercing the feet and protecting the ankles against slips, trips and falls, which are common when ground clearance takes place or heavy pruning is undertaken, especially with stumps and uneven ground being produced.

Maintenance and aftercare – this is an area that is often overlooked and can result in serious damage to everything concerned in the task being undertaken: the plants, people and equipment. Careful note should be taken of any instructions given, especially when they relate to the maintenance instructions and the time periods involved.

One of the main areas to consider is that of loose nuts, bolts and blades, together with following the manufacturer's instructions for both use of the equipment and the required maintenance.

Handtools

Knives – the classic item, with a solid curved blade, for formative pruning of trees and shrubs, now less used and less popular, due have selected. If the one bud is facing the wrong way then cut higher to a bud facing the correct way, unless the stem is very weak, in which case cut it to the basal rosette and encourage the dormant buds to break.

Other more involved shapes can be used to achieve a more decorative effect and a visit to an established garden growing a range of fruit for inspiration would be recommended: places such as any of the four Royal Horticultural Society gardens, West Dean Gardens, near Chichester and Kasteel Hex Gardens in Belgium.

Fan – commonly used with cherries, plums, gages, apricots, nectarines and peaches, this classical pruning method is again more about training and the understanding of the movement of plant hormones, e.g. auxin. The main branches are spaced 30 to 45 cm (12 to 18 inches) apart and their number will be dictated by the vigour of your plant and the intended space that is available. When wall training, it is important in spacing the wires that full support is provided for the young plant and also for the more mature one under the full weight of fruit. The recommendation of the first wire being set at 45 to 60 cm (18 to 24 inches) has been found to be too high for the space provided. Therefore set all the wires at a spacing of 30 cm (12 inches) apart and remember that it is best to have each one tied and made taut independently of the others and if possible adjustable, rather than running one piece of wire through all the levels. This makes it very difficult to make it taut and also when it breaks, usually with a full set of potential prize-winning fruit present, the whole support system is ineffective.

Finally on this area, please present the branch to the support wire and tie it on the front with strong garden twine. Do not tuck it behind the wire, even on a temporary basis, as this will result in a damaged branch, allowing an entry point for fungal and bacterial diseases, which the stone fruit group are particularly prone to and eventually also to a damaged support system.

Delayed open-centre – is used to attain a more traditional growth system, to produce a rounded 'brandy glass' shape of bush, on a short single stem between 30 and 45 cm (12 to 18 inches). Again the main undertaking during the first five years is the formation of a semi-permanent framework, with a series of main stems evenly spaced around the whole circle of available space, with a spacing of 30 to 45 cm (12 to 18 inches) apart. One recommended pruning system that you may have heard of is the renewal system, where whole branches are removed after fruiting or if poorly placed to allow the growth of replacements, which will, in turn, fruit in two to three years' time.

Pyramid – this training system was originally developed for apple trees that are tip-bearing in their fruiting method, e.g. *Malus domestica* 'Worchester Pearman'. They have a central main stem with rings of fruiting branches set 30 to 45 cm (12 to 18 inches) apart. Imagine, if you will, a number of bicycle wheels set on top of each other: commercially this has developed into a spindle bush. This method provides the maximum opportunity

to allow the growth of short lateral fruiting growths, with the highest-quality length being up to 15 cm (6 inches) to allow for fruit formation without overcrowding, which can lead to bad outbreaks of diseases such as brown rot. See Chapter 12 for further details and other control methods.

GROWTH TRAINING FOR STONE FRUITS

In modern times, with the use of plum trees on more dwarfing rootstocks, a technique called **festooning** is being used to promote maximum fruiting while at the same time controlling the new vigorous growth that the plum group are well known for. This involves tying down the long young main branches in June back to the main stem with strong garden twine, for a minimum of six weeks, as after this period of time the branches stiffen and then remain in their new position. Twine is also recommended, as great damage in the way of unwanted pruning, can be undertaken if the plastic twine is not removed before cutting into the stem. This can result in the breakage of branches, due to the weather or weight of fruit. Once fruited, in a year's time, the whole branch is removed to stimulate new replacement growth.

Figure 10.7 The fan training technique is being used on the growth of plum, *Prunus domestica*.

Stepover – an excellent training method for any small space and especially used to good effect as an informal low border or edging to divide one area of garden from another. A single set of espalier arms or, even in some situations, one arm is grown. Think about their use around a patio or terrace area or where a change of level within a garden is required or planned. These are both decorative and productive using the same space and work especially well in modern gardens as well as traditional situations.

Traditional shapes – these include **standards** and half-standards, which are grown on a clear main stem of 1.83 m (6 feet)

Figure 10.8 The stepover training method can be used very effectively as a low barrier to divide different growing areas.

Figure 10.10 When grown on their own roots fruit trees can dominate a garden situation.

and 1.2 m (4 feet) respectively, historically to allow animals to be run beneath them, cattle and sheep, depending on the stem height, to keep the developing fruit out of eating range.

Figure 10.9 The height of the fruit trees allowed animals to graze underneath in a traditional growing situation.

Usually these sizes are now grown to order, on more vigorous rootstocks, so unless you need these shapes for historical context, keep to the more compact delayed open-centre, pyramid (for tip-bearing plants) or trained forms, e.g. cordon. If, however, you have a set of these growing in your garden, then please read the next section for a self-help guide.

Pruning established trees

One of the most challenging tasks when people move into a new garden is to inherit the mature plants, be they ornamental or fruit trees, shrubs, climbers or herbaceous. Some are overawed by the responsibility of these stately specimens, thinking of them, correctly so, as a living link to our horticultural and social history; others see them as problems getting in the way of their 'blank

canvas' approach. Do try, where possible, to retain them as they could be unique examples of your local area, also see Chapter 12 for comments on the value of heritage cultivars.

Techniques used for their recovery include the 'grade one' approach, which entails reducing the tree back to a main framework.

Figure 10.11 A dramatic method of rejuvenation, reducing all the branches back to a lower level.

This is the way to go if the specimen is very badly affected by diseases, such as apple and pear canker or bacterial canker (affecting *Prunus* spp., especially cherries and the plum group), see Chapter 12. If major rebuilding works are planned this will also provide greater ease of access and reduce the physical and chemical damage to the trees caused by the building process, e.g. concrete dust and splash, which is extremely alkaline and can burn the foliage.

However, the problems can start with the vigorous, uncontrolled regrowth the following spring, which can result in masses of unproductive **epicormic** growth, which will then require intensive thinning the following summer to select out the new branches for subsequent fruiting. This may mean a 'take one, leave one' policy to allow room for the fruiting stems to develop over the next two to three years; remember that it may take up to four years to set fruit buds, especially with **triploid** apples, e.g. *Malus domestica* 'Blenheim Orange', so this should be considered in the process of pruning adopted for these plants.

One other technique, preferable if possible, is 'partial dehorning', which involves removing up to one third of the mature branches, back to the main framework, but allowing the rest to continue to grow, tipping them if required to keep them in check until removal over the next two years. Therefore a total of three years is taken to reduce the overall size and shape of the tree. In the first year, emphasis is placed on removal of any dead, diseased, dying, misplaced or crossing branches. This allows improved airflow and light penetration, resulting in higher-quality, controlled regrowth. The other advantage is that with more stormy weather, including higher, more gusting winds, being forecast and experienced, the new canopy will be much more open, allowing the wind to pass through, without causing any major damage. This technique can also be employed with overgrown evergreen shrubs, e.g. *Elaeagnus* × *ebbingei* 'Limelight'. Whichever method may suit you best and you select, the method of major branch removal is important, together with the timing.

The widely recommended and practised method is called a **jump cut** or step cut – this involves undercutting the required stem a minimum of 15 cm (6 inches) above the final cut, through by one third, then on the top side, moving up another 3 cm (1.5 inches) cutting totally through. This takes the main weight of the stem and prevents the bark or rind splitting into the major framework. Finally, to finish, remove the stump or snag with a clean cut, just off the horizontal (to allow water runoff), but do not go beyond the basal ridges as this will slow down or prevent healing.

Fruit type:	Apple, Malus domestica	Pear, Pyrus communis	Cherries, Prunus avium and P. cerasus	Plums, Prunus domestica	Apricots, Prunus armeniaca	Peaches, Prunus persica	Grape, Vitis vinifera	Fig, Ficus carica
Recommended months:								
January	▲	▲			▲		▲	▲
February	▲	▲			▲	▲	▲	
March	▲					▲		
April	▲		■	■				
May					■	■	●	
June								
July	●	●	▲	▲				▲
August	●	●	▲	▲	▲			▲
September	●	●		▲	▲	▲		
October		▲						
November	▲	▲						
December	▲	▲					▲	

Key:

▲ = Annual maintenance pruning

■ = Thin or remove damaged branches only

● = Annual summer pruning for restricted cropping systems

Table 10.2 Recommended pruning times for named species or group types.

Figure 10.12(a, b) The result of a well-undertaken pruning cut: growth of callus tissue to naturally seal the wound.

Autumn pruning recommendations

Some ornamental and fruit trees, e.g. the tulip tree, *Liriodendron tulipifera*, walnut, e.g. *Juglans nigra*, and mulberry, e.g. *Morus nigra*, have special requirements as regards the timing of when they are best pruned, as if this is undertaken too late in the spring any wounds made will weep uncontrollably. This situation is best avoided by pruning in the late autumn, after the leaves have dropped; this makes it much easier to see the branch structure and identify any that require attention or total removal.

CLIMBERS

This is another area of specialist pruning where gardeners usually leave well alone until the climber falls off, outgrows the space provided, or the support system fails. The more demanding plant groups include those decribed below.

Clematis

The pruning of this very popular and enduring genus has been made more and more complex by the introduction of a wide range of interbred species and the increasing popularity of the less well-known species, to give an all-year-round flowering impact. Taking things back to the absolute basics, the factor that we need to remember is flowering time and the key month for this is July: if your clematis flowers before July then prune it, after flowering, removing old flowered stems and tying in young stems to supports for next year or for an improvement to the overall shape of the plant. This can be undertaken by the use of a hedge trimmer, to remove the bulk, take care with the level of dust that is produced, so a dust mask and some form of eye protection would be recommended. Then finish off with a pair of secateurs, always cutting to a bud, to avoid **snags**. These are sections of nodeless stems that subsequently die back to the position of the next node and allow the entry of diseases. Clematis species in this group include *Clematis montana, C. armandii* and *C. alpina* (Figure 10.13a, b and c).

If, however, your clematis flowers in July or later then prune it in the spring, removing the dead and winter-damaged parts by up to two thirds. The month when spring occurs, of course, depends on many factors, but look out for the return of swallows and the formation of new worm casts.

Clematis species in this group include *C. texensis, C. rehderiana* and *C. tangutica*. Plus all the late large-flowered cultivars, e.g.

Figure 10.13(a) *Clematis montana* is a vigorously growing species containing a range of cultivars.

Figure 10.13(b) *Clematis armandii* is a large-leaved species with scented flowers.

Figure 10.13(c) *Clematis alpina* is a species containing cultivars resulting in a wide range of colours being available.

Figure 10.14(a) *Clematis texensis* has a long flowering season with architectural flowers.

Figure 10.14(b) *Clematis rehderiana* is an excellent choice for late summer colour, flowering through into the autumn.

C. 'Hagley Hybrid', *C.* 'Jackmanii' and *C.* 'Polish Spirit' (Figure 10.14a, b, c and d).

Remember that a larger selection of the scrambling herbaceous species are now more commonly available and these die down completely in the late autumn, leaving the old seed heads for the birds, and should be cut to the ground at the end of February each year, such as, *C. recta*.

Figure 10.14(c) *Clematis tangutica* produces long-lasting bells of bright yellow, followed by persistent seed heads.

Figure 10.14(d) A collection of modern large-flowered *Clematis* cultivars.

Wisteria

For established plants – July is the classical month for pruning after flowering: allow the full extension of new growth before pruning the new elongated growths back to four buds; think about the direction of the next generation of new growths when selecting your buds – not into the support frame or towards a window. If after an excellent growing year then regrowth is seen or, even more commonly, a second flowering is seen, then repeat this operation again but take the regrowth back to only one bud.

Once the leaves have fallen, then a maintenance prune is required, to remove or adjust the spacing and position of the most mature stems and also to review the quality of the support method. Replacement is much easier during the winter months than after major support failure – usually while you're away on holiday, resulting in a mass of free-growing wisteria for you to cope with on your return.

Remove dead, dying and diseased material. In very mature examples removal of one or more of the most mature stems may

be undertaken, this will reduce the weight on your support system, which can be quite considerable when the whole plant is in full leaf and is wet.

Figure 10.15 A mature plant of *Wisteria* spp. can take some room and will require good support when in full leaf

For newly planted and non-flowering subjects – Wisteria will always grow vegetatively, that is producing stems and leaves, until it has completely filled the space provided for it or it runs out of nutrients. This can be a period of some years and is often exacerbated by the kindness of the proud owner.

Treat them mean once the plant has made good growth to provide coverage within the modest space that you have provided. The aim for the first three to five years is to establish a main framework for maximum flower display. It is important at this time not to tuck any new growths behind the support wires, as they will remain in that position for all time and as they enlarge will remove the supports.

OTHER EXAMPLES

Rambling roses

For established plants – First of all note the different thicknesses and ages of main stems presented to you, start from the ground upwards, a change of colour will sometimes be clearly seen and is very helpful in deciding the stems to remove, allowing regrowth to replace the oldest stems. For rambling roses a system of total stem removal of the selected oldest stems is best and these should be reduced to 10 cm (4 inches), the timing is usually best after the first, or only, flush of flowering. If, however, the plant in question has set excellent hips, which are one of its special decorative features, then after these have been consumed by the local bird population is the best time, as also these could have been covered by a hard frost providing a pleasing winter bonus.

If the stems are badly tangled then removal of the selected oldest stem should be undertaken in sections, but please take care as the stem you cut may not be the one you intended. Check twice and cut once, or the bonfire will benefit once more. An odd number of stems should be left, as this provides a more visually pleasing result, and the final number will be chosen depending on the space available and the starting number. Remember that a maximum of one third should be removed at any one time. Regular replacement of stems will result in growth of new stems at more even intervals and on more vigorous plants may result in the selection of the best replacements being required. If an excessive number of new stems is seen, do check that some of these are not rootstock growths called 'suckers', as these all should be removed as soon as they are seen, the best way to do this is by twisting them off or even better by rubbing them off with your thumb, when they regrow. One of the classic methods of sucker identification was always to count the pairs of leaflets: more than three pairs meant a suckering stem, less than this meant a cultivar. However, now with the advent of modern shrub roses and different propagation techniques this test is no longer reliable and strong vigorous growth from the base should be prized and may be tipped as part of pruning back up to one third, to encourage the growth of three replacements of more controllable height and mature the growth to improve the flowering potential.

Figure 10.16 A mature plant of *Rosa* spp., where the oldest third of stems should be removed annually.

For newly planted and non-flowering subjects – For the highest quality rose plants these have a minimum of three main stems and would be of greater thickness than a pencil. Space the stems on your selected support system and remove any dead, dying and diseased stems, also removing any very thin growths. Cut back all of the main stems removing two thirds of the growth height to induce vigorous regrowth providing the basis for your main framework for future years. Select buds that will provide growth in your desired directions for the space available. Feed any plants that have not flowered after three years with a high potassium fertiliser to encourage the wood to mature. Do not continue to prune the plant as this will result in further juvenile growth and delay flowering even longer. If the non-flowering persists then it may be that the plant is too shaded and requires moving into a position that receives greater light.

Hydrangeas

Also increasing in complexity, due to modern breeding and size reduction, is the pruning of these superb mid- to late-summer-flowering (if not over-pruned) shrubs. The selection of method and timing depends on where, and when, the flower buds are formed. This is, for the main *Hydrangea hortensia* cultivars, on last year's growth, so if in the spring period you give an all-over plant tidy, then a number of flowering shoots will be removed. Selection of the oldest flowered stems should be made in spring to allow the stimulation of new replacements, which will flower the following season, thereby maintaining a balance in the age of the stems while moderating the overall size of the plant. If very vigorous new shoots do arise and spoil the overall shape of the plant then these can again be tipped by one third to encourage a division of stems that will flower within the desired shape of the plant. The old flowering stems are left unpruned until late spring, when the new buds have broken, that is started into growth, as protection against cold-weather damage over the winter period.

Figure 10.17 A mature plant of *Hydrangea aspera* Villosa Group in a woodland glade setting.

Camellias

Often thought of as slow-growing shrubs they can become unwieldy and outgrow their space more quickly than is realised. When discussing the pruning of this genus in past times the recommendation of minimal pruning was stated, just lightly, undertaking to keep a balanced shape. However, with the advent of climate change and a wider range of cultivars being available, major thinning or height reduction can be required. After flowering has finished the plant will have a growth flush and so this is the

Painting over of the pruning cuts is not necessary, as it may slow the plant's natural healing processes, if a clean wound is made to a node without splitting the stem.

Other evergreen shrubs

The timing of branch removal after flowering is complete can be used across a whole range of evergreen shrubs, including *Photinia × fraseri* 'Red Robin', which will also reward your efforts with rich red young growths. *Hebe* spp. and cultivars together with *Ceanothus* spp. and cultivars are other examples of where pruning practices have changed, with the norm now being the annual removal of up to one third of the oldest or misplaced branches, allowing the regeneration of the plant, rather than leaving young specimens totally untouched and then replacing the whole plant after five to seven years. This more modern technique results in extending the useful life of the plant, in fact, up to doubling it in some cases.

However, exceptions to this timing include *Pieris* spp. and cultivars, e.g. *Pieris* 'Forest Flame', where after flowering the new young flush of material is initially bright red, fading to pink, then white before eventually becoming green. Once the new material is green then is the time to reshape your plant: removing whole branches back to a joint and ensuring that no snags (sections of stems with no buds) are left. Also remember that female plants of evergreen shrubs, e.g. spotted laurel, *Aucuba japonica*, cultivars, need their pruning to be delayed until the fruit has finished, usually enjoyed by the birds, and so early spring is favoured for major reduction or reshaping.

Figure 10.18 A mature plant of *Camellia japonica* in need of reshaping after flowering, with some branches removed to show the adjusted shape.

recommended time to adjust the shape or size of your plants before this flush occurs. Also take the opportunity to remove any winter damage, e.g. split stems, by making clean cuts using the appropriate tool for the size of the branch being removed. A folding saw is the ideal tool for branches too thick for secateurs, even the more modern ratchet type, as you can more easily gain access to the pruning points causing minimum damage. If you are tempted to reach for a pair of loppers, consider the room you will require for access and the potential for accidental damage to the adjacent branches. On this point, do remember to ease off the pressure on your folding saw as the cut is nearing completion to avoid nicking the next stems.

Some other safety points

Care should be taken when pruning or reshaping some plants as they have a layer of fine hairs or dust-like material, which when disturbed can cause an irritation to the eyes, skin and in extreme cases breathing problems. Examples of plants to take care with include Jerusalem sage, *Phlomis fruticosa*, and *Elaeagnus × ebbingei* 'Limelight', where one recommendation to offset the effects is to prune them when the foliage is wet. Wearing the appropriate PPE should be standard when undertaking any practical work and a dust mask and safety glasses may be required. Dust and dry material can also be a problem when reducing mature hedges during dry periods.

CASE STUDY Wiltshire College, Lackham campus

Set up in the mid 1940s as a county agricultural and horticultural training college, the Lackham site has seen many changes since those early days. It now provides full- and part-time courses at many levels for a wide range of student requirements. The range of protected structures includes modern aluminium glasshouses, polythene tunnels and cold frames, each of which are used to produce commercial crops for sale and plants used in the grounds; most importantly they are for students to receive theory and practical sessions in plant knowledge, propagation and crop production. These teaching outcomes require a variety of different growing environments, from cold unheated zones with high light levels to hot and humid for the 'tropical' display collection. Heating is provided by fixed hot-air units. Propagation is carried out using mist units and covered benches with bottom heating. Many glasshouses have rolling benches in order to maximise the growing area while minimising path surfaces. Good access is achieved via well-positioned doorways and concrete paths. Semi-automated thermal screens are operated in some sections to reduce heating and provide shade for some crops.

The Curvilinear Range, National Botanic Gardens, Glasnevin, Ireland. Opened in 1849.

Chapter 11

Protected cultivation

Chris Allen

INTRODUCTION

This chapter will show that growing plants in protective structures can be enormously beneficial for commercial growers and gardeners who are producing crops all year round or trying to extend the growing season in the spring and autumn. These extra crops can attract premium prices for producers, an important factor for those whose cash flow is reduced during the winter months.

The positioning, building and management of any structure requires careful planning and preparation of the site in order to maximise the benefits and justify the additional costs involved compared to growing in the open ground.

Protective structures are made from many materials and constructed in many shapes and sizes: from the home-made polythene lean-to on an allotment, a cedar wood amateur greenhouse in a private garden to a modern, fully automated aluminium glasshouse covering many hectares. Polythene and netting tunnels, conservatories, cold frames and cloches are also considered as protected structures. What they all have in common is the ability to create an environment that induces improved plant growth compared to those growing in an open environment outside.

The uses for protective growing structures are varied. Many are used to produce edible crops grown direct in the ground or modern methods can be used where plants are grown in artificial growing media and have no contact with the soil such as hydroponic and nutrient film units.

Growers can use protected structures for a wide range of crops throughout the complete production cycle or for a specific growing period, such as propagation and establishing newly potted plants. Some houseplants and cut flower crops require carefully controlled light regimes to ensure they have colour and flower at a certain time in order to be saleable for a specific market; other plants require protection from the extremes of the weather to ensure they grow to the desired size and quality within strict time periods.

The Fundamentals of Horticulture: Theory and Practice, ed. C. Bird. Published by Cambridge University Press. © The Royal Horticultural Society 2014.

Figure 11.1 Commercial glasshouses in the old walled garden on the Lackham campus.

Figure 11.2 A poinsettia crop growing at Lackham campus for the Christmas market sales.

THE MAIN TYPES OF PROTECTIVE STRUCTURES

This section will look at the main types of structure available to commercial growers for large-scale plant production, and amateur gardeners for growing a range of plants for home consumption, gate sales and wholesale distribution. It will cover:

- size and construction methods for glasshouses, polythene and netting tunnels, frames and cloches and cloche tunnels
- cladding materials, including glass, polycarbonate, polythene, plastics and netting, horticultural fleece
- the benefits and limitations of these materials.

The scope and complexity of units is extensive and potentially quite daunting for both the keen amateur gardener growing a range of ornamental and edible plants in their own garden or allotment, or a commercial producer growing intensive crops for sale all year round.

Before purchasing a new structure careful research should include factors such as the main construction and cladding materials used and how they affect the growing conditions within the structure; also the ongoing maintenance both inside and outside of the structure plus the capital and running costs to keep production running efficiently. The amount of available space is often the governing factor when it comes to deciding the final size for any new structure. A small garden will necessitate a smaller structure, an allotment or larger garden may accommodate a medium-sized unit, but local bylaws may restrict final roof height and therefore the overall dimensions. On a commercial production unit the amount of actual growing space will dictate the volume and value of the saleable crops to be harvested and therefore the viability of the business. It is essential to carry out this research as part of any business plan to ensure the correct structure and resources are obtained at the start so the viability of the venture is more secure. Major structural changes at a later date are not only expensive but can cause disruption and delay to cropping programmes.

Glasshouses

The smaller, amateur greenhouses are useful for growing and propagating produce for home consumption for many months of the year. The choice between a timber or aluminium structure will be based on the capital costs and the visual impact it will have in a small area rather than the fact that the timber frame will transmit heat more slowly than a metal frame, plus the regular maintenance for the timber will be higher, on a small scale these factors have less impact on the viability of the project.

Other important points worth considering with a smaller structure are that the smaller volume of air within it will heat up and cool down more rapidly than in large structures, ventilation will need to be adjusted more frequently to maintain a uniform growing environment along with maintaining the correct levels of irrigation, both of these topics are covered in a later section.

A visit to a local garden centre with a greenhouse franchise site will show good examples of models available. The choice is large with variations to suit every site and budget. Many companies offer a supply-only deal or the complete supply and installation service. It is worth visiting several companies and looking through the local and national media to compare specifications and prices, ex-display models are often advertised at reasonable prices in order to make space for displaying new models.

The early commercial glasshouses were predominantly constructed using large sheets of horticultural glass with western red cedar timber frames connecting them together; these were known as Dutch-light houses and provided many growers with economical growing space for all-year-round production.

Now aluminium is the main material used. It is light in weight while having great strength and load-bearing capacity when extruded into surprisingly thin glazing bars with complex

Figure 11.3 A range of consumer glasshouses on show.

Figure 11.5 A modern glasshouse with minimum framework allowing maximum light levels within the structure.

Figure 11.4 A traditional Dutch-light glasshouse.

profiles; these being essential in the construction of a modern glasshouse to allow maximum light transmission into the house and minimum shade falling on the growing crops. With the help of computer aided design (CAD) programs the designers and engineers plan every detail into making the structure as efficient as possible.

Timber frames are often more aesthetically pleasing in a garden setting. However, the larger glazing bars do create more shade across the crop and also have a high maintenance requirement to reduce them rotting.

Horticultural glass has been the traditional cladding material due to its high light transmission qualities. Other advantages include long lifespan and better heat retention compared to polythene coverings. However, the disadvantages are that it requires a strong load-bearing frame due to its weight, it is also fragile and if broken creates potential health and safety issues, including hazardous material within the soil and crops.

Figure 11.6 A timber glasshouse makes an attractive feature in a garden setting.

Polycarbonate is available as twin- and triple-walled sheets. Being lightweight and thermally efficient it is a useful alternative to glass. Its limitations are that it can discolour and reduce light transmission into the structure.

Greenwood, P. & Halstead, A. (2009). *RHS Pests & Diseases*. Dorling Kindersley.
This publication has been updated to cover adjustments in the problems now experienced by, and the controls available to, the private gardener.

Hessayon, D. G. (2008). *The Greenhouse Expert*. Expert Books/Transworld Publishers.
The accent here is on quick access to information through a range of illustrated methods supported with bullet-pointed text.

Walls, I. G. (2001). *The Complete Book of the Greenhouse*. Cassell Illustrated.
An excellent balance is provided between text and images to cover the establishment and management of the home greenhouse.

RHS level	Section heading	Page no.
2 1	The main types of protective structures	224
2 2	Controlling the environment within different structures	229
2 3	The main types of protective structures	224
2 4	Controlling the environment within different structures	229
2 6	Pests and diseases	231
2 7	Managing crop production and ornamental plant displays in protected environments all year round	232

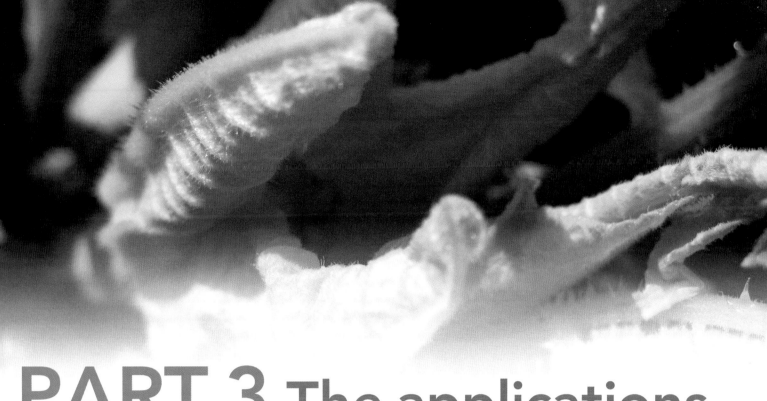

PART 3 The applications

INTRODUCTION

Once the foundations and adjustments of plants and their environment have been studied, and reflected upon, then our attention turns to the application of this knowledge to meet, or sometimes to exceed, our needs and expectations, however challenging, humble or extravagant they may be. Moving forwards with new products and ideas need not mean forgetting tried and tested methods but achieving a balance that we are comfortable with and that we know is robust enough to withstand any changes or challenges that may occur in the future, be they climatically, environmentally or financially induced.

Gardening for food must be one of the success stories of the last few years, with a massive increase in interest shown within this area by all ages and all interest groups. The allotment has come of age and has developed into a prized possession. Everyone is starting to 'grow their own' and being encouraged to do so by a number of the large UK charitable organisations, including the Royal Horticultural Society, the National Trust and Garden Organic. The supermarket chains are also providing support and impetus to the whole movement, underpinned by the government's 'Five a day' campaign.

Remember that vegetables and fruit are very decorative and can be also used as focal points and garden features, such as hedges or garden dividers or to clothe arbours and pergolas.

Features such as these lead us into another area of public interest and leisure activity: designing gardens and landscapes, encompassing everything from 'credit crunch' DIY to the London 2012 Olympics. This chapter introduces or re-acquaints us with the art and skill of design, with themes and ideas influenced,

adjusted or re-used from ancient times but with a whole plethora of emerging materials, both living and non-living, keeping in mind the call for minimum impact on the environment and all things to be sustainable (more of this in Chapter 18).

Using plants in the garden provides us with the opportunity to re-evaluate the constantly emerging plant palette becoming available to us through introductions from overseas, controlled breeding and home selection. Remember that the so-called 'classic planting combinations' are just that, because they work and are associated with memorable images of childhood or yesteryear.

The science behind gardens has been proving more useful and supportive in recent years, as a greater understanding of the links between all living things across the whole world has emerged, whether you follow a creational or evolutionary path. Certainly this closer understanding has provided changes and adjustments of growing methods and materials to the world of commercial horticulture, with massive reductions in energy use, especially by tapping into combined heat and power units, together with the reduction of waste materials and the development of recycling methods and reuse, where possible.

Another area of development, which includes a high science input, is pests, diseases and other problems, also including the ecological aspects of interrelationship, and identification of problem, host and predator combinations. The establishment of a balance, resulting in a holistic state, leads us on to another emerging area: conservation and sustainability.

The establishment of this state may have taken us right round to the beginning again, where all things are providing for our needs as they change through time.

CASE STUDY   Hampshire Potato Weekend with Garden Organic

Over 180 cultivars are available and seed sales, a seed exchange, culinary demonstrations and potato talks all take place. This underlines the developing interest in the widest range of cultivars and the fact that they are now becoming more easily available, including heritage cultivars and organic seed.

In January 2012, over 2,000 people of all ages attended the two-day event, which is held annually on the last full weekend of January in Whitchurch, Hampshire, UK.

Hampshire 'Potato Weekend': some of the range of cultivars available.

Chapter 12

Gardening for food

Chris Bird

INTRODUCTION

One of the most rewarding elements of gardening is that of harvesting and consuming what you have just grown, the taste is outstanding, whatever the crop, and the sense of achievement immense. The range of crops becoming available, both for new species and cultivars, is on a world scale; the variation of colours, textures and tastes is increasing with each new growing season.

The opportunity to grow edible crops across a wide range of sites is increasingly being taken by the widest range of people imaginable, fuelled on one level by concerns over international biosecurity: uncontrolled use or high levels of chemicals, bacterial or fungal infections and unbalanced nutrient levels. Add to this the general perception of modern food having bland flavouring and low nutritional levels, and the drive and stimulus to grow your own crops has captured most people's imagination and heightened their interest as to what can fit into a smaller modern space.

All-year-round cropping, regional variation and a return to seasonal continuity have all fuelled this drive to grow your own. Not to mention the wider benefits: the reduction in food miles and the potential improvement in general health. Also seen is the social side of allotments and community projects to develop or rekindle neighbourhood relationships, a true activity for all the family, together with the positive therapeutic advantages.

This chapter encompasses the range of food crops becoming available to the grower, including herbs, nuts and the developing fruit and vegetables, whether old favourites that have fallen out of favour or newer cultivars. Included are tips and recommendations as to their selection, arrangement, suitability, timings, growing, management, maintenance, harvest and storage. Also covered are the practicalities of producing these crops: dealing with diseases, disorders and weeds. The control of pests is covered more broadly in Chapter 17.

Key concepts

✿ Plant selection for vegetables, fruits and herbs

✿ Their cultivation, in both soil and containers

✿ Management for maximum success of production

✿ Outlining storage methods and timing

✿ Encompassing both growing in the open and under cover

The Fundamentals of Horticulture: Theory and Practice, ed. C. Bird. Published by Cambridge University Press. © The Royal Horticultural Society 2014.

HUMAN NUTRITION

This whole topic area has now become very high on the economic, social and political agendas of not only the UK and the wider Europe but on a world scale. However, for very different reasons. In broad terms the developed countries, especially Western Europe and North America, increasingly suffer from overproduction and very high levels of food wastage; for developing and Third World countries, the opposite is seen, malnutrition and starvation. If we take a look at the numbers involved, the most common reference point is that of the United Nations Food and Agriculture Organization (FAO), which measures 'undernutrition'. The figures for 2010 are that 925 million people were undernourished. If we take a world population of 7 billion people, this means 13.1%, or nearly 1 in 7 people, are hungry.

Nutrients and food preparation

Encapsulated by the UK government's 'Five a day' campaign, the issues of the quality and quantity of modern foodstuffs has been brought to a head in recent years, with a number of national, if not international, players heading up the specific and wider health benefits of a balanced and broad-based diet. Together with the accent on the Royal Horticultural Society's 'Grow your own' and the National Trust's 'From plot to plate' initiatives the home production of fruit and vegetables is very much back on the agenda – with increasing interest in regional and local plant species and cultivars, both basic traditional standards and imports on a world scale.

Added to this is the emerging concern over 'food miles' and the wider use of energy for items, such as cleaning, grading and controlled atmosphere storage, to provide the same crops on an all-year-round basis, e.g. cucumber, *Cucumis sativus*, and tomato, *Lycopersicon esculentum*.

The well-publicised example of the contrasting vitamin C level of a pre-Second World War orange, *Citrus sinensis*, and its present day equivalent (a seven to one reduction) has been well used to show up the drive for looks over nutritional content. This is now beginning to change with the advent of local farmers' markets, and an increasing number of organic growers' and vegetable box schemes where a local producer provides a box of mixed seasonal vegetables on a weekly or fortnightly basis for a set cost, usually also within a set radius, e.g. 25 miles.

Another area that has caught the attention of the general public is that of 'superfoods'. From the re-examination of old favourites, such as broccoli, *Brassica oleracea* Italica Group, to the increasing popularity of blueberries, *Vaccinium corymbosum*, and the shift into wider berry use from a wider geographical area, such as the demand of the fruit of the red chokeberry, *Aronia* x *prunifolia*, which has the highest level of antioxidants of any temperate berry-bearing fruit. However, do not overlook the humble blackberry, *Rubus fruticosa*, as this also has very high levels of vitamin C, antioxidants and fibre.

GROWING VEGETABLES

With increasing concerns over global food security, quantity and quality, the general public has refocused an increasing area of land to the local or home production of edible crops, with the demand of allotments standing at a waiting list of 83,000 names across the UK alone recorded in July 2011 by the National Society of Allotment and Leisure Gardeners, based in Corby, Northamptonshire, UK.

The classical conditions required for growing vegetable crops are a well-structured light clay loam soil with a pH of 6.5 and a minimum depth of 30 cm (12 inches) in a sheltered but well-lit situation, and these provide the best balance to grow the widest range of vegetables. However, some crops do have special requirements in terms of pH and light levels; Table 12.1 provides an overview.

Type	Full sun	Semi-shade	Shade	Water	pH range
Asparagus *Asparagus officinalis*	●●	●		▲▲	5.5–8.5
Beetroot *Beta vulgaris* subsp. *vulgaris*	●●	●		▲▲▲ due to bolting	6.5–9.0
Broad bean *Vicia faba*		●●	●	▲▲	6.5–9.0
Carrot *Daucus carota*	●●	●		▲▲	6.5–8.5
Cabbage *Brassica oleracea* Capitata Group		●●	●	▲▲	6.5–9.0

(cont.)

Cauliflower *Brassica oleracea* Botrytis Group		●●	●	▲▲	6.5–9.0
Celery *Apium graveolens* var. *dulce*	●●	●		▲▲▲	6.5–8.5
Chard *Beta vulgaris* subsp. *cicla* var. *flavescens*		●●	●	▲▲	6.5–8.5
Courgette *Cucurbita pepo*	●●	●		▲▲▲	5.5–8.0
Leek *Allium porrum*	●●	●		▲▲	6.0–9.0
Lettuce *Lactuca sativa*	●●	●		▲▲▲ due to bolting	6.5–9.0
Onion *Allium cepa*	●●	●		▲▲	6.5–9.0
Mizuna *Brassica rapa* var. *nipposinica*		●●	●	▲▲	6.5–9.0
Pak choi and su choi *Brassica rapa* var. *chinensis*		●●	●	▲▲	6.5–9.0
Parsnip *Pastinaca sativa*		●●	●	▲▲	6.0–9.0
Potato *Solanum tuberosum*		●●		▲▲▲	5.5–8.0
Radish *Raphanus sativa*	●●	●		▲▲▲ due to bolting	6.5–9.0
Salsify *Tragopogon porrifolius*		●●	●	▲▲	6.0–9.0
Sweet potato *Ipomoea batatas*	●●			▲▲▲	6.0–9.0
Sweet corn *Zea mays*	●●			▲ when flowers show	6.0–9.0
Tomato *Lycopersicon esculentum*	●●	●		▲▲▲	5.5–8.5

Key:
●● = tolerable light level; ● = reduced cropping light level.
▲▲▲ = full watering required; ▲▲ = reduced watering tolerated; ▲ = dry periods tolerated.
For pH range the figures given are the range for tolerable growth, depending on growing media type.

Table 12.1 Preferred pH and light requirements of common vegetables.

Diversity of garden vegetables: including exotics

The scope for vegetables grown or available to the UK in the twenty-first century has never been wider, and are taken from right across the plant kingdom, from the standard Brassica family to the exotic and increasingly common yam or sweet potato, *Ipomoea batatas*. Another shift that has been seen in the last decade is that towards a more oriental taste, following the popularity of cooking methods such as stir fry and the wider use of the wok. This has opened up the selection, cultivation and consumption of Chinese leaf vegetables, e.g. pak choi, *Brassica rapa* var.

chinensis, su choi and mizuna, *Brassica rapa* var. *nipposinica*, which also extend the growing season and open up an increasing range of autumn and early-winter vegetables. One of the leading authors on this subject has been Joy Larkcom with her seminal work *Oriental Vegetables*, which was totally revised and updated in 2009.

Equally, the taste buds have been excited from other geographical sources: the Caribbean, India and the wider Asian continent including Vietnam and Singapore.

With the proposed increasing changes in the UK climate then the use of these newcomers to expand and complement our stable favourites is causing some exciting times, especially with the growing of home produce becoming, once more, a multigenerational activity, also undertaken in family units or across local communities, to provide mutual benefits to all those involved. Transition Network (transitionnetwork.org) is springing up in many different geographical areas and one of their keynote activities is that of garden sharing: they offer a conduit or forum so that those people with spare land or garden space are put in contact with those that require space to grow their produce and all within a local area. This therefore reduces food miles to a minimum and provides social regeneration or neighbourhood contacts.

CASE STUDY   **Totnes, Devon**

The original 'Transition Town' of the UK, Totnes has stepped up to the challenges of climate change: reducing carbon use, and reusing materials or products.

Features have included the listing of those people within the local area that wish to grow their own edible crops, and linking them with another list of people with available land but who are unable for a number of reasons to make the most of the opportunities available. With both the parties involved then sharing the produce over the season, either through consumption or sales.

Another positive initiative has been the recycling of vegetable oil to power a rickshaw taxi to transport people from the railway station to the town centre.

CULTIVATION OF MAJOR VEGETABLES

Tables A 2.1, A 2.2 and A 2.3 (in Appendix 2) give the recommended sowing, growing and use periods for many seasonal crops. For the purposes of our discussion: spring is taken to be the months of March, April and May; summer, June, July and August; autumn, September, October and November; and winter, December, January and February. Any references to the 'early', 'middle' or 'late' periods relate to the first, second or third month of each listed season. Remember that this provides a guide for the central part of the UK; please subtract a fortnight for the extreme south and west and add a fortnight for the extreme north and east.

As you will quickly find, your own plot, whatever its size, will have its own microclimate and some crops will be much more successful than others; usually totally independent of your personal skill level and experience. Also each year is unique, so keeping a personal record of crop and cultivar selection, together with sowing, transplanting, cultivation and harvesting dates, is key to building up a detailed knowledge of what your plot will grow well without a high level of input from you; inevitably this will be a different set of crops to that you wish to grow.

Also the soil type, its current physical state and general fertility will all have a role to play in determining the range and season of your crops. With heavier clay-based media then a shorter growing season is likely, but less watering will be required during any drier periods. By contrast, soils with a high sand content will allow a longer season, especially in the early spring period, but much higher levels of irrigation and feeding will be needed.

Seed sowing

The selection of top quality seed is vital to successful vegetable production, this not only relates to the crops themselves but more importantly, these days, to the cultivars (commonly but usually incorrectly called varieties) available. A general understanding of some of the common words and phrases used is very helpful in matching what you intend to grow and what you end up with. Seed packets and packaging have changed for the better: with seed counts as to how many seeds are actually in the packet. Don't be dismayed if you only have two cucumber, *Cucumis sativus*, seeds in an expensive packet, this is to pay the breeder and it should have the phrase 'F^1' or 'F^2' in the name.

Some label marks to look out for include NIAB (the National Institute for Agricultural Botany), which selects out crops and cultivars that are good for large-scale production, with characteristics such as to be cut in one pass, but also, increasingly, pest and disease resistance. Also look out for AGM (the Award of Garden Merit) – awarded by the Royal Horticultural Society after being trialled at Wisley Gardens in Surrey, England; this mark is also used for fruit and ornamental plants and indicates a high level of quality in vigour of growth and reliable cropping.

Seed packets also now have a clearly stated 'sow-by' date, which is a vast improvement from the days of 'packeted in year ending…', and is very helpful when a change of season is required in larger shops or garden centres. Check the dates at the end-of-season sales, as you can pick up the majority of your favourite staple vegetables at discount prices, and as they are now vacuum packed there will be no loss of quality or viability. This area is also discussed in Chapter 7.

Another very helpful aspect now seen on most modern packaging is that of a sowing, planting and harvesting bar or pie chart; often this is also colour coded for ease of use, with the colours selected still shaded to be of use to those that are colour blind.

The temperature of the soil when sowing is one of the main reasons for patchy or a failure of germination; this is especially true of those vegetables from the tropical or subtropical regions of the world, e.g. French bean, *Phaseolus vulgaris*, or runner bean, *Phaseolus coccineus*.

A minimum soil temperature of 10°C (50°F) should be looked for to ensure a good level of germination. To check whether your soil has obtained such a temperature there are a number of options: check it with a soil thermometer (or sit on it with your bare bottom), if the soil feels comfortable then sow, or, and this is much better on well-populated allotment sites, check whether there are any fresh wormcasts about. When the earthworm, *Lumbricus terrestris*, casts then the soil has reached 10 to 12°C (50 to 54°F), and it is therefore suitable to sow even the most difficult seeds.

However, it is interesting to note that with beetroot, *Beta vulgaris* subsp. *vulgaris*, a crop that should be in everyone's starter vegetable growing pack, a botanical difference is seen: what looks like a seed is, in fact, a multiseeded fruit. So that no matter how thinly you sow them a group of three to five seedlings will germinate in one place – this means that thinning of the seedlings will be required if an even size of root is required. The upside of this is that the thinnings can be used as part of a mixed salad and in fact some people use the whole first crop of the year in this way. Another way to prevent overcrowding is to select and grow the monogerm group of cultivars, e.g. *B. vulgaris* subsp. *v*. 'Monodet' or 'Monopoly', which have been bred to have single-seeded fruits. Spacing should be at 30 cm (12 inches) between the rows, with the final plant spacing to be 10 cm (4 inches) between plants and the point of harvest being when the roots are the size of a tennis ball.

FLUID DRILLING

This is a technique used for carrot, *Daucus carota*, and especially parsnip, *Pastinaca sativa*, to aid where a very low germination percentage is seen (40%).

Place some blotting paper or kitchen roll in a container, moisten, then sow your seeds. Place the container into a warm, dark place, such as an airing cupboard providing a temperature of 21°C and check every day for the first signs of germination. The seeds only need to **chit**, which means that they have taken up water (imbibition) and the testa (seed coat) has split with the radicle (first root) just emerging.

Then wash off the seeds from the blotting paper through a kitchen sieve (after getting permission from its owner) and after mixing flour and water into a paste, half fill a clear plastic freezer bag with the mixture, transfer the chitted seeds into the mixure and stir slowly – taking care not to damage the seeds. Then using the bag like an icing bag, cut one of the lower corners off and pipe the mixture into a pre-prepared seed drill.

If you can find a wallpaper paste without a fungicide in it, then that can be used instead of the flour paste.

Transplanting

Some types of vegetables, such as leek, *Allium porrum*, or the members of the cabbage tribe (cabbage, *Brassica oleracea* Capitata Group, cauliflower, *B. oleracea* Botrytis Group, and broccoli, *B. oleracea* Italica Group), need to be sown and grown on to a set size in another area of soil or in propagational modules due to the fact that the space required is still being used by the previous crops or they take up a large volume of space in the later stages of their life cycle but take a long period of time to get there. To avoid these problems then the method of sowing them in prepared seed beds has been used. For the cabbage tribe, when the plants get to the size of five true leaves then it is time to transplant them into their growing-on positions at their final spacings. To offset the waste of space, a method of cropping called interplanting has developed. This is where a short-term crop, e.g. lettuce, *Lactuca sativa*, is grown between the widely spaced young plants for the early part of their growth.

A similar method, called intersowing, is also used in seed drills, where a longer term crop, e.g. parsnip, *Pastinaca sativa*, and a short-term crop, e.g. radish, *Raphanus sativa*, are sown in the same drill, so that a crop can be harvested while the longer term crop is left at the correct wider spacing for winter cropping, after the taste is improved by the action of the frost.

In allotment situations many people take the opportunity to swap plants grown on in seed beds so that an increased range is available to each person. Therefore each person only grows from seed one or two species or cultivars, but can then have available a wide range to grow throughout the winter season.

However, one of the major problems that can develop with any member of the cabbage family, *Brassicaceae*, is a fungal disease called club root, *Plasmodiophora brassicae*, which can last in a soil for a minimum of 15 years.

Cultivation

One of the most difficult areas to master is that of timing, you start with a well thought-out plan, but then the weather plays its part, together with competition from weeds, pests and diseases to throw you off course. So do keep things a little loose and not a computer-controlled program.

Do think about using the full range of cultivars available in crops such as lettuce, *Lactuca sativa*, through the seasons: kos, crisphead, butterhead, loose-leaf and cut-and-come-again types.

Onion, *Allium cepa*, is another crop group that spans the seasons providing all-year-round production: through main crop, Japanese, salad, and Welsh onions, *A. fistulosum*. These can be grown together with their close relations to extend the range: shallot, *Allium cepa* Aggregatum Group, and garlic, *Allium sativum*, from cloves, sets or seed.

Leek, *Allium porrum*, can be grown in household downpipes to keep them free of soiling from the media and encouraging blanching (whitening and softening) of the lower leaves.

CROP PROFILES

ENDIVE

- ✿ Botanical name: *Chicorium endive*
- ✿ Common names: endive, escarole

Salad leaves, similar to lettuce but with a bitter taste, grow in a rosette. The leaves may be curled (Frisbee type) or broad-leaved (Batavian type), which are hardier for late crops. Bitterness can be reduced by excluding light and blanching for two weeks. Endive can also be used as a cooked vegetable.

Figure 12.1 Endive, *Chicorium endive*, growing as part of a salad crop.

Common cultivars

- ✿ 'Grobo' – broad-leaved, bolt resistant
- ✿ 'Jeti' AGM – curled, high-quality foliage
- ✿ 'Moss Curled' – curled type, summer or autumn crops
- ✿ 'Pancalieri' AGM – curled, bolt resistant

Cropping schedule

Season	Spring		Summer			Autumn		Winter	
Sow		•	•	•	•	•			
Transplant				•	•				
Harvest					•	•	•	•	•	•

Site and soil: An open sunny site with fertile, moisture-retentive soil and low levels of nitrogen. Midsummer crops can tolerate light shade.

Sowing and planting: Either sow 1 cm (3/8 inch) deep in modules under cover in spring for transplanting 23 cm (9½ inches) apart in early to mid summer, or sow direct outdoors in early summer. Rows 30 to 35 cm (12 to 14 inches) apart. Early crops are liable to bolt if exposed to less than 5°C (41°F) for several days, but bolt-resistant cultivars can be grown, e.g. *C. endive* 'Pancalieri' AGM.

Endives can be grown as cut-and-come-again crops under protection from spring to late summer.

Aftercare: Endives can withstand light frosts and so remain usable into the autumn period. Also using cloches or greenhouse protection extends the season well into winter.

Harvesting: Endives usually mature over a period of 12 weeks, and when the heads reach full size they are blanched by covering each plant for 10 days (20 in cold weather) with an inverted light-proof container (e.g. a bucket) or by laying an inverted dinner plate over the centre of the plant. Plants are susceptible to rot at this stage and covering with a cloche can help. Alternatively use twine to tie the head into a tight bunch and exclude light from the inner leaves.

Storage: Heads should be used as soon as they are blanched, or greenness and bitterness return with light.

Problems

Slugs and snails: non-chemical control methods include physical removal, barriers, beer traps and grapefruit skins. The parasitic nematode *Phasmarhabditis hermaphrodita* can be used where soils are well drained and the temperature is above 5°C (41°F), though this is more effective on slugs than snails. Aluminium sulphate and metaldehyde pellets are used for chemical control.

Aphids (sap-sucking insects 2 mm (1/12 inch) long) may cause reduced growth and leaf distortion. Organic treatments include pyrethrum, rotenone and insecticidal soaps; bifenthrin can be used if the manufacturer's guidelines indicate.

Lettuce root aphid, *Pemphigus bursarius*, can be difficult to control, but watering with an insecticide containing bifenthrin at spray strength may be effective.

Caterpillars: a range of species can damage foliage and roots; the bacterial control *Bacillus thuringiensis* can be used against them, or rotenone, bifenthrin or pyrethrum when signs of feeding are seen, or the caterpillars can be removed by hand – they are generally night-feeders and can be found easily by torchlight.

Diseases and disorders

Tip burn, where leaf margins are scorched and brown, is usually caused by calcium deficiency, particularly on light dry soils. For calcium deficiency add lime to acid soils to raise the pH.

BROAD BEAN

- ✿ Botanical name: *Vicia faba*
- ✿ Common names: broad bean or Fava bean

Very hardy annual, grown usually for the immature seeds but young pods and shoot tips can also be eaten.

Figure 12.2 Broad bean, *Vicia faba*, growing as part of a crop rotation system.

Cultivars

Longpod beans have eight seeds per pod, Windsors have four large seeds. Modern cultivars are intermediate and have shorter stems and tender seeds.

- ✿ 'Aquadulce Claudia' AGM – Longpod, one of the best; overwinters for early use
- ✿ 'Imperial Green Longpod' AGM – disease and weather resistant, exhibition cultivar
- ✿ 'Jubilee Hysor' AGM – Windsor type, good flavour
- ✿ 'The Sutton' AGM – dwarf, suitable for containers and cloches; do not overwinter
- ✿ 'Topic' AGM – modern cultivar with very good yields

Cropping schedule

Season	Spring		Summer		Autumn		Winter	
Sow	•	•				•	•	•	•
Transplant	•	•							
Harvest			•	•	•	•			

Site and soil: Any moderately fertile, well-drained soil. Soil should be deeply dug (broad beans have a long tap root) and well drained. Best yields are on heavy soils, but on lighter soils early crops will give a good yield if watered when in flower. Average yield is 1 kg (2 lb) per metre (3ft 3 inches) of row.

Sowing and planting: Successional sowing in spring – sow when seedlings from the previous sowing reach 8cm (3½ inches). Dwarf cultivars are best for later crops.

If the soil is not waterlogged and the temperature of the soil is over 5°C (41°F), broad beans can be sown out of doors in late winter. In sheltered areas autumn sowings are possible, though the crop can be lost in a severe winter.

Autumn and winter sowings can be made under cover and transplanted in early to mid spring.

Sow 8 cm (7½ inches) deep, 23 cm (9½ inches) apart. Single rows should be 45 cm (18 inches) apart, double rows 23 cm (9½ inches) apart with 60 cm (2 feet) between the rows.

Aftercare: Control weeds, drawing soil around base of the plant with a hoe to protect and support it. Stake tall cultivars as they emerge with old shrub prunings or hazel, *Corylus avellana*, sticks for extra support.

When the lowest blossom has set, pinch out the top to promote early cropping and remove any blackfly, *Aphis fabae*. Watering during dry spells when the crop is in flower will increase the yield considerably.

Harvesting: Spring-sown crops mature in three to four months, autumn and winter sowings will take longer.

Beans mature in succession starting with the lower pods – pick regularly before the pods get too old. Once the point of attachment of the seed within the pod has turned brown or black the bean is becoming tough.

Problems

Black bean aphid, *Aphis fabae*, sucks sap, causing stunting of leaves and stems. Pyrethrum, soft soaps and rotenone may be effective before infestations get really heavy. Pinch out infested tips.

Pea and bean weevil, *Sitona lineatus*, damages leaves on young plants by nicking the edge but does no great harm to the crop.

Seeds are attractive to **mice**, e.g. *Mus musculus*, and **bean seed beetle**, *Bruchus rufimanus* – try scattering dry holly, *Ilex aquifolium*, leaves in the planting area to dissuade the mice.

Chocolate spot, *Botrytis fabae*, can infect the plants in wet seasons, particularly overwintering crops. Avoid high nitrogen fertilisers, which promote soft growth and the application of sulphate of potash may help harden plants. Spray with carbendazim, or increase spacings within and between rows for better air circulation. Maintain good control of weeds.

COURGETTE AND SUMMER SQUASH

- ✿ Botanical name: *Cucurbita pepo*
- ✿ Common names: courgette, summer squash and zucchini

Figure 12.3 Courgette, *Cucurbita pepo*, growing as part of the high summer cropping to provide total soil coverage preventing weeds problems.

supply is known to exacerbate the condition and so it is commonly seen on the return from a holiday period.

A shortage of calcium is also the cause of bitter pit found on fruit, especially apple, *Malus domestica*, with some cultivars being more susceptible than others, e.g. *M. domestica* 'Howgate Wonder' is a common indicator cultivar being one of the first to show the disorder. Brown slightly sunken spots, starting at 1 mm (1/25 inch) in size and developing into 5 mm (1/5 inch) over a matter of weeks, are seen on the skin of the fruit. If peeled these brown areas are only just beneath the skin but make the whole of the flesh bitter to taste, also the problem develops further in storage and is an entry port to fungal rots.

Further details of trace element problems are given in Chapter 6.

Physical problems include weather damage caused by such events as frost, hail, sun and wind, together with the excess or lack of water.

Figure 12.4 Weather damage to vegetables: bolting.

INTEGRATED CONTROLS

The way forward, commonly highlighted in recent years, has been to adopt a more holistic approach to growing, that is, an all-encompassing method of growing providing a balance of all

the plant's needs. This has included things such as the correct spacing for each plant and the overall density of the crop for a given area. Current research is actively investigating the impact of different growing arrangements and plant configurations.

Balance is the word to remember, as an excess or deficiency of anything will result in stress on the plant, slowing down growth and providing an opportunity for pests and diseases to become established.

CASE STUDY | **Equidistant planting**

One of the developments in tandem with the more common use of the raised bed is that of equidistant planting, where the plants are set out on the square, with the same distance being used for between the rows as in the row.

The aim of this method being to provide total crop cover at the earliest opportunity, that is, to cover the soil or growing media with the growing crop, thereby allowing no space for weeds to grow. This has caused an increase of crop weight for a given area being recorded and easier maintenance, such as hoeing less often, resulting in less plant disturbance and no thinning.

Vegetable storage

More modern methods of traditional storage include bottling, canning or freezing, but all these require some form of preparation or processing. However, the area of fresh storage methods is being re-examined as they have low cost with very low energy inputs. A garage or outhouse, e.g. a garden shed, even a Second World War air raid shelter can be brought into use.

The main criteria required for successful storage of vegetable or fruit crops are:

- ✿ moist, not dry, atmosphere
- ✿ even temperatures around 4°C, but up to 10°, over the summer period
- ✿ frost free, aim for a minimum of 2°C, during the winter period
- ✿ free from rodent access to avoid damage and contamination
- ✿ good air flow around the store and between trays or shelves of produce, including the use of a fan or heater running on cold
- ✿ space between items of produce, e.g. apple, *Malus domestica* and pear, *Pyrus communis*
- ✿ store only undamaged produce – fruit with just the stalk (pedicel), missing will have a reduced storage time and be an entry port for the introduction of rotting
- ✿ store in usable amounts, in several small containers rather than one large one, to avoid the risk of infection, especially with common fungal problems, such as brown rot, *Monilinia fructigena*, on fruit
- ✿ regular, e.g. weekly, inspection of the stored produce, including turning or picking through and removing any diseased items.

One of the major problems or limitations for producing an excessive harvest is that of storing it until it is required. In many modern homes there are no places left to store fresh produce – space under the stairs or the walk-in pantry have been removed. So other methods need to be used: a return to the traditional methods of outdoor storage such as clamping is now occurring. Clamping involves digging out a shallow hole of a suitable size; remember that several smaller clamps are much better than one large one as once opened the produce must be used quickly. The hole is then lined with a 5 cm (2 inch) layer of dry straw and on the surface dry produce is added then topped with more straw of the same depth and then covered with a 5 cm (2 inch) layer of topsoil. The only other thing to remember is that each clamp requires a 10 cm (4 inch) diameter vertical 'chimney' of straw, linking the produce to the outside atmosphere; this is to allow the exchange of gases and to keep the temperature even. Vegetables that are suitable for this method of 'in field' storage are root crops, such as carrot, *Daucus carota*, or parsnip, *Pastinaca sativa*, and stem tubers such as main crop potatoes, *Solanum tuberosum*.

HERBS AND HERB GARDENING

Diversity of herbs, culinary and other uses, and historical associations

Figure 12.5 A classical medieval-style herb garden.

Most people associate herbs and their uses with two main classical historical peoples that have left a lasting influence on our selection, arrangement and use of herbs: the Romans and the medieval religious orders. Both of these peoples brought with them not only their knowledge but also the plants themselves. Monks, especially in their extended time within the UK, including in the Middle Ages, provided a very wide range of plants from all across the plant kingdom.

Remember that the plants provided not only medicines, culinary flavourings and dyes but also great symbolism and the lasting impression of protection. These lasting impressions still echo to this day and some folklaw recommendations have subsequently been proven to supply the protection that have been associated with them since before medieval times, including anti-bacterial effects. Well-documented examples include: lavender, *Lavendula spica*; thyme, *Thymus sempervirens* and English marigold, *Calendula officinalis*.

One of the greatest benefits provided by growing herbs in association with vegetables and fruit is that of supplying the continuity of nectar to introduce and maintain the level of pollinating insects across a wide seasonal period. Together with supplying home to a wide range of beneficial organisms, these provide protection against pests, including aphids and whiteflies. Also the root exudates, chemical products released into the surrounding soil, are being used to clean up soils, e.g. Mexican tarragon, *Tagetes lucida*.

Pests and other problems

With the advent of a potentially warmer climate a number of new pests have been found more commonly across the range of the herb group of plants. Rosemary beetle, *Chryolina americana*, has now become widespread round the UK, after being first recorded at the Royal Horticultural Society Garden at Wisley, in Surrey during 1994, a commonly found beetle around the banks of the Mediterranean has now been introduced with imported plant material. The very attractive adult is up to 8 mm (2/8 inch) long and is a metallic green with purple stripes. However, the eggs are sausage shaped and 2 mm (1/16 inch) long, held under the foliage from September until April, once hatched they become soft-bodied grubs, an off-white colour with five dark lines along their bodies, developing to around 5 to 8 mm (1/8 to 2/8 inch) long. For organic control pick off pests at all stages or place a collecting sheet under the plant and tap the plant. Chemical controls are available, such as products containing pyrethrum, which is approved for use on edible herbs, but a harvest interval is still recommended. Bay sucker, *Trioza alacris*, has also been more of a problem, after a series of mild winters, this is seen as an enlarged yellow or brown edge to some leaves over the summer period and is best controlled by leaf picking as soon as any damage is identified.

A greater area of problems is found with diseases on herbs, especially with the mildews, both powdery, *Erysiphe* spp., and downy, species of *Peronospora*, *Plasmopara* and *Bremia*, and rust, *Puccinia menthae*, which is one of the few things that can limit the growth rate of mints, *Mentha* spp. The best method of instant control is leaf removal, as soon as the problem is noted; however, this may mean total removal of all the foliage at the same time, over the summer period. This breaks through the cycle of infection by spores and allows the free flow of air, promoting the quick regrowth of a new set of healthy foliage within a few weeks.

CASE STUDY The Winter Garden, Sir Harold Hillier Gardens, Romsey, Hampshire, UK

The Winter Garden at the Sir Harold Hillier Gardens exemplifies the results of a successful mix of horticultural expertise and design skills. The Winter Garden was established in 1996 and officially opened in 1998. Issues of sustainability were taken into account, and recycled and reclaimed materials were used in its construction. The situation near to the entrance and all-weather paths makes the Winter Garden easily accessible for all. Both the structure of the planting and the subsequent maintenance techniques employed, have created a garden for inspiration throughout the seasons, becoming particularly vibrant in the winter when many others gardens are past their best.

A striking and uplifting design in the Winter Garden, Sir Harold Hillier Gardens, Romsey, Hampshire, UK.

Chapter 13

Designing gardens and landscapes

Jenny Shukman

INTRODUCTION

An understanding of the processes, principles and history of garden design is important for anyone involved in garden and landscape planning. It also increases enjoyment from simply visiting gardens.

A well-designed garden or landscape will fulfil its function at all levels. It will be aesthetically pleasing, inspiring, and yet practical and easy to use, such as the Winter Garden at Sir Harold Hillier Gardens, Romsey, described in the case study and illustrated above. The starting point for good design is an understanding of the principles of design, which apply equally to small domestic gardens and large parks. Linked with an understanding of design principles is an understanding of garden history. Styles have changed throughout history, associated with fashions for a variety of formal and informal designs. Great influxes of new plants into cultivation in particular eras have also influenced these changing styles.

Gardens and parks need to be accessible for everyone. The educational potential for children is immense. Gardens, parks and green spaces provide fresh air, exercise and stimulation. The value of these throughout history has not always been recognised to the same degree, but it is again being brought to the forefront with links to the benefits of a healthy lifestyle.

It is also vital that garden planning is carried out with sustainability in mind (see Chapter 18). A garden designed for wildlife will make a valuable contribution, providing different habitats, shelter and food sources for our native fauna. It does not matter what size the garden is, it is an effective way of gardening on any scale.

The design process allows the garden to come into being. This involves initial surveying, discussions, and research into suitable materials and plants, leading to the production of a working plan.

The features to include may be divided into living and non-living materials, i.e. soft and hard landscaping respectively. An assessment of the climate and microclimate of a garden or landscape should be included as part of the survey. An understanding of the problems of a changing climate needs to be taken into account in the subsequent planning (see Chapter 5).

Key concepts

✿ Understand the principles of design

✿ Describe key design styles and features

✿ Describe how to design a garden to encourage wildlife

✿ Understand the requirements of a range of garden users

✿ Explain the different roles for urban parks

✿ Evaluate a range of historical styles and the role of plant introductions

The Fundamentals of Horticulture: Theory and Practice, ed. C. Bird. Published by CAMBRIDGE UNIVERSITY PRESS. © The Royal Horticultural Society 2014.

ORNAMENTAL GARDEN DESIGN

… the best kind of garden should arise out of its site and conditions as happily as a primrose out of a cool bank

(Robinson, p.viii, 1883)

The successful design of ornamental gardens involves linking an understanding of the site conditions and requirements of the garden users with the principles of design.

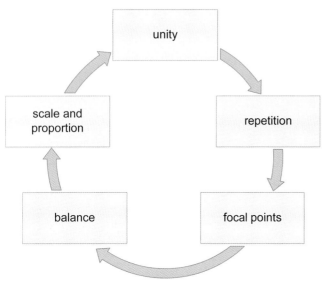

Figure 13.1 Interlinking principles of design.

Principles of design

To create a garden that is aesthetically pleasing, it is important to design the basic layout around certain design principles. Many of these have their roots in the most ancient of gardens, although they are equally relevant for a contemporary style.

Unity is a major principle, ensuring that the garden works as a whole. All parts of the garden need to tie in with each other, which can be achieved by choosing an overall style or theme for the garden design. Other methods of achieving unity include:

❀ linking hard landscape materials of buildings and paths or patios, for example, repeating a brick from the house as an edging for a patio
❀ a colour theme
❀ repetition
❀ using the shape of the dominant building, usually the house, to inform the layout of the garden

Repetition, a key element for providing unity, can be achieved in many ways, by repeating:

❀ shapes, forms, colours and textures
❀ specimen plants or plant groups
❀ features

Repetition may be used to create a sense of rhythm within a garden.

Two major design principles are those of symmetry and asymmetry, and form the basis of distinct garden styles throughout history.

Symmetry: a symmetrical design is one that forms a mirror image along a central line or axis. There may be one, two or several lines of symmetry. Radial symmetry may also be used, dividing the area into similar parts from a central point rather than line.

The symmetrical design in Figure 13.2 demonstrates two lines of symmetry, dividing the garden into identical quarters.

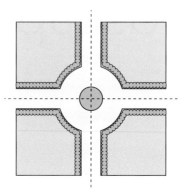

Figure 13.2 A symmetrical design with two lines of symmetry.

Asymmetry: an asymmetrical design does not form a mirror image along a central axis.

The asymmetrical design in Figure 13.3 has no lines of symmetry; it does, however, have the property of balance.

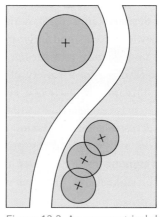

Figure 13.3 An asymmetrical design with no lines of symmetry.

Balance is vital to create a garden that feels pleasant to use. A symmetrical design may be more straightforward to balance, particularly if there are two lines of symmetry. The same shapes, features and plants will be positioned in each quarter of the garden to reflect each other. However, care still needs to be taken to position all these features to create a balanced feel. The easiest way to consider the question of balance is to imagine a seesaw, each side of the central pivot being sides of the garden. If one side is too heavy the garden will feel out of balance.

Balance is equally important in an asymmetrical design, and again, the seesaw image can be used. Three smaller features, e.g. shrubs, may be used to balance one larger feature, e.g. a tree.

Movement: a sense of movement may be achieved by the careful placing of focal points or the use of repetition.

Scale and proportion: to achieve balance, unity and harmony within a garden all the elements must be of the appropriate scale and proportion. A massive statue would look out of place in, for example, a small urban garden, and a small pebble fountain equally out of place in a large country estate.

Pleasing proportions may be achieved by using the **golden rectangle**. This ratio has long been used over the ages by artists and architects, such as Leonardo da Vinci, to provide aesthetically pleasing effects.

CASE STUDY	A4 paper and bank cards = common shape?

They are both practical examples of the use of the 'golden rectangle': a rectangle with sides in the ratio of 1:*phi*, or approximately 1:1.618. When a square section is removed, the remainder forms another golden rectangle.

Focal points are widely used to create points of interest in the garden and lead the eye. They can help create movement through a garden or make a dramatic full stop. Strong architectural shapes can create focal points, such as spiky plants, upright forms and obelisks, as well as bold colours.

To achieve a balanced garden, the proportions of mass and void need to be considered. Mass refers to three-dimensional features, such as buildings, trees and shrub borders. Void refers to two-dimensional features, such as lawns, ponds, paths and patios. A pleasing ratio is 1:2, mass to void, i.e. a third of the garden as mass and two thirds as void. Too much mass will make a garden feel heavy and closed in, whereas too much void will give an extremely open and maybe featureless feel. There are, of course, exceptions. A wildflower meadow could be regarded as void, but may be exceptionally pleasing. Conversely, the aim might be for an enclosed, tropical, urban back garden, with a far greater proportion of mass.

Perspective is a useful tool to use to manipulate the garden space. A garden can be lengthened by the use of a narrowing path, or cool colours placed at the boundary. By placing pots, plants or other features of decreasing height either side of the path, the effect is further enhanced. Figure 13.4 demonstrates how simple this effect can be. This may also be termed a **trompe l'oeil**, or 'trick of the eye', which can be used in the garden to deceive and suggest greater depth. This method of creating an illusion is often created by means of a painting, mirrors or trellis work.

Borrowed views may be used to enhance the garden or make it appear larger. Unsightly views should be screened by carefully placed features, whereas an attractive view should be framed, or

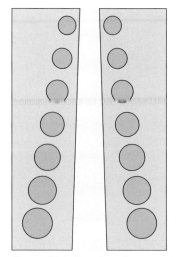

Figure 13.4 Creation of perspective.

blended into the garden to make the garden and wider landscape appear seamless.

Colour, texture and **form** all play a vital role. Colour can be used to create different moods. Cool colours, such as green and blue, create a calming effect, as shown in Figure 13.5. Greens and blues also provide a feeling of distance. To make a garden appear longer or wider, cool colours can be used at the edges. Hot colours, such as red, orange and yellow, give a bright, uplifting feel. They can be used to create focal points and draw the eye. As with all aspects of design, it is also important not to adopt too rigid an approach. Christopher Lloyd's wonderful and exuberant colour schemes at Great Dixter, East Sussex, UK, were achieved with scant regard for the colour wheel.

Figure 13.5 The blue of *Ajuga reptans* among green foliage provides a cooling effect in the dappled woodland.

Texture refers to the surface qualities of a plant or feature. This may range from glossy, smooth and reflective to rough and matt. Figure 13.6 shows an example of glossy texture in the bark of *Prunus serrula*. This contrasts with the matt, peeling bark of *Betula papyrifera* (Figure 13.7).

Form refers to the shape of a feature or plant. Upright forms include plants such as *Phormium tenax* and *Taxus baccata* 'Fastigiata', and hard landscape features such as obelisks. These make punctuation marks, or focal points in a landscape, garden or individual border. Rounded forms create a softer effect. Horizontal forms can lead the eye to another feature or plant. They can be used to link upright and rounded forms in a planting scheme. Weeping forms, such as *Itea ilicifolia*, illustrated in Figure 13.8, can also create focal points, or provide linking features. Spiky architectural foliage can be used to create dramatic focal points. Many plants and features combine several elements to provide striking focal points. *Kniphofia rooperi*, illustrated in Figure 13.9 combines hot colours, an upright form and architectural foliage.

Figure 13.6 Glossy bark of *Prunus serrula*.

Figure 13.8 *Itea ilicifolia* provides a weeping form.

Figure 13.7 Matt, peeling bark of *Betula papyrifera*.

Time is an element that is crucial to consider for garden design, and sets it aside from many other schools of design. Gardens are living and evolving, and need to be planned with a view to their development. Plant life cycles are varied, from ephemeral to perennial. An annual border may be created to achieve its potential in one season, but a tree may live for hundreds of years. A

Figure 13.9 *Kniphofia rooperi* combines several features to provide a strong focal point.

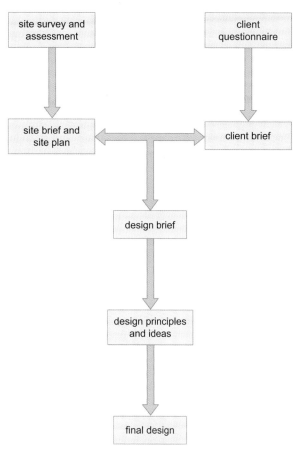

Figure 13.10 The design process.

decision has to be made as to how much space to allow the developing plants. The plan may allow space for the estimated growth of a plant within five years, or maybe ten years. Many gardens and landscapes are planted for instant effect, and therefore need thinning out as they mature.

It can be seen that all the design principles are interrelated and need to work together to produce a satisfying design.

The design process

A logical process needs to be worked through to achieve a design that suits both the site and its users. Figure 13.10 outlines the main stages that need to be brought together in sequence to achieve a successful design.

The first step is to survey the site to accurately plot and measure the boundaries and features. Simple methods may be used to survey a site without any complex equipment, including **triangulation** and **offsets**. These may be carried out using long tapes, e.g. 30-metre tapes. Running measurements may be taken directly from the tape, for example measuring along the side of a house to plot doors, drains and windows.

The tape may be used as a fixed baseline for offset measurements. These are taken at regular intervals at right angles from the baseline. This method is useful for shorter distances on relatively flat sites with few obstructions, and is useful for plotting curves. Figure 13.11 demonstrates the use of offsets for plotting a curved border.

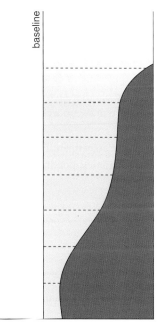

Figure 13.11 Using offset measurements to plot a curved border.

Triangulation involves accurately locating an unknown point using two fixed points. This method is useful on sloping and uneven sites with obstructions, and for longer distances. Triangulation may be used to plot the boundaries of a site, or important features such as trees and garden buildings. Angles are measured from the two fixed points to the feature required, and these are used to plot a triangle. Where the two lines cross accurately locates the unknown point. The same result may be achieved by measuring from the two fixed points with the tape measure (called trilateration). Alternatively, equipment may be used such as laser levels, dumpy levels and automatic levelling equipment such as total stations. For larger and more complex sites global positioning systems (GPS) are often used. Figure 13.12 demonstrates the use of triangulation to plot the far boundaries of a garden, using the corners of the house as fixed points.

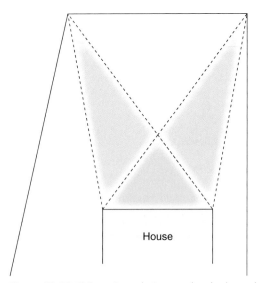

Figure 13.12 Using triangulation to plot the boundaries of a garden.

The survey measurements are drawn up on a survey plan, which is a rough plan, but with accurate measurements recorded on it. From this an accurately scaled site plan is drawn up. This may be drawn by hand or by using computer aided design (CAD).

As well as measurements, a wide range of other factors need to be recorded to assess the site. It is useful to draw up a checklist of factors to assess, as in Table 13.1.

All the information gathered from a site assessment is summarised in a site brief, setting out the characteristics, possibilities and potential problems of a site.

Boundaries	position
	type
	condition
	ownership

Soil details	pH
	texture
	structure
	depth of top soil
	organic content
	drainage
	contamination
Climate and microclimate	wind direction and exposure
	temperatures
	frost pockets
	rainfall
Topography	degree of any slopes
	direction of any slopes
Aspect	sun and shade
Views	pleasant views to enhance
	unpleasant views to screen
Existing services	overhead and underground power cables
	position of drains
Existing hard landscape features and buildings	style and materials for house and any outbuildings
	type and condition of garden structures
Existing soft landscape features	type and condition of lawns, trees, shrubs, herbaceous perennials
Existing water courses or features	natural or installed
	condition and site
Legal factors	planning issues
	Tree Preservation Orders
	protected species
Wildlife issues	existing wildlife
Access	relating to machinery, delivery of materials

Table 13.1 Factors required for a site assessment.

Alongside the site assessment, equally important is the assessment of the client's or users' needs. This is often assessed by the use of interview, discussion and questionnaires. Examples of factors to determine for the garden users are shown in Table 13.2.

After careful analysis of the client questionnaire, a client brief should be drawn up, summarising the needs of the client.

The site brief and client brief are brought together in a design brief. This takes into account features of the site and wishes of the client, and suggests design outcomes to best fit the two. Sometimes the two may conflict, and the client may wish for a feature or type of planting scheme that does not fit the site well, for example, ericaceous plants on an alkaline soil.

Once a design brief has been agreed, the process of producing the new plan begins. The site plan is the starting point, and if drawing by hand, tracing overlays may be used to produce alternative designs. All the principles of design then need to be considered to produce a design that is aesthetically pleasing and fits the design brief.

The plan should include a scale, direction of north and key. Standard symbols are used to represent features and plants within the garden.

Hard landscape features

Hard landscape may be defined as the non-living elements in a garden or landscape. An ever-increasing range of materials is available to the garden designer, but several factors need to be taken into consideration before selecting materials. These include suitability for the purpose, durability, appearance and cost. The principles of design must, again, be considered. The materials need to fit with the general design and provide a sense of unity.

Hard landscape features may be divided into horizontal and vertical elements.

Horizontal elements – Paths may be constructed from a range of materials, including concrete, gravel and paving. Concrete is functional, providing a solid, hardwearing surface for easy access. It can be laid to create a variety of shapes and sizes desired, but may not fit aesthetically in more ornamental areas.

Gravel has a more natural appearance, and can again be laid to a variety of shapes and sizes. However, even if laid on solid foundations, it is more difficult for access, for example for wheelchair users and pushchairs.

Paving may be of natural stone, concrete or reconstituted stone. Larger units are difficult to use to create curves and awkward shapes, although preformed curved and circular shapes are available. Natural stone can be more aesthetically pleasing, although issues of sourcing and sustainability need to be considered. Reconstituted stone or concrete is generally less expensive, and an increasing range of products is available from purely functional to highly decorative.

Garden users	number of people
	ages
	disabilities
	allergies
	pets
Garden style preferred	contemporary
	informal
	formal
	naturalistic
Functional requirements	entertaining area
	play area
	food production
	utility area
	car parking area
	security
Soft landscape features required	lawn
	meadow
	vegetable garden
	fruit garden
	herb garden
	mixed borders
	specific types of planting
Hard landscape features required	patio
	paths
	garden furniture
	statues/ornaments
	arches, pergolas, gazebo
	water features
Garden buildings required	greenhouse
	shed
	summerhouse
Maintenance levels	
Budget	

Table 13.2 Client questionnaire.

Small unit paving is easier to use to create curves and fit awkward spaces. It is commonly used for driveways.

UNDERSTANDING THE SETTING OUT AND CONSTRUCTION OF LANDSCAPING ELEMENTS IN THE GARDEN

Flexible paving: paving units such as block pavers, granite setts, or paving slabs are laid on a soft material such as sand. The bedding layer remains flexible or fluid and does not set. Edging is required to prevent movement.

Inflexible paving: a fluid material is used, which sets solid and is therefore inflexible. An example is concrete, which requires expansion joints at intervals. Alternatively, individual units such as block pavers or paving slabs may be laid on a mortar base to provide inflexible paving. No edging is necessary.

Permeable paving: paving that allows water to infiltrate rather than run off. For example, water will infiltrate through the sand in the joints and bedding of inflexible paving, or through a gravel path with a geo-textile membrane. The use of permeable paving is important to prevent problems of water run-off and flooding. The use of permeable paving is part of the wider aim of sustainable urban drainage (SUDS), which aims to change many traditional methods of drainage and deal with the water at source rather than discharging into drainage systems or rivers. More sophisticated systems are also available to store the water for later use.

Decking can add a more natural feel to the garden. Choice of timber affects the longevity of the feature. Composite decking is also available, which uses recycled materials.

Steps and ramps provide access to different levels within a garden. A woodland-style garden could have steps constructed with log risers and bark treads. A more formal garden could have steps constructed from brick risers and paving treads.

Vertical elements – Walls and fences are common boundary features. Walls may be constructed of brick, maybe matching with the house, or rendered blockwork painted to create a particular atmosphere or theme. Examples of fencing include closeboard for greater privacy and security, or open picket fencing for a cottage garden feel. Woven willow hurdles would suit a naturalistic theme or temporary need.

Pergolas and arches create focal points and can be used to direct movement in the garden. Timber links with a natural style, or metal, or plastic-coated metal could be used in a formal style. Obelisks also create focal points, and can be used to create height in borders.

Statuary and garden furniture should be selected to suit the style of the garden, whether contemporary or traditional. The former, in particular, creates the opportunity to personalise a garden.

Water features include ponds, rills, fountains and canals. These are often also linked with rock gardens. Materials and design should be selected to fit with an overall theme or style. Ponds may be formal or informal, symmetrical or asymmetrical. Examples of rills include the symmetrical layout at Hestercombe Gardens, Somerset, UK and the informal design at the National Botanic Garden of Wales, Carmarthenshire, UK. Fountains may be grand and impressive, for example in the formal layout of the extravagant gardens of the Palace of Versailles, France, or smaller and gentler, such as pebble fountains. Examples of canals can be seen in the historic Dutch-style gardens of Westbury Court, Gloucestershire, UK.

Rock gardens may be created using natural stone, reconstituted stone or artificial stone. The option of natural stone may be the most aesthetically pleasing. However, the stone needs to fit with the site, preferably being locally sourced. Issues of conservation and sustainability come into play here. If imported stone is used, not only do environmental factors have to be taken into account, but also the important consideration of ethical production. Alternatives to natural stone have long been experimented with. Many historical gardens have rock gardens created from Pulhamite, an artificial rock made by cement render, which was used by the firm, James Pulham from the 1830s to 1940s. The range of artificial materials available today has been extended to include rocks constructed of polyurethane.

Soft landscape features

Soft landscape may be defined as the living elements within a garden or landscape. Some may be providing the same function as hard landscape alternatives, such as boundaries and paths; others are providing an entirely different function. See also Chapter 14.

Hedges provide living boundaries, creating habitats for wildlife and effective noise barriers. Taller hedges tend to be used as boundary markers, lower hedges as divisions within a garden. Hedging plants for a formal garden need to be able to respond to regular close clipping, and form dense cover. **Topiary**, **pleaching**, **coppicing** and **pollarding** may be used for special effects.

Other vertical elements or masses within a garden are provided by a range of plants, from trees and shrubs to herbaceous plants. Themes may be selected with particular plants, creating rose gardens, herb gardens, and shrub borders or bedding themes. Alternatively, borders may be mixed with a variety of layers and types of plants.

Lawns, meadows and grass paths provide horizontal elements, or voids in a garden or landscape.

Garden styles

It is crucial to decide on an overall style at the outset. This may be determined by the setting of the garden, e.g. a rural setting may suggest a cottage style or an urban setting may lend itself to an enclosed courtyard style.

Formal styles have several distinct features. They are symmetrical and planned around geometric shapes, such as squares, rectangles, hexagons, ovals and circles. Straight lines are commonly used, and curves only included as part of a geometric shape. Formal features are included such as tightly clipped hedges, topiary, and formal water features such as canals and fountains. Plants may be used to provide patterns or symmetrical features. The interest in the plants tends to be based around their architectural value or blocks of colour.

Informal styles are characterised by a more flowing style, consisting of organic shapes and curving lines. They are asymmetrical, but still need to be balanced to be aesthetically pleasing. A wider palette of plants may be used, and the style may be developed to include naturalistic planting.

Cottage gardens are characterised by a wonderful mix of different plants. The layout may be fairly simple and unstructured, perhaps with just a single path leading to a front door between two borders. Edible plants and ornamentals are combined. Rustic, natural materials should be used, such as hazel hurdles or picket fencing.

Contemporary gardens are those of the current time. An interesting contemporary style is to combine the formal and informal, with the use of tightly clipped hedging surrounding a naturalistic, exuberant and free-flowing planting style.

Minimalist gardens place an emphasis on form and style. Colour palettes should be limited and sleek lines and materials utilised.

Utility/domestic gardens generally need to combine aesthetics with practicality. For most domestic gardens a range of users need to be considered. Features need to be incorporated into the design such as play areas for children, entertaining areas, an area for the wheelie bin and clothes dryer, maybe somewhere for a rabbit or guinea pig hutch. The function of the garden will need to change as children grow up or people find the maintenance harder. Alongside these practicalities, the garden should provide a pleasant and inspiring environment.

Naturalistic gardens are carefully planned to appear as if they have developed naturally. They may place an emphasis on native plants, but can equally include exotic plants carefully selected for particular characteristics. Herbaceous perennials, for example, should require minimum maintenance in terms of feeding and staking.

Wildlife gardens, as discussed in the next section, are designed to maximise the habitats and food sources available for a variety of insects, mammals, birds and reptiles.

Although, as stated earlier, an overall style is important to achieve unity and cohesion, it is sometimes possible to include a change and element of surprise. Larger gardens may have distinct

UNDERSTANDING GARDEN SURVEY TECHNIQUES AND DESIGN PRINCIPLES

Moorish: Religious symbolism features highly, exemplified by the Persian paradise garden layout of a square quartered by canals. Enclosed courtyard gardens with high walls, and walls and paths with coloured mosaic tiles, Islamic motifs used for decoration and pattern.
Italian: Italian Renaissance gardens are exemplified by formality, flamboyancy of style, symmetrical gardens with elaborate water features such as fountains and cascades, topiary, parterres, terraces with grand steps, balustrades, statuary and vistas.
French: French Renaissance gardens adapted the Italian style for a flatter landscape. They are exemplified by formality, grand symmetrical designs, parterres, vistas, elaborate pools, topiary, restricted use of colour and avenues radiating into the surrounding countryside. The parterres further developed in intricacy, creating parterres de broderie, resembling intricate embroidery.
Japanese: Religious symbolism features highly, typified by contemplative Zen gardens. Informal, asymmetrical but well-balanced gardens, including dry gardens representing landscapes, using rocks to represent mountains and raked gravel or sand to represent water; symbolic routes to tea houses. Simplicity and harmony, albeit carefully contrived, dominate the style.

Figure 13.13 Naturalistic planting style, Knoll Gardens, Wimborne, Dorset, UK.

areas within them, where a change of style is effective. This may be a more formal enclosed garden within a larger woodland or naturalistic garden, for example at Knoll Gardens, Wimborne, illustrated in Figures 13.13 and 13.14, and described in the case study.

Figure 13.14 The Dragon Garden, Knoll Gardens, Wimborne, Dorset, UK.

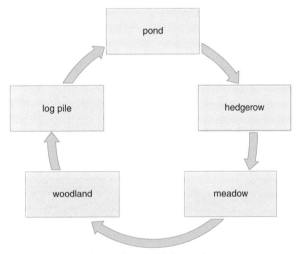

Figure 13.15 Habitats for a wildlife garden.

CASE STUDY **Knoll Gardens, Wimborne, Dorset, UK**

Knoll Gardens is an example of a garden that epitomises the naturalistic style, created by an ornamental grass specialist, Neil Lucas. The overall style is informal, developed around an existing mature framework of trees and shrubs. Naturalistic plantings of ornamental grasses and herbaceous perennials, curving borders and a gravel garden have enhanced the site. All-year interest is provided by this changing framework, complemented by spring-flowering shrubs, summer interest from the herbaceous perennials, autumn fruits and winter stems and seed heads. A separate garden room, the Dragon Garden, is designed in a more formal style, with symmetrical beds and a central pond and dragon sculpture. The planting in the beds is, however, exuberant and overflowing, carrying on the theme of ornamental grasses and herbaceous perennials. The garden is being further developed with the aims of being sustainable and wildlife friendly. Evidence of this is easy to see if visiting on a warm summer's day, when the garden is alive with birds, butterflies, bees and a myriad of other insects.

WILDLIFE GARDENS

…the mutual dependence of the animal and vegetable kingdoms.   (Wilson, 1920, p.ix)

Figure 13.16 Obvious benefits of a wildlife garden – the migrant painted lady butterfly, *Cynthia cardui*, is attracted to the flowers of *Allium schoenoprasum*.

All gardens are generally good for wildlife, but one designed with wildlife specifically in mind can really enhance and benefit wildlife populations, and also create far more interest in the garden. Some of the benefits may be obvious, such as a garden alive with feeding birds, bees, butterflies and hoverflies (see Figures 13.16 and 13.17).

Figure 13.17 Bees, *Apis mellifera*, attracted to the flowers of *Sedum telephium*.

Figure 13.18 The single flowers of *Caltha palustris*.

Figure 13.19 The double flowers of *Caltha palustris* 'Flore Pleno'.

Others are less noticeable but equally important, such as beetles and a myriad of soil-dwelling creatures.

Garden ecology

Ecology may be defined as the interrelationships of living organisms, including plants and animals, both with each other and their environment.

Much of horticultural practice can be seen to be disturbing these natural relationships that have established throughout the process of evolution, so the aim of a wildlife garden is to restore this balance. One of the most important points is therefore to maintain a natural balance within the garden. This may seem difficult when most gardens are aiming to grow plants that would not have naturally occurred in them. However, there are many things that can be done to improve the ecological value of a garden:

- minimise the use of pesticides, allowing a natural balance to build up between pests and their natural predators
- avoid monoculture
- provide a range of **habitats** to encourage **biodiversity**
- select plants to attract wildlife, including a wide range of flowering and fruiting plants
- use native plants of local provenance
- adopt a sustainable approach and recycle where possible
- avoid plants highly bred to have double flowers as these will often have developed more petals at the expense of nectar – plants such as the native *Caltha palustris* (Figure 13.18) will be more beneficial to wildlife than the cultivated *Caltha palustris* 'Flore Pleno' (Figure 13.19)

Designing a wildlife garden

Although this will depend on other factors, the aim is to provide the widest range of habitats possible. Most gardens do naturally provide a variety of habitats, and these can be enhanced for the benefit of wildlife.

The existence of a tree in the garden has been said to be the most significant factor for wildlife. It is important to think in three dimensions for garden habitats, and consider the vast volume that a tree will occupy. The very best is our native oak, *Quercus robur*, but not all gardens will have the space for this. Other natives, such as *Betula pendula*, all provide homes for a great range of native fauna.

Boundaries need to be considered for their wildlife value. Hedges provide the best habitats, and a wildlife hedge all the better. A mix of native hedgerow plants such as *Crataegus monogyna*, *Prunus spinosa*, *Corylus avellana*, *Acer campestre*, *Viburnum lantana*, *V.opulus* and *Ilex aquifolium* will provide nesting sites and food sources for birds and maybe a **wildlife corridor**, linking borders and hedgerows across gardens. Hedges are of great value for passerine birds. Passerine birds are small, perching birds such as sparrows. If the only option is for a fence or wall, this can also provide valuable habitats. Climbers and wall shrubs can enhance the habitat, and a wall itself can be home for bees. Baines (2000) suggests training climbing plants a few centimetres away from the wall or fence, by using battens to secure the support. In this way, a shady space is created for hibernating butterflies and nesting birds.

The pond is a key element for a wildlife garden, although the inclusion of this in a design needs to take the safety of garden users into account, in particular the risks for small children. If space is at a premium, a half barrel can encourage a surprising number of inhabitants. A pond will provide a permanent water source for birds and mammals. If this is not possible in the design, then alternative water sources, such as bird baths, should be considered.

A typical domestic garden will already provide something akin to woodland edge and meadow habitats. They may just need adapting to be more suitable and attract a wider range of inhabitants. Many garden borders are designed with a similar structure to a woodland edge, with taller shrubs to the back, maybe including small trees, and smaller shrubs, herbaceous perennials and perhaps biennials and annuals to the front. Layers of planting are provided, which mimic the woodland edge habitat by providing more dense cover to hide or nest in, gradually merging to a more open site with a range of food sources. This border is then often next to a lawn, providing an open habitat. There are many ways of adapting a lawn to increase its wildlife value. Allowing the grass to grow longer and encouraging native wild flowers to grow in among the turf grasses will greatly encourage biodiversity. All of the lawn could be cut to a height of about 8 to 10 cm, and certain areas left uncut. Maintaining three different heights of lawn provides the maximum range of wildlife attraction in smaller spaces.

If space allows, a wildflower meadow will make a valuable contribution. The aim is for low fertility, in contrast with a traditional high maintenance and low cut lawn. A summer wildflower meadow could include native plants such as *Leucanthemum vulgare*, *Centaurea nigra*, *Achillea millefolium* and *Knautia arvensis*. A spring-flowering meadow could include *Cardamine pratensis*, *Primula vulgaris*, *P.veris* and *Stellaria graminea*.

Log piles are a classic example of features to benefit wildlife. Along with compost areas, these are examples of allowing the natural processes of decay to take place, and encouraging the recycling of nutrients. At the same time, the vast numbers of organisms involved in this recycling process are encouraged.

Managing and maintaining a wildlife garden

The management of a wildlife garden needs to be based on an understanding of the life cycles and needs of its vast array of inhabitants.

Timing is crucial, and pruning and hedge cutting must be not be carried out while birds are nesting, commonly between March and August. For a wildlife hedge, the plants that provide food sources must not be cut back too soon. Traditional methods of cutting back herbaceous perennials at the end of their flowering period, often autumn, deprive many creatures of valuable seed heads. (Not to mention the aesthetic value of winter frosts on stems and seed heads.)

It is not a natural state for soil to be left bare, so management techniques should be used to ensure there is always cover. In nature there will generally be a cover of plants or leaf litter. Aim to cover the ground with plants and use mulches, such as composted bark, garden compost or leaf litter, as a temporary cover until the plant cover matures.

Construction of wildlife ponds

The siting of a wildlife pond is crucial, ensuring that the pond receives adequate sunlight.

To be effective for encouraging wildlife, a pond needs to meet certain criteria. It should have a minimum depth of 60 cm to protect hibernating creatures. The sides should be sloping to allow access for wildlife. Vertical sides result in animals such as hedgehogs falling into the pond with no escape route. Shelves incorporated for **marginal planting** will increase the range of habitats provided.

Preformed liners are available that already include these features; alternatively, butyl liners or a good-quality, UV-stabilised PVC may be used. Puddled clay is a traditional alternative.

A bog garden is a useful addition to a wildlife pond, extending the habitat range and planting possibilities.

Planting a wildlife pond

The crucial point is to include the correct balance of aquatic plants. This should include a mix of **oxygenators**, floaters, deep water aquatics and marginals as illustrated in Figure 13.20. Native oxygenators include *Ceratophyllum demersum*, *Myriophyllum spicatum* and *Potamogeton crispus*. Marginals include *Iris pseudacorus*, *Myosotis scorpiodes* and *Veronica beccabunga*. *Ranunculus aquatilis* has both submerged and floating leaves. If there is plenty of space, the native waterlily, *Nymphaea alba*, will provide an attractive deep-water plant with floating leaves. It is very important to select plants carefully, avoiding invasive aliens such as *Crassula helmsii* and *Myriophyllum aquaticum*.

Common problems with ponds include algal bloom and the spread of blanket weed. If the right balance of plants is established and a pond cover established of approximately one third, these problems will be minimised. Take care, also, to check plants before planting, to make sure that there is no blanket weed being introduced.

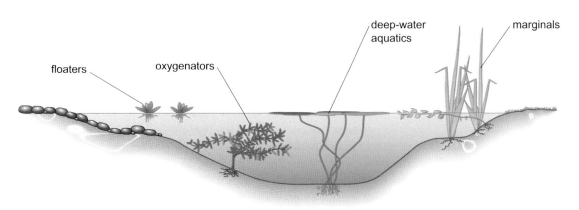

Figure 13.20 Mix of plants for a wildlife pond.

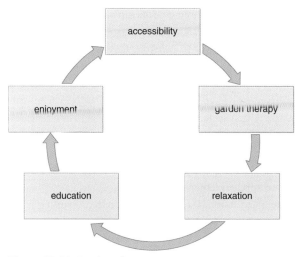

Figure 13.21 Gardens for everyone.

Wildlife gardens and conservation

Much valuable research has been carried out by Sheffield University, as part of the Biodiversity in Urban Gardens Projects (BUGS Projects) and has highlighted the importance of all gardens for wildlife. Other useful contacts include wildlife organisations such as the Wildlife Trust, Plantlife and the Royal Society for the Protection of Birds (RSPB).

DESIGNING GARDENS FOR EVERYONE

There are no happier folk than plant-lovers and none more generous than those who garden. There is a delightful free-masonry among them, they mingle on a common plane…
                                        (Wilson, 1927: preface)

Figure 13.22 *Dianthus barbatus.*

Gardens should be accessible for all to enjoy, and this factor should form a crucial part of the design brief.

Gardens for the visually impaired

Plants may be experienced in a multitude of ways, sight being just one of these aspects. Scent is a powerful sensory aspect. It can evoke different feelings, memories and responses. The most obvious part of the plant providing scent is the flower, but other aspects of a garden also need to be considered. A lawn with dew early in the morning has a particular freshness, chipped bark paths or leaf litter mulches may evoke a feeling of being in a woodland environment. The aroma of the needles of conifers is more invigorating, foliage of herbs such as lavender and thyme, relaxing. Although important, scent should not be overdone in a garden to the point of making it disorientating. A few key plants will be more effective than a mass of confusing scents.

- ✿ Scented plants for the winter: *Sarcococca hookeriana* var. *digyna*, *Viburnum × bodnantense* 'Dawn', *Lonicera fragrantissima*, *Daphne bholua* 'Jacqueline Postill'
- ✿ Spring: *Hyacinthus orientalis*, *Convallaria majalis*, *Viburnum carlesii*, *Osmanthus × burkwoodii*
- ✿ Summer: *Lonicera periclymenum*, *Rosa* 'Buff Beauty', *Lavandula angustifolia*, *Philadelphus* 'Virginal', *Dianthus barbatus* (Figure 13.22), *Lathyrus odoratus*, *Heliotropium arborescens*, (Figure 13.23)
- ✿ Autumn: *Cercidiphyllum japonicum* with an unusual aroma of burned sugar from the foliage, *Elaeagnus × ebbingei*, *Crinum × powellii*, *Clerodendron trichotomum*

Some plants need to be planted close to a path for the scent to be fully appreciated, such as an edging of lavender to a border, which can be brushed against while walking past. Others, such as some of the larger shrubs, e.g. *Daphne bholua*, will waft their scent across greater distances.

Figure 13.23 *Heliotropium arborescens.*

Sound is another sensory aspect of plants to be considered. Plants that move and rustle in the wind can add a further dimension. Many plants are more rigid and static, but others provide less resistance, such as *Miscanthus sinensis* (Figure 13.24), *Phyllostachys aurea* and *Betula pendula*. Seed heads at certain times of the year can provide surprising noise, such as the **dehiscent** pods of *Ulex europaeus* and *Cytisus scoparius*, and the dehiscent capsules of *Papaver orientale* and *Nigella damascena*.

Figure 13.24 *Miscanthus sinensis*.

Sound may also be introduced with hard landscape features, such as wind chimes. Care must be taken not to overdo these features, however, and to consider the overall style or theme of the garden. Running water can be an effective addition, and may be achieved on a variety of scales, from a full blown waterfall to a trickling pebble fountain. Touch is a further sensory experience, but this also needs to be considered alongside access. Scent and sound may carry, but plants to be touched need to be carefully positioned.

Raised beds may be useful in this case to increase accessibility to smaller plants. Those plants that could cause harm when touched should be avoided. This includes those that produce irritant sap, such as euphorbias, thorns, spines and prickles, such as pyracantha, sharp-edged foliage such as *Cortaderia selloana* and spiky plants such as *Yucca filimentosa*. Texture and form of plants have been discussed for garden design, mainly considering visual appearance, but touch adds a further dimension. Examples of plants with interesting textures and forms include:

- ❧ glossy foliage, such as *Fatsia japonica*
- ❧ soft, feathery foliage, such as *Foeniculum vulgare*
- ❧ soft, hairy foliage, such as *Stachys byzantina*
- ❧ twisted stems, such as *Corylus avellana* 'Contorta' and *Salix babylonica* var. *pekinensis* 'Tortuosa' AGM
- ❧ bark may be shiny, such as *Prunus serrula*;, corky, such as *Euonymus alatus*; or peeling, such as *Betula papyrifera* and *B.nigra*.

Hard landscape features can also be conspicuous in a **sensory garden**. Sculptures may be carefully positioned and designed for touch, maybe with warm, smooth, curving surfaces of polished wood, grooved surfaces, rough textured stone or cold steel. As with soft landscape features, sharp edges need to be avoided.

For the partially sighted, bold colours and contrasting textures should be used.

A key feature for design is to consider easy access and orientation. The plan should be kept as simple as possible, using straight paths and avoiding curves.

Trip hazards must be avoided. These include badly placed seats and ornaments, which should be set back in appropriate recesses, adjacent to pathways. Plant selection is vital, and overhanging branches and trailing stems avoided. The overflowing herbaceous border, softening the edges of a pathway is not suitable. Planning for the ultimate spread of plants is also vital.

Path surfaces need to avoid trip hazards, such as protruding edges of paving. Good construction is essential. Different path surfaces can be used to indicate a change of direction, or interesting feature to pause by. Slightly raised edges for the paths enable an easier route around the garden; different areas need to be clearly distinguished. The route needs to be straightforward and not confusing. Landmarks or reference points around the pathways will guide the user.

The width of borders should enable easy access. Where access is available on both sides, the width should be no more than 1,200 mm, narrower if from one side only.

Organisations involved with the blind and partially sighted include Thrive and the Royal National Institute of Blind People (RNIB), who jointly launched the National Blind Gardeners' Club (NBGC).

Gardens facilitating access

To plan a garden for people with limited mobility, the design needs to be kept relatively simple, as with gardens designed for the visually impaired. However, curved paths are an option, and may be preferable in some instances, as sharp, tight corners may be difficult to negotiate.

The width and surface of paths needs to be considered for those using wheelchairs or walking frames. Loose aggregates, such as gravel and bark chippings, for surfaces are unsuitable for wheelchair use. Gradients for ramps should be no more than 1 in 12. Many of these factors also apply to parents with pushchairs.

Sometimes steps are actually easier to negotiate than ramps, if a wheelchair is not being used. In this case, sturdy handrails need to be installed to facilitate use. Landings should be constructed at appropriate intervals, with no greater than 9 metre (29 feet 6 inches) lengths between landings. Steps should be constructed using the correct ratio between riser and tread. The riser should have a maximum height of 150 mm (6 inches) and a minimum height of 75 mm (3 inches), with an optimum between 100 and 125 mm (4 and 5 inches).

Design features for therapy

Gardening and visiting gardens is of immense value to people of all abilities. Horticultural therapy can be used to improve both physical and mental wellbeing, from exercise, fresh air, stimulation, relaxation and social interaction.

For those suffering with **dementia**, gardens can be designed in specific ways to aid memory. Way-marked routes are important, and the easiest format is to create a garden with a circular route. Scents may be used to evoke pleasant memories.

Shade and seating need to be incorporated into the design. Seats should be carefully positioned that are both easy to access, yet do not create trip hazards.

The value of gardens in aiding recovery is becoming increasingly recognised, and hospital gardens have an important role. In some situations, calm and relaxing colours and forms should be used. Other situations require the use of stimulating features.

Interior landscaping has been shown to reduce stress in the workplace by providing a more natural and green environment. Several plants are particularly effective at removing toxins and pollutants, such as the peace lily, *Spathiphyllum wallisii*, and weeping fig, *Ficus benjamina*.

The charity Thrive runs and supports horticultural therapy projects for a wide range of people, in allotments, community gardens, prisons and hospitals, and its own projects such as the Trunkwell Garden Project, Reading, Berkshire, UK.

Design features for easier gardening

Raised beds are a useful addition for those with limited mobility. A well-constructed path needs to provide easy access, preferably to all sides of the bed. Thrive provides useful recommendations for raised bed construction and path dimensions (www.Thrive.org.uk).

Although an option is to create a garden with plants as low maintenance as possible, this may not always be the best option. The aim may be to stimulate interest and create a garden for interaction. Plants that require nurturing, tying in, pinching out, thinning or deadheading may add to the pleasure of gardening.

Many tools are now available to make gardening easier for those with differing abilities.

Lightweight and long-handled tools may be used in preference to working on raised beds. An increasing range of ergonomic and flexible tools is also available.

Gardens for children

Gardens for children need to be engaging and interactive. As part of an overall garden design, certain areas may be planned specifically with children in mind, for example play areas. These can be planned to be adapted as children grow older. Alternatively, some gardens are designed to always be used by children, such as educational gardens.

Health and safety factors need to be taken into account, but it is also being increasingly realised that play areas need to be able to stretch and stimulate. Consideration needs to be made of whether children will be using the garden unattended. If this is the case, for younger children, open water needs to be avoided or securely fenced off. Water can be included in a design in other ways, such as pebble fountains.

Many common garden plants are poisonous; some may cause skin allergies or skin or eye irritation. A further category can cause the skin to become particularly sensitive to sunlight after contact with the sap, e.g. rue, *Ruta graveolens*. Whereas it is important to avoid those that are highly toxic, education also plays a role for both adults and children, in understanding which plants are safe to touch and maybe eat, and those that can cause harm. The Horticultural Trades Association (HTA) has drawn up a list in conjunction with the Poisons Unit at Guy's Hospital, the Royal Horticultural Society and the Royal Botanic Gardens, Kew, detailing categories of potentially harmful plants.

Some examples of potential harm from common garden plants are given in Table 13.3. It must be stressed that this list is a selection and by no means an exhaustive list.

Aconitum (monkshood)	Poisonous, skin irritant and harmful via skin
Alstroemeria (Peruvian lily)	Skin irritant
Arum (cuckoo pint, lords and ladies)	Poisonous, skin and eye irritant
Chrysanthemum	Skin irritant
Convallaria majalis (lily of the valley)	Poisonous
Euphorbia (spurge)	Poisonous, skin and eye irritant
Daphne (mezereon, spurge laurel)	Poisonous, skin irritant
Digitalis (foxglove)	Poisonous
Fremontodendron	Skin and eye irritant
Hedera	Poisonous, skin irritant
Helleborus	Poisonous, skin irritant
Heracleum mantegazzianum (giant hogweed)	Sunlight causes severe irritation
Lupinus	Poisonous
Laburnum	Poisonous
Narcissus (daffodil)	Poisonous
Prunus laurocerasus (cherry laurel)	Poisonous
Ricinus communis (castor oil plant)	Poisonous
Ruta	Sunlight causes severe irritation
Taxus	Poisonous
Wisteria	Poisonous

Table 13.3 A selection of potentially harmful plants commonly found in gardens.

The role of education needs to be considered alongside the basic principles of design, together with consideration given to selecting plants and materials for colour, shape, form and texture.

A garden that changes distinctly with the seasons provides a valuable experience, watching a deciduous tree or shrub shed its leaves, become dormant and then burst to life again in the spring needs to be experienced outdoors. Incorporating edible gardens in the design, planting vegetables, fruit and herbs, provides valuable lessons in where our food comes from (see also Chapter 12).

A garden for wildlife teaches the value of biodiversity from an early age. Exploring the range of habitats in a garden can provide an insight into the world of **food webs**, inter-relationships between flora and fauna and complex animal communities.

Considering the garden from the height of a child is a valuable reference point. Most gardens are planned considering the height of an average adult, but the perspective changes dramatically with a lower viewpoint.

A range of different play surfaces is available for use with play equipment. Selection of surface needs to consider the critical fall height.

Figure 13.25 *Calendula officinalis.*

UNDERSTANDING THE SELECTION AND USE OF LANDSCAPING ELEMENTS IN THE GARDEN

Factors that need to be considered include the critical fall height, i.e. the maximum height a child can fall from the equipment, and the impact-absorbing qualities of the material used. There are three main types: grass, loose fill and synthetic surfaces.

Grass: useful for ball games, or play areas with critical fall height up to 1 metre. Advantages include the fact that it is a natural material with good developmental value for children, particularly if mixed with wildflowers and left to grow longer. Problems occur in dry or freezing conditions, particularly on clay soils.

Loose impact attenuating surfaces include:

Bark: minimum depth of 300 mm required. A natural appearance is achieved with bark. Barriers are required to limit drift to other areas of garden. Topping up is required, particularly under equipment such as swings where it is liable to be kicked away.

Sand: useful for small areas, e.g. under a climbing frame or at the end of a slide. Disadvantages include problems with cats and when it becomes wet.

Impact absorbing synthetic surfaces include:

Tiles or *wet pour rubber*: usually efficient, long-lasting, but requires skill to lay and is expensive. Wet pour rubber provides a seamless surface.

Plants for children's gardens should be considered from the viewpoint of both interaction and subsequent interest. Seeds that are easy to handle and germinate are ideal. Many hardy annuals fall into this category. *Calendula officinalis* and *Helianthus annuus* are classic examples, illustrated in Figures 13.25 and 13.26. Half-

Figure 13.26 *Helianthus annuus.*

hardy annuals such as *Tagetes patula*, with large seeds, are ideal for children to sow indoors in seed trays or pots.

The collection of seeds is another element to consider. This could involve shaking the capsules of seed heads such as *Nigella damascena* or waiting for the pods of *Lathyrus odoratus* to twist open.

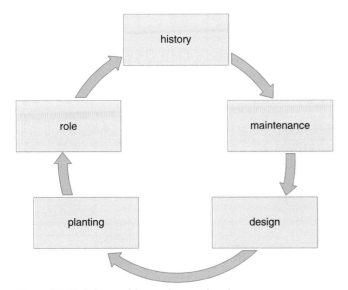

Figure 13.27 Urban public gardens and parks.

Many vegetable seeds are suitable, either sowing in trays or modules, or *in situ*. Seeds that germinate quickly, such as radish and lettuce, and can be harvested relatively quickly are useful.

An increasing range of tools has become available, suitable for children to use. Also health and safety factors need to be taken into account, with regard to the age of the child.

URBAN PUBLIC GARDENS AND PARKS

> But many of us live too near the city to possess a real garden.
> (Kingdon-Ward, 1935)

Green space may play a different role in urban areas where space is more limited. Innovative means can be used to create gardens in small spaces or even vertical spaces. The large open spaces that have fortunately been preserved as public parks provide valuable breathing spaces within cities, but also have extra demands placed on them.

Brief history

The majority of Britain's public gardens and parks were created in the nineteenth and early twentieth centuries. This came about largely as a response to increasing industrialisation, leading to rapid population growth in urban areas and poor living conditions. Britain was the first to establish municipal parks, with the aim of improving the health and wellbeing of the urban populations, providing an opportunity for fresh air and exercise.

This importance can be seen echoed today, in the increasing awareness of the value of exercise and healthier living styles.

These municipal parks were free for anyone to use, although opening times were sometimes restricted. The Royal Parks at first had limited entry, but subsequently opened to a wider population, today enjoying free access. Regent's Park is such an example, being designed by the architect John Nash between 1811 and 1826.

Botanical and zoological gardens were also developed in the early nineteenth century, but these have always charged fees or required a subscription. **Pleasure gardens** also were popular at this time, again charging a fee, but providing a different function of entertainment rather than education.

Many of the public parks were designed by local engineers and surveyors; however, certain influential designers had a key role. Conway (1996) identifies these to include Humphrey Repton, with his influence of the sweeping landscape style, John Claudius Loudon, John Nash, Joseph Paxton and later, figures such as Thomas Mawson and Geoffrey Jellicoe.

Loudon's parks included Derby Arboretum; Joseph Paxton was a key figure for the Crystal Palace Park. Thomas Mawson designed several parks between 1890 and 1933, working on a theme of a fairly formal centre set in landscaped surroundings.

The late nineteenth century saw the development of many parks in seaside resorts, such as Brighton and Blackpool, with the aim of creating attractions. The role of sport also became increasingly important from the end of the nineteenth century.

In recent history, parks have been subject to economic pressures and re-organisation, including contracting out of much of the maintenance. The large-scale bedding schemes of Victorian parks has, in some cases, been replaced by more sustainable and lower maintenance alternatives. The Heritage Lottery Fund has proved beneficial, injecting some much-needed capital into many parks. The **Green Flag Award** was launched in 1996 to encourage high-quality green spaces, considering factors from its welcoming nature to sustainability.

Suitability of plants and materials

A different range of factors need to be considered for the design of gardens open to the public. Maintenance levels need to be taken into account, which has often led to a very restricted range of plants being used. Planning needs to consider **desire lines** and directing movement.

Vandalism is, unfortunately, another factor alongside plant theft, and plants may have to cope with pollution from traffic. Plant selection should also be planned around the seasons of interest required. This may well be all year, or may be for a seasonal resort.

The range of users affects the design. Play areas for young children are ideally fenced off.

Skateboard parks are a feature for young people, and these also need to be located in separate areas. Adequate seating needs to be provided in easily accessible positions.

Planting and maintenance of amenity gardens

Although bedding is not so much in vogue as in Victorian times, it still features in many public gardens. Some schemes are sponsored by local businesses and are planted as picture bedding in a form that creates an advertisement.

Maintenance requirements for herbaceous schemes should be taken into consideration, avoiding those needing excessive staking, tying in and feeding.

Trees are subject to extra pressures for successful establishment. Vandalism is a major problem with newly planted trees. Tree guards should be used to protect the trunk against damage from animals, machinery and people. Adequate staking is essential. Tree selection needs to take account of safety factors, and maintenance and regular inspection is vital in public areas.

A particular problem associated with growing trees in public gardens and heavily trafficked areas is that of compaction. Trees start to suffer as oxygen becomes limited in the **rhizosphere**. One option is to redirect traffic, maybe altering paths so that well-used walkways do not interfere with the health of the tree. A remedial treatment is to inject the affected soil with compressed air, using specialist equipment. General maintenance of shrubs is similar to that in private gardens (see Chapter 14); however, the role of shrubs and subsequent management may differ slightly.

Shrubberies were important features of Victorian parks. Low-maintenance shrubs feature highly today in many modern parks and landscapes. Planting is often carried out to a higher density for amenity planting, in order to provide a quick and efficient ground cover. Subsequent maintenance needs to take this into account and thinning out may be required. Concern in the 1990s over public safety and threats from the provision of shelter for criminals in dense undergrowth, led to some clearance and cutting back of these features to provide clear views.

Lawns are often subject to much harder wear than domestic lawns. Areas for heavy traffic or sports use will need particular attention paid to drainage and adequate aeration.

Meadows have been previously confined to rural areas, but are increasingly being planted in urban areas with much success.

The development of green spaces has a key role to play in urban regeneration today. Organisations involved include the Design Council, Cabe, the government's advisor on public spaces, and the charity GreenSpace, which aims to promote and provide information on parks and green spaces.

LEVEL 3 BOX

UNDERSTANDING THE USE OF THEMED ELEMENTS IN THE PLANTING OF GREEN SPACES

UTILISING THE VERTICAL ELEMENT, LIVING WALLS

Climbing plants and wall-trained plants have long been used to clothe walls of buildings and ameliorate the environment. An extension of this is to use a wider range of plants in permanent structures to create screens. Hydroponic systems are used to create a growing environment for vertical gardens, such as the façade at Foundation Cartier, Paris, designed by Patrick Blanc.

UTILISING ROOF SPACE

Roof gardens or green roofs can enhance high-rise horizontal spaces. The former refers to planting in containers and placing on a roof. Green roofs are those that cover the whole roof space. The roof is covered with a drainage layer, growing medium and a layer of vegetation. In this way, for example, meadows can be created on the top of roofs. Pre-grown vegetation mats are available for green roofs.

In all cases, the roof needs to be structurally suitable for the weight, which increases substantially when the growing media absorbs water, and protected against water seeping through.

In general, the greening of urban spaces in such ways will have a multitude of benefits. These include:

- increasing biodiversity
- therapeutic effects with a view of green plants
- regulatory effects on climate – cooling buildings in summer and warming them in winter
- reducing water run-off
- reducing pollution
- providing areas for outdoor recreation

GARDENS THROUGH TIME

In history books too much is told of man's destructive quarrels and too little about his constructive work in developing the arts of peace.   (Wilson, 1920: p.x)

The history of gardens can be seen to be irrevocably bound with periods of peace throughout history. In times of peace and prosperity, the ornamentation of gardens could develop, and many different styles emerged. The UK is fortunate in having a wealth of historic gardens, but this also raises difficult questions as to how and whether they should be conserved.

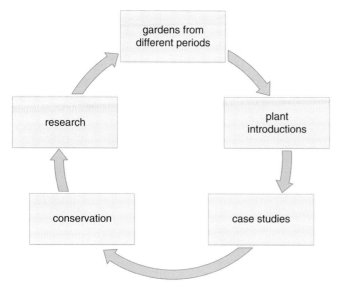

Figure 13.28 Gardens through time.

Researching, designing and planting an historical garden

There are many sources of information for research into historical gardens. These include historical maps, estate records, letters, literature, archaeological evidence and aerial surveys. Local county archives hold a wealth of information freely accessible to all.

A decision needs to be made whether to conserve what is there in the garden, to restore certain features or recreate the garden. An example of a garden recreation on a grand scale is that of Kenilworth Castle, Warwickshire, UK, as described in the case study below.

CASE STUDY   **The garden recreation at Kenilworth Castle, Warwickshire, UK**

An example of a garden recreation is that of the Elizabethan garden at Kenilworth Castle by English Heritage. It was decided to recreate a lavish garden in the footprint of one implemented by Robert Dudley, Earl of Leicester, in 1575, in order to impress and entertain Queen Elizabeth I. Part of the evidence for this reconstruction was provided by archaeological excavations carried out by Northamptonshire Archaeology from 2005 to 2006. This was linked with other academic research, including a letter written by Robert Laneham, usher to Robert Dudley, describing the garden in great detail.

The garden is formal in style, with a combination of Tudor and Renaissance style and Italianate features. A grand terrace overlooks a garden divided into quarters, with an elaborate fountain as a centrepiece. Other features include obelisks and spheres, and an ornate aviary. The planting is based on research into planting of the era.

An important feature of the garden, which would not have been present in Elizabethan times, is that of a stair lift at the end of the terrace to facilitate access. The view from the terrace is an important part of the garden experience, and improving access enables a wider range of visitors to appreciate this aspect.

To plant an historical garden with authenticity, research needs to be carried out on plants that were likely to have been used at the time. This very much links with the eras of different plant introductions.

Chronology of plant introductions

Britain has a limited palette of native plants, albeit delightful and of great ecological importance. This paucity resulted from the last great Ice Age, some eight thousand years ago, which wiped out much of the native flora. However, the gardens of today have an immense palette of plants. This is a result of a combination of the endeavours and fortitude of the great plant collectors, and the temperate climate of the British Isles, which has allowed a vast range of plants from all over the world to establish. Some have even established too well, to become troublesome invasives, e.g. Japanese knotweed, *Fallopia japonica*. It is interesting to consider the future trends, and how much our palette of plants will be affected by climate change, and perhaps, also, by changing legislation concerning the list of plants it is illegal to import or allow to spread into the countryside.

Many important gardens have evolved directly as a result of these plant introductions, as the owners sponsored many of the plant hunting trips. Examples include Bodnant Garden, Conwy, Wales, UK and Caerhays Castle, Cornwall, UK. With such a vast selection of plants, the skills of the garden designer are important to use this palette effectively.

The earliest plants to be introduced came from continental Europe. Roman introductions include plants such as *Acanthus mollis*, *Vitis vinifera*, and others that may not be considered as welcome today, such as *Aegopodium podagraria*.

Plant hunters began exploring further afield. Interesting links can be seen with the names of plant hunters and botanical nomenclature.

Philip F. von Siebold (1796–1866) was an early plant hunter in Japan. He introduced plants such as *Malus sieboldii*, *Wisteria floribunda* and *Trachycarpus fortunei*.

David Douglas (1798–1834) introduced many important plants from North America. Shrubs introduced by Douglas that are widely used in gardens today include *Garrya elliptica* and *Ribes sanguineum*. He also introduced annuals, such as *Eschscholtzia californica* and *Limnanthes douglasii*, and is famous for conifer introductions such as *Abies grandis* and *Pseudotsuga menziesii*.

William Lobb (1809–1863) was an influential plant hunter in South America, introducing shrubs and conifers such as *Berberis darwinii* and *Seqouiadendron giganteum*.

Robert Fortune (1812–1880) was an early plant explorer in China and Japan. Influential plant introductions include *Weigela florida*, *Jasminum nudiflorum* and *Akebia quinata*.

Ernest Wilson (1876–1930) is regarded by many as the greatest of plant collectors, with expeditions to China and Japan. Few gardens in Britain will not have a plant that was introduced by Wilson. His first two expeditions were for the Chelsea branch of the great Veitch's Nursery and subsequent trips were made for the

Arnold Arboretum in Boston, USA. Originally searching for *Davidia involucrata*, Wilson collected plants such as *Acer griseum* (Figure 13.29), *Ceratostigma willmottianum, Kolkwitzia amabilis, Lilium regale* (Figure 13.30) and *Liriodendron chinense* (Figure 13.31).

Figure 13.29 *Acer griseum.*

Figure 13.30 *Lilium regale.*

George Forrest (1873–1932) was a further prolific plant hunter in China, responsible for many of our rhododendrons, including *Rhododendron sino-grande*.

Frank Kingdon-Ward (1885–1958) carried on the tradition of collecting in China, visiting South West China, Upper Burma and Tibet. His introductions include *Primula vialii, Cotoneaster conspicuous* and *Meconopsis integrifolia*.

Modern-day plant hunters continue the theme, but with regard to modern-day legislation and restrictions on plant traffic.

Figure 13.31 *Liriodendron chinense.*

Gardens from different periods

A brief introduction to historical British garden styles shows a changing preference for formal and informal styles, and indicates the origins of many features in contemporary gardens.

CASE STUDY | **The High Beeches, Handcross, Sussex, UK**

The High Beeches woodland and water gardens, set in the High Weald Area of Outstanding Natural Beauty, exemplify good practice in the conservation of a landscaped woodland garden of the early twentieth century. This is an example of a garden that developed as a result of the increasing palette of exotic plants being introduced into Britain.

Colonel Giles Loder inherited the garden in 1906, and was responsible for developing the gardens in this naturalistic, woodland style, using exotic introductions among native plants. He subscribed to some of the great plant-hunting expeditions, and there are plants in the garden from the expeditions of George Forrest and Frank Kingdon-Ward. Introductions of Ernest Wilson arrived in truckloads from the closing down sale of Veitch's Nursery in 1914. Detailed plant records have been kept since 1906, and these have been updated and are available on the website for High Beeches. The garden has also developed with nature conservation in mind with a natural, acid hay meadow and maintenance regimes to encourage wildlife and biodiversity.

Roman gardens – Roman gardens were essentially formal, with formal ponds, topiary and architectural features such as colonnades.

An interesting example of a Roman garden to be seen in Britain is at the Roman Palace at Fishbourne, West Sussex, UK. Archaeological evidence has been used to recreate a formal

Roman garden, with hedges of *Buxus sempervirens* and espaliers of apple trees. It is even thought that the formal gardens surrounding the villa gave way to less formal plantings further away, to link with the landscape. Thus the idea of a **borrowed landscape** and themes used in contemporary naturalistic planting styles were possibly inherent in Roman gardens

Medieval gardens – Monastic gardens were very influential, with formal designs and the widespread use of herbs and edible crops. Orchards were typically laid out in a **quincunx** pattern.

Features of medieval gardens include the flowery mead and turf seats. Outdoor dining and raised beds feature, again popular in gardens of today. The **outdoor room** has long been in existence. Queen Eleanor's Garden, Winchester, UK is an interesting medieval garden recreation using plants typical of the era such as *Calendula officinalis* and *Laurus nobilis*, and includes typical features such as a tunnel **arbour**, turf seat, fountain and chamomile lawn.

The formal garden carried on through Tudor and Elizabethan times, being very much influenced by the Italian and French formal styles.

Knot gardens were popular from Tudor times, enclosed in a square frame and originally consisting of simple geometric patterns based on squares and circles of low hedges. These gradually became more complex, evolving into parterres. **Parterres** were designed to be viewed from above. They consist of symmetrical designs of regularly shaped flowerbeds or designs cut into turf. Elaborate patterns developed for these, some of the most intricate being parterres de broderie, with rich patterns resembling embroidery.

The formal era was largely swept away by the English landscape movement, with an important shift towards the naturalistic. Landscape designers such as Lancelot Brown, William Kent and Humphrey Repton created parklands, vistas, woodland walks, lakes and ha-has. Gardens such as Stowe, Buckinghamshire, UK and Stourhead, Wiltshire, UK evolved with **iconographic** temples and grottoes.

Victorian gardens flourished with all the exotic new plant introductions. Large walled gardens, conservatories and bedding displays were popular. There was again a backlash against the trend to formality and influential figures, such as William Robinson, advocated a return to a more naturalistic style. Other important figures emerged, such as Gertrude Jekyll, famous for her herbaceous borders and colour schemes.

Naturalistic styles have become popular in recent times, with planting styles exemplified by Piet Oudulf. Increasing interest in conservation and sustainability has increased interest in wildlife gardening and meadows.

Throughout history garden styles have oscillated between formality and informality. A contemporary theme includes both formality and informality, with the use of formal features such as topiary and clipped hedges, enclosing naturalistic swathes of perennials.

Summary

It can be seen that there is a multitude of facets to consider for the design of gardens and landscapes.

The principles of design provide consistency in understanding how to create a pleasing and workable design. The whole garden should have a sense of unity and balance. The use of principles such as focal points, borrowed views and knowledge of the impact of colour, texture and form of hard and soft landscape elements add to the interest and success of the design.

The process of design begins with an assessment of the site and requirements of the client. The site survey should be accurately drawn to produce a site plan, and this should be analysed along with the client brief to produce a design brief.

An overall garden style is important in achieving a sense of unity. This may be formal or informal, from a particular historic period or contemporary.

Knowledge of a range of hard and soft landscape features is important to fulfil the design and create a garden that is both practically and aesthetically functional.

Designing for wildlife is vital to encourage biodiversity. Knowledge of suitable habitats to create or exploit will enhance the wildlife value and sustainability of the garden. All gardens have the potential to become valuable wildlife havens.

An understanding of the different needs of all garden users is essential in providing accessible and enjoyable gardens for all. Different design features can be used to facilitate this access and interest, and different tools and methods of gardening.

Urban parks have played an important role for recreation and health throughout their history, and continue to do so today. Different requirements exist to successfully manage, maintain and develop them.

A study of the history of gardens and plant introductions brings a greater understanding of the gardens and landscapes that exist today. Common themes for garden styles have run through garden history. Historic gardens that have been restored, conserved or recreated provide both an understanding of the past and an inspiration for gardens of today.

REFERENCES

Baines, C. (2000). *How to Make a Wildlife Garden*. London: Frances Lincoln.

Conway, H. (1996). *Public Parks*. Princes Risborough: Shire Publications Ltd.

Robinson, W. (1883). *The English Flower Garden and Home Grounds: Design and Arrangement Followed by a Description of the Plants, Shrubs and Trees for the Open-air Garden and their Culture*. Charles Scribner's Sons.

Wilson, E. H. (1920). *The Romance of our Trees*. Doubleday, Page & Company.

Wilson, E. H. (1927). *Plant Hunting, Volume 1*. The Stratford Company.

Revision questions

1. Describe how to **achieve** balance and unity in a garden.
2. Explain the **differences** between formal and informal garden styles.
3. List the **factors** to be assessed in a site survey of a garden.
4. Describe how to **encourage** wildlife into the garden.
5. Suggest different features that could be used in a garden to **facilitate** access and enjoyment for all ages and abilities.
6. Describe the different **roles** that urban parks play.
7. Suggest some **design features** that can be seen in historic garden styles that are still used **successfully** today.
8. Explain the **role** of the plant hunters in developing the gardens of today.

FURTHER READING

Alexander, R. (2004). *The Essential Garden Design Workbook*. Cambridge: Timber Press.
Rosemary Alexander's workbook provides a practical and easily accessible guide to garden
    design, covering equipment required, surveying, drawing skills and symbols, and
    presentation of plans.

Baines, C. (2000). *How to Make a Wildlife Garden*. London: Frances Lincoln.
Chris Baines' wildlife gardening book provides a wealth of information on creating and
    enjoying gardens to encourage wildlife, for gardens of any size. Different habitats are
    described and there are many useful tables of both native and exotic plants suitable for
    wildlife gardens.

Campbell Culver, M. (2001). *The Origins of Plants*. London: Headline Book Publishing.
Maggie Campbell Culver's work on plant introductions provides a chronological account of
    thousands of plants that came to Britain through many different routes, and describes the
    major influence these have had on the botanical richness and diversity of the gardens of
    Britain.

Conway, H. (1996). *Public Parks*. Princes Risborough: Shire Publications Ltd.
Hazel Conway is a key figure for research into public parks, and this Shire Garden History
    publication covers the history of public parks in Britain from the early nineteenth to the late
    twentieth century.

Uglow, J. (2005). *A Little History of British Gardening*. London: Pimlico.
Jenny Uglow's 'potted history' of British gardening provides a very informative and readable
    account, covering topics from Roman gardens to the modern day.

RHS level	Section heading	Page no.
2 1.1–1.5	The design process	271
2 2.1	The design process	271
2 3.1	Principles of design	268
2 3.2	Principles of design	268
2 4.1	Garden styles	274
2 4.2	Gardens from different periods	286
2 5.1	Hard landscape features	273
2 5.2	Hard landscape features	273
3 1.1–1.7	Hard landscape features	273
3 1.1	Level 3 box: Understanding garden survey techniques and design principles	274

CASE STUDY Beth Chatto Gardens, Colchester, Essex, UK

Beth Chatto continues to have an inspirational impact on plant use related to a given site or theme. She has created sensational garden areas on a difficult site near Colchester in Essex, one of the driest counties in the UK. Beth's experiences and recommendations have been recorded in a series of books, each related to an area of extreme growing conditions.

Also Beth's prolonged correspondence with her great gardening friend, the late Christopher Lloyd of Great Dixter, published in the *Guardian*, provided guidance and reassurance of the current garden problems and themes to the readership for many years.

Beth Chatto Gardens, Colchester, Essex, UK is also the site of an exceptional nursery, with a broad range of over two thousand types of plant being available at any one time.

The pond planting in July, part of the Beth Chatto Gardens, showing the use of plants suitable for the site and soil conditions.

Chapter 14

Using plants in the garden

Jenny Shukman and Rosie Yeomans

INTRODUCTION

Plants provide the living elements in the garden, with their immense variety of life cycles, forms, colours, textures and scents. They add vitality and bring change, both within the cycles of the seasons and their own life cycles. Annuals may be used to provide a short buzz of colour in one growing season, whereas woody perennials may outlive the garden creators. An understanding of these life cycles is paramount in deciding how to place and use them within the garden.

The decision of which particular plants to select should be reached after considering overall factors of garden style and size, functional purposes, aesthetic value and environmental factors. These factors that come into play when designing gardens and landscapes are further discussed in Chapter 13. It is possible for the gardener to alter and ameliorate the environment in order to increase the range of plants that may be grown, or to enable them to thrive and develop more successfully. Container growing provides an extreme example, where the growing media can be selected or mixed specifically for the desired plants. Raised beds also offer an opportunity of providing suitable conditions, including enhancing the natural drainage of the selected site. Structures can be erected to enable climbers to be grown more easily; windbreaks allow less hardy plants a chance. Drainage can be improved, soil improvers and fertilisers added, or aquatic environments created.

An approach that should always be borne in mind is that of selecting plants suitable for existing site conditions. This may not always be possible, but does offer a more sustainable alternative. An appreciation of the natural habitats of plants is vital here, both considering their countries or regions of origin and the habitat they occupy within their regions. Naturalistic styles of gardening and wildlife gardens fit well with this theme.

Key concepts

❀ Understanding the impact of plant life cycles

❀ Introduction to natural habitats and the growth habits of plants

❀ Hardiness and creating microclimates to promote growth

❀ Manipulating growth for maximum ornamental display

❀ Scheduling maintenance with seasonal timings

ESTABLISHMENT AND MAINTENANCE OF PLANTED GARDEN FEATURES: HEDGES, PLEACHING AND TOPIARY

These terms refer to plants that have been trimmed intensively to thicken the lateral growth, either to create a barrier in the case of hedging or define a distinct shape in the case of topiary.

Hedges

Hedges are plants that are growing in a continuous line and create a boundary, a microclimate by providing wind shelter, visual or noise barrier, structural enclosure or design pattern in the garden. The function of the hedge will dictate the species chosen, and it is not unusual to have several species mixed to create a hedge. Hedges vary in height depending on the species and purpose.

Hedges are broadly categorised as 'formal' and 'informal'. Formal hedges are trimmed to a definite shape and are either used as high boundary hedging for creating enclosures in the garden, or for patterning in **knot gardens**, **parterres** and **potagers**.

Figure 14.1 Formal hedge shapes in a parterre setting.

Plants suitable for formal hedges include:

- ✿ *Taxus baccata*
- ✿ *Buxus sempervirens*
- ✿ *Fagus sylvatica*
- ✿ *Carpinus betulus*
- ✿ *Ilex aquifolium*
- ✿ *Griselinia littoralis*
- ✿ *Elaeagnus pungens* 'Maculata'
- ✿ *Teucrium chamaedrys*.

Informal hedges have a less formal shape with species pruned in the way an individual specimen would be maintained. The exception to this is a native hedge, grown to encourage wildlife, which would be trimmed annually to control height and spread. Informal hedges are used to encourage wildlife, create a boundary

Figure 14.2 Lavender hedge in flower with standard roses.

that looks like an extension of the border, be a flowering feature or create an attractive desire line.

Plants suitable for informal hedges include:

- ✿ *Crataegus monogyna*
- ✿ *Viburnum opulus*
- ✿ *Euonymus europaeus*
- ✿ *Lavandula angustifolia*
- ✿ *Rosmarinus officinalis*
- ✿ *Escallonia rubra* var. macrantha
- ✿ *Cotoneaster simonsii*
- ✿ *Berberis* × *stenophylla*.

Choice of hedge – The plants chosen to create a hedge will depend on height, style, function, geographic location, soil type and cost.

Site – Hedges are often a necessary garden feature and the species must be chosen to tolerate the site conditions. For example, *Fagus sylvatica* hedges are popular but will not succeed in moisture-retentive soils; *Carpinus betulus* is a better choice and will achieve a similar effect.

Where hedges are on an exposed boundary it is important to plant young specimens that will acclimatise to the site with ease. Mature plants will give a more instant effect and can be used in sheltered conditions but are vulnerable to scorching in extreme conditions.

Site preparation – The site for planting must be cultivated and cleared of all vegetation, paying particular attention to perennial weed control. The area cleared must be the same width as the intended width of the established hedge. The soil improvement will depend on the soil type (see Chapter 6 for a full discussion). Organic matter will improve the structure and fertility of all soils. A general fertiliser should be incorporated into the ground prior to planting.

Planting – Plants are best sourced between November and February when they can be bought as bare-root specimens. These are much cheaper than pot-grown plants. However, some species such as lavender will only be available as pot-grown stock and can be sourced all year round. Plants are either put in at **nursery level** against a line at a given spacing or planted in two staggered rows to give a thicker and faster establishing hedge.

Spacing will depend on the site conditions and the speed of cover needed, but may also be influenced by vigour of the species chosen and budget. See Table 14.1 for a general guide.

Figure 14.3 (a) Newly planted, single line and (b) double-staggered hedge line.

Formal hedging plants	Description	Spacing
Taxus baccata	Dense evergreen that makes a strong, long-lived hedge 1–5 m height. Well-drained soils.	450–600 mm
Fagus sylvatica	Deciduous hedge 2–5 m height with dried leaves persisting over winter. Well-drained soils.	400–700 mm
Buxus sempervirens	Small-leaved evergreen with compact growth 0.3–3 m. Tolerant of most soil types.	300–600 mm
Carpinus betulus	Deciduous hedge 2–5 m height with leaves persisting over winter. Good on wet soils. Suitable for pleached hedges.	400–750 mm
Prunus laurocerasus	Vigorous evergreen hedge 3–5 m height. Large-leaf canopy makes a good noise barrier.	700–1000 mm
Informal hedging plants	**Description**	**Spacing**
Lavandula angustifolia	Low evergreen flowering hedge of 0.4–0.6 m height. Well-drained and light soils.	300–450 mm
Fuchsia magellanica	Deciduous summer-flowering hedge of 1–1.5 m height. Needs well-drained soil and a mild climate.	500–1000 mm
Escallonia rubra var. macrantha	Evergreen, summer-flowering hedge of 1–2 m height. Best on well-drained soil.	600–900 mm
Prunus cerasifera 'Pissardii'	Deciduous purple-leaved cherry with pink flowers in the spring. Hedge 2–3 m height. Tolerant of most soils.	600–750 mm
Crataegus monogyna	Strong upright, native deciduous hedge with thorns, white spring flowers and red berries. Tolerant of most soil types. Height 2–3 m.	400–600 mm
Berberis darwinii	Small-leaved evergreen with orange flowers in late spring and purple berries. Height 1–1.5 m. Well-drained soils.	600–700 mm

Table 14.1 Hedge spacing guide.

It is useful to plant through polythene mulch for weed suppression. On exposed sites it is also prudent to use rabbit guards for protection from vermin damage.

Some hedging plants will need some support and can be individually staked with a cane or tied to a taught wire line that runs the length of the hedge.

Establishment – Newly planted hedges must be well watered during the first year of establishment.

Formative pruning to encourage lateral development is crucial (see Chapter 10). All new growth should be trimmed by two thirds once it has made 10 to 15 cm (4 to 6 inches) of new growth. This may be two or three times in the first growing season. Conifers are trimmed carefully so that no woody material is exposed and the centre leader is left intact until the desired hedge height is reached.

Maintenance – The hedge should be given a spring *top dressing* of a compound fertiliser or rotted farmyard manure and kept weed free throughout the growing season.

Hedges must be trimmed to encourage thick lateral development. They are trimmed with shears by hand, or electrical or petrol-driven hedge trimmers with reciprocating blades. Safety equipment must be used, particularly when working at height.

Figure 14.4 An operator wearing full PPE on a supported step ladder.

Hedge trimming depends on the species and effect required. Twice a year during the late autumn and mid summer maintains the shape for formal hedges. Formal hedges such as *Buxus sempervirens* used in border garden features may need to be trimmed up to four times a year to keep them looking neat. Informal hedges are pruned once a year depending on the maintenance pruning requirement of the species.

Pleaching

This is a hedge grown above a clear stem, rather like a hedge on stilts. The principles of establishment are similar to formal hedges.

Figure 14.5 A pleached hedge.

The most suitable trees for this type of feature are *Carpinus, Tilia, Sorbus, Malus* and *Pyrus*.

Topiary

This is where a hedging specimen is trained and clipped intensively to maintain a particular shape. The best topiary

Figure 14.6 Inspiring topiary specimens found at Levens Hall, Keswick, Cumbria, UK.

specimens are *Buxus sempervirens* and *Taxus baccata*. Complicated shapes need to be trained on wire frames to fix their intended form during the establishment period. Topiary must be clipped several times during the growing season to maintain a shape.

CASE STUDY   Levens Hall, Kendal, Cumbria

Originally laid out in 1690 for Colonel Grahame by Monsieur Beaumont, who as gardener to King James II also had an input to the layout of Hampton Court Palace Gardens. This garden, within the Lake District of the UK, exhibits incredible topiary specimens of English yew, *Taxus baccata*, several hundred years old in a range of shapes, both geometric and novelty, including hats of the time. The garden also has outstanding blocks of single-colour bedding, to complement the scale and colour of the topiary, herbaceous borders and vegetable areas. It is open for five days a week during the main growing season and for special events during the winter. Please check the seasonal dates and details before visiting.

TREES

Trees in the garden enhance wildlife and provide height, wind shelter, screening and shade. Garden features such as avenues are created with lines of boldly spaced trees. Single specimens are dramatic and planted for their visual beauty, while groups of trees create a habitat for shade-loving flora and fauna.

Tree species must be chosen carefully to suit the site and avoid visual problems of scale. Practical problems of structural damage caused by roots to buildings and garden structures must be considered. Vigorous trees such as *Populus, Alnus* and *Salix* must not be planted near buildings.

Planting and establishment

Trees are sourced as **standards** or **feathered maidens** and are pot grown or supplied bare root in the winter. Planting is best in the spring or autumn for pot-grown stock and in the winter for bare-root specimens.

Trees produce a few strong roots for stability but need to mass a lot of root near the soil surface for nutrient and moisture absorption. These roots will travel beyond the tree canopy and must be considered when preparing the soil for planting.

The ground must be cleared of perennial weeds. Soil improvement with organic matter or grit may be necessary depending on the soil type. Soil from the excavated planting hole is improved before back-filling, for further detail on soil types see Chapter 6. A planting hole is excavated and should be a third or a half wider and deeper than the root ball of the tree to be planted. A high phosphate fertiliser is incorporated into the base of the hole and into the soil used to backfill the hole. A tree stake is driven into the hole before the tree is planted. There are many staking

(a)

(b)

Figure 14.7 Trees with (a) a short vertical stake, (b) an angled stake and (c) a guyed transplant.

Figure 14.7 (cont.)

Figure 14.8 Planting a standard tree.

systems used. The principle is that the root system must be held securely while the tree canopy moves in the wind. The canopy movement encourages root development and helps strengthen the trunk. The roots of trees that have been supplied bare rooted can be pruned by up to one third in order to stimulate more lateral root development. The tree is planted at nursery level, firmed well and tied to the stake before watering and mulching. The newly planted tree must be kept well watered through the first year of establishment. Tree ties are checked and loosened as necessary.

	Name	Ornamental value
Small trees, 5–10 m	*Acer griseum* (paperbark maple)	Deciduous Rich brown winter bark Red autumn leaf colour.
	Amelanchier lamarckii	Deciduous White spring flowers Copper-coloured spring leaves Red autumn leaf colour.
	Arbutus unedo (strawberry tree)	Evergreen White flowers and red fruit in late autumn Winter foliage and deep-brown bark Best on acid soils but shows some tolerance to chalk.
	Cercidiphyllum japonicum (katsura tree)	Deciduous Smokey-pink autumn leaf colour, colours best on acid soil.
	Ilex aquifolium (holly)	Evergreen Winter foliage and red berries (dioecious) Good on chalk.
	Laburnum × *watereri* 'Vossii'	Deciduous Pendulous racemes of yellow flowers in early summer Poisonous.

(cont.)

	Magnolia × *soulangeana*	Deciduous Large white, flushed pink, tulip-shaped flowers in the spring Best on acid soils.	
	Malus 'John Downie' (crab apple)	Deciduous White spring flowers Orange, red autumn fruit.	
	Pittosporum tenuifolium	Evergreen Winter foliage and strong pyramidal shape Honey-scented flowers in spring.	
	Sorbus aucuparia (mountain ash or rowan)	Deciduous Red autumn berries Tolerant of chalk and very acid soils.	
Medium trees, 10–15 m	*Betula pendula* (silver birch)	Deciduous Pendulous habit with white winter bark.	
	Catalpa bignonioides (Indian bean tree)	Deciduous Large green leaves in late spring White summer flowers in panicles Late summer bean-like fruits.	
	Juniperus scopulorum (juniper)	Evergreen conifer Very narrow conical shape Grey-green fine foliage all year round.	
	Parrotia persica (Persian ironwood)	Deciduous Coppery autumn colour Acid soils.	
	Thuja occidentalis	Evergreen conifer Strong, narrow conical shape Glossy, fine, aromatic foliage all year round.	
Large trees, >15 m	*Eucalyptus gunnii*	Evergreen Silvery-green leaves, round when juvenile and lanceolate on mature shoots White flowers in summer Peeling grey-green bark.	
	Liriodendron tulipifera (tulip tree)	Deciduous Distinctive lobed leaves that colour yellow in autumn Summer green/yellow tulip-shaped flowers.	
	Pinus sylvestris (Scots pine)	Evergreen conifer Fine needle-like, grey-green foliage all year round Acid soils.	
	Quercus ilex (holm oak)	Evergreen Glossy dark-green leaves all year round Dry soils.	
	Tilia cordata (small-leaved lime)	Deciduous Fragrant yellow flowers in summer.	

Table 14.2 Garden trees and their vigour.

Maintenance

Some **formative pruning** may be necessary in the first few years to shape the crown development. Please see Chapter 10 for further details. Once the tree has established, there will be little need for pruning and it would be prudent to employ the services of an arboriculturalist for any high-level work.

The tree may need water in very dry weather. Tree ties are checked and loosened annually.

SHRUBS

Shrubs in the garden provide structure, all-year interest and a range of decorative features such as flowers, fruit, foliage colour, scent and evergreen foliage. Shrub borders are low maintenance and low, spreading shrubs can be used as ground-cover planting.

Figure 14.9 Shrubs suitable for dry conditions.

There is a massive range of shrubs for use in the garden and their choice is dependent on ornamental value but also site conditions, vigour, longevity and cost.

When planning a shrub border, one must account for the spread and vigour of the shrub together with site conditions to assess the correct spacing. Densely planted shrubs require more maintenance once the border is established. Shrubs are also used for structural interest in mixed borders.

Figure 14.10 Woodland shrubs in rich soil.

Planting

Shrubs are pot grown and can be planted at any time but are best planted in autumn or spring when conditions are mild. The ground must be cleared of weeds and prepared to suit the conditions needed by the chosen shrubs. Mediterranean shrubs such as *Cistus*, *Brachyglottis* and *Lavandula* do not grow well where the soil is rich so adding grit to heavy soil is necessary to aid establishment. Shrubs such as *Daphne*, *Rhododendron* and *Hydrangea* are best in an organically enriched soil. Pots should be well watered prior to planting. The shrub is knocked gently from the pot and roots loosened at the base to prevent spiralling. The shrub is planted at the depth in which it is grown in the pot. The plant should be firmed well, watered and mulched to conserve moisture.

	Name	Height and spread	Ornamental value
Winter interest	*Cornus sanguinea*	2 × 2 m	Deciduous Bright orange-red winter stems Moist soils.
	Erica carnea	0.3× 0.4 m	Evergreen Winter flowers, white, pink or purple cultivars grown Lime tolerant.
	Garrya elliptica	3 × 2.5 m	Evergreen Long racemes of catkin-like, grey-green flowers in mid winter (dioecious).
	Mahonia japonica	2 × 1 m	Evergreen Scented yellow flowers in late winter.

(cont.)

	Viburnum tinus	3 × 2 m	Evergreen Dark-green glossy leaves all year round White flowers from pink buds in clusters all winter.
Spring interest	*Ceanothus thyrsiflorus*	2 × 2 m	Evergreen Bright-blue flowers in spring Well-drained soils.
	Chaenomeles speciosa	3 × 2 m	Deciduous Cultivars range from white to pink flowering in spring Fruits in summer.
	Magnolia stellata	1.5 × 1.5 m	Deciduous Compact and slow growing White star-shaped flowers in spring Acid soils.
	Pachysandra terminalis	0.3 × 0.4 m	Evergreen Spreading growth bears white flowers in the spring Acid soils and will also tolerate slightly alkaline soils.
	Skimmia japonica 'Rubella'	1 × 1 m	Evergreen Rounded shrub with glossy leaves and pink flowers in the spring (dioecious, male) Acid loving.
Summer interest	*Buddleja davidii*	3 × 2 m	Deciduous Late summer flowers, cultivars range from white to blue Tolerant of poor soils.
	Hydrangea macrophylla	1.5 × 1.5 m	Deciduous Large flower heads colour blue in acid soils but pink in alkaline soils Sheltered sites and moist soils.
	Rhododendron spp.	2 × 2 m	Evergreen Acid loving Vast range of cultivars flowering early summer, pink, white, mauve, purple.
	Rosa glauca	2 × 1 m	Deciduous Grey-green foliage persists in mild winters Pink, single flowers in summer Red hips over winter.
	Yucca filamentosa	0.8 × 0.9 m	Evergreen Rosette of spiked leaves, the centre from which appears a strong spike of white flowers in mid summer Dry soils.
Autumn interest	*Acer palmatum*	3.5 × 3.5 m	Deciduous Fine foliage held in horizontal layers Coppery autumn colours Acid soils.
	Ceratostigma willmottianum	0.8 × 0.6 m	Deciduous Bright-blue flowers from mid summer to October with red foliage colour in autumn.
	Cotoneaster horizontalis	1.5 × 1.5 m	Deciduous Spreading shrub with red berries and red foliage colour in autumn Tolerates poor soils and will grow up a wall or spread along the ground.
	Pyracantha rogersiana	2 × 1.5 m	Evergreen Thorned shrub with white flowers in summer and red berries in the autumn Useful for hedging.
	Rhus typhina	4 × 3 m	Deciduous Large compound leaves that turn bright red-orange in the autumn.

Table 14.3 Seasonal shrubs.

Maintenance

The newly planted shrub should be watered as necessary in the first year of establishment. A regime of mulching the shrub border after wet weather will maintain moisture without further irrigation. The shrub border must be kept weed free and top dressed each spring with organic matter or a compound fertiliser to maintain healthy growth. Shrubs do not all need annual pruning but see Chapter 10 for pruning techniques. After any activity among a shrub border the foot compaction should be lightly forked to keep the surface roots aerated.

ROSES

Roses have traditionally been used in separate, formal rose gardens, but a variety of other uses are also possible. Species and shrub roses can create structure and interest in a more naturalistic style, incorporated into mixed borders and wildlife gardens. Some may be used as hedging such as *Rosa rugosa* for an informal hedge and *R. canina* as part of a wildlife hedge. Climbers and ramblers add vertical interest, clothing walls, pergolas and pillars in a range of settings. Most roses are grown for their flowers, such as the large-flowered and cluster-flowered roses (formerly hybrid teas and floribundas), examples of which are shown in Table 14.4.

Large flowered	*Rosa* Alexander	Bright-red double flowers
	R. Freedom	Bright-yellow, double flowers
	R. 'Just Joey'	Copper-pink, fully double flowers
	R. Lovely Lady	Salmon-pink, fully double flowers
	R. Peace	Pale yellow flowers with pink tinge, fully double
Cluster flowered	*R.* Escapade	Pink/violet flowers with white eyes, semi double
	R. Iceberg	White, fully double flowers
	R. 'Korresia'	Yellow, double flowers
	R. Sexy Rexy	Fully double, pink flowers
	R. The Times Rose	Crimson, double flowers

NB Those plant names written without quotation marks are trade designations rather than cultivar names. These are the names under which the plants are sold rather than their original cultivar name.

Table 14.4 Large-flowered and cluster-flowered roses.

Many roses are highly scented, for example the old climbing Bourbon rose, *R.* 'Madame Isaac Pereire', and modern shrub roses such as *R.* Gertrude Jekyll. Others also have attractive foliage, such as *R. glauca*, interesting stems, such as *R. sericea* subsp. *omeiensis* f. *pteracantha* or attractive hips such as *R. moyesii*. Patio roses are suitable for container growing, and roses have also been bred to provide ground cover, such as the county series.

CASE STUDY **English Rose gardens**

Gardens to visit for inspiration about the use of roses include Mottisfont Abbey, near Romsey, Hampshire, UK and the Royal National Rose Society Gardens, Chiswell Green, Hertfordshire, UK. Mottisfont Abbey, a National Trust property, has a walled garden designed by Graham Stuart Thomas to include a National Collection of old fashioned roses (pre 1900), with rich herbaceous underplanting. The Royal National Rose Society Gardens displays thousands of roses, both old and modern, among companion planting. Both are open to the public and have longer hours of access at peak flowering times – it is always best to check the seasonal details with the property before visiting.

Most roses prefer a slightly acidic soil, pH 6.5, and an open, sunny but sheltered position. There are some exceptions to this, for example, *R.* 'Madame Alfred Carrière' (noisette type) will tolerate a north-facing aspect unlike most climbing roses. Most roses will show signs of iron deficiency if planted in alkaline soils, and most modern roses are best avoided in such conditions.

It is important to select a site that has not grown roses before. If possible, prepare the site several months before planting, incorporating organic matter and thoroughly clearing weeds.

Roses establish well from bare-root plants if these are available, but planting times are more restricted. Bare-root plants should be planted in late autumn/early winter or early spring. These are also the optimum times for planting container-grown roses, but the time period can be extended as long as sufficient aftercare is provided with regard to watering in dry conditions, and planting is not carried out in extremes of temperature or soil water conditions.

Spacing varies according to the type of rose. Overcrowded roses with a lack of airflow are far more prone to fungal diseases. Bedding roses should be planted about 45 to 60 cm (18 to 24 inches) apart, and 30 cm (12 inches) from the edge of the bed. Hedging roses should be planted at about 1 metre (39 inches) apart.

Select plants with at least two to three good strong shoots, and avoid any pot-bound plants. Make sure the roots are not dry when planting, by soaking the plant in a bucket of water for a couple of hours if necessary. The hole should be wide enough to spread the roots out well and deep enough so the bud union for grafted roses is 2.5 cm (1 inch) below the ground. Mycorrhizal products are available for planting roses. If planting climbers against a wall, prepare the planting pit at least 45 cm (18 inches) away from the wall, spreading the roots out in a fan shape away from the wall to avoid the effects of a rain shadow.

Carefully backfill with soil, ensuring there are no large air pockets, and firm well while avoiding damage to the stem of the rose. The soil for backfill can be enriched by mixing with some well-rotted compost and seaweed meal.

Ongoing annual maintenance tasks consist of weeding, mulching and feeding, checking for pests and diseases, deadheading and pruning. Mulching should be carried out annually in the spring, preferably just after pruning. A 5 to 8 cm (2 to 3 inch) layer of well-rotted manure, garden compost or recycled green waste all make excellent mulches. Always keep the mulch clear of the stems to avoid them being damaged by rotting.

Large-flowered and cluster-flowered roses flower on the current season's growth, and should be pruned in March. If the site is prone to wind exposure, stems should be shortened in the autumn to prevent windrock. Harder pruning can be used to produce fewer but larger blooms, for example for exhibition purposes, but in general, stems should be cut back just above outward facing nodes to about 20 to 25cm (8 to 10 inches) for large-flowered roses, and 35 to 40 cm (14 to 16 inches) for cluster-flowered roses. Bush roses are pruned with the aim of keeping the centre open and achieving a balanced goblet shape.

Regular deadheading will promote flowering, and should be carried out unless the rose has been planted to produce a display of autumn hips.

CLIMBERS AND WALL SHRUBS

Climbing plants are those that have specific adaptations, such as tendrils, twining stems, adventitious roots or a stem outgrowth of prickles. These require support of a structure, either artificial or living, and may need tying in or may be self-supporting. Wall shrubs are those that may be trained against or through a structure, but do not have any specific adaptations. An interesting plant is *Euonymus fortunei*, which is a shrub, but can develop the climbing adaptation of adventitious roots against a wall or other structure in its adult form.

The ability of these plants to rapidly gain height makes them very useful in garden design terms.

Vertical elements can be provided relatively quickly in a garden. Climbers may be used to clothe unsightly walls or fences, or complement attractive ones. They can be used over arches, pergolas and pillars. Naturalistic effects can be produced by growing through trees or shrubs, and seasons of interest can be extended in this way. Some can be used to provide effective ground cover. It is important to select the appropriate plant for the situation required, considering the aspect. Many climbers are particularly rampant, and whereas this can be used to advantage to produce a quick screen or feature, careful planning is required to ensure they are positioned carefully.

Consider the eventual height of a climber in relation to the space available before selection, or how well the climber responds to pruning and the time you have available for seasonal maintenance.

Climbers may be used as long-term features, selecting woody perennials such as *Hydrangea anomala* subsp. *petiolaris*, or for short-term display, such as the annual *Lathyrus odoratus*.

Evergreen climbers, such as *Hedera colchica*, create useful screens. Ivy is often used to create an instant green screen. Deciduous climbers create seasonal interest, such as the autumn foliage of *Parthenocissus quinquefolia* and *Vitis vinifera*. Some provide scent, such as *Akebia quinata* and *Lonicera periclymenum*, others provide wonderful floral displays, such as the late spring-flowering *Clematis montana* var. *rubens*, showy early summer hybrids such as *C.* 'Nelly Moser', and late summer *C. viticella*. The use of climbers provides extra wildlife habitats and shelter, for nesting birds and a variety of insect life.

Climbers: south-facing aspect, sunny site	*Actinidia kolomikta*	Woody, deciduous, twining climber. Interesting foliage with pink and white tipped leaves.	Height to 4 m
	Humulus lupulus 'Aureus'	Herbaceous, twining climber with yellow/green foliage.	Height to 6 m
	Lathyrus latifolius	Herbaceous, tendril climber with pink/purple flowers in summer to early autumn. Toxic seeds.	Height to 2 m
	Trachelospermum jasminoides	Woody, evergreen climber with scented white flowers in summer.	Height to 9 m
	Wisteria sinensis	Woody, deciduous, twining climber with lilac flowers in early summer. Toxic seeds and bark.	Height to 30 m

(cont.)

Wall shrubs: south-facing aspect, sunny site	*Abutilon megapotamicum*	Evergreen or semi-evergreen shrub with red and yellow flowers in summer and early autumn. Hardy to −5°C.	Height to 2 m
	Carpentaria californica	Evergreen shrub with glossy foliage and white summer flowers. Hardy to −5°C.	Height to 2 m
	Ceanothus 'Autumnal Blue'	Evergreen shrub with blue flowers from summer to autumn.	Height to 3 m
	Fremontodendron californicum	Evergreen or semi-evergreen shrub with bright yellow flowers from late spring to autumn. Hardy to 5°C. Highly irritant.	Height to 4 m
	Itea ilicifolia	Evergreen shrub with catkin-like flowers in summer and early autumn. Hardy to 5°C. Highly irritant.	Height to 3 m
Climbers: north-facing aspect, shady site	*Celastrus scandens*	Woody, deciduous, twining climber with orange fruits. Dioecious.	Height to 10 m
	Hedera helix 'Parsley Crested'	Evergreen, woody climber with adventitious roots.	Height to 2 m
	Hydrangea anomala subsp. *petiolaris*	Woody, deciduous climber with adventitious roots. White flowers in summer. Poisonous plant.	Height to 15 m
	Lonicera × tellmanniana	Woody, deciduous climber with twining stems. Orange flowers from late spring to mid summer.	Height to 5 m
	Parthenocissus henryana	Woody, deciduous climber with tendrils. Good autumn leaf colour. Poisonous berries.	Height to 10 m
Wall shrubs: north-facing aspect, shady site	*Chaenomeles speciosa* 'Moerloosei'	Deciduous shrub with white flowers flushed with pink in spring followed by green/yellow autumn fruits.	Height to 3 m
	Cotoneaster horizontalis	Deciduous shrubs with pink/white flowers loved by bees in the spring, followed by red autumn fruits.	Height to 1 m
	Garrya elliptica 'James Roof'	Evergreen shrub with winter catkins.	Height to 4 m
	Jasminum nudiflorum	Deciduous shrub with yellow winter flowers.	Height to 3 m
	Pyracantha 'Orange Glow'	Evergreen shrub with white flowers in the spring and orange berries that last well from autumn to winter.	Height to 3 m

Table 14.5 Climbers and wall shrubs for different aspects.

Maintenance, propagation and planting regimes vary greatly with the different types of plants.

Annual climbers are useful for clothing obelisks in herbaceous or mixed borders, or growing as part of a cutting or ornamental kitchen garden. *Lathyrus odoratus* is a hardy annual, so may be sown *in situ* in the spring. For earlier flowers in mild areas, an autumn sowing can be made. Other climbers used to provide temporary summer displays include the frost tender *Ipomoea lobata* and *I. tricolor*.

The honeysuckles include many useful climbing species and cultivars for different situations. The fragrant, native *Lonicera periclymenum* can be used to twine through a native hedgerow in a wildlife garden. Many cultivars exist, such as *L. periclymenum* 'Graham Thomas', which has a longer flowering period. These will all quickly clothe arches and pergolas. Any pruning should be carried out after flowering, cutting back the older growths. An evergreen, or semi-evergreen, species is *L. japonica*. This climber may be pruned in spring, as it flowers on the current season's growth. *L. japonica* 'Halliana' is a useful cultivar.

Clematis are useful for a wide range of situations, often planted as a companion to another shrub, tree or climber to extend seasonal interest. The more rampant, such as *Clematis montana*, may be grown through trees and left unpruned. *Clematis montana* var. *rubens* has deeper pink flowers and a red tinge to the foliage; this is a less rampant climber and may be trained against support such as trellis on a wall. Pruning should be carried out after flowering, cutting back the flowered shoots to vigorous new growth.

Clematis should be planted with their roots in the shade, but the stems reaching towards the sun.

Planting clematis at a greater depth than the norm helps prevent damage from clematis wilt. *C. viticella* is less prone to this disease. A variety of cultivars produce a range of colours, from the purple *C.* 'Étoile Violette' to the wine-red *C.* 'Madame Julia Correvon'. These should be hard pruned in March, cutting all stems back to within about 30 cm (12 inches) of the ground.

Wisteria, *Wisteria sinensis*, is a classic climber for a pergola. Pruning needs to be carried out twice a year, cutting back the laterals to about 15 cm (6 inches) in the summer, and harder, back to two or three buds, in the winter.

Wall shrubs need to be selected according to how well they respond to pruning and training. A strong system of wires on straining bolts or trellis supported on battens is required for both wall shrubs and climbers that are not self-supporting. Keeping an air space behind the climber will benefit plants such as roses, which are prone to fungal diseases such as powdery mildew. Early pruning and training is essential to establish a framework. Further detail on pruning techniques and tools is found in Chapter 10.

LAWNS

As with any soft landscape feature, the function of the lawn should be assessed before decisions are made as to the type of lawn. Although long regarded as a quintessential element of British gardens, the ways in which lawns are managed now need to reflect the changing climate and issues of sustainability. Nevertheless, the lawn provides a valuable design and functional element within many gardens. The green expanse provides a calm setting for shrub or herbaceous borders, a valuable play space for children, sitting or entertaining areas, sports areas, and can provide the setting for formal gardens, surrounding formal bedding.

Lawns may be established from seed or turf, and there are benefits and limitations for both options.

Lawns from seed
Benefits:

- a wider range of seed mixes is available, and a suitable mix for a range of situations more easily achieved
- costs of purchasing seed are generally lower than turf

Limitations:

- times are more limited when seed can be sown
- ground preparation needs to be more exacting
- the lawn takes longer to be ready for use
- protection from birds is required.

Lawns from turf
Benefits:

- an instant effect is achieved, which can be used more quickly
- good preparation is still vital, but the soil does not need to be prepared as finely
- the time period for laying turf is wider, although weather conditions still need to be taken into account.

Limitations:

- less choice is generally available
- turf is generally a more expensive option
- for a high-quality ornamental lawn, fine grasses need to be used to create a sward that can be mown more closely
- a hardwearing utility lawn will also include some of these grasses, but will also incorporate tougher ones.

The optimum time for sowing lawn seed is late summer or in spring, usually in late March to April.

Preparation needs to be thorough. Any levelling required should be carried out at the start. Should the ground be very uneven, the subsoil may need to be levelled at the outset, removing topsoil, levelling subsoil, and subsequently replacing the topsoil. Drainage problems need to be sorted out at this stage as well, breaking up any compacted subsoil, or installing drainage systems if necessary.

Primary cultivation should be carried out several months before sowing, digging the area or using a rotary cultivator. Soil ameliorants should be incorporated as necessary, such as composted green waste, compost or well-rotted organic matter. Avoid applying organic matter that will bring in unwanted weed seed, or excessive amounts of organic matter, which will then cause the ground to settle.

Prior to sowing, secondary cultivation is carried out. Fork the area to a depth of a few centimetres, breaking up the soil to produce a fine tilth, raking is subsequently carried out to produce a finer tilth.

Tread the soil before raking for consolidation.

Remove stones larger than about 1 cm (½ inch) in diameter during raking. Allow the seed bed to settle for three to four weeks. A few days before sowing a balanced, compound fertiliser should be applied such as pelleted chicken manure or a granular fertiliser. Any weeds that have germinated can be hoed off before sowing.

A suitable seed mix for a high-quality lawn would be:

✿ 80% by weight chewings fescue, *Festuca rubra* var. *commutata*
✿ 20% by weight brown top bent, *Agrostis tenuis*.

A suitable mix for a hardwearing utility lawn would be:

✿ 50% by weight, perennial ryegrass, *Lolium perenne*
✿ 30% by weight chewings fescue, *Festuca rubra* var. *commutata*
✿ 10% by weight smooth-stalked meadow grass, *Poa pratensis*
✿ 10% by weight brown top bent, *Agrostis tenuis*.

Sow at 35 g/m^2 (1.5 oz/yd^2). Mix the seed with fine sand to facilitate an even spread, and broadcast the seed evenly over the prepared plot. For large areas it is advisable to divide the plot up and calculate the amount of seed for each section. A mechanical broadcaster may be used to facilitate sowing.

Finally, rake the seed lightly into the surface. Protection from birds is advisable.

Should conditions turn dry after sowing, irrigation is vital.

Creating a lawn from turf

Preparation follows guidelines as for seed, but the tilth does not need to be as fine and stones over 2.5 cm (1 inch) should be removed.

A firm and level surface is essential. Turf is available in sections or rolls. Start laying the turf at one side or corner, working forwards and treading on planks set across the turf. Bond the turves like bricks in a wall and butt joins well, sifting sand between the joints. One of the most common faults is for gaps to appear between the turves. Firm after laying by tamping down with a 'bishop'.

Irrigation in dry periods after laying is essential.

Maintenance tasks

Mowing – Cylinder or rotary mowers may be used. Cylinder mowers are suitable for ornamental lawns and will achieve a finer finish. The roller produces the striped effect. Rotary mowers are suitable for general purpose lawns. Mulching mowers cut the arisings finer and redistribute them to mulch and feed the lawn.

Traditional mowing dates were between March and October. Climate change is affecting these dates, and in milder areas the season for mowing is much extended.

Heights of cut – For a lawn grown from seed, wait until the grass is about 50 mm (2 inches) before cutting to 25 mm (1 inch) initially. Established utility lawns may be cut to 25 mm (1 inch). Different heights of cut can be used to create interesting effects, such as mown grass paths through longer areas, and a higher general cut will preserve the lawn better in conditions of drought.

High-quality ornamental lawns are cut lower than utility lawns. The former may be cut as low as 5 mm (1/5 inch), although this needs to be done every two to three days, and involves largely unsustainable practices. A height of 10 mm (2/5 inch) in summer, and slightly higher in the spring and autumn will create a fine appearance suitable for most high-quality lawns.

Health and safety factors need to be borne in mind for any turf machinery. These include the use of personal protective equipment such as steel-toe-capped boots, regular maintenance of machinery, and existence and proper use of the dead man's handle. Also ear defenders should be worn if required.

Edging – Long-handled edging shears can be used to trim the lawn edges after mowing. To establish a straighter or cleaner line, the edges can be cut periodically with a half-moon edging iron or powered edger.

Scarification – Scarification is carried out in the autumn to remove thatch from a lawn. This can be done by hand with a spring tine rake, or with a mechanical scarifier.

Aeration – Lawns inevitably become compacted if subjected to regular wear. Aeration can help relieve this and hand and mechanical methods are again possible. Hollow tine aeration is particularly effective as a core of soil is removed. Slitting machines are also used, and a manual method of spiking with a hand fork is another option for a smaller scale. This should be carried out in the autumn.

Top dressing – Top dressing can be carried out following hollow tine aeration, and a mixture of loam and sand brushed into the holes created. This will keep the structure more open.

Fertilisers – Different fertiliser regimes are appropriate at different times of year. A fertiliser with higher nitrogen is appropriate for the spring when growth is beginning, and higher potassium in the autumn to 'harden' the turf over the winter.

Weed control – The degree of weed control desired depends on the function of the lawn. A high-quality fine ornamental lawn will require more attention than a general purpose utility lawn. Clover, *Trifolium* spp., often regarded as a turf weed, is now often included in some seed mixtures. It aids in nitrogen fixation.

On a small scale, hand weeding is appropriate for larger weeds, such as plantains, *Plantago* species. Overall, proper site preparation, appropriate selection of grass mix, and good maintenance cutting to appropriate heights, all contribute to eliminating the problem of weeds.

Selective and translocated herbicides are an option for a fine lawn.

HERBACEOUS PERENNIALS

Herbaceous perennials may be defined as those plants with soft, i.e. non-woody, growth that complete their life cycles in a period greater than two years.

Many die back at the end of the growing season, and have perennating organs, such as rhizomes or corms, to enable them to survive over the dormant season. Others will remain evergreen, providing more permanent features in a design.

MEADOWS

A different regime exists for creating and maintaining meadows. Traditional lawns require the build-up of soil fertility, whereas meadows rely on low fertility for a rich biodiversity. If the site for a proposed meadow has a rich and fertile topsoil, this will need to be removed, or fertility lowered by constant removal of crops grown on the area. Meadows may be created by using native plants or exotics suited for the purpose. The use of the native, semiparasitic plant, yellow rattle, *Rhinanthus minor*, helps lower the fertility of the sward. Different types of grass are used compared with traditional lawns, the fine grasses being more suitable, and bulbs are often incorporated.

Specialist wildflower and grass seed mixes are available with correct proportions of the different seeds, prepared for different environmental situations and aesthetic purposes, such as colour schemes.

Ground preparation for a meadow is similar to that for a lawn, with the exception of improving fertility. A moist site could also be used for those plants adapted to such conditions, such as snake's head fritillary, *Fritillaria meleagris*; whereas for a lawn, the drainage would need to be improved.

Another option of creating a meadow is to adapt an existing lawn. The lawn should be mown with arisings repeatedly removed. Plug plants can then be used as appropriate. Plants are generally selected for spring- or summer-flowering meadows, examples of which are given in Table 14.6.

Maintenance regimes vary according to the season of interest. For a spring-flowering meadow, cut from mid summer onwards. For a summer-flowering meadow, cut from early autumn onwards. The height of cut is higher than that for a traditional lawn, at about 7.5 cm (3 inches).

Grasses	Agrostis tenuis
	Alopecuris pratensis
	Festuca rubra
	Holcus lanatus
	Phleum pratense
	Poa pratensis
Spring flowering	Crocus tommasinianus
	Narcissus pseudonarcissus
	Primula veris
	Prunella vulgaris
	Veronica chamaedrys
Summer flowering	Camassia quamash
	Centaurea nigra
	Geranium pratense
	Knautia arvensis
	Leucanthemum vulgare

Table 14.6 Plants for meadows.

Paths can be permanently mown through the meadow to allow access through and enjoyment of the feature.

An alternative method of creating the interest and effect of meadows is to leave areas of long grass within the garden. This practice has long been carried out with spring-flowering bulbs, where the foliage should be left uncut long after the flowers have finished. The practice can be extended to include a wider range of plants through the seasons. The wildlife value of the garden will be much enhanced whichever option is taken.

Herbaceous perennials may be used in a variety of ways, to suit a range of garden styles. They may be used in herbaceous borders, to provide ground cover or selected for shade.

The traditional herbaceous border, epitomised by Gertrude Jekyll, is planted purely with herbaceous perennials. It often consists of one or two long rectangular borders with plants tiered according to height, providing a summer and early autumn display. Formal hedging, such as *Taxus baccata* or *Carpinus betulus*, provides a suitable backdrop to best display the plants. Ensure that space is left at the back of the hedge to access for hedge cutting. The front of the border may be edged with grass or paving. A hard edging such as paving facilitates maintenance, particularly in heavily trafficked areas. Borders should be at least 2.5 metres (8 feet 2 inches) deep to achieve the structure, height and variety of planting.

The concept of island beds was developed in the 1950s by Alan Bloom. The arrangement of the plants needed to be altered as the beds are viewed from all angles, and informal curves suited the style.

Herbaceous perennials are widely used in mixed borders. Disadvantages of seasonal interest can be overcome by planting among a framework of woody perennials, which provide a permanent structure. Herbaceous perennials also make valuable ground cover plants, and may be used as the lower layer of planting in the forest canopy-style planting.

Naturalistic methods of using herbaceous perennials avoid the highly structured border style, and plants are grouped together considering natural plant communities. By considering the soil, climate and topographical situation of their origins, a more natural effect can be achieved with plants suited to their environment. The plants are not tiered according to height. Plants are selected that do not require staking and heavy feeding. They are also often selected for attractive seed heads and structure of the stems, which can provide interest over the winter months. Ornamental grasses fit well with this type of planting scheme.

Ground preparation for herbaceous border

A few months before planting, thoroughly clear the ground, removing all traces of perennial weeds.

Once established, deep digging to remove weeds is not advisable around herbaceous perennials.

Dig the border, incorporating a 5 to 10 cm (2 to 4 inch) layer of organic matter, such as well-rotted manure, garden compost or composted green waste, to ameliorate the soil.

Planting should be carried out in the autumn or spring, the latter being preferable for colder conditions. Ensure containers are

Biennials should be sown in May/early June, either outdoors in a nursery bed in drills, or under protection in modules or seed trays. Those grown under protection will require hardening off before planting out. Those sown in nursery beds can either be thinned or transplanted to a wider spacing of 10 cm (4 inches). Before planting out in early autumn to replace the summer bedding, ensure that the plants have been well watered. Lift with a fork, taking care not to damage roots, and make sure that the roots do not dry out before planting. Plant out at a spacing of about 25 to 30 cm (10 to 12 inches) using a trowel, or a dibber may be suitable for wallflower, *Erysimum cheiri*, grown in a nursery bed. Firm the plants in well.

PLANTS IN CONTAINERS

Plants grown in pots or large containers can add decorative display to patios, balconies and are very useful for roof gardens. Plants that have special requirements, for example acid-loving plants, can be grown in pots where the garden has alkaline soil. Tender trees and shrubs can be moved from protection in the winter to outdoor display over the summer.

The range of plants suitable for containers is very large but can be categorised into permanent planting and seasonal or temporary display.

Seasonal display

Containers such as hanging baskets, tubs and window boxes are used to display plants. Summer displays include annuals and tender perennials, while for winter displays low-cost evergreens and spring-flowering plants are used. Spring displays will also contain bulbs. Also it is possible to grow vegetables in containers.

Figure 14.11 Hanging baskets.

Figure 14.12 Vegetables growing in a container.

Seasonal displays are temporary, which means that the cultural requirements for these plants are simple. They need enough volume of compost to develop a root system, high-nutrient compost and plenty of water. Therefore the container choice is wide and can vary from old sanitary fittings to worn-out boots as well as the more traditional manufactured pots and baskets. The material from which the container is made will influence the water requirements of the display. Porous materials like terracotta need much more water than solid materials like plastics.

Compost choice will also be influenced by the container type. For a temporary display, soil-less compost, the use of which is discussed in Chapter 6, is suitable and as it is light, crucial for hanging baskets. It is useful to use a water-retaining gel with soil-less composts, particularly for water-demanding plants like vegetables. A slow-release fertiliser is incorporated into the compost.

Plants are potted depending on the season of display and hardiness of the plant chosen. Summer displays are normally potted under glass and established with frost protection until it is safe to put them into their display position for the summer. Winter and spring displays are potted in October and immediately put in their display position until May. Temporary displays are potted with very little space left between plants as the aim of the display is to provide a burgeoning display of growth. It is normal to have a central, dominant plant with infill around it and trailing plants cascading over the edge of the container.

Maintenance

The container is watered as required, daily for summer displays, and compost kept moist for winter and spring displays. A weekly liquid feed through the summer will greatly improve the quality of growth. Deadheading of flowering plants will extend the period of display.

Permanent containers

Plants grown in containers for a long period of time require more careful planning than seasonal displays. When selecting a container, the vigour of the plant and site preference must be considered to ensure the material is suitable and the size is physically realistic. The surface area of the container should be as large as possible to facilitate root aeration.

Compost for permanent containers is also crucial. Soil-less compost will dry out quickly and its decomposition will present a problem of deteriorating root volume over time. It is therefore sensible to use a loam-based formulation or at least a mixture of the two. Loam-based compost offers the weight, mineral stability and provides long-term trace nutrients, which will maximise the potential for trouble-free display over time.

The plant, see Table 14.11, when first obtained is often much smaller than the chosen container. If this is the case, the plant should be potted in the centre with some annuals or temporary plants put around the edge to stop the unused compost stagnating before the roots are able to grow to the edge of the container. These temporary fillers are removed once the main plant has established and the compost replenished around the edge of the pot.

The plant should be potted to **nursery level** in the spring or autumn when conditions are mild. It is firmed well in the compost, watered and mulched.

Figure 14.13 *Acer palmatum* growing in a container.

Plants for permanent containers and raised beds		Plants for seasonal display in baskets and pots	
Shrubs:	*Acer palmatum*	Tender perennials for summer display:	*Pelargonium zonale*
	Fatsia japonica		*Canna indica*
	Ficus carica		*Kalanchoe blossfeldiana*
	Laurus nobilis		*Chlorophytum comosum*
	Olea europaea		*Argyranthemum frutescens*
	Buxus sempervirens	Half-hardy annuals for summer display:	*Impatiens walleriana*
	Rhododendron luteum		*Lobelia erinus*
	Pinus mugo		*Petunia* × *hybrida*
Climbers:	*Wisteria sinensis*		*Ageratum houstonianum*
	Clematis florida		*Salvia splendens*
Grasses:	*Helictotrichon sempervirens*	Plants for winter and spring display:	*Viola* × *wittrockiana*
	Carex buchananii		*Erica carnea*
Herbaceous:	*Hosta sieboldiana*		*Hedera helix*
	Agapanthus africanus		*Primula elatior*
	Penstemon barbatus		*Tulipa tarda*

Table 14.11 **Plants for containers.**

Maintenance

Plants must be kept watered, in the summer this may be weekly, even daily in dry periods, in winter monitored and never allowed to completely dry out. A visual assessment of water requirement is not enough as composts will quickly dry out at the surface but may still be moist at the root. Use weight or a cane pressed into the compost, you can also use your fingers, to assess the moisture content of the compost.

An annual top dressing of slow-release compound fertiliser is required in the spring with liquid feeds through the summer if the species is demanding. The compost level can be replenished as required unless it is possible to repot the plant into a larger container.

Other maintenance techniques such as pruning or deadheading will depend on plant choice.

HERB GARDENS

The plants chosen for herb gardens are based on plants that are of some use to us. The earliest human writings from China, Egypt and later Greece and Rome all mention the use of plants for medicinal purposes. There is a vast history to research on this area and in Great Britain the medieval monastic herb gardens were followed by seventeenth century knot gardens where herbs were grown inside patterns of low box hedging with topiary accents.

Figure 14.14 A knot garden.

It is possible to put a use to most plants and therefore when designing a herb garden one has to be more specific about its purpose or use history to refine the plant selection.

Culinary herbs are the most obvious for the modern garden but it is also possible to have beds of dye plants, scented plants for pot pourri, those for cosmetics or an endless choice of plants for medicinal purposes.

Figure 14.15 Modern culinary herb garden.

Herb gardens should be carefully planned because of the huge mixture of plants that may range from native annuals to tender perennials and established trees. It is particularly successful to use shrub structure or low hedging as a permanent backdrop to the annual and herbaceous growth of many useful herbs.

The planting, establishment and maintenance techniques are dependent on the type of plants used. Refer to the sections in this chapter on trees, shrubs, herbaceous and annual bedding. Also see Chapter 12 for further details.

WATER GARDENS

Water gardens present an opportunity to plant a wide range of aquatic plants that would not be successful in normal borders. Aquatic plants are predominantly herbaceous perennials, although some are evergreen and plants that grow on the margins of the pond can be woody perennials.

Water gardens can be native, varied and ornamental or focus on a small range of species such as water lilies, *Nymphaea* spp.

Aquatic plants are categorised in the following way:

Submerged aquatics: these plants grow under water, are rooted into the substrate at the bottom of the pond and are excellent for aerating the water (and are also known as oxygenating plants) and creating habitat for the fauna in a pond. This group of plants can be plunged straight into the pond at depth, and will quickly acclimatise to light and water quality.

Floating leaved aquatics: these plants are rooted into the substrate in the bottom of a pond and rapidly produce growth in the spring, which reaches the water surface and

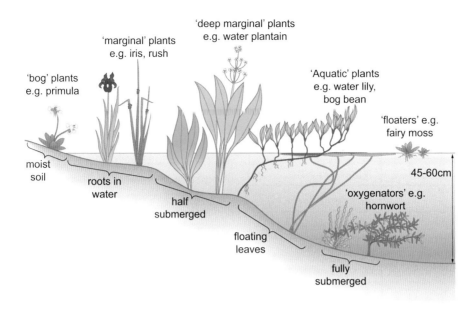

Figure 14.16 A transverse section of a pond with planting depths.

reveals leaves followed by flowers. They are useful for providing shade in the water for fauna and also reducing light transmission for algae control. Care must be taken to establish young plants gradually as leaf petioles expand to the depth required of them. Surface leaves will not survive being immediately submerged.

Marginal aquatics: this group of plants grow with the roots in water but the main shoot or flowering system emerges above the surface. Some plants are deep water emergents and others do better in the shallow margins of the pond.

Bog plants: these plants grow well in moisture-retentive soil and will tolerate periodic flooding but will not grow well in permanently wet margins.

Aquatic plants can be planted into baskets or planted direct into a soil substrate. Organic compost is not suitable as it will float in the water. Bog plants are always planted directly into the soil, which can be improved with organic matter. When planning a water garden it is essential that the planting shelves in the water are completely horizontal for the stability of the plants.

Maintenance

Nutrient availability is key to managing a water garden. If there is too much free nutrient in the water, the algae will grow at the expense of the other plants. Algae should be allowed to bloom as the water warms up in the spring, which in turn will encourage the build-up of insect larvae to feed on it.

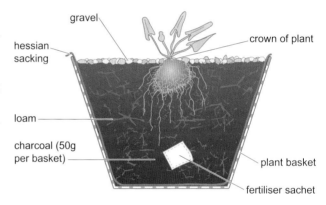

note: always ensure that the plants are at their correct depth

Figure 14.17 Aquatic basket showing soil, gravel and mesh.

However, the larger plants will be emerging and growing vigorously, thereby using a share of the nutrient. By the time the insect larvae have hatched, the algal bloom will recede leaving the water clear. The introduction of nutrient-rich composts, larger creatures such as fish, and particularly fish food will all disrupt this natural balance and give algae an opportunity to cloud the water.

Biological and UV filters can be used to keep the water clear where algae is a big problem or fish are an important component of a pond.

Aeration of the water can also be a maintenance issue and can be improved by the use of submerged aquatics or oxygenating plants, listed in Table 14.12, the annual removal of leaf and plant debris, the use of pumps to create fountains or water flow, and the addition of cool water in the summer when oxygen levels drop with rising water temperatures. The use of a ball or polystyrene block floating on the water surface in winter allows air holes to be

Description	Name
Submerged, oxygenating plants.	*Ceratophyllum demersum*
	Lagarosiphon major
	Myriophyllum spicatum
Floating leaved plants	*Nymphaea alba*
	Aponogeton distachys
	Nymphoides peltata
Marginal plants (0–15 cm water depth)	*Alisma plantago-aquatica*
	Butomus umbellatus
	Caltha palustris
	Iris pseudacorus
	Mentha aquatica
	Typha angustifolia
Bog plants	*Primula japonica*
	Astilbe x arendsii
	Gunnera manicata
	Hosta fortunei
	Ligularia dentata
	Rodgersia aesculifolia

Table 14.12 Aquatic plants.

Figure 14.18 Water lily, *Nymphaea* sp.

made easily when ice is formed with no disturbance to the wildlife in the pond.

Annual feeding of plants in the water is necessary when they are in baskets and can be done by repotting or top dressing with slow-release fertilisers. The dose rate should be minimal due to the algae response to nutrient described above.

The dying herbaceous stems of marginal aquatics are cleared from the pond at the end of the winter. In the spring, aquatic plants begin to shoot a short time after plants in garden borders due to the cool water temperature. Once plants are growing actively, if necessary, they can be lifted and divided (see herbaceous section above). When trimming plants, lifting baskets or weeding out vigorous species, the debris should be left on the side of the pond before being composted so that insects and amphibians can escape back into the water.

ALPINES AND ROCK GARDENS

Alpines are grown on rock gardens, in stone sinks or dry stone walls.

Alpines are plants that grow at high altitudes. This group of plants have physical characteristics which mean that they are well

Figure 14.19 An alpine feature.

Figure 14.20 A rock garden.

adapted to grow in conditions found high up in mountainous regions.

The conditions found are typically short spring and summer seasons, periodic watering, high wind movement, poor soil and extremes of temperature. These alpine plants are therefore spring and early summer flowering, are compact and often succulent. Alpine features are rocky, have very little soil substrate and are mulched with grit or stone chippings. Alpine plants include *Saxifraga aizoon*, *Draba rigida*, *Gypsophila repens*, *Erodium corsicum*, *Globularia meridionalis*, *Sedum acre* and *Sempervivum arachnoideum*.

Alpine regions also include lower meadows where the conditions are more temperate and wetter. Plants from these regions are more tolerant of rich garden soil and need more moisture in the summer. Alpine meadow plants include *Gentiana sino-ornata*, *Prunella grandiflora*, *Salix reticulata* and *Ranunculus alpestris*.

The term 'rock garden' is generally used to describe a feature that includes rocks with ground-creeping perennials, many of which are much too vigorous to be included in an alpine collection.

Rock garden plants include: *Aubrieta deltoidea*, *Cerastium tomentosum*, *Dianthus deltoides* and *Iberis sempervirens*.

Planting

Alpines must have good drainage around the crown to prevent rotting, and little water retained around the roots.

Plant in the spring into a soil mix that is free draining. Sand or grit added to a loam-based compost is ideal and is worked into the crevices of the rock feature. Alpines must be placed in sites that closely reflect their natural environment.

Bulbs such as *Iris reticulata* and tulips should be planted in areas that get full sun. *Crocus*, *Scilla* and *Chionodoxa* can be placed under mat-forming plants. Vigorous plants for rock gardens must be kept away from small crevice plants.

Maintenance

1. *Feeding*: alpines need very little nutrients and can be given an annual dressing of a balanced compound fertiliser at a quarter of the normal rate in the spring. Mulch as necessary with grit or chippings.
2. *Protection from cold*: alpines vary a lot in their tolerance of frosts and while they naturally cope with extremes of temperature at different seasonal stages, they are best not sited in a frost pocket, which would also encourage stagnant, humid air. For the least hardy alpines, a cold frame or alpine house is recommended. It may also be necessary to protect some plants from rain with temporary covers to avoid rotting.
3. *Protection from drought*: alpines need water when they come into spring growth. Alpine plants that grow in meadows will also need summer irrigation.
4. *Weeding*: perennial weeds are very difficult to eradicate from established alpine gardens. Hand weeding is the only option and must be done continually through the growing season.
5. *Pruning*: only needed for those plants that make a lot of growth each year such as aubrietia, *Aubrieta deltoidea*, which is cut hard back after flowering. Deadheading will lengthen the display.

PLANNING MAINTENANCE

To plan the maintenance of a garden, the combination of planted features and their requirements must be taken into account. Most gardens will have at the least mixed borders, hedges and lawns but many will have more specialised features.

A simple maintenance plan is outlined in Table 14.13.

Some tasks such as leaf sweeping are simple and fast, others such as pruning require some expertise and will take some time. Specialist machinery may be required for tasks such as hedge trimming and mowing.

Garden feature and operations	J	F	M	A	M	J	J	A	S	O	N	D	Tools and materials
	Visits per month												
Shrub border:													
Weeding			1	1	2	2	2	2	2	1			Fork, trowel, wheelbarrow.
Feeding			1										Farmyard manure or compound slow-release fertiliser.
Mulching			1							1			Organic matter, wheelbarrow, shovel, fork, rake.
Pruning		1		1			1			1			Secateurs, loppers, pruning saw, gloves.
Pest and disease observation	1	1	1	1	1	1	1	1	1	1	1	1	Diagnosis and treatment as appropriate.
Herbaceous border:													
Weeding			1	1	2	2	2	2	2	1			Fork, trowel, wheelbarrow.
Feeding			1										Farmyard manure or compound slow-release fertiliser.
Mulching			1							1			Organic matter, wheelbarrow, shovel, fork, rake.
Trimming		1				1							Shears, secateurs, fork, wheelbarrow.
Deadheading				1	1	1	1	1	1				Secateurs, wheelbarrow.
Division			1				1			1			Forks, knife, spade, planting compost, wheelbarrow, water.
Lawn:													
Weeding	1	1	1	1	1	1	1	1	1	1	1	1	Daisy grubber, wheelbarrow.
Feeding			1					1					Compound slow-release lawn feed.
Mowing and edging			1	2	3	4	4	4	3	2	1		Mower, edging tool, wheelbarrow.
Irrigation						1	1	1					Lawn sprinkler if necessary.
Leaf sweeping									1	2	2		Lawn rake, besom or spring tine rake, wheelbarrow.
Scarification										1			Spring time rake or scarifier, wheelbarrow.
Aeration										1			Hollow tine aerator, lawn rake, wheelbarrow.
Top dressing										1			Top-dressing media, wheelbarrow, besom or **lute**.

Table 14.13 Annual maintenance plan.

Summary

Gardening is very much concerned with manipulation of the environment to provide ideal growing conditions for those plants desired in the garden. The understanding and creation of different microclimates is an important tool.

To aid the matching of the microclimate a vast range of plants is available for the gardener to choose from. Rather than constantly fighting against existing soil and climatic conditions, wonderful effects that are easier to maintain may be created by working with the site.

Manipulation of the plants as well as the growing environment is a further important skill. Shrubs and trees may be pruned and trained to fulfil a variety of different practical and aesthetic functions, such as hedging and topiary. Pruning of shrubs can enhance flowering and fruiting or encourage bright winter stems. Techniques such as stopping and pinching out can control shapes of plants and size and quantity of blooms.

However, thorough site preparation is paramount for the future success of any planting scheme, and will vary according to existing conditions and the amount of manipulation of the growing environment required. Following selection and the appropriate site preparation, the establishment and maintenance of plants needs to be considered, but the area of maintenance is often seen as a 'soft option' to be the first area to be reduced or ignored.

Another method of establishment that has enjoyed a resurgence is that of container growing, where you are independent of the soil and can adjust the microclimate as the day progresses: providing increasing sun or shade as the sun tracks across your site. Some of the main advantages of this method of growing are the use of space and ease of movement, allowing you the opportunity of recreation of a design, including increasing the space given, using the same plants, depending on your mood or the prevailing season.

This again brings in a need to understand the life cycles of the plants linked with the desired aesthetic and functional effects. Attention paid to correct planting techniques and times, followed by clear maintenance schedules will provide the best opportunity for success. Also in these more relaxed times, with an eye to wildlife attraction, enjoy the seed heads and mature stems of herbaceous perennials as they move into senescence and welcome the frosty days of another season.

FURTHER READING

Christopher, M. (2006). *Late Summer Flowers*. London: Frances Lincoln.
A good round up of the plant species used to extend the season of interest.

Court, S. (2004). *Roses in Modern Gardens*. London: Mitchell Beazley.
Ideas abound in this publication for plant siting with a modern twist.

Lloyd, C. (2004). *Meadows*. London: Cassell Illustrated.
Experience the delights and pitfalls of establishment and maintenance of these now popular features.

Oudolf, P. & Gerritsen, H. (2003). *Planting the Natural Garden*. Cambridge: Timber Press.
Clear ideas are outlined to ease the problems commonly encountered when moving towards a more natural look.

Thomas, H. (2008). *The Complete Planting Design Course: Plans and Styles for Every Garden*. London: Mitchell Beazley.
A broad range of ideas and planting examples are discussed in this publication.

Revision questions

1. Explain how the **purpose** of a hedge will influence the choice of species.
2. Name **two** shrubs for organic rich soils and **two** shrubs for well-drained soils.
3. Describe the specific site conditions required when planting and siting *Clematis* in the garden.
4. Outline the practical tasks required in the **autumn** to keep a lawn in good condition.
5. Explain the benefits of using **ornamental** grasses in herbaceous planting schemes.
6. Describe the maintenance needed for a permanently containerised **shrub**.

Figure 15.1 Vegetation survey paraphernalia used for the rare fen orchid, *Liparis loeselii*, study, one of the plant species studied by UK botanical science students.

Figure 15.2 Tomatoes undergoing biotechnology research in Belgium.

PLANT COLLECTIONS AND DATABASES

There are many organisations whose primary aim is to conserve and maintain the integrity of genes, individual plants, cultivars and species of plants pertinent to horticulture.

Conservation (of any kind) is a watchword of many but, in reality, is very difficult to achieve. In order to conserve plants that are, by definition, rare or of vital importance, several factors have to be considered: is the size of the initial population of plants to be conserved viable, is there the physical space available to store/grow the plants, is the technology available to carry out micro-propagation if it's required and are the financial resources available in order to effectively carry out such conservation work?

Habitat conservation must also be considered, is it worth using limited resources to save a small number of threatened individuals in a glasshouse, if their habitat is lost and the plants never returned?

In a perfect world, all plant cultivars and species would be conserved. We all know this is not a perfect world so difficult decisions have to be made as to how the resources are allocated to which plants. Who has the moral right to make such decisions?

The NCCPG seeks to conserve, document, promote & make available Britain and Ireland's rich biodiversity of garden plants for the benefit of everyone through horticulture, education & science.

Figure 15.3 Plant Heritage (formally known as the NCCPG) works tirelessly in order to conserve endangered garden plants.

One such organisation is Plant Heritage (formerly known as the National Council for the Conservation of Plants and Gardens, NCCPG). Plant Heritage's aims are to encourage the propagation and conservation of endangered garden plants in the British Isles, both species and cultivars; encourage and conduct research into cultivated plants, their origins, their historical and cultural importance, and their environments; and to encourage the education of the public in garden plant conservation.

Through its membership and the National Collection holders, Plant Heritage seeks to rediscover and re-introduce endangered garden plants by encouraging their collection, propagation, maintenance and distribution of as many taxa as possible.

Plant Heritage works closely with other conservation bodies as well as botanic gardens, the National Trust, the National Trust for Scotland, English Heritage, the Royal Horticultural Society and many specialist horticultural societies.

Figure 15.4 Botanic Gardens Conservation International (BGCI) helps conserve plants within a network of botanic gardens around the world.

One way to conserve endangered species of plants is to secure their survival within botanic garden collections (this is termed *ex situ* conservation, i.e. conserving organisms away from their natural habitat in a controlled environment). Botanic Gardens Conservation International (BGCI) assists in the implementation of one of the targets of the Global Strategy for Plant Conservation (GSPC) where 60% of threatened plant species are to be conserved in *ex situ* collections, thus helping to preserve those species in greatest danger. The BGCI also works with partners such as ENSCONET (European Native Seed Conservation Network) in an attempt to conserve the seeds of Europe's most threatened species.

Botanic Gardens Conservation International is a worldwide organisation that is dedicated to supporting the development of 1,600 botanic gardens along with gathering and disseminating information relating to plant conservation. The BGCI focuses on five categories of plants in need of conservation. These being; rare and endangered plants, economically important plants, species required for the restoration of ecosystems, keystone species (plant species that are important in keeping an ecosystem healthy) and taxonomically isolated species with scientific value. The BGCI also supports *in situ* conservation of habitat restoration throughout the globe (*in situ* conservation is where organisms are conserved within their natural environment).

Figure 15.5 The Palm House at the Royal Botanic Gardens at Kew. Kew's conservation programmes include *in situ* and *ex situ* projects.

The Royal Botanic Gardens at Kew is a world leader in the conservation of plants. Not only do they hold many specimens of rare or endangered plants, but they are also involved in finding unknown plant species, researching the possible benefits of phytochemicals not yet understood, *in situ* conservation and seed preservation.

Figure 15.6 Wakehurst Place is a National Trust property managed by the Royal Botanic Gardens, Kew.

Why save seeds?

The conservation of plants is immensely important to the survival and wellbeing of human society. Plants are the basis of all terrestrial ecosystems, and as such, if plant species were to decline in numbers, animals including we humans would suffer too. At present it is thought that 60,000 to 100,000 plant species are under threat from climate change, habitat loss and over-exploitation. By collecting and storing seeds, it is hoped that the genetic information contained within will be available for possible future release and/or research.

The Millennium Seed Bank already holds seeds from species thought to be extinct in the wild and has recently celebrated banking its 24,200th plant species; 10% of the world's plant species! The project's main aims are to collect seeds, herbarium specimens and data from 24,200 plant species, not only at the Millennium Seed Bank but also in the countries of origin. The project also works to increase public awareness of the need for plant conservation and its importance in this era of changing climate and urban expanse.

The project's work is not as simple as collecting seeds and keeping them in airtight containers. The seeds have to be dried to a moisture content of below 7% and sealed in moisture-proof containers such as laminated foil bags, aluminium cans or glass jars. The processed seeds are then stored at −18°C (−0.4°F) for a period of time, possibly as long as several years. Of course, seeds will not survive indefinitely using this storage method and need to be periodically germinated in order for seeds to be recollected and stored for the future.

For new plant species that have not yet undergone this preservation treatment, or are recalcitrant seeds (these are seeds that do

CASE STUDY Hillier Nurseries, Romsey, UK

This world-famous nursery has been growing hardy nursery stock since 1864 and has over 225 successful new plant introductions produced by its propagation department. It now consists of several wholesale nursery sites, a chain of garden centres and the highly acclaimed landscape divisions. Their broad range of horticultural enterprises ensures Hillier Nurseries are at the forefront with new developments, marketing and promotion campaigns. The high standards demanded by the company are reflected in all its operations: from the 31-hectare (75-acre) container nursery to the extensive 208 hectares (500 acres) of open ground sites with trees up to semi-mature specimen sizes.

The well-stocked retail garden centres offer top-quality plants with stimulating displays to satisfy the demand of ever-increasing numbers of discerning consumers. The highly skilled landscape teams implement the designs for commercial and private clients using the quality plants from the Hillier range. Plants are also supplied wholesale to other growers, retailers and landscape companies.

All these business activities are enhanced by Hillier's many promotional activities, including the world famous *Hillier Manual of Trees and Shrubs* plus the numerous books and publications aimed at both the professional plant user and consumers for use in their own private gardens. Hillier's has an enviable history of winning awards at major flowers shows, national landscape ceremonies and business consortia events. In 2012, their amazing show display won the nursery their 67th consecutive gold medal at the RHS Chelsea Flower Show.

Maximise cropping space to increase efficient production and sales.

Chapter 16

Commercial horticulture

Chris Allen

INTRODUCTION

This chapter will show that commercial horticulture is a diverse and multidisciplinary global industry with thousands of commercial companies and organisations producing crops or offering their services to commercial customers, clients and retail consumers. These activities are carried out to satisfy local or regional demand, multinational operations or trading as vast global networks. These can vary from specialised plant growers to multiple retail chain stores; high street florist shops to seed producers and landscape contractors to tree surgeons, all supported by a wealth of research and development networks. With the world markets becoming more accessible and competitive it is essential that growers and suppliers are aware of how global issues will affect their existing customers and potential markets in the future. These factors could include fluctuating exchange rates within Europe or the sharp rise in global fuel prices, both dramatically affecting the supply and demand for crops. These variables can cause extreme difficulties for growers within an overcrowded, overstocked seasonal marketplace, especially where time-sensitive edible crops are being produced as any disruption in the supply chain will affect crop quality and its value. Many horticultural businesses can be divided into two main categories: those that sell their products or services to other commercial companies within the industry, these being wholesalers, and those that sell direct to consumers in retail nurseries, garden centres or multiple chain stores, these operate as retailers. There are companies that trade in two or more sectors, for example a wholesale production nursery supplying direct to its own onsite garden centre, or a landscape contractor growing plants for direct use in their private clients contracts.

This chapter will highlight some of the key considerations for anyone involved in commercial horticulture whether at the start of their career, progressing within an existing company or setting out on a brand new venture with limited experience. This chapter will hopefully improve the chances for success and help generate profitable returns.

> ### Key concepts
>
> ✿ Understanding horticultural markets and managing a commercial business
>
> ✿ The importance of planning
>
> ✿ How to keep financially viable
>
> ✿ Keeping on the right side of the law

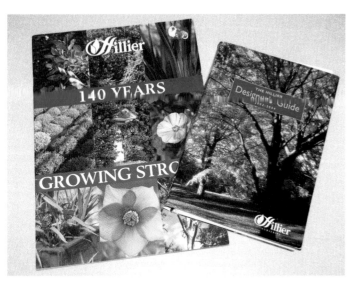

Figure 16.1 A small selection of Hillier Nurseries publications to support plant selection and use.

UNDERSTANDING THE HORTICULTURAL MARKETS AND MANAGING A COMMERCIAL BUSINESS

Commercial horticulture can appear to be a fragmented and complex industry. Finding a new job, creating a new business or developing an existing company is a daunting prospect for many people and will demand a lot of time and energy. However the job satisfaction and sense of achievement can be great, and last a lifetime.

This section will look at the following topics:

- ❀ different horticultural businesses
- ❀ local, regional and global markets
- ❀ selecting business activities and the right markets
- ❀ the importance of good management
- ❀ developing and increasing production and sales
- ❀ telling everyone about the business
- ❀ business infrastructure and development
- ❀ management of resources, production systems and provision of services
- ❀ production methods and operating services
- ❀ health and safety issues
- ❀ efficiency, sustainability and environmental good practice
- ❀ creating a professional, green, ethical business image.

Different business types

In order to begin to understand the horticultural industry and how it operates it is essential to visualise just how many different components there are within it.

From edible and ornamental crop production, sports and leisure sites to research and media there are businesses with professional horticulturists providing the highest standards of service and products to satisfy regional, national and international

markets. A visit to the library to look at reference books, journals and magazines will begin to build up the picture. Some publications will be aimed at the consumer and offer helpful information relevant to their own gardens such as *Amateur Gardener*, others such as *The Garden* published by the RHS will give a lot more detailed information about specific plants or places of interest plus the people involved. Then there are the trade publications aimed at professional horticulturalists who need technical information on latest developments and up-to-date industry news on a regular basis, a good example being *Horticulture Week*. There are also many specialist trade publications covering landscaping, glasshouse production, sporting and leisure sites.

Use of the internet can locate specific details for a specific market sector; thousands of websites around the world can be accessed, selected and used for sourcing information at all levels. Other sources of useful information include the many trade bodies whose members are involved in every sector of the industry; the internet will provide access to their websites and the details of how they support their members. Most companies produce literature describing their business activities and product ranges; this information can be obtained by visiting their websites or their stands at the annual trade shows that take place around the world. These regional, national and international shows provide great opportunities to gather information, meet people and gain valuable insight into how the industry operates, the scale of operations being carried out and how many of them link suppliers, customers and consumers together by providing the essential supply chains and communication networks.

CASE STUDY **The British Association of Landscape Industries**

The British Association of Landscape Industries (BALI) was set up in 1972 to promote communication and to maintain and increase industry standards across the UK as the country's representative trade association for the landscape sector.

Their vision statement is: 'An association for all landscape professionals that **supports**, **promotes** and **inspires** its members to be leaders of an environmentally, ethically and commercially sustainable landscape industry'.

This is enacted through annual awards, members' events and trade shows.

The marketplace

Modern horticulture is a global business operating around the clock on every continent. A visit to a supermarket or market stall selling fresh produce will provide the evidence. A close look at the labelling for many items may show they have been imported from another country. There are many reasons for this level of imported produce, competitive prices are a main factor, another is the fact that in warmer countries many crops can be grown all year round

Figure 16.2 The British Association of Landscape Industries is usually abbreviated to BALI and all members can display this logo.

unlike in the UK where growers may only have a short season unless they invest heavily in protected cropping structures and heating (see Chapter 11 for further details). Consumers are now familiar with certain crops being available all year round (AYR) and are out of touch with the seasonality and availability, therefore the demand is high so commercial growers, suppliers and retailers are keen to satisfy it. It is not only edible crops that are being shipped around the world; ornamental growers are maximising the growing seasons in different countries to get extra growth on their plants in the shortest timescale. Plants may be propagated in one country with higher temperatures earlier in the year; the young plants are then shipped to another continent to extend their growing season then to the European markets for dispatch for final growing on or sent directly for retail sale. The major benefits of these production systems are the additional plant growth achieved compared to that possible in a single UK season plus the time and labour saved in producing saleable plants makes quicker financial returns for the companies involved. By working together many individual specialist plant breeders and commercial growers link their production programmes and their output to

provide buyers with a wider range to choose from, this in turn offers a continuous, more efficient supply route for retailers to buy from so helping to ensure their customers have an uninterrupted supply to purchase. There are many examples where different companies work together for all or part of the production, supply and sales operations. This could be as a local co-operative benefiting from joint bulk purchasing agreements and sharing specialist resources or an international consortia pre-ordering and buying their production requirements on a global basis then selling their produce via the large international markets for distribution around the world. In recent years the introduction and global use of the standard Danish trolley system has dramatically improved the physical movement of living plants. It has played a major part in the development of this highly efficient and rapidly developing global horticultural marketplace where production and supply chains really do link together to form smooth and reliable transport routes around the world.

Whatever the scale of operation there are many advantages for linking with other companies, the benefits must result in cost savings, improved efficiency, quality products available to wider customer groups and, most importantly, increased financial returns to secure the future of the business with potentially higher profits for all involved.

Right business, right markets

Whatever the type, size and location of any operation it is essential to be producing saleable crops, or supplying professional services at the right price at the right time. Without customers it is obvious there will be no sales and no income generated.

Ongoing market research should identify what crops are to be grown or which services offered and how new ones could be planned for the future, where potential markets are located and what existing or future customer requirements may be.

The geographical location of the business will dictate and influence which crops are selected and how they are to be produced. Key factors that need to be considered include the topography of the land, its aspect and the microclimate within the site itself and how these will influence the length of the local growing season and future plant growth. This may dictate the possible need for protected cropping structures on exposed sites. If growing plants directly in the open ground it is essential to know the soil type and its suitability for the proposed crops plus what condition it is in: a mistake at this point would be extremely costly. The ease and efficiency for getting crops to market, delivery services or attracting customers when operating in remote locations where access to and from the site is difficult can lead to potential problems, these could range from dispatching edible crops that are particularly time sensitive or accessing additional staff at peak production times. The choice of crops to be grown or the services provided will depend on many factors including the horticultural knowledge and expertise of available staff; the financial capital available for setting up or the upgrading of an enterprise with specialised equipment or the income required between the start and finish of a service contract or the time between the production and harvesting of a crop. The time differences for the payments on a saleable lettuce, *Lactuca sativa*, plant and those for a mature specimen tree will be years. These vastly different time periods must be considered at the earliest planning stages to avoid potential problems when generating income through different production cycles.

Good management

One definition of management is 'to have effective control of the administration of the business concerns'. All businesses require consistently good management systems operating across all levels. Owners, managers, supervisors and employees all play an important part in making any business successful, or not. The size and scale of the enterprise will dictate the numbers of staff or amount of resources required; however, the services being offered or the types of crops being grown will be a major factor in the recruitment of horticultural staff, they will require appropriate knowledge and experience to carry out the specific operations. They will also require good supervision and leadership to ensure they are available to work efficiently and safely when required. A failure by staff at a key time can have a serious effect not only on cash flow with delayed payments but also on the reputation of the business. Letting clients or customers down with excuses is not good for ongoing business relationships. Another aspect that puts demands on any management system is the weather. Poor seasonal weather can have both immediate and long-term effects on any horticultural business. It is therefore even more important that staff are encouraged to participate positively in management activities and decision making to overcome existing or potential problems. Each sector of the horticultural industry has become a huge subject on its own and very few individuals have sufficient expertise to cover all aspects. With the continuous flow of updated specialist information it becomes even more important to have good staff who are able to carry out tasks at different levels while being a part of the business team and aware of the common aims and objectives.

The competition for selling products and services is rarely confined to a local region, with up-to-date communication systems managers can access information and talk to suppliers and potential customers anywhere in the world. High-speed transport networks enable haulage companies and specialist couriers to collect and distribute products efficiently to the required destination. The benefits for growers adopting these flexible systems include a reduction in owning delivery vehicles, less driving time and expense. These benefits would allow a manager more time to manage the important job of producing the crops or providing the services essential to the financial success of the business.

Developing the production and sales

In order to make a business grow and become more successful it is important to tell as many people about it so they will buy the

Figure 16.3 There are many publications to keep the team up to date with industry news and developments.

products or use the services, or recommend their contacts to do so. Improving income generation from increased sales is key to business growth and future development. Keeping up to date with horticultural news and development will be useful when looking to extend business activities and sales. It is essential to know what is happening currently within the trading sectors and monitoring what the future trends may be, this will allow better forward planning in order to develop new business opportunities and pick up market share before competing companies step in. The internet, trade publications and shows can play an important part in researching useful information about the horticultural industry and find useful business links to follow up. The consumer press and gardening shows can also give good indications of what the amateur market is buying and any possible future trends that are worth investigating in order to increase sales of plants and products.

Telling the world about it

It is important to remember that there are opportunities for generating new business around every corner; however, it will often require quick thinking to see them and not lose too many valuable opportunities. For plant breeders and growers the introduction of new cultivars is an ideal opportunity to attract some positive publicity especially if closely linked with a planned advertising campaign. Landscape companies that have completed major projects may be encouraged to enter for industry awards, these can attract the attention of the media who will welcome a press release about the project resulting in good editorial coverage. There are numerous ways of promoting a business, if paid advertising is to be used a timetable should be in place with the most relevant publications, then a spending budget can be agreed with the advertising department. It is essential that the anticipated response to any adverts can be met, any customers placing orders then being told their order cannot be completed will not be impressed and are likely to shop elsewhere in the future.

Attracting new customers is vital for business growth; however, it must not be at the expense of existing customers. The time taken searching for new customers means time not spent making certain established customers are being looked after and their orders are being processed as they would expect. In local, national or international marketplaces competition is getting tougher, it is therefore more important to get the company's message out to the maximum number of existing and potential customers and clients. The business plan should contain details of how and when advertising campaigns are being run, who is responsible for carrying them out and that the necessary finances are available to pay the bills.

INFRASTRUCTURE AND DEVELOPMENT

Many horticultural enterprises have evolved over many years, some over many generations of the same family. In family businesses the opinions of an individual may have dominated and prevented new capital investment or the development of new techniques to improve efficiency and so halting business progress. There are of course many good examples of family run enterprises that have seen continuous development and success. These are often owned or managed by people who have worked for different companies that had a clear, structured staff team who all understood the importance of continually creating new business opportunities at all levels. As a business evolves and grows in size more people and resources will be required, this in turn requires more careful management and support. Clear staffing structures and management guidelines with good supervision will help to ensure efficient work programmes, avoid confusion and give responsibilities to employees.

All staff should be aware of the staffing structures to help them understand their own role within the company; it can also provide a positive incentive for those wishing to develop their careers within the company. The management team must also be aware of additional resources required for any planned development: these could range from small items of equipment to large machines, or additional buildings or land for growing crops. With this information the budgets and timescales can be drawn up. It is essential that all business planning processes identify how, when and who will manage the developments, including all the strengths and any potential weaknesses that could cause future problems. A strong staff team supported with good resources and facilities will be best equipped to respond to new demands and deal with the anticipated business growth.

Managing the business

The management and the efficient use of all business resources has never been more important. A company seen to be wasting materials and working inefficiently will lose credibility on environmental and **sustainability** grounds with their reputation lost in a world where the consumer has become environmentally aware and not afraid to voice their opinion. It is therefore important that all staff have a clear understanding of the company policy on safe and efficient working procedures, handling and using materials and equipment, plus their own responsibility towards environmental issues such as minimising waste and its appropriate disposal.

A business using a lot of materials while producing their products or services should be aware of what percentage of waste is generated as this will have a direct effect on the ability to generate a profit or sustain losses. This can only be achieved by routine observations during the operations and recording the materials purchased, the labour and other costs incurred plus the money generated from the sales. This is where the viability of specific operations or the profitability of the whole business is identified.

Production methods and operating services

Where required some production methods or services may need upgrading to make them more efficient and cost effective. Any that are no longer viable should be scrapped and not kept going because they are being subsidised by profitable operations; over time this may lead to serious **cash flow** problems. These major decisions should only be taken after investigating all the options available. On occasions it may be necessary to inject some investment money into the business to allow for these proposed changes, once completed the overall efficiency and viability of the business should improve dramatically. From propagation to dispatch, a production nursery must aim to create the ideal growing conditions at all stages of the crop cycle. Across the site the demands will vary, from undercover protected structures (see Chapter 11 for further details) to container stock beds and open-ground fields. Where plants are moved from one area to another at the end of each growing stage it is essential to make each operation as efficient as possible with the use of mechanised equipment. Provided they have suitable hard surface paths and roadways many nurseries now use Danish trolleys, these are now a worldwide integrated handling system for moving plants easily and safely.

CASE STUDY   Danish trolley development

This light-weight, collapsible transport system is also called the 'Dutch trolley' or the CC container. It is a wheeled returnable transport item (RTI) and is the horticultural standard method for transporting plant material around the whole of Europe.

The main system is run as an international pool, where trollies are exchanged when materials are delivered or collected. These can be hired on a short- or long-term basis.

The use of mobile conveyor belts to move and load plants can also save many hours of staff time walking backwards and forwards, they also reduce the risk of plant damage during handling.

On larger nurseries four-wheel trailers pulled in tandem by tractors speed up plant handling over greater distances. Careful planning is needed to ensure that any new systems fit in with, and

Figure 16.4 The Danish trolley system.

Figure 16.5 Installing this conveyor belt within the dispatch department will improve the preparation and loading times dramatically when fully staffed.

improve the cost effectiveness of each operation. Research will show there are many different production systems and pieces of equipment to aid plant growth and handling, the manufacturers will be keen to provide technical details and advice on their products.

Health and safety issues

All members of staff in every type of business have a responsibility for many aspects of health and safety in their workplace, including an understanding of first aid procedures. Horticultural businesses can provide many opportunities of dangerous environments for staff, customers and visitors. When walking around any site it is possible to see situations where an accident or incident could occur, this could involve a reversing delivery lorry in the car park because nobody was guiding the driver back or contamination by pesticides used in a glasshouse because warning signs were not displayed on the doors. Both of these incidents would be

due to human error. Other examples could be glass sheets blown from the sides of the garden centre shop on a windy day and hitting customers: would this be classed as an 'act of God', a lack of maintenance on the building, or a lack of decision making to close the shop during the stormy weather? The results of each could be the same – people being injured. There are now so many rules and regulations connected with health and safety that it can be difficult to keep up to date. It is important that someone within the company is appointed to take the lead in order to help co-ordinate and adopt the recommended guidelines. Every business must produce its own health and safety policy outlining all aspects of how it proposes to deliver and maintain safe working practice. **Risk assessments** must also be written for all operations carried out by the business; these must be relevant to when and where each operation takes place and identify the process from start to finish. There is plenty of help and information available from central and local government sources, plus many specialist consultants who advise and help implement the necessary legislation.

Good practice

The horticultural industry is perceived to be closely integrated within the 'green' environment. It is a fact that much of the output does improve the environment: the plants and products sold to consumers and commercial organisations do help improve their local surroundings. However, what is the global environmental cost of all this production? The use of peat in commercial horticulture and gardening products has raised many issues relating to the 'green' credentials of the industry, add to this the fact that vast quantities of non-recyclable plastic pots and polystyrene trays and large volumes of pesticides are still used. It is important to reduce this environmental impact. There have been significant outcomes relating to the sustainable alternatives used in composts and new natural-fibre pots plus the ongoing withdrawal of many pesticides and chemicals for use on crops. Other uses of sustainable materials include processed wood chips as a biofuel in modern, highly efficient heating boilers and the many different biological control systems now used as replacements to pesticides (see Chapter 17 for further details). In the nursery it is possible to make savings in most areas. Installing thermal screens, efficient boilers and thermostats will reduce heating demands in growing structures.

Irrigation and rainwater run-off can be collected and processed for use in more efficient low-level or drip-irrigation systems that replace very wasteful overhead spray lines. All vehicles and machines should have routine maintenance checks to ensure maximum fuel efficiency, savings on fuel bills and reductions in exhaust emissions.

Good storage facilities for materials should prevent wastage through spillage and reduce the risk of accidental contamination. Ongoing training will make staff aware of their responsibility towards safe working practice, the potential risks to others and the environment, plus the benefits to the business by working efficiently, reducing wastage therefore improving its viability.

Figure 16.6 Thermal screen will reduce heating requirements during cold periods and provide shade in summer.

Figure 16.7 All aspects of the project should be looked at in detail, including the site.

LEVEL 3 BOX

AN ETHICAL IMAGE

The horticulture industry has become a highly technical and mechanised, multidisciplinary profession with huge potential for research and new developments in the future. It operates in a global marketplace worth billions in the currency of each country while providing employment to millions of people and, most importantly, supplying food and raw materials to most of the world's population.

To ensure this continues it is essential that all levels of business activities are carried out following professional and ethical codes of conduct. This is important for many reasons: horticultural products and services play an essential part in the wellbeing of so many people both as producers and consumers, the possible threats to human health and the global environment are great if wrong decisions or mistakes are made, plus the future supplies of new, essential resources could be threatened without sufficient investment in research and development, often abbreviated to R&D, programmes. It is also essential to ensure that the growing, supply and distribution of horticultural crops and services does not lead to the exploitation of those involved wherever that may be in the world.

THE IMPORTANCE OF PLANNING

This section will look at how businesses decide what to do: who is going to be involved and where operations will take place. Without careful planning and ongoing monitoring a business can drift into new areas of work at the expense of their key operations, this can be detrimental to positive growth and development. Some topics to consider are:

- ✿ the business plan
- ✿ seeking advice from professional organisations
- ✿ working with the authorities
- ✿ identifying the right type of business to develop

- ✿ the importance of a good business location
- ✿ identifying the key business objectives and the resources required
- ✿ processing and selling the products and services
- ✿ staff recruitment – the importance of getting it right
- ✿ monitoring progress and growing the business.

The business plan

A **business plan** can be a most useful document provided it is used as an ongoing reference document and not filed away and forgotten about. The plan is used to define both existing and potential objectives, and how they contribute to the organisation, and the processes required in achieving and implementing them, also the timetable when the planned developments are to take place and when decisions will be taken to allow necessary resources to be made available. Identifying the people who are available with the skills to carry out the tasks is also essential, especially if considering the use of specialist outside agencies. The production schedules and marketing strategy should also be included plus the financial requirements to carry out all the proposals.

The plan should be reviewed regularly or when new proposals are discussed, and any additions or amendments recorded. This will ensure that new developments are clarified and the benefits to the business clearly identified so all those involved are aware of the changes.

Asking the professional

As the horticultural industry has become more and more complex it has become more unlikely for individual members of staff to have sufficient knowledge, a full understanding, or the experience to manage every aspect of any business. It is therefore necessary to seek expert advice from outside agencies. This takes place routinely with the bank manager and accountant or with the appointment of a specialist consultant who provides professional advice for a specific aspect within the business who then produces a report and charges for their services. Another option is to become a member of a trade organisation that offers a wide range of business support as part of the membership package. There are several land-based bodies to choose from. It is important to know what these organisations offer their members and what will be of specific use to an individual business operation, many include advice on legal, marketing and employment issues plus special rates with insurance, healthcare and vehicle leasing and purchasing agencies. These services could prove to be an efficient and cost-effective method of acquiring specialist expertise. There are many benefits to joining the right organisation but prior research is required, their websites have a lot of information, making contact with other known members and attending regional meetings could be beneficial before committing to joining.

Working with the authorities

Wherever a business is situated, and whatever it does, sooner or later it will come to the attention of staff working in a department from the local, regional or national government agencies. This can be a daunting prospect especially if it is related to an incident that has contravened legal procedures or official guidelines. However, these departments can also offer positive and constructive support for helping business development. It is important to understand that these official departments have the full weight of the law behind them, so it is far better to get the relevant people on side as soon as possible with any business developments so they can have a positive input at the early stages and possibly save a lot of time and possible expense. The main departments to approach would be the local planning office for all new structures or changes to existing facilities; the highways department for anything to do with access in and out of the premises; environmental health department for all aspects of hygiene, food handling and health and safety, trading standards officers for inspection of products and services; the local authority open spaces department tree officers or countryside wardens on aspects of trees and habitat. Many local authorities produce a small directory giving all the contact details for the different departments, this can save time when trying to find a specific person to process your enquiry. Other official bodies include the National Rivers Authority who keep a close watch on all waterways to prevent, and identify, any pollution incidents; the Health and Safety Executive who can monitor and inspect everything, anywhere that is or could be a hazard to health. HM Revenue & Customs have enormous powers to look at most things linked to finance or the handling of illegal substances. All the above organisations have websites and produce datasheets containing vast quantities of information, some will be relevant to the business and it is important to read and understand the contents, none more so than links to health and safety and the legal obligations placed on us all.

The right business development

When starting a new enterprise or taking on an existing business it is essential to fully understand what the commitment is going to be.

Figure 16.8 Details for all operations should be recorded.

At the outset all the necessary documentation must be available for all parties to see, this can reduce the risk of any legal or financial obstacles cropping up later in the proceedings. If the ownership or renting of land or property is involved then professional legal advice must be used, the costs for these services may appear expensive but will be money well spent in the event of any future legal disputes regarding land ownership, tenancy, boundaries or access, or the activities allowed or banned on the site.

Agricultural covenants

These are also known as an 'agricultural restriction' or 'agricultural planning restriction' and are a freehold covenant that restricts the occupancy of a property to anyone in agriculture who fulfils one of three criteria, namely:

✿ be employed in horticulture or agriculture within a 30-mile radius of the property

✿ be retired from horticulture or agriculture within a 30-mile radius of the property

✿ be using the land at the property intensively to provide your main source of income.

This restriction can be linked back to the post-war years of the late 1940s when many properties built on farmland were given planning permission subject to an agricultural restriction. This may limit buyers of these properties, allowing an opportunity for starting up a new enterprise. However, in some cases the covenanted restriction may be lifted or modified by application to the courts or the local council's planning authorities.

The type of business development will depend on many factors such as the knowledge, skills, experience and commitment of the key people involved and the amount of finance available to invest; the geographical location of the business site is important, especially in relation to the main customer base. If a retail site is difficult for consumers to find it should not just rely on regular passing trade; therefore the product range, customer service and facilities will have to be exceptionally good to encourage people to make a special journey. However, a wholesale nursery business that delivers all its plants with its own vehicles and has fewer trade visitors calling in could manage on the same location.

Figure 16.9 The business site must allow for future expansion.

Location, location, location

The importance of selecting a good site location for any business cannot be overstated. The ease of moving staff, customers, visitors and resources in and out of the site is essential to the smooth running of a company. A poorly located site could be one in the middle of a rural area with endless narrow lanes to navigate along, or situated on a busy road junction making it difficult and dangerous to gain access, both of these situations could create endless problems such as recruiting staff unable to get to the site or frustrated visitors having to ask for directions. Another aspect to be considered is the additional costs involved in transporting people, products and materials to remote sites. A well-located site should be easy to find, have good access from a main road to allow for the ever-increasing size of commercial vehicles, plus the ease of getting all the necessary services connected from the mains supply lines. In the age of high-tech communication it is important to check that mobile phones, the internet and other wireless systems can be received throughout the site – it is surprising where loss of signals can occur. A disadvantage of selecting a central prime site is the higher value and purchase price or rent payments plus the regular costs of business rates and services compared with a more remote situation. Additional security arrangements may also be required if located in more populated areas, together with the possibility of restricted working hours or vehicle access due to noise regulations.

Business objectives and resources

An integral part of the business planning process is to establish the key objectives for the proposed new business, or a project development within an existing company.

After many meetings and discussions there should be a business plan that outlines the key objectives with details as to who is involved, where the activities are to be carried out, a breakdown of the costs and finance arrangements, the products or services and the returns they will make and, importantly, when each stage is going to start producing an income. This revenue is the cash flow that is essential for paying the bills. The list of resources required and readily available to help operate any modern business could fill an entire book and could empty the business bank account in doing so. An established business may be wishing to expand into a new area and require additional land for increased cropping, or a larger vehicle to transport more materials more efficiently; these new resources should meet the criteria for each key objective and budget set out in the business plan. It is essential that everyone involved in the business is familiar with and has agreed to the content of the business plan, the actions to be carried out plus the financial and legal implications when they are followed through.

Selling services and products

Establishing how the business will process, sell and dispatch its products or services efficiently and profitably is fundamental to the whole success or failure of the project. Inefficiency within any part of the system will have a negative impact on other sections of the organisation, especially financially.

In the early stages of a new company any minor malfunctions can be quickly sorted out and everything and everybody can get on with their work; however, as a business grows in size so does the complexity of each section or department within it and the ability to implement change becomes more cumbersome and costly. Each department will require its own work schedules highlighting how and when things should happen, these will then link with other stages in the overall programme to ensure a manageable flow along the production line, or service provision with no single activity creating problems along the way. A well-motivated and communicating management team is essential in keeping this a reality.

Staff recruitment

Figure 16.10 Recruiting and keeping good staff will be essential.

Selecting the right staff at the right time and in the correct numbers is important especially within the horticultural sectors where environmental conditions play such an important part in how and when many operations are carried out. Many companies operate with seasonal, part-time staff to satisfy the peak production times and a small team of full-time employees to operate throughout the year. Staffing costs are often the largest expense for many businesses and it is important to establish accurate numbers from the outset. It is also essential that the business can actually afford to pay the employees plus the other overheads including tax and National Insurance contributions. The question should be asked as to how soon can these new employees generate sufficient income to cover their own costs plus contribute towards the company's profit; if it is too long a period the risk may be too high and a reduced plan implemented. When the need for new staff is established it is important to identify the skills required and the personal qualities that would fit within the existing business structure. The level of experience and qualifications required must be sufficient to do the current job and to develop it further.

Proposed salaries should be comparable with the same levels of employment within the industry in order to attract good people to apply and take account of the national minimum wage at the time. Contracts and work schedules should be written prior to advertising the posts. Placing adverts in local or national media, using the services of an employment recruitment agency or the company's own website are effective ways of getting adverts seen by a wide audience. The use of emails has revolutionised how employers and candidates pass details to each other, however the person-to-person meeting and interview is still important in the assessment and selection process to help establish if all sides wish to work with each other. New staff will require contracts, terms and conditions, induction training plus the resources to carry out their new work. All employees play an important part within any company; however, it is the managers that often keep the different sections working towards the main aims and goals. When recruiting new management it is essential they have broad experience not only with extensive horticultural knowledge but also people skills, motivation and enthusiasm to make the business perform well and develop into the future.

Growing the business

For many practical horticulturalists the office administration systems are chores that need to be done as infrequently as possible. In today's high-speed, high-tech world computers can do a lot of the boring number crunching. The need to carry out stock control by counting all the pots on the container beds or measure a site for the amount of lawn turf required is still important for providing the key information. This detailed information can then be used to monitor everything to do with a specific operation. This could be a crop grown from propagation stage to a three-litre plant and determining its saleable price; knowing the amount of turf required on a landscape site that will dictate the size of lorry needed for transporting it and the number of people needed to lay it, all of which costs money that must be included in the costs paid by the client. Without this type of information it is difficult for any business to calculate the true operational and production costs and therefore the true prices to charge. Without careful and accurate monitoring of the business activities it is not possible to know how much money is being made, or lost, and how to produce detailed plans for the future development of the business. As a business grows in size this aspect becomes even more important because as the overheads increase so do the risks.

HOW TO KEEP FINANCIALLY VIABLE

Borrowing money can be expensive and result in placing a considerable burden on any business. Increases in the interest rates offered by financial institutions can have dramatic effects on the cost of keeping a business financially viable; the monthly payments for business loans on property or essential equipment can put a severe strain on cash flow and budget forecasts require constant amendments. All this will place added pressure on the sales team to increase the income by improving sales to existing customers and prioritise the need to find new customers.

This section will demonstrate some actions that could help a business to maintain its financial viability:

- ✿ investigating possible sources of money
- ✿ the importance of professional financial advisors
- ✿ maintaining accurate records and systems
- ✿ ensuring the tax man is informed
- ✿ establishing which capital items are required
- ✿ growing the business for good cash flow
- ✿ keeping control and monitoring costs
- ✿ establishing the correct pricing and fee structures.

Possible sources of finance

Figure 16.11 The high street banks provide plenty of literature and advice on finance for businesses.

Reading through the national newspapers or surfing the internet can provide a list of money lending outlets; many of these are after the large international corporations looking for millions of pounds, dollars or euros. Most small enterprises will, however, look for a more regional source that may prove to be more commercially attractive and understand the local marketplace. The high street banks are still a logical place to start gathering information to research current terms and conditions, interest rates and types of loan that are available, this will also improve your understanding of the terminology and small print that is used in the financial world. A key member for any borrowing proposal plan should be the business accountant who prepares the annual accounts for HM Revenue & Customs; they will be familiar with how your business is run, the current position and forecasts of the financial situation plus advice on the implications for borrowing money, and more importantly how it will be paid back within an agreed time period. The accountant may also have knowledge of private individuals, or small consortia of business people who are looking for progressive companies to invest in and prepared to lend at more adventitious rates. However, and wherever, a financial loan is secured it is essential that the processes and papers are checked by financial and legal advisors, preferably those belonging to a professional body, this can reduce the risk of complications and legal difficulties at a later date.

Using professional advisors

Professional advisors can be found to cover every aspect of professional life. When running a horticultural business it is possible to hire an advisor to give guidance on the land itself, the positioning of services, the sizes and types of structures and what to grow in them. The production, harvesting and selling, all your transport and machinery needs, the staff and their working conditions plus all the financial, safety, legal implications and, planning for your retirement possibilities! The fact is that it would be uneconomic for a small business to spend money on fees for all these different consultants on a regular basis. Each business will need different consultants at different times and frequencies depending on their needs, this flexibility is one of the benefits of appointing independent consultants, you pay them when you need them.

Selecting the best advisors for the business needs careful research, start by looking at the websites and contacting the relevant industry bodies such as the Institute of Horticulture (IOH), the Horticultural Trades Association (HTA) or the British Association of Landscape Industries (BALI), where members may be advertising their professional services. Visiting industry trade shows will also give good opportunities to meet advisors who have a trade stand, also ask other exhibitors within the sector if they could recommend suitable people.

Suppliers and manufacturers should also be given the opportunity to provide information; bear in mind that this will be biased towards their products or services but will contain details that will prove useful. There are some aspects of business where it is necessary to buy in advice because technical reports are required specific to the business, these include health and safety, risk assessments, construction and planning proposals for the site, financial accounting and employment law; up-to-date information is critical when dealing with these specialist sectors.

Financial systems and records

The demands for business information come in from all directions, the company accountant, HM Revenue & Customs, VAT, the banks and local authorities, all these expecting accurate details within given time periods with penalties for non-compliance. This may appear doom and gloom; however, if this information is readily available as part of the normal business operations system then it can be used as an integral part of the business plan to grow the business in a structured, profitable way.

Information will include production figures, service operating costs and overheads, staffing costs and most importantly details of the sales over given periods of time.

With careful cross-referencing and the help from financial advisors valuable information will show which parts of a business are viable and, with further market research, could be developed; also the times of the year when business is at a peak and the slack times when bills still need paying. Modern technology has enabled the gathering and storage of vast quantities of data, then to gain access and cross-reference within a fraction of a second. Record keeping is necessary and useful provided the information is accurate, up to date and in a format that is usable, this is relevant to the number of plants in the tunnel or pounds in the bank account.

Keeping the tax man informed

Figure 16.12 Ensure all financial records are kept up to date to comply with current legislation.

Keeping the financial records up to date for any company is a legal requirement. The law is incredibly complex even for the smallest of businesses and a lack of knowledge or understanding of these complexities appears to be no excuse for not complying with HM Revenue & Customs' requests. An important point is to ensure all business receipts and invoices are carefully filed along with bank statements showing money going in and out from relevant accounts. There are many publications available from local bookshops and libraries giving guidance on book keeping and business accountancy and what is required to produce the end-of-year accounts. For many horticulturists the thought of having to spend regular amounts of time producing the accounts is not the most stimulating activity, therefore the appointment of a good accountant can be cost effective and worth considering, especially as a business grows and becomes more complicated. Allowing the accountant to have access to accurate information on time will provide a more efficient and cost-effective service for all concerned. The consequences of getting the accounts wrong, sending late tax returns or late payments can result in demands from the tax inspector for payment of fixed penalty sums, added to this will be the trauma of sorting out any disputes and visits by the inspection team.

Capital expenditure

Figure 16.13 New fixture and fittings will be required as the business develops.

As a business develops and grows so do its demands: new resources are needed to produce the products or provide the services that customers and clients wish to buy.

The cost of providing the everyday materials and sundries should be covered by the regular income; however, when a major piece of new equipment, vehicle, building or additional land is required then additional funding will need to be found.

This level of spending is known as capital expenditure. It is very important that if major new projects are to be carried out the end results are viable for the business and all its activities and operations. A new glasshouse without sufficient water to irrigate it all will not be viable unless more money is spent on improving the water supply systems; a shiny new rotavator that will not fit in the vehicle safely will need a trailer and a tow hitch fitted on the van. These two examples highlight the need for careful planning to establish the specifications, the practical and financial outcomes and importantly the impact and benefits to the business.

Keeping the cash flowing

Cash flow refers to all money coming in and going out of the business on a regular basis, it includes cheques, bank transfers, ebanking transactions and of course real cash. All businesses need good cash flow to keep going. The difficult part can be keeping the money coming in from the sales faster than it is going out to pay the bills.

The type of business will effect how and when money comes into the business: selling produce at the farm gate or charging visitors to tour the garden will produce cash on the spot, a wholesale grower or landscape contractor may not receive payment for several weeks after an invoice has been sent to the customer. These time delays between supply and payment can prove financially difficult for many businesses. The producers will have

Figure 16.14 Keeping the cash flowing is essential for every business – whatever the scale of the operation.

invested in expensive materials and time to supply the goods or services to the customer with no immediate payment made in return to allow paying their own suppliers or staff, or themselves! If the supplier does not have enough reserve money in their bank account to last until they receive their payment then serious difficulties can arise. This becomes worse when late payments from customers or clients occur on a regular basis. For this reason it is essential that businesses have clear and efficient systems for obtaining payments for products or services supplied. Invoices should be sent out soon after delivery or completion of services stating the payment periods and contacting any late payers requesting payment, also do not allow large amounts of credit to build up with individual customers while still supplying them with more products or services.

Keeping it all under control

Regular reviews on product and plant sales will highlight what is selling and, equally importantly, what is not selling. This information will help produce accurate figures for the future production output. The services being offered must also be reviewed regularly to ensure they are being sold at rates based on realistic fee structures. With the ever-increasing costs for transport, labour and materials it is easy for the profit margin to be reduced to such an extent that it becomes no longer viable to continue offering certain services. The way that work is carried out must also be monitored, it is not essential to carry out elaborate time and motion studies or hire expensive consultants to investigate what is going on. With the support of staff and managers it is possible to identify working practices where improved efficiency and cost saving can be implemented. By reducing wastage of time and resources and replacing inefficient with efficient systems the overall profit margins should be improved. An example of an inefficient system is where the potting machine is continuously stopping and starting due to the lack of small plants coming in for potting, or the congestion building up around the machine

with the newly potted plants not being taken away fast enough. Both reasons lead to loss of potting time, daily targets not being achieved and staff not working efficiently, this results in higher running costs; the extended potting period results in loss of plant growth, which could result in smaller plants not being saleable and requiring over wintering, all these factors will reduce profit margins and the viability of the business. This example raises questions such as does the nursery have the resources to cope with a potting machine running efficiently or would a well-planned hand-potting shed be more efficient and cost effective. The same question-and-answer process should be applied to all activities carried out before expenditure is committed to.

Charging enough

Having monitored the business activities and cross-referenced with the financial records it should now be possible to establish the true costs involved in running the business. With the help of a good calculator and some financial support the unit costs for each item or service can be established. A percentage figure must also be added to allow for a realistic profit margin. If these sales prices appear high compared with similar items offered by other companies it is time to analyse why this is; it could be the company's running costs and overheads are too great or the profit margins are set too high, further analysis and cost savings will be required to make the unit costs competitive while maintaining a realistic profit. The opposite will be if the sales prices appears too low, this will certainly require rechecking of the data used in the calculations to ensure it is accurate plus checking that quality and standards are being maintained. However, if it is genuinely possible to produce and sell sufficient products and services, and still make a good profit, then increase production and expand the market carefully so as not to break the winning formula.

KEEPING ON THE RIGHT SIDE OF THE LAW

This section will act as a reminder that all commercial operations must comply with current legislation in all its various forms. The risks of getting on the wrong side of the law could have serious consequences for employers, employees and customers. Covered within this section are:

- ✿ staff training
- ✿ operating in safe conditions
- ✿ financial record keeping
- ✿ working with the local authorities
- ✿ selling quality products and services.

Staff training

This can take place in the workplace with visits from trainers and assessors or at a college with enrolment on full-time or part-time courses. Whichever system is available and suits both the business and the staff, there are benefits for all by improving the knowledge and understanding of the chosen subjects. The range of subjects offered by awarding bodies is wide and includes many specialist topics within each horticultural sector. Ongoing training for staff is often referred to as continuing professional development (CPD) and can provide a way of upgrading staff knowledge and skills levels to enhance the employment opportunities. Many professional organisations and larger companies expect their staff to complete a certain amount of CPD over set periods of time; this being a good way to ensure staff are keeping up to date with developments in their specialist fields. All staff should receive training in aspects of health and safety and first aid relevant to their workplace; this should take place during the induction period for new members of staff. Correct use of personal protective equipment (PPE) and the handling and safe use of tools, machinery and materials must also be covered before staff can carry out such work. There is a legal requirement that an employee must obtain a certificate for the safe use of many types of machines before being able to operate them, some examples include the use of chainsaws, pesticides and forklift trucks. Food hygiene regulations also require staff involved with food processes to receive appropriate training and certification. Failure to comply with any current legislation can lead to prosecutions and possible cancellation of payments from insurance companies in the event of an accident.

Keeping everyone safe

Figure 16.15 People at work, and don't forget the customers!

Most businesses operate with the support from full-time or part-time employees, subcontractors, directors, volunteers and family members. Where health and safety issues are concerned the law is clear that everyone carrying out work activities on behalf of a business is classed as an employee and therefore is covered by the multitude of regulations to establish safe working conditions and practices. It should appear obvious that everyone should carry out their work in a professional, efficient and safe manner; however, we all know this not the reality. The media shows frequent evidence where the law has not been observed and accidents have occurred. The law also covers the customers and the consumers of the products and services being sold and used.

Keeping the records straight

As has already been indicated in the financial section it is vitally important to keep business documentation in order and accessible. All health and safety records such as accident books and Reporting of Injuries, Diseases and Dangerous Occurrences Regulations 1995 (RIDDOR) forms must be retained; the same applies to staff contracts and records of employment, all of which require secure storage to maintain confidentiality. Files containing information relating to vehicle ownership, property, people, insurance details and chemical product datasheets must all be kept safely to ensure they are readily available in case of an accident or emergency. With the increasing use of electronic recording and storage it is equally important to keep information secure with a reliable backup system in place to retrieve information in the event of a technical failure with the office system.

Keeping the neighbours informed

Wherever a business is located and whatever activities a business carries out sooner or later the local authorities will come knocking on the door to clarify the situation, and possibly make a request for an application of some sort to be completed and sent to a committee with a cheque attached. Rather than this outcome occurring a better plan is to keep all the relevant departments at the local town hall informed of potential activities so they can make their comments in advance. It is far cheaper to make changes on a paper plan than rebuilding, or demolishing, part of a project already underway. The main departments to be contacted are the local planning office where a planning officer may be appointed to liaise with the project, the highways agency regarding road use and access, the conservation and listed building team if working on or near a sensitive site plus the river authority if there are any implications regarding visible and underground natural water courses. The local residents will also take a keen interest in any development proposals and getting them interested and showing their support could be crucial during any formal planning applications.

Are you fit for purpose?

The providers of services and the manufacturers of products are required by various laws to ensure that what is being offered is fit for the purpose that it is intended for.

It is also a legal requirement to give clear descriptions of what is being offered for sale, any misleading or inaccurate information given could be unlawful under the Trade Descriptions Act. Examples that contravene the regulations often appear as media reports about imitation, fake goods made with poor-quality materials and to a poor standard but sold to give the impression of a higher quality item. Staff from the local trading standards department have the power to confiscate and destroy any such dangerous items plus the option of issuing legal proceedings where injury has been caused as a result of their use. Services carried out that do not meet the specifications identified in any sales information can also be investigated to establish potential shortfalls in performance, quality or quantity. It is therefore important to produce, promote and supply goods and services to the highest standards expected by customers. This requires the management team and all production and service staff receiving appropriate training and being aware of the company systems and policies for quality control. Continual monitoring and testing will help safeguard against any business output being called into question.

Summary

Ensure you have the right team around you to operate safely and efficiently, this will prove invaluable if times get tough. Careful recruitment, selection and induction to the business are required, especially where a small team is employed, with complementary skills and personalities.

Seek good professional advice from the outset of any business venture, poor decisions could be very expensive to sort out at a later date. Make good use of any local knowledge and work with the authorities to avoid expensive pitfalls in the design and layout of your site and intended operating methods.

Understand your markets by continually researching the facts and responding when new trends or opportunities appear. Maintain high standards of production and customer service, if these aspects start to fail valued customers may move away. 'Underpromise and overprovide' is an excellent maxim to work with.

One of the biggest challenges for any business is maintaining sufficient cash flow to pay all the monthly bills and financial outgoings. With so many external factors to be considered by a horticultural business continuous financial planning becomes even more important to maintain all-year-round viability. If financial shortages are looming always think twice before borrowing any money; the recent financial crisis has put enormous pressure on all businesses with many not able to pay back borrowings within the imposed terms and timescales. Before borrowing money for major capital investment projects it is essential to seek professional guidance before committing to any agreements.

Everyone has a legal and moral obligation to keeping everyone and everything safe in all areas of the company, and to be aware of changes in regulations and recording requirements. There is plenty of information and advice available from many sources, ignorance of the law is not considered an excuse!

Whichever method of selling or enterprise is developed, the plants or services can be produced using traditional methods or embracing the latest technology, scientific knowledge and up-to-date modern facilities; the desired outcome should always be the same – the production of quality saleable plants or services at a realistic price to develop and maintain a viable business.

In today's tough economic climate all businesses are encouraged to have clear guidelines set out in well-documented business plans that will help determine their future development and growth.

Keep your business plan as a working document and not gathering dust on the shelf, the initial investment in preparing it would be wasted plus the business could drift into other directions away from its core operations. Review the business plan on a regular basis – it is much better to set an agreed time period – whether this is the need to make small incremental changes on an annual cycle, for quick cropping items, together with a longer term, say five-year, projection.

Revision questions

1. Outline **four** horticultural markets serving different sectors of the industry, including the volume of sales and employment numbers for each.
2. Describe **four** items that need to be considered as part of planning a new horticultural business.
3. Discuss the importance of **cash flow** to an expanding horticultural enterprise.
4. State **four** possible sources of finance for a new landscaping business, including the benefits and limitations of each.
5. List **four** items of legislation where records need to be kept, including the level of detail and time period required for each.

FURTHER READING

Buchanan, D. & Huczynski, A. (1997). *Organizational Behaviour an Introductory Text*, 3rd edn. Hemel Hempstead: Prentice-Hall.
This publication is very wide ranging and provides an introduction to this important area, covering larger, more complex structures and also businesses with few staff.

Federation of Small Businesses (2010). *Good Small Business Guide 2010: How to Start and Grow your Own Business*, 4th edn. London: A. & C. Black.
This well-structured publication covers all the required areas with excellent case studies.

Power, P. (2007). *How to Start Your Own Gardening Business: an Insider Guide to Setting Yourself up as a Professional Gardener*, revised edn. How To Books.
A more focused book related to horticulture following a tried-and-tested route to self-employment.

Ramsden, P. (2008). *Finance for Non-financial Managers*, 3rd edn. Sevenoaks: Hodder & Stoughton.
A practical look at this vital area for small business operation.

Whiteling, I. & Welstead, S. (eds.) (2011). *Start Your Own Business 2011: How to Plan, Fund and Set Up your Business*, 3rd edn. Oxford: Vacation Work.
A comprehensive book written by those that have started their own businesses and covering from having an original idea through to getting the maximum from your operational business.

Other sources of information include:

* Food and Environment Research Agency (FERA) – the government department carrying out scientific and field research on all aspects of land-based operations and activities
* Health & Safety Executive (HSE) – many up-to-date handouts covering all aspects of health and safety
* High Street Banks – up-to-date details for business finance at all levels
* Horticultural Trade Association (HTA) – up-to-date information, and support available to members
* *Horticulture Week* magazine – a Haymarket publication
* National Farmers Union (NFU) – wide range of information relevant to land-based industries.

RHS level	Section heading	Page no.
2 1	Health and safety issues	338
	Working with the authorities	340
	Using professional advisors	343
2 2	Working with the authorities	340
2 3	Keeping everyone safe	346
	Keeping the records straight	346
2 4	Managing the business	337
2 5	Capital expenditure	344
2 6	Keeping everyone safe	346
2 7	Location, location, location	341
2 8	Good practice	338
	Location, location, location	341
2 9	Location, location, location	341
3 1	Location, location, location	341
	Using professional advisors	343
3 1.5	Health and safety issues	338
3 1.6	Keeping everyone safe	346
3 1.7	Managing the business	337

CASE STUDY Harlequin ladybird for aphid control

The harlequin ladybird, *Harmonia axyridis*, is a non-native species that was commercially released in France and Holland (Van Lenteren *et al.*, 2006) for aphid control in protected crops. Although they feed predominantly on aphids they can also feed on a wide range of other insects including other ladybird larvae. Van Lenteren *et al.* (2008) have done a risk assessment on the impact of *H. axyridis* in Northwest Europe designating this species as an 'invasive alien species' with great potential to establish and out-compete many native natural enemies.

The harlequin ladybird hunting aphids.

Chapter 17

Integrated pest management

Neil Helyer

INTRODUCTION

The majority of plants may suffer with pests or disease outbreaks at some time or other. Integrated pest management (IPM) brings together all aspects of pest and disease control: initially using cultural techniques, biological control, environmental control and then finally selective pesticides as a backup. This has developed as a more sustainable method of pest and disease control, and as a pesticide-resistance management tool. The vast majority of UK-grown edible crops are produced within strict guidelines that have been developed between advisors, growers and the retail outlets (www.assuredproduce.co.uk). The non-governmental organisation GlobalG.A.P. (www.globalgap.org) sets voluntary standards for growth and production of many crops including edibles, ornamentals and livestock production. Both organisations specify the importance of using IPM techniques in crop production.

Controllable environmental conditions can influence pest and disease pressure by inhibiting plant pathogens and by providing more suitable conditions for biological control agents to work more effectively. Pesticide regulation in Europe and the impact due to loss of key active ingredients on agriculture and horticulture has recently been reviewed by Chandler (2008) with the conclusion that IPM should be at the centre of European Union (EU) crop protection policy.

Pesticide use, particularly with selective products, should be used only when necessary (Thacker, 2002). As a guide it is best to use a pesticide when pests can be clearly seen on a plant or when leaf or yield damage is evident. By using a **short persistent** or **selective pesticide** to reduce an existing pest population natural enemies can be introduced, usually within a few days, to maintain the control. Biological control always works best when pest numbers and plant damage are low. This chapter will describe the IPM practices for many pests including insects, mites, molluscs, fungal and bacterial diseases, weeds etc. Currently this method of pest control is used on a wide range of crops by many commercial nurseries in the UK; over time it may transfer to be used in a garden situation.

Details of the use and disposal of chemical and biological pesticides, licences and the requirement for certified training are given below. Successful IPM requires a degree of knowledge, and details are given on various sources of information, cultural IPM techniques and pest-monitoring techniques.

Key concepts

* Range and control of plant pathogens
* Cultural, biological and integrated pest control
* Selective pesticides and further sources of information
* Types and uses of chemical and biological pesticides
* Sources of information, disposal, licences, registration and certificates

The Fundamentals of Horticulture: Theory and Practice, ed. C. Bird. Published by Cambridge University Press. © The Royal Horticultural Society 2014.

BIOLOGICAL AND INTEGRATED PEST CONTROL

The principal pest organisms infesting plants in many gardens and greenhouses are aphids, caterpillar, leaf hopper, leaf miner, mealybug, scale insects, slugs, spider mite, thrips, whitefly and vine weevil. Integrated pest management techniques are available to manage most pest organisms and are used by the majority of professional growers, many amateur growers and contractors responsible for private and public gardens, conservatories, interior atriums etc.

Biological control, as part of an IPM programme, infers that the technique may control an existing pest outbreak; whereas in the majority of instances it works best when pest levels are low and may be best thought of as biological maintenance. The majority of biological control agents and a few IPM compatible pesticides are available for use by anyone (see below in registration, licences and certificates). Sap-sucking pests such as aphids, mealybug, psyllids, soft scale insects and whitefly can excrete vast quantities of sticky, sugar-rich **honeydew** that, in addition to direct plant damage, will further disfigure plants with associated growths of sooty mould and loss of plant vigour.

Spider mites (see below) are also extremely common and severely damaging pests with a rapid rate of reproduction when conditions are hot and dry. Very effective predatory mites are available that frequently eradicate the pest completely. However, it is usually advisable to intervene where spider mites have already established with a selective pesticide to reduce further plant damage before introducing the predators. Various species of thrips can cause differing types of leaf damage as well as transmitting plant viruses to a wide range of plants. Sciarid and scatella flies are generally only a problem in plant propagation areas, whereas slugs and snails can be a serious problem to young and mature plants in both protected and outdoor-planted areas. Vine weevil can be a major problem on protected and outdoor nursery stock production units, granular pesticides (chemical and biological) can be incorporated into the growing medium to provide long-term protection. Remedial treatments with entomopathogenic nematodes may be required for unprotected plants or existing infestations. There are **natural enemies** for almost all the above pests (Hajek, 2004), the vast majority are, as the name suggests, completely natural but only seasonally present. An increasing range of commercially mass-produced natural enemies, also known as **beneficial organisms**, are commercially available for release all year round, providing the required environmental conditions are present. Various **biopesticides** are now a major component of many IPM programmes to control a wide range of plant diseases, insect and mite pests (Shah & Pell, 2003). See pesticides.gov.uk for the latest approved products.

Most commercial nurseries producing edible and ornamental plants as well as many private gardens and those on public display use a full IPM programme to control and prevent outbreaks of the common pests and diseases. These include the major insect and mite pests such as aphids, caterpillar, spider mite, thrips, vine weevil and whitefly. Plant diseases such as *Botrytis*, powdery mildew and most damping-off root pathogens including *Alternaria*, *Pythium*, *Phytophthora* and *Rhizoctonia* may all be controlled by biocontrol agents. A programmed approach to pest control with regular applications of beneficials for major pest organisms is the best approach, particularly in protected structures. A BASIS® (www.basis-reg.com) qualified advisor with experience in biological control and IPM should be called upon to prepare a programme of beneficial introductions and give advice on IPM-compatible pesticides. **Selective pesticides** are available that can be integrated to counter a larger pest attack that may cause plant damage before the beneficials gain control. This can be done as spot treatments for individual plants as well as larger planted areas. However, where large infestations are found a rapid response is essential to prevent further damage and spread of the pest.

The following section covers some of the major pests, examples of their biological control agents, and selective pesticides as used on a range of protected and outdoor plants. Further details of these and many other organisms along with colour photographs of each organism are in Helyer *et al.* (2014) *Biological Control in Plant Protection: A Colour Handbook*, 2nd edn.

APHIDS

- ✿ Parasitoid wasps: as a guide use *Aphelinus abdominalis*, *Aphidius ervi* and *Praon* spp. for large aphid species while *Aphidius colemani* and *A. matricariae* are more suitable for smaller round-bodied aphids.
- ✿ Predatory midge: *Aphidoletes aphidimyza* for aphid colonies.
- ✿ Predatory lacewing larvae: *Chrysoperla carnea*, generalist predator of aphids and other soft-bodied prey; can be used inside and outside.
- ✿ Fungal pathogen: *Beauveria bassiana* (Naturalis-L).
- ✿ Selective pesticides: pirimicarb (Aphox), pymetrozine (Chess), imidacloprid (Imidasect).
- ✿ Short persistence, contact pesticides: pyrethrins (Pyrethrum 5EC, Spruzit).
- ✿ Physical control: Majestik (plant extracts), SB Plant Invigorator (foliar latice and plant feed), sticky traps.

Order: *Hemiptera*; superfamily: *Aphidoidea*

Commonly known as blackfly or greenfly, aphids are small soft-bodied insects ranging in size from 0.5 mm (1/64 inch) to about 6.5 mm ($\frac{1}{4}$ inch) in length. Colours can vary between different species and even between strains of the same species, ranging from shades of green through to black, yellow, chestnut brown, pink, grey and waxy white. Several species have darker markings on a lighter toned body, camouflaging them against their plant background. Aphids may be differentiated from other *Hemiptera* by the ability to produce active young (viviparous birth) from unimpregnated females (parthenogenesis); the same species may be present as winged (**alate**) and wingless (**apterous**) stages on the same leaf. Alate aphids are frequently produced in response to adverse conditions such as overcrowding, poor plant condition, the onset of winter conditions or for migratory purposes.

The life cycle of aphids can be one of three types.

1 **Monoecious holocycle**, which is the simplest and most common on outdoor crops, where all the generations develop on the same host plant species. Overwintered eggs hatch to **parthenogenetic** females in the spring. These produce both **alate** and **apterous** aphids through the summer months and a sexual generation in the autumn that after mating lay eggs to survive the winter.

2. **Dioecious holocycle**, where the aphid has two different host plants, a woody winter host in which eggs are laid to over-winter and a non-woody summer host that supports the bulk of generations. Usually two generations occur after the eggs hatch in the spring followed by a migration to the main summer host plant on which multiple generations (both winged and wingless) are produced. As above, a sexual generation develops in the autumn that migrates back to the winter plant for egg production.

3. **Anholocyclic**, in which no sexual stages are produced and so no eggs are laid, all the aphids being parthenogenetic females as alate or apterae depending on plant condition and colony density. This type is common in the Mediterranean, subtropical and tropical regions where winter conditions are very mild to non-existent; they also occur on many glasshouse crops where they can remain active throughout the year.

Aphids feed by penetrating the plant tissue with four ultra-thin stylets that are held within a protective rostrum, which at rest is folded against the underside of the insect. Only the stylets enter the plant as the insect feeds, injecting saliva by way of a small canal formed between the hypodermic-like stylets and sucking up the partially digested sap. During feeding, aphids (and many other sap-feeding insects) can transmit numerous plant viruses, cause galls to form or leaves to blister or even roll around the pest colony. Large volumes of sap are taken in during feeding and almost equal amounts of sugary water excreted as honeydew. This is flicked away from the aphid to fall on the upper surface of leaves and stems. High numbers of aphids can give rise to copious quantities of sticky honeydew, which in turn can become infected with a fungus turning the whole area black with sooty mould. Ants are frequently associated with aphids and may be assumed to be feeding on the insect when they are in fact feeding on the sugary honeydew and will defend aphids from attack by predators.

In most years the aphid population crashes to very low numbers, usually in July to mid August (Bellows & Fisher, 1999), when the majority of the aphids' natural enemies are at their peak, these include naturally occurring parasitoid wasps, hoverflies, ladybirds and other predatory insects. Commercially produced aphid parasitoids include *Aphelinus abdominalis*, *Aphidius colemani*, *A. ervi*, *A. matricariae*, *Ephedrus cerasicola* and *Praon* spp. While predators include the cecid midge, *Aphidoletes aphidimyza*, green lacewing, *Chrysoperla carnea*, two spotted ladybird, *Adalia bipunctata*, and hoverfly species.

Parasitoid wasps

Numerous aphid species can be found throughout the world, fortunately for the purposes of biological control with parasitoid wasps they can be conveniently split into two types. The smaller, round-bodied species such as the melon-cotton aphid, *Aphis gossypii*, black bean aphid, *Aphis fabae*, and the peach-potato aphid, *Myzus persicae*, are parasitised by *Aphidius colemani* and *A. matricariae*. Whereas the larger elliptical-shaped aphids such as the glasshouse-potato aphid, *Aulacorthum solani*, the potato aphid, *Macrosiphum euphorbiae*, and the pea aphid, *Acyrthosiphon pisum*, are best controlled by *A. ervi*, *Aphelinus abdominalis* or *Praon* spp. These small, 3 to 5 mm (8/64 to 13/64 inch) parasitoid wasps lay a single egg inside the host aphid that hatches to a minute larva, which feeds inside the aphid body. During this process the host aphid remains alive but feeding and reproduction is severely impaired. Once fully developed, the larva spins a silken cocoon inside the now dead aphid cadaver, which is called the 'mummy' stage (Helyer *et al.*, 2014). Adult wasps emerge by cutting a circular hole through the 'mummy', always at the rear end. An irregular-shaped hole found near the aphid mummy head or on its back indicates hyper-parasitisation, usually by a cynipoid wasp. These insects lay their egg into the developing parasitoid that is inside the aphid, feeding on its body and killing it after the mummy has formed. *Aphelinus* and *Aphidius* wasps can also kill one to two aphids per day through direct host feeding providing an additional level of pest control. This is done by the adult female inserting her ovipositor several times and slicing the internal organs with an up and down motion, the adult then feeds from the wound frequently sucking the aphid dry. To broaden the scope of aphid parasitoids, which can be fairly specific in the aphids they control, various mixes of these wasps are available in different ratios for different crops and aphid pest species. (See Table A 3.2 in Appendix 3.)

Mixed species of parasitoids are recommended for use in greenhouses with mixed plants as they will control most aphid species, particularly in the spring and early summer months.

Aphid predatory midge

The midge *Aphidoletes aphidimyza* is introduced as cocoons wrapped in fine sand and distributed in damp vermiculite either in a bottle or in small blister packs from which the adults emerge. Females can deposit over 100 eggs close to aphid colonies, these hatch to orange-coloured larvae, which insert a toxin that paralyses the aphid allowing it to suck out the body fluids. Each predatory larva, 4 mm (5/32 inch), feeds on 5 to 35 individual aphids depending on host size, before dropping off the plant to spin a silken cocoon in which it pupates. The last larval instar within the cocoon requires a minimum of 15.5 hours of light per day to re-emerge as an adult within two to three weeks, with fewer hours of light the cocoon enters diapause and it may take several months before the adult emerges. The optimum period for use of *A. aphidimyza* to establish and produce subsequent generations is early spring to early autumn, unless supplementary lighting is provided for the plants. This predatory midge is ideal for use in both protected and open gardens, particularly where vegetables are grown, as over a period of a few years they will establish, returning each year.

Figure 17.1 *Aphidoletes aphidimyza* – a predatory midge larva.

Lacewings

Lacewing larvae, *Chrysoperla carnea*, 1 to 8 mm (3/64 to 20/64 inch) when mature, will devour most aphid species and can be used to control quite large infestations; they are particularly useful in organic systems where selective pesticides are unavailable. They will feed on other soft-bodied organisms including leafhopper nymphs, mealybug, scale insects, spider mite, thrips and developing whitefly. Lacewing larvae may be introduced to outdoor growing areas as well as protected structures. They have also been introduced to hedges close to glasshouses and tunnels in mid spring and again in early summer. This has helped reduce the number of pests in the local environment before they migrate into the production area and establish a breeding population of predators on site. *Chrysoperla carnea* are excellent generalist predators and can be used in all gardens, greenhouses and interior atriums to control a wide range of pests such as aphids, mealybug, moth eggs and young caterpillar, thrips, etc. However, lacewing larvae can be cannibalistic if food supplies run short; so they need to be distributed thinly, rather than many in one place.

Entomopathogenic fungi

The fungal pathogen *Beauveria bassiana* (sold in the UK as Naturalis-L) is a registered biopesticide that can infect a wide range of insects (and spider mites). This stable liquid formulation is applied as a high-volume spray to contact as much of the target pest population as possible, spore germination and production of a germ tube that penetrates the cuticle requires a relative humidity of 60 to 80%. Once hyphae have penetrated the target body, further fungal growth throughout the body kills the host after a few days.

Under favourable conditions the infected bodies shrink as their nutrients are utilised by the fungus for reproductive spore development. Many hundreds of thousands of spores may be produced from each infected cadaver; spores may also develop saprophytically on other dead insects that can spread to entire colonies as an epizootic infection.

CASE STUDY | **Selective pesticides**

There are several pesticides with activity against aphids, some of which can be integrated while others have a much broader range of activity, killing most natural enemies as well as (hopefully) the pest. The aphicide pymetrozine (Chess WG) will integrate well with most biological controls and can be used to control any difficult or large outbreaks without harming beneficials. Aphox (pirimicarb) is moderately harmful to most of the beneficials above but is safe to hoverflies and ladybirds. Soil incorporated applications of imidacloprid (Imidasect 5GR or Intercept 5GR), or thiacloprid (Exemptor) as well as Intercept 70 WG for mixing with water and applying as a drench treatment, can prevent attack by aphids and other sucking pests for several months. Physically acting sprays are generally of short persistence and can be used as spot treatments without too many side effects on the biological controls. Physically acting products such as SB Plant Invigorator (growth stimulant and foliar lattice surfactants) can kill most small insects (aphids, mealybug, scale insect, whitefly etc.) and mites as well as having activity against powdery mildew spores. These are non-selective (they generally kill what they hit), but are of very short persistence and hence will integrate reasonably well with biologicals. Frequently, with short persistence contact sprays, several applications are necessary over a relatively short period to fully control a pest outbreak. SB Plant Invigorator can be applied at up to three-day intervals without harmful side effects to plants or the environment and is ideal to treat 'hot spots' of a pest infestation and to correct any imbalances in an IPM programme. SB Plant Invigorator appears to be reasonably safe to most beneficials, particularly the larger, more robust species and those protected within the host's body (*Encarsia*, black scale, *Aphidius*, mummy etc.). After reducing the pest population with a couple of sprays the biocontrols can be re-introduced or they will transfer naturally from unsprayed areas to maintain control. Other physically acting insecticides include Eradicote/Majestik (plant extract) and Savona (fatty acid soap concentrate); all have a zero harvest interval for edible crops but will kill most biocontrols on contact. It is therefore recommended to introduce a combination of biological control agents, particularly against major pests such as aphids (Van Lenteren, 2000) rather than relying on one single species.

(See Table A 3.2 in Appendix 3.)

WHITEFLY

- ✿ Parasitoid wasps: *Encarsia formosa* and *Eretmocerus eremicus* attack and kill larval stages. *E. eremicus* requires a grower licence for release onto glasshouse crops (UK) (see below).
- ✿ Predatory mite: *Amblyseius montdorensis* feeds on whitefly eggs and young larvae, young thrips, rust mites and tarsonemid mites. *A. swirskii* attacks eggs and young larvae. Both predatory mites require a grower licence for release onto glasshouse crops (UK) (see below).
- ✿ Predatory bug: *Macrolophus pygmaeus*, Mediterranean insect licensed for release onto glasshouse crops (UK).
- ✿ Pathogenic fungi: *Beauveria bassiana* (Naturalis-L).
- ✿ IPM-compatible pesticides: pymetrozine (Chess WG, SOLA/EAMU), imidacloprid (Imidasect 5GR), *Beauveria bassiana* (Naturalis-L), pyrethrins (Pyrethrum 5EC and Spruzit), physical mode of action pesticides such as SB Plant Invigorator, Majestik and Savona.

Order: *Hemiptera*; family: *Aleyrodidae*

The presence of whitefly detracts from the plant value; some species can transmit plant viruses and in high numbers can produce copious quantities of honeydew. The sugary honeydew lands on the upper surface of leaves and fruit on which a black, sooty-mould fungus can grow.

Adults are small, 1.25 to 2 mm (3/64 to 5/64 inch), grey to white winged flies. Some species have bands or spots on their wings, adults are usually found at rest on the underside of apical leaves (near the head or top of plants). Eggs are initially white and laid singly or in complete or partial circles; after a few days they melanise to an almost black colour before hatching to a six-legged nymph. This 'crawler' stage moves a short distance to locate a suitable site for larval development. The four larval instars are commonly known as whitefly scales; the last instar is often referred to as a pupal stage, although as no moult occurs it is actually a false pupal stage. Development of *Trialeurodes vaporariorum* from egg to adult is approximately 32 days at 20°C (68°F) to 26 days at 30°C (86°F). Oviposition (egg laying) usually begins one to two days later, with each female capable of depositing several hundred eggs, they may live for several weeks. In some species the 'pupa' may be covered in a waxy secretion making it difficult to see clearly. Integrated pest management methods can provide excellent control of whitefly at all stages of plant growth using various combinations of biological control agents and selective pesticides.

Parasitoid wasps

The whitefly parasitoid, *Encarsia formosa*, is well known and widely used on many nurseries, particularly those producing fruiting, salad and ornamental flower/shrub crops. *Encarsia formosa* is mainly used to control the glasshouse whitefly, *T. vaporariorum*, and has been commercially used in the UK since the 1970s. For ornamental plant production introduce the parasitoid wasps at 2 to 5 per m^2 (2 to 4 per yd^2) weekly for about eight weeks to 'seed' the area with parasites as a preventive measure. Further releases may be required through the season if whitefly increase or invade from elsewhere (purchased plant material can be a common source of pests). Alternatively a programme of regular introductions through the growing season offers greater flexibility and a quicker response for control of minor hot spots as the biocontrols are delivered on a regular basis alleviating the time delay between ordering and receipt.

The black and yellow adult females (males are all black and in very low numbers) are 1 mm (3/64 inch) in length and lay a single egg into a third instar whitefly larva (commonly called whitefly scale) that hatches to a minute parasitoid larva. The *E. formosa* larva develops along with the host whitefly larva, changing from third instar through to pupation inside the whitefly 'pupa'. Parasitised whitefly turn black while unparasitised pupae remain creamy white and emerge seven to ten days later as the adult whitefly. Parasitoid development takes a further three to six days (temperature dependent) before the wasp cuts a 'D' shaped hole through the black pupal case to emerge as the fully developed adult parasitoid. Mass production of *E. formosa* is usually done using glasshouse whitefly *T. vaporariorum* on tobacco plants, *Nicotiana tabacum*, which have large flat leaves, each leaf can produce up to

Figure 17.2 Severe whitefly damage on a tomato, *Lycopersicon esculentum*, crop showing sticky honeydew, sooty mould and dead adults stuck on fruit.

conditions return. Under glasshouse conditions, where good plant growth and reasonable temperatures above 16°C (61°F) can be maintained, the mites are active through the year, although their life cycle is considerably slower at lower temperatures. Females produce up to a hundred spherical eggs, 0.14 mm (0.0055 inch) in diameter, that are initially transparent but turn white to light yellow just prior to egg hatch. A six-legged larva emerges and feeds for a short while before settling on the leaf to form a protonymph; this may last a few days depending on temperature before developing into a deutonymph. At this stage females are distinguishable from males by their larger size and rounded shape, while males have a pointed rear and are much more active.

Predatory mites

Prevention of spider mite attack is much better than trying to control an outbreak. The predatory mite *Amblyseius* (*Neosieulus*) *andersoni* is a UK sourced and produced mite (Syngenta Bioline product guide). They are available in tubes or bags with bran and vermiculite carrier as well as controlled release system (CRS) sachets that hang on plants and release mature mites continuously over a six- to eight-week period. They are ideal for production nurseries where plants may remain on site for several months and for display plants as in botanic gardens, conservatories or interior atriums. *Amblyseius andersoni* can also provide control of fruit tree spider mite, russet mites and other small prey but do not perform well where there is spider mite webbing (Enkegaard, 2005). However, they survive well in surrounding areas, where they ambush straying spider mites limiting their spread. Due to their ability to survive on other food sources such as extrafloral nectaries, glandular hairs and pollen they are ideal to use as a preventive measure on susceptible plants (van der Linden, 2008). Like most *Amblyseid* mites they prefer to deposit their eggs among pubescent hairs as shown in Figure 17.5; they are also commonly found on several wild plants such as red campion, *Silene dioica*.

The more voracious and widely used predator *Phytoseiulus persimilis* can be used alongside *Amblyseius* spp. as a curative treatment (Buxton *et al.*, 2006). These predatory mites are introduced in a vermiculite carrier that is sprinkled directly over infected plants. For use on display plants or those with an open, feathery-type leaf canopy (palms) it is suggested to place the predators and their carrier on a section of shade netting or in a small fleece pouch (Helyer, 2008). These platforms can be placed on the plants for a few hours or left *in situ*, allowing the mites to disperse. *Phytoseiulus persimilis* are most commonly used as a curative treatment for spider mites, frequently being introduced after an initial knockdown spray with a physically acting product such as SB Plant Invigorator. Although they will survive at 5°C (41°F), optimum conditions are 22°C (72°F) and 75 to 85% relative humidity (Malais & Ravensberg, 2003). Under these conditions the predator produces more eggs and its development is quicker than its prey. Predators should be introduced at the first sign of spider mite damage and a repeat application made a fortnight later. *Phytoseiulus persimilis* is highly recommended and the most likely biological control to be used for control and prevention of spider mites.

Figure 17.6 Adult *Phytoseiulus persimilis*, a spider mite predator – the egg is a spider mite egg.

The licensed predatory mite *Amblyseius* (*Neosieulus*) *californicus* (see below) is tolerant to higher temperatures and lower humidity. They can also feed on pollen, other mites and small prey making them ideal to 'seed' susceptible plants such as protected strawberry before spider mites appear.

Spider mite predatory midge

Larvae of the predatory midge, *Feltiella acarisuga*, can kill several mites per day and provide good control of large pest populations; adults fly to locate spider mite colonies where they lay several eggs. *Feltiella acarisuga* requires a reasonable population of spider mites before egg laying begins, thus they may be better suited for use on edible crops where leaf damage is less of an issue. Further introductions of this or other predators may be necessary during the summer months.

Figure 17.5 *Amblyseius andersoni*, a predatory mite – these minute mites hide among pubescent hairs where they also deposit their eggs.

Macrolophus pygmaeus – Adults and nymphs feed on all stages of spider mites as well as several other pests; they are introduced under the guidance of a grower licence that permits its use on glasshouse grown crops in production.

Entomopathogenic fungi

The fungal pathogen *Beauveria bassiana* (Naturalis-L) provides excellent results on most crops providing good spray coverage and a relative humidity of 60 to 80% can be achieved.

CASE STUDY   **Integrated pest management compatible pesticides**

If spider mites are not noticed until plant damage is evident, a rapid response is required to minimise further damage and pest spread. Spray with abamectin (Dynamec) followed seven to ten days after the last spray with an introduction of *Phytoseiulus* at 10 to 50 mites per m² (8 to 45 per yd²) of infested crop. Dynamec has translaminar activity and may be used for any early season outbreaks or as an emergency treatment in summer, allow up to seven to ten days before predator introduction. It also makes an excellent end of season clean-up treatment to prevent mites entering diapause. This pesticide is not photostable and will break down in bright sunlight conditions (May to August) in about five to seven days or fewer if continuously hot and bright. When the high rate is used (50 ml per 100 l) Dynamec has activity against leaf miner, thrips, other mites and leaf nematodes. It can be harmful to biological agents for up to two weeks in the winter but fewer in the summer or in high-light conditions. Other IPM-compatible acaricides include clofentezine (Apollo) and etoxazole (Borneo) that both have good activity against eggs and young nymphs. The acaricides tebufenpyrad (Masai), spiromesifen (Oberon) and fenpyroximate (Sequel) have activity against most stages of spider mites and, depending on the cropping situation, may be used under SOLA/EAMU notification.

The bioinsecticide *Beauveria bassiana* (Naturalis-L) has good activity against spider mites as well as a range of other pests. SB Plant Invigorator is also useful in situations where the predatory mites have been introduced and require additional help to control a severe outbreak. It is worth remembering that the spider mite life cycle from egg to egg-laying adult is no more than seven days at 30°C (86°F) (Malais & Ravensberg, 2003), and under the same conditions *Phytoseiulus* becomes less active, preferring to stay lower in the crop canopy where it is cooler. Also under conditions of high pest density the predatory mites can be quickly overfed (Sunderland *et al.*, 1992) and will take time before their numbers increase to provide the necessary degree of control. All predatory mites struggle to move between well-spaced leaves and plants, to improve their activity strips of fleece can be draped over and between plants that improve mite mobility by acting as a bridge. Increased humidity around plants helps decrease the rate of spider mite reproduction and also improve predator activity. Under ambient temperature and natural day-length surviving spider mites may begin to leave the plants in the autumn to overwinter in canes and the greenhouse structure as they enter diapause. An end of season clean-up spray with Dynamec has proved very effective over several years of use on many nurseries.

(See Table A 3.4 in Appendix 3.)

THRIPS

- ✿ Predatory mites: *Amblyseius cucumeris*, *A. montdorensis* (licensed predator), *A. swirskii* (licensed predator).
- ✿ Predatory bug: *Orius laevigatus*.
- ✿ Pathogenic nematodes: *Steinernema feltiae*.
- ✿ IPM-compatible products: Thripline ams pheromone, spinosad (Conserve), abamectin (Dynamec), *Metarhizium anisopliae* (Met52 Granular Bioinsecticide), *Beauveria bassiana* (Naturalis-L), natural pyrethrins (Pyrethrum 5EC and Spruzit).

Order: *Thysanoptera*; family: *Thripidae*

Adults are tiny, winged insects ranging in colour from light golden yellow to brown (with and without stripes) to black with and without stripes. They have various common names including thunder bugs and thrips, named for the plant or geographical region from which they originated. Several thrips species can cause widespread plant damage, particularly western flower thrips (WFT), *Franklineliia occidentalis*, and onion thrips, *Thrips tabaci*. Cereal thrips, *Limothrips cerealium*, damage many plants in late summer when they migrate in from senescing fields. The black *Echinothrips americanus* may be found on many houseplants, sweet peppers and in interior landscaped atriums.

Thrips damage leaves, fruit and flowers by piercing and sucking out the cells of the surface membranes, leaving silvery/grey patches littered with small black faecal pellets, see Figure 17.7, leading to premature senescence (Van Driesche, 1998). Leaf and fruit distortion can occur when thrips feed on very young, developing parts of the plant, leading to severe economic losses. Several thrips species are also responsible for transmission of plant viruses (tomato spotted wilt and impatiens necrotic spot virus are of particular importance in the UK) to a wide range of plants potentially causing severe infection symptoms on the whole plant. Certain species of thrips are notifiable quarantine pests and their presence may lead to crop destruction or more usually a chemical spray programme. Adult females oviposit into plant tissue and may leave a small raised blister where the egg was laid. Eggs hatch after two to twelve days to minute first instar larvae, which last only a few days before moulting to second instar larvae (4 to 10 days), these develop to prepupa and pupa. Pupal stages of thrips may occur on the plant or in the growing media below and can last for a few days to several months depending on temperatures. Adults have a pre-oviposition period of two to ten days before they commence their next generation.

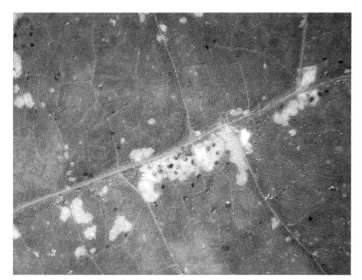

Figure 17.7 Feeding damage from thrips on cucumber, *Cucumis sativus*, leaf.

CASE STUDY   **Types of thrips**

Cereal thrips, *Limothrips cerealium*, live on cereals and grasses, in most years they can increase to high numbers and will migrate off their host plants as they ripen in mid to late summer. The migration of adult thrips can be detected on yellow traps leading to a dramatic increase in thrips counts leading to concerns by growers of an explosion of thrips on their crops. Most thrips species will damage plants as they feed; however, cereal thrips are black in colour whereas western flower thrips are golden brown to dark brown. Western flower thrips (WFT) have a wide range of host plants on which they can feed and reproduce, whereas cereal thrips can only reproduce on cereals, grasses and other monocotyledon plants.

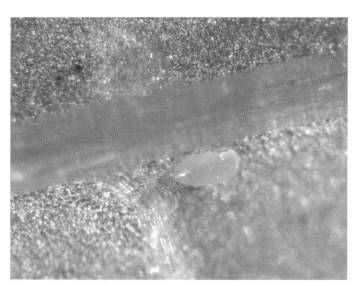

Figure 17.8 Adult *Amblyseius cucumeris* – a predatory mite of young thrips larvae.

Predatory mites

Thrips are best tackled preventatively by regular introductions of the predatory mite *Amblyseius* (*Neoseiulus*) *cucumeris*, particularly to susceptible plants, either in loose vermiculite carrier material weekly or in controlled release system (CRS) sachets. This predatory mite feeds mainly on the first larval stage of thrips with little activity against second stage larvae, for best results they should be introduced as a preventive measure at 100 to 200 mites per m² (90 to 180 per yd²) on a regular cycle (every two to three weeks from loose material). The CRS contains a breeding population of predatory mites in various forms of sachet that last between six to eight weeks and continuously release mated females that distribute themselves over an area of about 1 m² (1.2 yd²). The CRS sachets are very good for use across a range of crops, in particular bench or floor grown crops where leaf contact between plants can be maintained. The waterproof sachets can also be hung on larger plants, they come in various unit numbers ranging from 200 to 500 CRS single sachets, twin (Gemini) sachets to hang as an inverted 'V' (Figure 17.9) and in continuous lengths of up to 140 m (153 yards) called Bugline (Lambert, 2009). *Amblyseius* (*Typhlodromips*) *montdorensis* (sold in the UK as Montyline am) is a licensed, generalist predator of first and second stage larvae of thrips and young whitefly (see whitefly above). *A. swirskii* is also a licensed predator that is useful where there is a mixed population of thrips and whitefly. However, this southern Mediterranean mite requires a constant temperature above 20°C (68°F), having little development below 15°C (59°F) and preferring 25°C plus (77°F). These mites are available either loose with vermiculite carrier or in sachet form as above.

Figure 17.9 *Amblyseius* continuous release system (CRS) Gemini sachet in raspberry, *Rubus idaeus*, crop.

Predatory insects

The predatory bug *Orius laevigatus* can feed on all active stages of thrips and kill several individuals each day. These are introduced as larvae and adults to established populations of thrips, a source of pollen is recommended to maintain this predator if thrip numbers

Figure 17.10 Adult *Orius laevigatus* – a thrips predatory bug.

THRIPS PHEROMONES AND SELECTIVE PESTICIDES

A specific sex aggregation pheromone called Thripline ams is available for both male and female WFT adults to monitor and improve pesticide activity (GreatRex, 2009). Should thrips become a problem spinosad (Conserve) will control all thrips stages and is safe to predatory biocontrol agents and nematodes; two sprays at a seven-day interval will usually correct an outbreak. Conserve has contact and translaminar activity and will also provide control of caterpillar, including Tortrix moth and sawfly larvae, leaf miner and dipterous insects such as sciarid and scatella flies. When using Conserve the whole crop should be treated so as not to allow any refuge for survivors. For most ornamental crops a maximum of six applications per structure per year is permitted (some edible crops may be allowed fewer applications), as part of a resistance-management strategy.

Due to the feeding activity of thrips the addition of an appetent in the form of natural sugars can greatly improve the efficacy of Conserve and other pesticides that have ingested stomach poison activity. Attracker® is liquid fructose sugar with a preservative that is approved for use with insecticides as an appetent, it also works well with Dynamec when sprayed for thrips and spider mite control. Pyrethrum 5EC is a natural pyrethrum insecticide from chrysanthemum and formulated with a synergist called piperonyl butoxide. Synergists are chemicals that have no direct pesticidal effects on their own but enhance the activity of some, mostly naturally derived, pesticides such as pyrethrum. Pyrethrum 5EC, tank mixed with Conserve and Attracker (at their respective rates per litre) has been found to dramatically improve efficacy against thrips. Nemasys can also be tank mixed with several pesticides; a full and updated list is available from the manufacturer BASF Ltd.

are low. They can be used on all protected crops that require a biological curative agent for thrips.

Entomopathogenic fungi

The fungal pathogen *Beauveria bassiana* (Naturalis-L), see above, may provide excellent results on most crops providing good spray coverage and a relative humidity of 60 to 80% can be achieved. This fungus works well on many plant species but if thrips are hiding within flowers and contact is limited results may be impaired.

Metarhizium anisopliae as Met52 Granular Bioinsecticide is a fungal pathogen principally used to control soil-dwelling pests such as vine weevil. However, several other pest species such as thrips have a soil-inhabiting pupal stage, which is highly susceptible to fungal infection. Met52 is incorporated in the compost growing media and can remain active for 18 months to 2 years. The fungus has good activity from 15°C (59°F) up to 32°C (90°F) providing the compost remains moist. Its use against thrips is not regarded as a total control agent but rather as an additional element in an integrated approach to control this frequently difficult pest.

Entomopathogenic nematodes

Nemasys, *Steinernema feltiae*, is formulated in a clear gel carrier that can be applied as a foliar spray and has given excellent and reliable control of WFT on chrysanthemum and other flowering plants such as *Gerbera* and *Saintpaulia*. The nematodes should be applied weekly as a high-volume spray to leave a wet residue on the plant surface (Piggott *et al.*, 2000). Nematodes swim through this film of water to attack and kill various prey insects. Once the spray has dried the nematode activity ceases so the application should be timed to allow a film of water to remain on the leaf for as many hours as possible. Drench treatments of Nemasys control soil pests such as sciarid fly larvae and the pupal stage of soil-pupating insects, including many thrips species.

Figure 17.11 Thripline pheromone lure for adult western flower thrips, *Frankliniella occidentalis*, on blue sticky trap.

(See Table A 3.5 in Appendix 3.)

LEAF MINER

- ✿ Parasitoid wasps: *Dacnusa sibirica* and *Diglyphus isaea* for larvae within the leaf.
- ✿ Pathogenic nematodes: *Steinernema feltiae* for larvae within the leaf.
- ✿ Predatory bug: *Macrolophus pygmaeus* (licensed predator) larval predator.
- ✿ IPM-compatible sprays: abamectin (Dynamec), spinosad (Conserve).
- ✿ Yellow sticky traps for adult flies.

Order: *Diptera*; family: *Agromyzidae*

Adult females are 2.5 to 3 mm (3/32 to 4/32 inch) long and wound leaves with their ovipositor leaving open pits from which both sexes feed on plant juices; in some pits an egg is deposited. This hatches to a minute larva, which begins tunnelling in the leaf forming a characteristic whitish, snake-like mine. Low numbers are of cosmetic importance but high numbers can devastate a crop leading to yield loss in many edible crops. In some species the mine is spiral or serpentine while others produce mines that follow the leaf veins. The mine shape and frass trail within can help differentiate species of leaf miner, some of which are notifiable pests.

Figure 17.12 Leaf miner tunnel in tomato, *Lycopersicon esculentum*, leaf.

Mature larva up to 3.5 mm (9/64 inch) long can either pupate in the leaf as a small blister, usually on the underside of mined leaves, or drop out of the leaf to produce a puparium, depending on species. Those that pupate externally may land on the upper surface of lower leaves or drop through to the soil where they may remain for several months while they overwinter. Leaf miner larvae may be parasitised by specialised wasps such as *Dacnusa sibirica* and *Diglyphus isaea* that attack the larval stage of the pest. They may also be predated on by heteropteran bugs such as *Macrolophus pygmaeus* or eaten as pupae by ground beetles.

Nemasys, *Steinernema feltiae*, applied as a spray can enter leaves through the oviposition wound and swim through the tunnel to kill larvae. A period of free water on the leaf surface is required for the nematodes to swim and locate the entry hole for best activity.

Leaf miner is most likely to be a localised pest but can cause serious plant damage if left uncontrolled. At the first sign of mines in leaves place horizontal yellow sticky traps, sticky side up, close to the affected plants. If the outbreak is severe, a spray with a pesticide such as spinosad (Conserve) or abamectin (Dynamec) is very useful; however, this will kill parasitoid wasps for up to two weeks. Parasitoids (as above) should be introduced to maintain leaf miner control; follow-up units are likely to be required for a few weeks to fully control.

(See Table A 3.6 in Appendix 3.)

MEALYBUG AND SCALE INSECTS

Mealybug – order: *Hemiptera*, family: *Pseudococcidae*; scale insects – order: *Hemiptera*, family: *Coccidae* and *Diaspididae*

Mealybug bodies are covered with dusty white wax filaments, some of which form a fringe around the insect. Large clumps of mealybug can become too heavy to support their own weight and frequently fall to lower leaves or the ground where they disperse to start new colonies. Male mealybugs are delicate winged flies and after repeatedly mating, die within 48 hours. Most reproduction is parthenogenetic with females of some species giving birth to live young while others produce masses of eggs covered in a protective white woolly wax. An unmated female mealybug can survive on an inert surface for several months, before mating with a winged male and returning to a plant to reproduce.

They are an extremely common pest, particularly on slow-growing plants and those used for display purposes. Mealybug

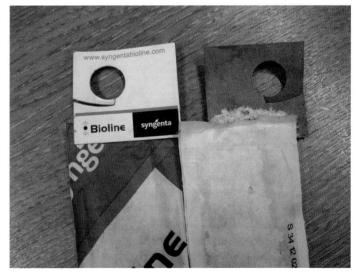

Figure 17.13 Mealybug hiding between layers of card and paper on *Amblyseius* CRS sachet, this insect will survive on inert surfaces for several weeks, including plant ties, sisal string and under pot rims.

are tenacious and can survive 'off the plants' for several months by hiding behind plant ties, within canes and under the rims of pots. One useful tip is to annually remove old plant ties and re-tie some 10 cm (4 inches) above or below the original site, in doing so large numbers of mealybug are also removed. Biological control with the Australian ladybird, *Cryptolaemus montrouzieri*, works well on egg-laying species of mealybug such as *Planococcus citri* and some *Pseudococcus* spp. but these ladybirds can be slow to establish on long-tailed mealybug, *Pseudococcus longispinus*. *Cryptolaemus* also feed on scale insects but will not establish without mealybug. Their white, woolly larvae can be very noticeable on display plants, which has led to the use of predatory lacewing larvae, *Chrysoperla carnea*, for mealybug control. Unlike *C. montrouzieri* adults that can easily fly away, lacewing larvae remain close to their release point, (see aphid section above for details of lacewings).

Scale insect adults range in size from 3 to 7 mm (4/32 to 9/32 inches) and can be circular or elongated; all have a waxy skin that protects the body. Soft-scale insects (*Coccidae*) produce honey-dew, often in large quantities giving rise to sooty mould, whereas armoured or hard-scale insects (*Diaspididae*) do not. Scale insects, particularly armoured, are hardier than mealybugs and have a larger geographical range, being found on several outdoor plants throughout Europe and North America. They are most frequently associated with conservatory plants. In outdoor situations they may be found on fruit trees and ornamental shrubs. Although mobile they tend to remain localised, frequently moving slowly along stems as the plant grows.

Specific parasitoid wasps and predatory beetles are seasonally available for mealybug and scale insects but their effective use requires expert advice. Pesticides with a physical mode of activity such as SB Plant Invigorator and spraying oil have given excellent control of mealybug and scale insects when applied as a contact spray with a second follow-up application no more than two days after the first.

The parasitic nematode *Steinernema feltiae* will control scale insects when sprayed on leaves; the nematodes swim through the film of water and enter under the scale to kill it. The systemic and translaminar insecticides pymetrozine (Chess), thiacloprid (Calypso) and acetamiprid (Gazelle) have activity against sap-sucking insects including mealybug and scale insects. Although these pesticides are approved on many crops, some areas of use such as amenity plants and interior landscapes may be restricted. (See Table A 3.7 in Appendix 3.)

CATERPILLARS

Order: *Lepidoptera* (moths and butterflies)

Caterpillars are the larval stages of butterflies and moths. They have powerful jaws capable of eating through most plant tissue, including roots, young stems, leaves, flowers and fruit. Tortricid (tortrix) caterpillars protect themselves by feeding within rolled leaves or growing tips that are stitched together with a silken thread, while others from this family feed inside fruit, such as the codling moth, *Cydia pomonella*, which is found in apples. Several caterpillars feed inside leaves and may be mistaken for leaf-mining flies but they can easily be identified by their capability of producing silk from spinnerets found behind the head, used to escape attack by parasitoids or predators.

The generally undifferentiated thorax (segments immediately behind the head) may bear a hard protective plate but always has three pairs of true legs (except in some of the leaf-mining caterpillars where the first instar is often legless). The abdomen has two to five pairs of fleshy prolegs, used for walking as well as clinging onto surfaces, and the rear of the larva has an anal plate and an anal clasper. The prolegs of lepidopteran caterpillar also have small hooks called crochets that aid in gripping when moving. Whereas sawfly caterpillar (order: *Hymenoptera*; suborder: *Symphyta*) have six or more pairs of prolegs and no crochets. This is important as some pesticides are only active against *Lepidoptera* (see below).

Caterpillar skin can be smooth, bristly or very hairy; the hairs are, in some species, irritating and act as a defence mechanism when the insect is attacked. Developing larvae must periodically shed their skins as they grow and once fully developed form a pupa or chrysalis. This is usually the overwintering stage and the pupa may remain hidden for several months before the adult moth or butterfly emerges.

Many predators and parasitoids attack and kill caterpillars including parasitoid wasps (*Hymenoptera*) and parasitoid flies (*Diptera*), which lay their eggs on or inside young larvae. The minute parasitoid, *Trichogramma* spp. lay their eggs into the eggs of several *Lepidoptera* species including tortrix. Predators include bugs (*Hemiptera*) such as *Anthocoridae* and lacewing larvae (*Neuroptera*). Birds and beetles will also devour caterpillars and pupae. Several pathogens attack *Lepidoptera*, these include bacteria of which *Bacillus thuringiensis* as Lepinox is widely used in the UK but may have limited activity against leaf-mining species. Fungi such as *Beauveria bassiana* can infect many species of caterpillar while specific viruses such as those within Carpovirusine are available for codling moth control.

Pheromone traps are a very useful addition to the IPM programme and will attract adult male moths from a wide radius, indicating when to begin a spray programme and often providing a degree of pest control. The company Exosect have developed a range of pheromone lures treated with an electrostatically charged powder that provides far greater control than normal sticky traps. Conserve (spinosad) will kill most caterpillars including sawfly larvae (*Hymenoptera*), which are not affected by *Bacillus thuringiensis* (Lepinox). Other insecticides for caterpillar include diflubenzuron (Dimilin Flo), indoxacarb (Steward) and pyrethrins (Pyrethrum 5EC and Spruzit).

SCIARID

Order: *Diptera*; family: *Sciaridae*

Commonly known as fungus gnats, the narrow black fly is 2.5 mm (3/32 inches) long and runs over the surface of compost depositing eggs on and just below the surface. The larvae are almost transparent with a shiny black head and can reach up to 6 mm (15/64 inch) in length before pupating close to the compost surface. Black and white threads inside the larva are the gut and fat deposits. Larvae feed on young root tissue, stunting the plant growth; cuttings can be tunnelled like a drinking straw often leading to their death. Both adults and larvae can exacerbate 'damping-off' root diseases such as *Pythium*, *Phytophthora* and *Rhizoctonia* (see below). Adults pick up fungal spores as they run over the compost surface and larvae create a wound for the disease to enter. Moist composts with high organic matter content are most favoured although it is not uncommon to find them in other media.

Sciarid larvae can be killed by the predatory mite *Hypoaspis miles* that lives in soil, compost or capillary matting and will also eat thrips larvae/pupae, springtails and other organisms in the growing media. The mite is best introduced as a preventative treatment to all plants, ideally within two weeks of potting or striking cuttings. They can also be introduced at the very first sign of sciarid or scatella fly adults on the compost. Horizontal yellow traps should also be used to control the adult flies (see below). Excellent results have been obtained by using *H. miles* at 100 to 150 mites per m^2 during propagation, as a single introduction on both seed- and cutting-raised material. Nemasys (see Figure 17.14), *Steinernema feltiae*, can provide good curative activity of sciarid fly larvae when drenched into compost and can be used in all situations where the pest is prevalent.

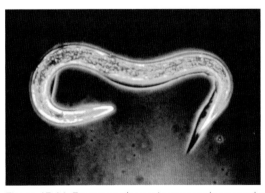

Figure 17.14 Entomopathogenic nematode as used to control sciarid fly larvae.

The predatory beetle *Atheta coriaria* can be introduced to control sciarid and scatella larvae, ADAS and HDC trials have developed a method of own production using sandwich boxes with peat compost and turkey feed (Bennison *et al.*, 2008). Entomopathogenic fungi such as *Metarhizium anisopliae* (Met52), *Beauveria bassiana* (Naturalis-L) and *Furia sciarae* can give high levels of control. The last is naturally occurring and leaves the dead larvae resembling threads of white cotton on the compost surface.

Many growers mistake fungus gnats for shore flies (*Scatella stagnalis*; family: *Ephydridae*), which are slightly longer, 3 mm (8/64 inch) long and broader, at 1.25 mm (3/64 inch) wide as adult flies. When at rest either on a plant, compost or other flat surfaces their folded wings appear to have three white spots across them. The larvae are greenish-white and segmented with a small black head; they feed in and on algae that may be growing on almost any surface including compost, benches, paths and floors where conditions allow algae to establish. A hard-skinned puparium develops in the algae and depending on temperatures the adult may emerge after only three to four days, giving rise to rapidly increasing populations. Control is best achieved by eliminating their algae food source with proprietary algicides. Biological control with *Atheta coriaria* is an effective predator against scatella flies.

VINE WEEVIL

Order: *Coleoptera*; family: *Curculionidae*

Otiorhynchus sulcatus is commonly known as the black vine weevil, cyclamen borer or strawberry root weevil and is found on many household, nursery and soft-fruit plants throughout the temperate regions of the world. Adults are nocturnal feeders of leaves and produce a characteristic edge notching, which in itself may cause economic damage to ornamental plants. In outdoor or unheated crops there is usually only one generation each year but in heated conservatories, botanic gardens and interior atriums there may be overlapping generations. All recorded adult vine weevils have been female; so reproduction is entirely parthenogenetic.

Adults are wingless, 8 to 10 mm (20/64 to 25/64 inch) in length, migration is limited when compared to many other insects although plant movement by humans accounts for most of its rapid colonisation. Eggs are laid on the compost surface, initially soft white and glistening they melanise to become rigid and reddish-brown in colour. The newly hatched or neonate larva, is white and legless with a brown head; older larvae normally have a 'C' shape and can acquire some colour from their host plant. Mature larvae may begin to burrow into corms and fleshy stems of several plant species where they are well protected from most forms of control, including pesticides. Larvae overwinter in cells produced in the compost and remain dormant until warmer conditions return and feeding activity resumes. After the final larval instar is reached a pupal cell is formed close to the surface and the weevil pupates, emerging some weeks later as an adult. Natural enemies of vine weevil include insectivorous mammals (hedgehogs, *Erinaceus europaeus*, and moles, *Talpa europaea*), several birds, frogs, toads, lizards and predatory insects such as carabid or ground beetles.

Entomopathogenic fungi

Commercial control agents include compost-incorporated insecticides; these can be in the form of conventional pesticides such as Imidasect (imidacloprid) or bioinsecticides such as *Metarhizium anisopliae* (Met52 Granular Bioinsecticide). Met52 contains a fungal pathogen produced and formulated on grains of rice. It is approved for use as incorporation in most growing media used for plant production. There are EAMUs for its use on a range of edible crops, open field applications, amenity situations and as a mulch treatment in a range of growing environments. The rice grains disintegrate leaving the green fungal spores that remain viable for 18 months to 2 years. Best results are achieved if the compost can be remixed two or more weeks after initial mixing as this distributes the fungal spores more evenly. Vine weevils deposit their eggs on the compost/soil surface, these hatch to neonate larvae that wriggle and burrow into the compost to reach roots on which they feed. In doing so, they invariably contact the fungal spores that germinate and penetrate the insect cuticle. Once inside the insect the fungus grows and it changes colour from creamy white to orange brown with white fungal growths that change to green as the spores mature. Met52 has good activity from 15°C (59°F) up to 32°C (90°F) providing the compost remains moist.

Entomopathogenic nematodes

If vine weevil larvae are found on any plants the insect parasitic nematode *Heterohabditis megidis*, as Exhibitline H and Nemasys H, is the fastest curative treatment when soil/compost temperatures exceed 14°C (57°F). This species of nematode can enter through the host cuticle as well as 'natural body openings', after entering the weevil larvae they release a small pellet of bacteria that kills the insect larvae changing them from creamy/white to pinkish red. *Steinernema krausseii*, as Nemasys L, can provide excellent preventative as well as curative activity from 5°C (41°F) up to 25°C (77°F) and is also applied as a drench treatment. This species is slightly slower acting but can be longer lived in soil or compost, providing good protection even when conditions are cold. None of the currently used soil-incorporated pesticides have any effect on these nematodes except Vydate (oxamyl), which could be used with a SOLA/EAMU to control non-indigenous leaf miner.

PLANT DISEASES

The section below details several of the most common plant diseases, their biology, cultural, chemical and biological control options; it is not an exhaustive list of all plant pathogens or their controls.

Horticultural disinfectants

Prevention is always better than trying to control any disease. Several horticultural disinfectants, also known as biocides, have broad-spectrum activity against most plant pathogens. These include peroxyacetic acid (available as various trade names), benzoic acid (as Menno Florades®), glutaraldehyde with quarternary ammonium compounds (as Horticide® and Unifect G®) etc. These will sterilise most surfaces such as benches, capillary matting, pots, tools and trolleys. Several plant diseases can be controlled biologically, either as part of an IPM schedule or as standalone products alongside conventional pesticides as part of resistance management. Biopesticides must all have an extant (current) **MAPP** (Ministerially Approved Pesticide Product) registration number. A **COSHH** (Control of Substances Hazardous to Health) risk assessment should also be completed before the use of these products.

Care should be taken when using biopesticides based on fungal preparations; the supplier's website should be consulted for the most up-to-date information. As a guide a period of up to 14 days should be allowed between treatments of a biofungicide and conventional (chemical) fungicide, although some are fully compatible and can even be tank mixed. The majority of biopesticides are approved for use on organically certified premises by Organic Farmers & Growers (OF&G) and the Soil Association. However, growers should check with their certification body before use.

GREY MOULD: *BOTRYTIS*

Botrytis cinerea, also known as grey mould, is a very common disease recognised by its fuzzy grey growth on the surface of buds, fruit, flowers, leaves and stems, see Figure 17.15.

Figure 17.15 Grey mould, *Botrytis cinerea*, on strawberry fruit.

It normally enters through damaged tissue and may be regarded as a secondary infection although it can infect healthy plants, particularly under humid conditions. Ghost spotting may be found as discrete spots on fruit, flowers and leaves, see Figure 17.16. For this to occur *Botrytis* must be present and spores must germinate on a wet surface; dry conditions later will kill the

Figure 17.16 Grey mould, *B. cinerea*; note collapsed flower heavily infected with the disease and darker pink spots on flowers known as ghost spotting.

spores, leaving the central spore and surrounding circle where its enzymes were beginning to work.

Botrytis may remain latent in and on many plants where disease symptoms may not develop for a considerable time. This is particularly noticeable on soft fruit where development occurs as the fruit ripens and sugar levels rise. Minute dusty, grey spores are released from mature colonies of *Botrytis* that carry in the air and water. Almost any movement of, or around, infected plants can trigger a release of spores. One technique that helps reduce spore spread is to place infected material in a plastic bag coated inside with soapy water, the 'soapy-bag' traps fungal spores and other mobile organisms such as adult whitefly, preventing their spread. Infection initially appears as brown, water-soaked lesions, the fungus develops to produce characteristic tan to grey hyphae and minute spores in grape-like clusters. Small black resting bodies called sclerotia develop on infected and sporulating tissue. They may be found inside dead and dying plant material where they can remain viable for a considerable time. Sclerotia may be isolated from many contaminated surfaces including benches, soil, tools and even seeds. Germinating sclerotia produce conidia that easily disperse to contact plant tissue and depending on conditions may develop ghost spotting or full disease symptoms.

Cultural and environmental controls for *Botrytis* include reducing crop humidity, this can be done by improving air movement through the crop either with air circulating fans or by additional venting (where possible). Many nurseries with the facility to heat the crop aim to have a heat boost and open vents to move air through the crop. Warm air rises and creates a gentle, drying air current that may be sufficient to control this and several other plant diseases.

There are several conventional (chemical) fungicides with good activity against *Botrytis* including iprodione (Rovral®), cyprodinil + fludioxonil (Switch®) and prochloraz (Octave®). The biopesticide *Bacillus subtilis* (Serenade ASO®) is active against *Botrytis* and several other plant diseases. It is applied as a spray to prevent and control early-stage infections. Having a zero-day harvest interval, no set maximum residue level (MRL) and up to 20 permitted applications per season it is ideal for use on many edible crops such as soft fruit. Prestop containing *Gliocladium catenulatum* is a biofungicide with good activity against most damping-off diseases as well as *Botrytis* stem lesions on tomato.

POWDERY MILDEW

Powdery mildews are common fungal diseases of many plants and normally, but not exclusively, show as chalky white spots that develop into larger patches, on the upper leaf surface, see Figure 17.17. If uncontrolled they can cover the entire leaf and stem severely weakening plants, leading to leaf and yield loss or even death of the plant.

Unlike most fungi, powdery mildew does not require a wet leaf surface for infection, although high air humidity is required for spore germination. In fact it is more prevalent in warm, dry conditions and when there are environmental fluctuations; high

Figure 17.17 Powdery mildew infection on honeysuckle leaf.

humidity favours asexual spore (conidia) production while low humidity favours spore maturation and release. Poor air circulation and damp shaded areas, as found with overcrowded plants, encourage this disease, so culturally, the opposite helps prevent or reduce the incidence of powdery mildews.

The majority of powdery mildews are restricted to just a few plants; these include most commercially grown plants and their associated weeds that act as alternate host plants. Fungal threads, called hyphae, grow on the plant surface forming mats of mycelium, taking nutrients from the plant with haustoria that penetrate the outer epidermal cells of the plant. As the fungus develops, minute spherical fruiting bodies may be produced called cleistothecia. Powdery mildews can overwinter as cleistothecia and mycelium on plant debris or dormant tissue (e.g. overwintering terminal buds of apple trees). In spring the cleistothecia rupture releasing sexual spores that can be transported by insects, water or wind, to initiate new infections.

Cultural control, as above, will help but in the event of powdery mildews being found the use of a fungicide is advised. There are several approved chemical fungicides including mycobutanil (Systhane) and bupirimate (Nimrod) that have good preventive and curative activity and are approved on many horticultural crops. Biologically, *Ampelomyces quisqualis* (AQ10), has good preventive and early-stage curative activity; this biofungicide is applied as a spray to contact the mildew's cell walls. High humidity at the plant surface is required for the beneficial fungus to penetrate and infect the powdery mildew hyphae, this usually occurs within 24 hours of initial contact. *Ampelomyces quisqualis* hyphae grow within the mildew and after seven to ten days minute fruiting bodies called pycnidia are produced. These can be found in the mildew hyphae, conidia and immature cleistothecia, killing all stages of the plant disease.

Potassium hydrogen carbonate as used in food preparation has a commodity substance approved for use as a fungicide against powdery mildews, to comply with CRD (Chemicals Regulation

Directorate) legislation it must be food-grade material. From experience good results have been achieved when it is mixed with SB Plant Invigorator. The mix has curative activity and may be considered an eradicant for the disease; although due to its naturally short persistence further instances of powdery mildew may appear some time later.

DAMPING OFF AND OTHER ROOT DISEASES

This complex of diseases is caused by several soil-borne fungi including *Fusarium*, *Pythium*, *Phytophthora*, *Rhizoctonia* and *Thielaviopsis*, which infect seedlings and young plants causing them to collapse and decay. These pathogens may also be isolated from mature plants, leading to severe losses. Cotyledons may break the soil surface only to wither and die, or healthy looking seedlings may suddenly collapse, which is known as post-emergence damping off. These diseases are most prevalent under levels of high humidity, poor air circulation and poor hygiene, overcrowding of seedlings and adverse germination conditions also exacerbate these pathogens. Although signs and symptoms of these diseases are mentioned below, microscopic determination by a specialist laboratory may be required. Damping off is easily confused with plant injury caused by insect feeding; sciarid flies, see above, are attracted to decaying plant tissue and will spread most fungal pathogens. Other symptoms similar to fungal damping off include excessive fertilisation, high levels of soluble salts, excessive heat or cold, and excessive or insufficient soil moisture, which also favour the diseases. It is therefore important to identify the cause of the disease/disorder so the correct course of action can be taken.

Fusarium oxysporum has several specialised forms – known as *formae specialis* (f.sp.) – that infect a variety of host plants causing various named diseases, thus *F. oxysporum* f.sp. *asparagi* is fusarium yellows on asparagus. Fusarium is a wilt disease and initially shows as transparent veins on the outer portion of younger leaves, followed by downward drooping of the older leaves called epinasty. As the fungus grows within the plant's vascular tissue, it causes a blockage leading to the characteristic wilt that can eventually kill the plant. At the seedling stage, infected plants may die soon after symptoms appear. In older plants, vein clearing and leaf epinasty are often followed by stunting and yellowing of the lower leaves. *Fusarium oxysporum* is also a common saprophyte in soil and organic matter where it can survive for long periods in the absence of host plants. The fungus can invade directly through root tips and open wounds in the roots as caused by physical damage and by sciarid fly larvae feeding, see above. Once inside the plant, the mycelium grows through the root cortex, reaching the xylem where it invades the vessels through the xylem pits. It can grow laterally to adjacent xylem vessels and upwards by way of the plant's sap.

Pythium (various species) occurs mostly in cool, wet, poorly drained soils and may be exacerbated by overwatering. Infection results in wet rot of roots that have no particular odour. When severe, the lower portion of the stem becomes black and slimy.

Usually, the soft, slimy outer portion of the root can be easily separated leaving a thread-like inner core. Most species of *Pythium* can survive for several years in soil and plant debris.

Phytophthora (various species) can be associated with root rots of established plants but are also part of the damping-off disease family; some species are involved with foliar diseases such as those that cause potato blight (*P. infestans*). These species have flagellate spores that can swim in wet compost towards root tips where they penetrate and cause a water-soaked brown to black rot similar to *Pythium*. Larger roots may show signs of soft rot and easily break, internally they may show a reddish brown to black core if scraped with a thumb nail. Arial symptoms include wilting, branch dieback and yellow or sparse foliage that progressively worsens until the plant dies. *Phytophthora* is usually regarded as a warm condition fungus as the symptoms are most commonly found in late spring to early autumn. Fungal spores can survive indefinitely in soil and plant debris.

Rhizoctonia solani is a fungal disease that causes rotting of cuttings, seedlings and infects mature plants. Infection is most prevalent when the growing media is overwatered and warm to hot, and shows as slightly sunken stem lesions at or just below the soil interface. The disease is found in all natural soils and can survive almost indefinitely. *Rhizoctonia* is easily transferred on infected hands, tools and from contaminated benches, plant trolleys, etc.

Thielaviopsis basicola is usually a problem of older, established plants. It infects the lateral roots where they grow from the main taproot. Diseased tissue turns dark brown and appears dry and shrivelled. The fungi survive for ten years or more in soil. It does not occur in strongly acid soils with a pH of 4.5 to 5.5.

Conventional fungicides such as metalaxyl-M (Subdue®) and propamocarb hydrochloride (Proplant) are active against *Pythium* and *Phytophthora* and also control downy mildew. The fungicide cyprodinil + fludioxonil (Switch WDG®) controls a wide range of plant diseases including *Botrytis* and *Fusarium* but has no approval for use as a drench into growing media and may indeed be phytotoxic if drenched.

Biofungicides

The biofungicide Prestop® is based on the beneficial fungus *Gliocladium catenulatum* and contains a nominal 2×10^8 cfu/g (colony forming units per gram), each of which can initiate an infection or start a colony in the soil. This fungus is a saprophyte in many environments but also pathogenic on a range of plant diseases. Fine, hair-like hyphae grow from germinating spores throughout the compost matrix and form an association with plant roots giving a good degree of protection against disease attack. *Gliocladium* hyphae, that contact a plant pathogen, kill it by enzyme activity, utilising its nutrients for its own growth, there being no penetration. *Gliocladium* is active at temperatures between 5 and 34°C (41 and 93°F) with best activity at 15 to 25°C (59 to 77°F), temperatures above 42°C (108°F) can kill the fungus, spores and mycelium. Prestop is approved on the majority of horticultural crops, both protected and outdoor, where it is used to control most damping-off diseases as well as some foliar and stem diseases such as *Botrytis*.

The biofungicide T34 Biocontrol® is based on the beneficial fungus *Trichoderma asperellum* and contains 1×10^{12} cfu per kg (dry weight). The antifungal abilities of *Trichoderma* have been known since the 1930s. The fungus can colonise soil and artificial growing media, forming an association with plant roots where they compete with other soil micro-organisms for nutrients and space. *Trichoderma* hyphae grow towards other fungi, coil around their hyphae and degrade the cell walls by secretion of various hydrolytic enzymes. The weakened cell wall allows *Trichoderma* to extract nutrients for its own growth and development. It may also penetrate the root epidermis and outer cortex of roots, which triggers the whole plant's immune system. This is known as induced systemic resistance (ISR) or systemic acquired resistance (SAR) and can provide protection throughout the plant against a range of foliar pathogens. Analysis of the signalling mechanism indicates the involvement of salicylic acid, jasmonic acid and ethylene as the most important active molecules.

CHEMICAL AND BIOLOGICAL PESTICIDES

A wide range of chemical plant-protection pesticides are available for professional use against most pest organisms found in horticultural situations. These include acaricides for spider mites, fungicides against various plant fungal pathogens (diseases), herbicides for weed control, insecticides for insects, molluscicides for slugs and snails, and rodenticides to control mice and rats. Other horticultural 'cides' include algicides to control algae, moss and liverworts. **Biocides** include algicides, many rodenticides and surface disinfectants etc. For UK use, plant-protection pesticides must have a current MAPP registration number.

A COSHH risk assessment must also be done to mitigate against any hazardous effects before a pesticide or other potentially harmful product is used. This involves obtaining a copy of the safety data sheet (**SDS**), sometimes called the manufacturer's SDS, material SDS or product SDS, which is a 16-section (older SDSs had between eight and ten sections) document specific for each product. All manufacturers of pesticides are legally obliged to issue the relevant SDS. However, the majority may be downloaded from manufacturers' or suppliers' websites.

Any specified **PPE** (personal protective equipment) must also be worn for pesticide use; this may include appropriate waterproof footwear, gloves, coverall, hood, respirator and filters, face or eye protection. The appropriate PPE should be specified on the product label with additional guidance from the pesticide code of practice or the SDS. All PPE must be assessed for the operation for which it is to be used, be compatible with the user and other items of PPE being used. Full information on the use of PPE (INDG174(rev2), which is a downloadable document) can be found on the Health and Safety Executive's website. The product label should contain all the specific requirements for safe and effective use the product. It is good practice to keep a copy of the product label safely in a file away from the pesticide store so it can be read before using a pesticide.

Notices placed close to treated plants or at the entrance to a treated area, warning unprotected people, may be required for some pesticides. Some pesticide treatments may require a specified re-entry period to be observed by anyone not wearing the appropriate PPE.

All professional plant-protection pesticides issued with a MAPP registration number require users to be trained, and in most situations to hold a certificate of competence specific to the method of application. Certificates are issued (following a training course and examination) by the NPTC (National Proficiency Test Council). The first test is the foundation module known as PA1 (Pesticide Application 1). Further training and tests allow holders to apply pesticides by various methods including hand-held applicators such as a knapsack sprayer (PA6A hydraulic nozzle or rotary atomiser type sprayers) the suffix 'W' includes application in or near water, thus PA6AW. There are 12 further assessment schedules that include PA2A for tractor-mounted or trailed ground-crop sprayers, PA4G for granular applicators and PA9 for fogging, misting and smokes. Amateur pesticides require a MAPP registration number similar to professional products; however, currently there is no requirement for users to hold a certificate of competence or to wear any PPE. However, if amateur products are used by a contractor or in a professional situation there would still be a requirement for appropriate PPE. This is mainly due to the degree of potential exposure; amateur home and garden use would normally be over a relatively small area. However, if the operator uses a pesticide over a long period of time the potential exposure is greater and PPE is likely to be required. See Pesticides: Code of practice for using plant-protection products.

Although pesticides sold for use in Europe must show **efficacy** as well as operator, environmental and plant safety, control against many pests can be difficult due to the pest life cycle. Many organisms have protected stages that may last from several days (eggs laid within leaves, such as thrips), weeks (pupal stages in soil, such as leaf miner) or even months, particularly during colder periods. The majority of pesticides disrupt the nervous system of the target organism requiring it to be actively metabolising to be effective; at low temperatures pesticide activity can be greatly reduced. Pests can become tolerant and later resistant to several pesticide groups, particularly **broad-spectrum** and **long-persistence** conventional pesticides that have an extended period of potentially sub-lethal activity. However, products having a physical mode of action tend to work well over a wide range of temperatures and can be particularly effective under cool conditions. Many conventional chemical pesticides have been in use for many years.

The aphicide pirimicarb, for instance, was first commercialised in 1970 (Tomlin, 2009), and although resistance is known for some aphid species on some agricultural crops it can still give good results in many situations. Other conventional pesticides, including synthetic pyrethroids such as cypermethrin, deltamethrin and lambda-cyhalothrin, have been in use since the 1980s but now have known resistance issues to many pests. This is thought to be due to the long persistence of this group of products (frequently several weeks to months) when there is a period, towards the end of the spray's active life on the plant, that a pest insect can reproduce while being exposed to a sub-lethal dose of pesticide. When this chemical group was developed in the early 1970s and marketed later that decade it only took a matter of a few years before widespread resistance to permethrin by glasshouse white-fly, *Trialeurodes vaporariorum*, appeared. To a slightly less extent we now have resistance to synthetic neonicotinoid products such as imidacloprid (Intercept), thiacloprid (Calypso) and acetamiprid (Gazelle). Interestingly, granular formulations of imidacloprid as Imidasect 5G or Intercept 5G that are incorporated in compost for root uptake remain active against most sap-sucking pests. These pesticides can provide good protection for several months, if plants are re-potted in treated compost.

Pesticides that are applied as a spray and have systemic or translaminar activity are best applied as a high-volume spray and may require up to three hours as a wet residue on the leaf before drying to allow penetration into the plant (evaporation reduces penetration). Pesticides with contact activity against a pest require less time for activity and are often best applied early in the morning. This period is usually cooler and the pests more sedentary giving a better chance of hitting the target organism.

The UK has a scheme that allows many plant protection pesticides to be applied at different rates and on different plants/crops than those specified on the product label. Known as an **EAMU (Extension of Authorisation for Minor Uses)** – previously called a **SOLA (Specific Off-Label Approval)** – these notices are legal documents that must be adhered to. They can be found on the CRD website (www.pesticides.gov.uk) by entering the phrase 'off label – search' for current approvals and rates.

There is also a section called commodity substances that can be found on the same website. This section gives authorisation to use non-registered products that have known pesticidal activity as a conventional pesticide. An example of this is the use of food-grade potassium hydrogen carbonate, more commonly known as potassium bicarbonate, as a powdery mildew fungicide. The stipulations within the approval include the caveat that the material must be a food-grade product and that the maximum dose does not exceed 20 g per litre of water. The use of any product under the EAMU/SOLA and commodity substance approval scheme is always at the grower's risk for efficacy and most importantly crop/plant safety. There being no compensation from the supplier for damage or inactivity.

To reduce the threat of pesticide resistance, growers should alternate products with different modes/sites of activity against the target pest. The Pesticide Resistance Action Group has various committees that have developed a code system for all pesticides. However, those with a purely physical mode of action are exempt from registration. Resistance codes are a numerical way of indicating the pesticide's mode of activity. Think of a clock face with

the numerals being the resistance code numbers, the centre as the pest, and the hands leading to centre being the various modes/sites of activity in/on the pest. Insecticides have an IRAC (Insecticide Resistance Action Committee) number, fungicides a FRAC number and herbicides a HRAC number.

(See Table A3.0 in Appendix 3.)

Due to the restriction in the number of permitted pesticide applications and requirement to use products with different modes or sites of activity to help reduce resistance pressure, several pesticides need to be available against each pest.

Biopesticides

Several biological agents are registered as pesticides and are also known as **biopesticides** – these are based on micro-organisms or natural products. These include bacteria, fungi and viruses as well as plant-derived products. Just as with conventional chemical pesticides, biopesticides must have a MAPP registration number, these are only issued if the product reaches set standards of efficacy, has good operator, environmental and crop safety, etc. The reason for this is mobility of the organism: the majority of biopesticides are based on **beneficial pathogens** (disease-causing organisms), which are applied by spray to plants or by incorporation into growing media.

Beneficial organisms such as predatory mites, parasitoid wasps and **entomopathogenic nematodes** that locate the host by their own locomotion are known as macro-organisms. As such they do not require registration and can be used by anyone. However, some non-indigenous beneficials require a licence for use. Entomopathogenic nematodes are minute nematodes that kill insects by infecting them with a pathogen (disease). They are mobile and swim in moist growing media or in a film of water on plant and flower surfaces, to actively seek their host prey. Once located, the nematodes enter the host body through natural body openings while some species can also enter through the cuticle. Once inside the host, a small pellet of symbiotic bacteria is released, which utilises the body's tissues to reproduce, causing a septicaemia and death of the host organism.

Pesticide disposal

All registered plant protection pesticides have a MAPP number printed on the product label. Biocides such as algicides and some rodenticides may have an HSE approval number. Only those products with a currently approved (extant) MAPP/HSE number are legal to be used by a qualified person. A list of approved products can be found on the CRD (Chemicals Regulation Directorate (pesticides)) and HSE website at www.pesticides.gov.uk. Pesticides having an old or expired registration number usually have a legal 'use-up' and storage period that can be up to a year after the final sale date. Any product not appropriately used before the final 'use-up' date becomes illegal to use and store, and should be disposed by an approved hazardous waste disposal contractor.

GETTING STARTED WITH IPM

✿ Sources of information. Identify any potential pest or disease problems as soon as possible, keep up to date.

✿ Regulations and licence requirements.

✿ Cultural control, general nursery hygiene, ground-cover materials, weed control, insect netting.

✿ Monitor on a regular basis, use sticky traps, keep records (and look at them!).

Sources of information

Pest identification and knowledge of basic biology is essential if an IPM programme is to be fully effective. Various reference books are available; CRC Press (www.crcpress.com) produce a wide range of highly illustrated books for study and reference. The Agriculture Development Advisory Service (**ADAS**), Agriculture and Horticulture Development Board (**AHDB**), Food and Environment Research Agency (**FERA**), and Horticulture Development Company (**HDC**) are responsible for most of the 'near market' research and development work in the UK. Much of this work is conducted on commercial nurseries and at East Malling Research (**EMR**) in Kent, Stockbridge Technology Centre (**STC**) in Yorkshire and Warwick Crop Centre or by ADAS. Illustrated monthly reports are published in the *HDC News*, which is available to AHDB levy payers, corporate members and funding members; giving results of the latest horticultural R&D. DEFRA has funded HDC and ADAS who have published several guides for UK growers including: *Integrated Pest Management in Protected Ornamental Crops* (Buxton *et al.*, 2006) and *Environmental Best Practice Guide for the Ornamentals Sector* (Hewson and Perkins, 2008). *The UK Pesticide Guide* commonly known as the 'green book' is an annual publication giving details of pesticides and their various approvals for use on a range of edible and ornamental plants. It has additional information on the product's chemical group, mode of action, code number, EAMU/SOLA notices, efficacy, restrictions, environmental safety, hazard classification and safety precautions. This is also available as an electronic version from www.plantprotection.co.uk, which includes online access to *IdentiPest* – a pictorial diagnostic tool that helps identify pests, diseases and weeds (mostly agricultural crops). Although the green book is an extremely useful guide as to what each product can be used for and on which plants, it is always advisable to check the latest approvals on the CRD website.

The vast majority of UK-grown edible crops are produced within strict guidelines that have been developed between advisors, growers and the retail outlets (www.assuredproduce.co.uk). The non-governmental organisation GlobalG.A.P. (www.globalgap.org) sets voluntary standards for growth and production of many crops including edibles, ornamentals and livestock production. Both organisations specify the importance of using IPM techniques in crop production.

BIOCONTROL LICENCES, REGISTRATION AND CERTIFICATES

Non-indigenous biological agents require a licence for release from DEFRA, specifying the conditions of use for growers to legally use them. These are obtained by the producing company and used to regulate where and when certain organisms are released. They are also used to monitor where each organism was dispatched and must be retained by the user.

The predatory bug *Macrolophus pygmaeus* is of Mediterranean origin and has a 'grower licence' restricting its use on glasshouse grown crops. Similarly the spider mite predator *Amblyseius californicus* is restricted for use on glasshouse crops in production.

Pathogens that attack insects, mites or plant diseases, although of natural/biological origin must be registered as a pesticide and have a MAPP registration number. There are several fungal pathogens with excellent potential to control a range of foliar and compost/soil-dwelling pests. A wide range of plant diseases can also be controlled by specific beneficial pathogens. Indigenous beneficials that can actively move of their own accord include mites, insects and nematodes, and do not require any licences or registration to use.

Cultural methods

General hygiene throughout plant production, dispatch, storage and sales areas is the first line of defence in any pest-control strategy. Weeds can harbour many pests and disease species; as well as being unsightly, they impact directly on the growth of bed- and pot-grown plants. Ground-cover materials can have several roles; suppression of weeds is probably the most obvious and important as they can harbour many insect and mite pests as well as several important diseases. For example, various powdery mildews that affect crops can also be found on weeds such as dandelion; this is important for any garden growing vegetables and salad crops. Weed control can be tackled in several ways, from hand pulling to chemical herbicide applications; several useful articles on biological control of weeds can be found on the internet. Some of the woven polypropylene materials are extremely strong and hard wearing, often having an active life of ten years or more. These materials can withstand most chemicals used for surface disinfection, cleaning by brush or sweeper and considerable physical wear and tear.

Thinner, spun **polypropylene**, materials known as fleece, usually white in colour, can be used to help protect plants from frost or provide light shade to protect from excessive sunlight. They can also provide protection from attack by flying insects such as aphids, flea beetle, butterfly/moths etc. Black, spun polypropylene materials are relatively cheap to purchase and are used as a mulch to control weeds under plant production benches as well as open-bed grown plants such as lettuce, *Lactuca sativa*, and strawberry, *Fragaria × ananassa*. Some ground-cover materials are pretreated with copper hydroxide as Tex-R® fabric, which effectively acts as a barrier against plant roots (used on sand beds to prevent rooting through), germination of wind-blown weed seeds (pot toppers), and as a highly effective barrier against slug and snail attack. Woven polypropylene netting is available in various mesh sizes as a physical barrier to protect plants from attack by flying insects and birds, most of these materials are UV stabilised and should last several years.

Many other IPM techniques are in common use (Cloyd *et al.,* 2004), including companion planting and the use of plant push-pull strategies (Cook *et al.,* 2007) where the attractiveness of particular plants can be used to protect others. Canadian researchers (Nakano & Matteoni, 2008) have set up 'bug gardens' to determine which plants are best suited to encourage natural enemies that will develop sustainable colonies. The wider aspects of biodiversity in agriculture have been discussed by Scherr and McNeely, (2008) looking at the whole 'eco-agricultural' landscape, a wide range of plants helps create more habitats for natural enemies (Fahrig, 2003). Traditional walled gardens such as Arundel Castle Gardens, Beningbrough Gardens, Helligan, Wakehurst and West Dean Gardens have a range of fruit trees, ornamental flowering plants, seasonal vegetables and permanently planted herbaceous plants providing excellent **biodiversity**.

Figure 17.18 Taking a photograph with a mobile phone through a 'linen tester' type hand lens.

Pest monitoring

Monitoring and recording pest outbreaks to build up a diary of trends and identify vulnerable locations helps to reduce the incidence of major pest problems by anticipating and responding to problems quickly. Nothing beats regular crop scouting by walking through a crop, lifting pots and turning leaves for finding and monitoring pest or disease levels. A hand lens is a vital piece of equipment for identifying potential pests or diseases. Linen tester type lenses, at 8 to 10× power magnification (as shown in Figure 17.18) are easy to carry and use. These lenses fold away and are prefocused to the target area; as such a photograph can be taken directly through the lens. The majority of mobile/cell phones take photographs that can be emailed and kept on a computer to build up a library of useful images.

Disease-infected plants must be isolated and the type and cause of infection identified as soon as possible to prevent further spread. For example: a collapsed, wilting plant may be infected with one of several pathogens, many of which can be transmitted by sciarid flies that feed on roots and even burrow into the stems of plant cuttings causing direct feeding damage.

Sticky traps are available in blue and yellow; they work by attracting flying insects to the light reflected from the sticky surface. For best results, in either monitoring pest levels or mass trapping, the positioning of the trap is important. Horizontal yellow traps, sticky side up, at just above soil/compost level will trap scatella and sciarid fly adults but should be just above the plant head for flea beetle and leaf miner adults. Vertical sticky traps, ideally in a fixed position and not swinging on a string, are better for winged aphids, leaf hopper, thrips and whitefly. Blue traps tend to catch more thrips, at least in the summer, but can also attract high numbers of predatory hoverflies (they are also useful to catch houseflies when hung near the end of a fluorescent light). Pheromone lures are available for a wide range of adult moths; however, most of these lures use a sex attractant to catch male moths and are specific to each species, so correct pest identification is critical. Monitoring thrips with sticky traps has been improved by the introduction of Thripline ams (advanced monitoring system) from Syngenta Bioline that uses a sex aggregation pheromone to activate both male and female adults. These are specific for the western flower thrip, *Frankliniella occidentalis*, and can be used to monitor or to excite adults out of their hiding places within flowers and improve pesticide spray efficacy (GreatRex, 2009). Trap counts can be entered on a computer spreadsheet program, these all have the ability to plot a simple chart or graph of data counts. These graphs give an easy-to-read visual impression of pest and disease fluctuations, which can be compared with the controls being applied.

Summary

Integrated pest management is a more sustainable method of pest control using a combination of cultural techniques, biological control agents and selective or short-persistence pesticides to provide long-term pest management. European Union legislation for the framework for community action to achieve a sustainable use of pesticides (P6_TA(2009)0010) has been passed to ensure that all professional growers implement the general practices of IPM techniques by 2014. All pesticides – conventional (chemical), biological and if applicable amateur products – must have an extant (current) MAPP/HSE registration number; a COSHH risk assessment should also be completed before their use. Volunteer staff are essential to the majority of public properties; current legislation would exclude many from using professional pesticides unless they were suitably qualified in pesticide use. However, cultural and biological control requires no specific training or protective equipment, and can be regarded as a commonsense approach to pest management (Olkowski et al., 1991).

It is worth remembering that a fully fed predator will take some time to digest its food before it starts eating again, in the meantime the pest population will continue to increase. Sticky traps, although unsightly, can be an effective method of monitoring and even control in 'behind the scenes' situations, e.g. production greenhouses that are screened from public viewing. Registration-exempt, physical-acting pesticide sprays, such as SB Plant Invigorator, can be used by almost anyone, making it ideal for use by volunteers, to provide good control of a wide range of insect and mite pests. Generalist predators such as *Chrysoperla* (lacewing larvae) can be used inside or out, providing average temperatures exceed 12°C (54°F). They will feed on most soft-bodied prey including aphids, mealybug, young caterpillar and moth eggs. Lacewing and/or ladybird larvae can be introduced to beech hedging for beech aphid (*Phyllaphis fagi*) or *Buxus* for box psyllid or box sucker (*Psylla buxi*) and may transfer to many other plants, overwintering boxes help with long-term establishment of these very useful predators.

Identification of plant damage symptoms (diseases, insects or mites) comes with experience and will lead to improved control; specific biocontrol agents that target pests can often provide better control than generalist predators. Further training and certification (PA1, PA6, etc.) will allow a full IPM programme to be successfully run, utilising a full range of biocontrol agents and IPM-compatible pesticides (see tables in Appendix 3). Some plant cultivars tend to be more susceptible to pest and disease attack while others are much less susceptible and may even show signs of resistance. Selecting resistant cultivars or growing plants grafted onto resistant rootstock to prevent disease attack is a valuable element of IPM. Cultural techniques include general hygiene, ground-cover materials for weed control and monitoring with sticky traps to predict pest occurrence. Alternative methods of pest control include selecting resistant plant cultivars either by natural selection or, where allowed, genetic modification; which is further discussed in Chapter 9. Biological control becomes the next line of defence either by introducing beneficial organisms (Collier & Van Steenwyck, 2004) or as a consequence of reduced broad-spectrum pesticide applications allowing natural pest control organisms to establish. IPM techniques are currently being used in the majority of protected crops, increasingly on outdoor landscape plants (Dreistadt, 1994) and orchard crops (Cross, 2002).

Review questions

1. Name **four** methods of controlling pests and diseases, including **one** benefit and **one** limitation of each.
2. Define the following terms related to chemical selection and use:
   i) pesticide resistance
   ii) broad-spectrum
   iii) sub-lethal dose
   iv) pheromone.
3. List **four** control methods for an infestation of western flower thrips, *Frankliniella occidentalis*, including the reasons for their selection.
4. Describe the damage caused by **two** plant pests and **two** plant diseases.
5. Identify **four** sources of information on current pesticide legislation, reviewing their relationship to selection, storage and use.

REFERENCES

Bellows, S. & Fisher, T. W. (1999). *Handbook of Biological Control*. California, USA: Academic Press, San Deigo.

Bennison, J., Maulden, K., Maher, H. & Tomiczek, M. (2008). Development of a grower rearing-release system for *Atheta coriaria*, for low cost biological control of ground dwelling pest life stages. *IOBC/WPRS Bulletin*, **32**.

Buxton, J. (2009). The biology and control of mites in pot and bedding plants. *HDC Factsheet 12/09*.

Buxton, J., Bennison, J., Brough, W. & Hewson, A. (2006). *Integrated Pest Management in Protected Ornamental Crops: A Best Practice Guide for UK Growers*. Cambridge, UK: DEFRA/ADAS.

Chandler, D. (2008). The consequences of the 'cut off' criteria for pesticides: alternative methods of cultivation. Policy Department Structural and Cohesion Policies. European Union, Brussels: Agriculture and Rural Development.

Cloyd, R. A., Nixon, P. L. & Pataky, N. R. (2004). *IPM for Gardeners: An Guide to Integrated Pest Management*. Portland, Oregon, USA: Timber Press.

Cook, S. M., Khan, Z. R. & Pickett, J. A. (2007). The use of push-pull strategies in integrated pest management. *Annual Review of Entomology*, **52**: 375–400.

Collier, T. & Van Steenwyck, R. (2004). A critical evaluation of augmentative biological control. *Biological Control*, **31**: 245–56.

Cross, J. V. (ed.) (2002). Guidelines for integrated production of pome fruits in Europe. Technical Guideline III. *IOBC/WPRS Bulletin* **25**(8).

Dreistadt, S. H. (1994). *Pests of Landscape Trees and Shrubs: An Integrated Pest Management Guide*. Oakland, California, USA: Statewide IPM Project, University of California, Division of Agriculture and Natural Resources Publication 3359.

Enkegaard, A. (2005). Integrated control in protected crops, temperate climate. *IOBC/WPRS Bulletin*, **28**(1).

Fahrig, L. (2003). Effects of habitat fragmentation on biodiversity. *Annual Review of Ecology, Evolution and Systematics*, **34**: 487–515.

GreatRex, R. (2009). 30% better control of WFT using Dynamec and Thripline. *Fargro Croptalk. Summer, 2009*. UK: Fargro Ltd.

Hajek, A. (2004). *Natural Enemies: An Introduction to Biological Control*. Cambridge, UK: Cambridge University Press.

Helyer, N. (2002). Abamectin plus pymetrozine; an extremely useful addition to the IPM armoury. Integrated control in protected crops. *IOBC/WPRS Bulletin*, **25**(1).

Helyer, N. (2008). Integrated pest management for interior atriums and office plants. *Planteria*. **24/25**.

Helyer, N., Cattlin, N. & Brown, K. (2014). *Biological Control in Plant Protection: A Colour Handbook*, 2nd edn. CRC Press.

Hewson, A. & Perkins, S. (2008). *Environmental Best Practice Guide for the Ornamentals Sector: A Guide for UK Growers*. Cambridge, UK: DEFRA/ADAS.

Lambert, L. (2009) Innovative technologies for the bio-control of greenhouse diseases and insects. *International Symposium on High Technology for Greenhouse Systems. GreenSys, 14–19 June 2009*. Quebec City, Canada.

Malais, M. & Ravensberg, W. (2003). *Knowing and Recognising: The Biology of Glasshouse Pests and their Natural Enemies*. Berkel en Rodenrijs, the Netherlands: Koppert BV.

Nakano, M. & Matteoni, J. A. (2008). Bug gardens for education and research in conservation biological control and sustainable horticulture. Integrated control in protected crops, temperate climate. *IOBC/WPRS Bulletin*, **32**.

Olkowski, W., Daar, S. & Olkowski, H. (1991). *Common-sense Pest Control: Least-toxic Solutions for Your Home, Garden, Pets and Community*. Newtown, Connecticut, USA: The Taunton Press.

Piggott, S. J., Clayton, J., Gwynn, R. *et al.* (2000). Improving foliar application technologies for entomopathogenic nematodes. *Workshop Proceedings; University of Ireland, Maynooth, 13–15 April 2000*, pp. 119–27.

Sabelis, M. W., Janssen, A. Lesna, I. *et al.* (2008). Developments in the use of predatory mites for biological pest control. Integrated control in protected crops, temperate climate. *IOBC/WPRS Bulletin*, **32**.

Scherr, S. J. & McNeely, J. A. (2008). Biodiversity conservation and agricultural sustainability : towards a new paradigm of 'ecoagriculture' landscapes. *Philosophical Transactions of the Royal Society B*, **363**: 477–95.

Shah, P. A. & Pell, J. K. (2003). Entomopathogenic fungi as biological control agents. *Applied Microbiology and Biotechnology*, **61**: 413–23.

Sunderland, K. D., Chambers, R. J., Helyer, N. L. & Sopp, P. I. (1992). Integrated pest management of greenhouse crops in Northern Europe. *Horticulture Reviews*, **13**: 1–47.

Thacker, J. R. M. (2002). *An Introduction to Arthropod Pest Control*. Cambridge, UK: Cambridge University Press.

Tomlin, C. D. S. (2009). *The Pesticide Manual: A World Compendium*, 15th edn. Alton, Hampshire, UK: BCPC Publications.

Van Driesche, R (1998). *Western Flower Thrips* (Frankliniella occidentalis) *in Greenhouses: A Review of its Biological Control and Other Measures*. Amhurst, USA: Department of Entomology, University of Massachusetts.

Van Lenteren, J. C. (2000). Success in biological control of arthropods by augmentation of natural enemies. In Gurr, G. & Wratten, S. (eds.), *Biological Control: Measures of Success*. Dordrecht, the Netherlands: Kluwer Academic Publishers, pp. 77–103.

Van Lenteren, J. C., Bale, J., Bigler, F., Hokkanen, H. M. T. & Loomans, A. J. M. (2006). Assessing risks of releasing exotic biological control agents of arthropod pests. *Annual Review of Entomology*, **51**: 609–34.

Van Lenteren, J. C., Loomans, A. J. M., Babendreier, D. & Bigler, F. (2008). *Harmonia axyridis*: an environmental risk assessment for Northwest Europe. *BioControl*, **53**: 37–54.

van der Linden, A. (2008). *Silene dioica* (Caryophyllaceae: Silenoidae) as a reservoir and a hibernation site for predatory mites (Acari: Phytoseiidae). Integrated control in protected crops, temperate climate. *IOBC/WPRS Bulletin*, **32**.

FURTHER READING

Albajes, R., Gullimo, M. L., van Lenteren, J. C. & Elad, Y. (eds) (1999). *Integrated Pest and Disease Management in Greenhouse Crops*. Dordrecht, the Netherlands: Kluwer Academic Publishers.

Alford, D. V. (2012). *Pests of Ornamental Trees, Shrubs and Flowers: A Color Handbook*, 2nd edn. Academic Press.

Brown, A. E. (2006). Mode of action of insecticides and related pest control chemicals for production agriculture, ornamentals and turf. Pesticide information leaflet no. 43. Maryland, USA: University of Maryland, Department of Entomology.

Budge S. P. & Whipps J. M. (2001). Potential for integrated control of *Sclerotinia sclerotiorum* in glasshouse lettuce using *Coniothyrium minitans* and reduced fungicide application. *Phytopathology*, **91**: 221–7.

Lainsbury, M. A. (2012). *The UK Pesticide Guide 2012*. Alton, Hants, UK: BCPC Publications.

Watson, A. K. (ed) (1993). *Biological Control of Weeds Handbook*. Champaign, Illinois, USA: Weed Science Society of America.

WEBSITES

Several of the websites listed contain hyperlinks to other useful sites.

Agriculture and Horticulture Development Board: www.ahdb.org.uk

American Phytopathological Society: www.apsnet.org/online/archive.asp

Commercial biological control producers and suppliers – worldwide listings: www.agrobiologicals.com/index.html

European and Mediterranean Plant Protection Organization (EPPO): www.eppo

Fargro Ltd: www.fargro.co.uk

Global non-governmental organisation setting growing standards and monitoring crop production: www.globalgap.org

Horticultural Development Company: www.hdc.org.uk

International Biocontrol Manufactures Association: www.ibma-global.org

International Organisation for Biological Control: www.iobc.global

Non-governmental organisation setting growing standards and monitoring edible crop production in the UK: www.assuredproduce.co.uk

Ohio Florists Association, USA: www.ofa.org

Photographic images: pictures@flpa-images.co.uk

Protected Herbs: www.protectedherbs.org.uk, also www.hdc.org.uk/herbs

Royal Botanic Gardens, Kew: www.rbgkew.org.uk

Royal Horticultural Society, UK: www.rhs.org.uk

RHS level	Section heading	Page no.
2 1	Biological and integrated pest control	354
2 3	Biological and integrated pest control	354
2 3.4	Horticultural disinfectants	367
	Aphids	354
2 3.8	Biological and integrated pest control	354
2 4	Biological and integrated pest control	354
2 4.1	Aphids	354
2 4.6	Biological and integrated pest control	354
2 4.12	Aphids	354
	Horticultural disinfectants	367
2 5	Aphids	354
2 8.2	Aphids	354

CASE STUDY The London Olympic Park

The design and planting of the London Olympic Park epitomises the ethos of sustainability and conservation in amenity horticulture. The aesthetics of the design is enhanced by consideration given to the principles of sustainability and conservation. The aim of the design was also as a showcase for the 2012 Olympic Games, which could then be converted to a valuable public park. The park was designed with two distinct areas. The North Park has been designed and planted to represent native UK habitats, including meadows, wetlands, bioswales, rain gardens and woodland underplanting. The South Park has stunning annual meadows, half a mile of naturalistic perennial planting and gardens that represent the biodiversity hot spots of the world. The design and realisation of the park is the work of Nigel Dunnett, James Hitchmough, Sarah Price, LDA Design and Hargreaves Associates.

Wildflower meadow planting at the Olympic Park, London 2012.

Chapter 18

Conservation and sustainability

Jenny Shukman

INTRODUCTION

Conservation and sustainability go hand in hand and embrace all aspects of horticulture, from amenity to commercial and small to large scale. The gardener has a role to play whatever the size of garden. It is being increasingly recognised how important ordinary gardens are in providing a network of habitats for conserving wildlife and improving the environment.

Botanic gardens have long played a global role in conserving plant species, dating from the sixteenth century, with the first botanic garden in the UK being completed in 1663, in Oxford, (Oldfield, p.13, 2007). This role is carried on today with traditional methods combined with the advantages of modern technology, for example at Kew's Millennium Seedbank at Wakehurst Place, Sussex.

Many important charities have been founded to aid the conservation of plants and gardens, either as their main remit or as part of a wider theme of conservation and education, such as the National Trust and Plant Heritage. Legislation has played an increasingly important role, for example through the Wildlife and Countryside Act 1981, and Convention on International Trade in Endangered Species (CITES). Local planning authorities implement important planning legislation, for example, administering Tree Preservation Orders (TPOs).

Sustainable practices are fundamental to conservation. Again, these are important on all levels, from the compost heap and water butt in a back garden, to large-scale water storage for commercial nurseries and effective management of recycling in large parks and gardens.

Design is a crucial factor in creating sustainable environments. This can involve the general layout and planning of gardens to encourage wildlife, and specific features such as green roofs and living walls. Linked with design is the selection of suitable hard and soft landscape materials.

Key concepts

✿ An introduction to conservation principles

✿ The fundamental role of recycling

✿ Design for sustainable gardens and environments

✿ Selection of soft and hard landscaping materials to minimise environmental impact

✿ Management of resources via sound horticultural practice

✿ The role of legislation and conservation bodies

PRINCIPLES FOR CONSERVATION AND SUSTAINABILITY

The general aims for conservation and sustainability are to create the best conditions possible for the survival of the widest number of species, to reduce any negative impacts on the environment of horticultural practices and promote the positive impacts.

Biodiversity is a key principle for conservation. To conserve a wide range of species, and maintain genetic diversity in the environment, methods of *in situ* and *ex situ* conservation are used. Also see Chapter 15. The creation of suitable habitats is a vital part of both *in situ* and *ex situ* conservation. Botanic gardens play a vital role, but the role of ordinary gardens on a range of scales must not be overlooked. This is discussed further in Chapter 13.

An overriding principle for conservation and sustainability is that of taking a holistic approach. Rather than considering individual items such as plants, animals, soil organisms or gardens as isolated features, they need to be considered as part of a whole. The Earth should be regarded as a single living organism, with all the individual parts being interrelated, as proposed by James Lovelock via the Gaia Theory. Whatever is done to one part affects the whole. Understanding the connections between all living organisms underpins sustainable gardening techniques and successful conservation of species.

Climate change is a prime example of the Earth as a single living organism, with change in one part of the world affecting others. Horticultural practice has a role to play in reducing carbon emissions and alleviating the adverse effects. Conservation of a wide range of plant species is also crucial as increasing pressure is put on different environments. Throughout history, plants have evolved and adapted to different climates. The current rate of change is too fast in general for plant adaptations to evolve, and migration of plants is hindered by the fragmentation of habitats. The provision of a network of habitats through all types of land use can help conserve different plant species.

A useful tool for assessing the impact of any activity is the carbon footprint. Carbon calculators have been devised to achieve defined measurements. A carbon footprint is a measurement of the amount of greenhouse gases that are produced, for example by burning fossil fuels, transport, and indirectly through items purchased and used. It is measured in units of tonnes or kilograms of carbon dioxide equivalent.

Carbon offsetting involves activities to offset the production of carbon dioxide, such as planting trees. On a larger scale this may include reforestation projects and using renewable energy. The aim is to be carbon neutral, which can be achieved by balancing outputs and inputs.

By planting a tree the carbon dioxide will be sequestered in the process of photosynthesis. It is stored as carbohydrates, eventually becoming the wood of the tree, and oxygen is released as a by product. The tree can therefore be termed a 'carbon sink'.

The peat debate is very much concerned with carbon sinks. Extraction of peat on a wide scale has resulted in the release of much carbon dioxide into the atmosphere from the vast carbon sinks of peatland habitats. Lowland bogs, from which much peat has been extracted, are very fragile and vulnerable habitats. Draining these for peat extraction has had the effects of not only reducing biodiversity by destroying these habitats, but also releasing carbon into the atmosphere as the peat is drained. Peat habitats also play a vital role in the hydrological cycle and help prevent flooding and ameliorate climate change. Peat takes years to accumulate, growing at only about 1 mm a year. Extraction removes it at up to 22 cm (8 inches) a year. Much valuable work has been carried out on restoring habitats after peat extraction, but this is limited compared to the damage that has already occurred.

A government-led voluntary target of peat use reduction by 2010 was not met. The aim was to increase the amount of peat alternatives used in growing media to 90%. Most peat used in the UK is now imported from Ireland and the Baltic countries.

New aims in the Natural Environment White Paper are to reduce peat use to zero by 2030. This includes a voluntary phase out for peat use in the amateur horticultural sector by 2020, and a voluntary phase out in the commercial horticultural sector by 2030. A Sustainable Growing Media Task Force has been set up to oversee this.

Standards exist for carbon neutrality – the British Standards PAS 2060 specification for the demonstration of carbon neutrality. These standards are applicable on all levels, including individuals, communities, businesses, and local or regional-level governments.

RECYCLING

Recycling can play a key part in reducing the carbon footprint of horticulture and balancing inputs and outputs. It is particularly relevant for growing plants, as the natural cycles of nutrients and organic matter in the soil can be followed and encouraged in gardening practices. Maximum use can be made of the organisms and micro-organisms that, for example, break down organic matter in the soil, capture atmospheric nitrogen, and convert ammonia to nitrites and in turn to nitrates for plant uptake. Please also see Chapter 6.

As part of recycling in the garden, any green waste can be composted and returned as a mulch or soil conditioner.

Composting

Home composting is the keystone for sustainable gardening. A valuable source of organic matter is provided for the garden with no material being brought into the site, and no material leaving the site. Suitable methods for home composting are discussed in Chapter 6.

On a larger scale, the recycling of green waste can also play a major role. Rather than transporting vast amounts of plant material such as hedge trimmings to landfill sites, the green waste can be turned into useful compost to improve soil structure, add nutrients and act as a mulch to conserve moisture and prevent weed growth. Different local authorities have different systems for collecting waste and different companies are involved in turning this green waste into valuable compost. The case study gives an insight into a local authority green waste recycling scheme in Hampshire used to produce Pro-Grow.

Veolia Environmental Services, Hampshire

The process of producing Pro-Grow from green waste:

❀ Green waste from households is typically collected in green bins with roadside collections, or taken to local household recycling centres

❀ From there, lorries transport the green waste to composting sites. Figure 18.1 shows the original green waste at the site.

❀ Before depositing, the green waste is passed through a large mechanical shredder, as shown in Figure 18.2. This both reduces the volume and speeds up the composting process.

❀ After shredding, the green waste is placed in long trapezoidal heaps, termed windrows.

❀ For two to three weeks, the compost builds up high temperatures in excess of 75°C, which sterilises the material, destroying any pests, diseases and weed seeds. During this sanitisation period, the windrows are regularly turned with a machine termed a windrow turner to introduce oxygen and release carbon dioxide that has built up, improving conditions for the microbes that break down the green waste.

❀ The next phase is that of stabilisation in which high temperatures are maintained and the compost matures. The windrows are turned every two to three weeks and irrigated if conditions are dry.

❀ After four to six months the compost is ready to be passed through a large sieve or screen, termed a Trommell screener as shown in Figure 18.3. Material that is too large to pass through is returned to be mixed with fresh green waste to start the composting process again.

❀ Contaminants such as metal, plastics and stones are removed by an air separator.

❀ The compost is then returned to the windrows for further maturation.

❀ A final screening then takes place through a finer grade Trommell screener, typically between 5 to 20 mm.

❀ Figure 18.4 shows the compost when it is moved to a storage shed where it is kept dry and then either bagged or sent for distribution.

❀ The finished product arrives at a garden or landscape for use as a valuable mulch or soil conditioner. Pro-Grow is peat free and Soil Association approved as well as having the British Standard PAS 100. It is therefore suitable for organic growing systems.

Courtesy of Veolia Environmental Services

Figure 18.1 Green waste arriving for composting.

Figure 18.2 The shredding process.

Figure 18.3 The screening process.

WRAP (Waste and Resources Action Programme) is a government-funded, non-profit-making company set up in 2000 in the UK with the aim of preventing waste and supporting recycling. WRAP works with all sectors, being involved with individuals, businesses and communities. Of particular interest for horticulture is the advice on the use of composts produced to the British Standard PAS 100.

Figure 18.4 The storage shed.

Master Composter Scheme – The Master Composter Scheme aims to promote home composting through a network of trained volunteers. The schemes are organised in different counties, often through the local council in conjunction with Garden Organic, some through Wildlife Trusts. They are all community-based schemes working with groups such as allotment societies, schools, youth groups and horticultural societies. Advice and information is passed on through talks, demonstrations and articles.

Reed beds

Naturally occurring reed beds form important wildlife habitats. Reed beds are wetlands dominated by a particular plant, such as common reed, *Phragmites australis*. They are found in river flood plains and low-lying coastal areas. They can also be artificially created to perform extra functions such as a sustainable method for filtration for waste water, thus providing an alternative to chemical treatments.

Artificial systems are self-contained units, which are lined and filled with soil, sand or gravel. The area is then planted with reeds, such as *Phragmites australis* or sometimes other wetland plants such as *Typha latifolia* and *Juncus effusus*. Wetland plants improve conditions for aerobic micro-organisms in the root zone as they carry oxygen from the atmosphere via their leaves to the roots. These micro-organisms then degrade nitrogenous compounds and other pollutants such as heavy metals. The plants have extensive root systems, which create channels for the waste water to flow through. They also take up excess nutrients in the natural growth process.

These wetland ecosystems provide a high degree of biological degradation, which can be artificially created to treat filtered sewage or as buffer strips to filter run-off.

Conserving water

There are many different methods for conserving the valuable resource of water:

- ✿ selecting suitable plants for the site, e.g. drought-resistant plants, or xerophytes – plants with particular adaptations for living in dry conditions, see also Chapter 5

- ✿ mulching
- ✿ shading, e.g. shade tunnels, shading in glasshouses
- ✿ improving soil structure by the addition of organic matter
- ✿ not cultivating too deeply in dry periods
- ✿ not mowing lawns too often or too low in dry periods
- ✿ not watering in hotter parts of the day when evaporation losses will be higher
- ✿ irrigating at appropriate stages
- ✿ methods of irrigation – frequent applications of small amounts of water can make the problem worse as plants may not develop deep root systems
- ✿ planting at appropriate times, e.g. bare root trees in the dormant season
- ✿ provision of windbreaks – wind has a desiccating effect on soil and the plant.

Technical knowledge of a plant's needs and when irrigation is more critical is a key point for irrigation. This can particularly apply to commercial growers, and also relates to economics. **Response periods** for plants are determined when irrigation is critical and will affect eventual crops. For example, irrigation is critical for the potato, *Solanum tuberosum*, from the time the tubers reach marble stage, and for brussels sprouts, *Brassica oleracea* Gemmifera Group, when the buttons are about 15 mm in diameter. Some crops are crucial at all stages, such as carrot, *Daucus carota*, and lettuce, *Lactuca sativa*.

For all plants, irrigation is more critical at a seedling stage or when newly planted. Container growing has increased the flexibility of planting times, but planting in hotter months generally necessitates far greater irrigation.

On a smaller scale methods can be used in greenhouses such as sinking a flower pot next to a tomato, *Lycopersicon esculentum*, grown in a border, to water directly at the roots.

Rainwater harvesting

Water butts are an excellent means of harvesting water on a domestic scale. These can be fitted to guttering around glasshouses, sheds, summerhouses and on the house itself. The cleanliness of the water butt is important and it should be sealed to prevent leaves and organic debris falling into the water. This is particularly important if using the water for seedlings as the organic matter can harbour the fungi that cause damping-off diseases. A quick turnaround of the water is also important to prevent the build-up of pathogens in the water, and annual thorough cleaning of the water butt. Filters can be fitted to prevent debris entering from gutters.

On a larger scale, suitable for nurseries for example, rainwater can be harvested in storage tanks. These may be above or below ground.

Sustainable urban drainage systems (SUDS)

In the past, methods of drainage have often just concentrated on moving unwanted water away from the site. A more sustainable approach is to deal with water on site. Much of the problems of flooding in recent times has been exacerbated or caused by a non-sustainable approach, for example, building on flood plains, not maintaining ditches and diverting rainwater far away from site. Widespread use on non-permeable hard landscaping materials on horizontal surfaces has also contributed to the problem. Linked with this are problems of pollution due to leaching of nutrients, such as **eutrophication**.

The traditional soakaway drains and French drains are more in line with a sustainable approach. Piped drainage systems, on the other hand, sometimes just act in diverting water. Drainage methods are discussed in detail in Chapter 6.

Sustainable urban drainage systems aim to minimise water run-off, reduce pollution and, in turn, increase amenity value and biodiversity.

A particular modern-day problem has been widespread flooding due to the paving of many front gardens in urban areas, often to provide parking spaces. Legislation introduced in 2008 requires all new drives over 5 m² to have planning permission if traditional non-permeable surfaces such as concrete are used. Planning permission is not required if the water is directed to drain naturally onto a border or lawn, or permeable paving is used.

The variety of permeable options available is increasing, with some examples provided in Table 18.1. These generally rely on

water filtering through a permeable sub-base, but there are also systems where there is no further infiltration and the water is stored as part of a rainwater harvesting system. The sub-base is a crucial part. There are the normal requirements from the sub-base for stability and load bearing, but the role of the sub-base for permeable paving may also be to act as a store and slow down the rate at which water is released back.

Rain gardens and swales

Both rain gardens and swales are features designed to reduce surface run-off and create attractive features and habitats. Swales are linear grass-covered channels, relatively shallow and wide compared with traditional ditches. They are dry in dry weather and fill with rain in wet weather. They are positioned to take in excess water, for example running along a pathway at the bottom of a slope. Some of the water will naturally infiltrate, but in times of heavier rain the water will be led off to a different area such as a rain garden. Swales may be left open or, in the case of bioswales, planted with vegetation for aesthetic benefit and habitat creation. The plants themselves will take up some of the excess water.

Rain gardens are relatively shallow, saucer-shaped depressions. They should be created with moisture-retentive yet free-draining soil and planted with a variety of plants able to withstand occasional flooding.

Material	Method of permeability	Description	Uses
Gravel	Water filters directly through surface	Locally quarried may be available. A range of types and colours. A relatively cheap option.	Driveways, paths. Not easy for wheelchairs.
Concrete block permeable paving (CBPP)	Water filters through gaps between the paving	Block paving available in a range of colours and types. Methods of construction are different. Joints are larger than standard block paving and filled with an angular aggregate rather than sand. The sub-base has a large proportion of voids enabling it to store the surface water that permeates through. Can be designed for infiltration or attenuation.	Driveways, paths, patios.
Permeable resin-bound aggregate	Water filters directly through surface	Wide range of colours available and design opportunities. Different materials such as natural aggregates and recycled glass. As the material is not loose it does not scatter around as with gravel.	Driveways, paths, tree pits (around trees in urban areas), patios.
Porous asphalt system	Water filters directly through surface	Can be designed for infiltration or attenuation. Traditionally black, but a range of other colours now available.	Utilitarian pathways and driveways, car parks.
Matrix or cellular paving	Water filters directly through surface	Moulded plastic, generally in a hexagonal pattern, filled with a permeable material such as gravel, or planted with grass. Pre-cast concrete cells or concrete cells cast *in situ*. Useful for erosion control.	Hard standing, car parks, paths.

Table 18.1 A selection of options for permeable paving.

Energy

Sustainable use of energy plays a key role in limiting the carbon footprint. Plants are the ultimate resource for this, capturing sunlight energy through the process of photosynthesis.

Glasshouses also make good use of this natural energy source, capturing solar energy. The siting of a glasshouse is important to capture maximum light. The detrimental result of the greenhouse effect is described on a global scale, but is generally of great value in a greenhouse. Light energy enters the glass as shortwave radiation, warming everything inside the structure. This then re-radiates as longwave heat. The glass does not allow the long waves out and heat builds up inside the structure, only escaping through ventilation or conduction. If you want a quick way to remember the ray lengths, then think: short waves from the sun.

Solar energy can provide an alternative to fossil fuels through the installation of solar panels. Photovoltaic (PV) or solar cells are used to capture energy. These cells are connected to form modules, and the modules make an array. They are mounted on structures orientated towards the sun. Other alternatives include wind energy, through the installation of wind turbines, and biomass energy. The latter uses wood or agricultural residues for heat and power generation.

Methods of conserving energy are equally important. These include the use of insulation, for example, pipe insulation and thermal screens, as described in Chapter 11.

An understanding of scientific principles is vital to make the best use of valuable resources. For example, when applying fertilisers, an understanding of amounts required by the plant and optimum timing will avoid waste. An understanding of the law of limiting factors is important for inputs of energy through heating and lighting, and carbon dioxide enrichment.

The carbon dioxide produced from combined heat and power units can be recycled and used for carbon dioxide enrichment in the glasshouse. This method of heating is further discussed in Chapter 11.

DESIGN

An early record of ecological consideration for planning dates back to Roman times. The work of the architect Vitruvius Pollio, 'De Architectura' circa 27 BC, includes some ideas for landscape architecture that are still relevant today. The importance of using local materials for economy is highlighted, and considerations of climate for selecting sites.

The ideas of Vitruvius Pollio are referred to by Ian Thompson who identifies three main sources of value in designed landscapes: ecology, community and delight (Thompson, 2000). Blake (1999) describes the approach of designing planting schemes to mimic natural patterns. This is combined with the need to choose plants according to their preferred microclimatic and environmental needs.

William Robinson is famed for turning his back on non-sustainable formal Victorian gardening styles, such as intensive bedding. He promoted a more naturalistic style and the use of hardier plants.

Numerous modern-day garden designers have promoted a naturalistic style and have taken an ecological approach. Beth Chatto started to develop a garden in Essex in 1960 along ecological principles. Plants were carefully selected for seemingly inhospitable conditions, including poor gravel soil and damp shady conditions. The plant selection was aided and inspired by the extensive research of her husband, Andrew Chatto. This knowledge of plant ecology was invaluable in the successful planting of the Beth Chatto Gardens, matching plant habitats and creating plant communities that fitted the site. Along with a strong eye for design from Beth Chatto, this combination has resulted in the creation of stunning gardens, including a dry gravel garden, illustrated in the case study in Chapter 14.

An understanding of plant communities is crucial for ecological planting design. A plant community is a group of plants that live together and interact with each other in a particular environment.

The Dutch designer and nurseryman, Piet Oudulf, gained inspiration from both Beth Chatto and German nurserymen such as Karl Foerster and Ernst Pagels. Oudulf developed a style of new-wave perennial planting in a naturalistic style in the 1990s, which has in turn influenced many new designers.

CASE STUDY　Wildfowl and Wetlands Trust London Wetland Centre

The London Wetland Centre in Barnes was created on the principles of conservation and sustainability. It is managed by the Wildfowl and Wetlands Trust (WWT). The reserve was created on the site of Victorian reservoirs at Barn Elms that became redundant in their original purpose. The aim of the WWT was to turn this site into a series of waterbodies of different sizes to create a range of aquatic habitats, and surround these with further habitats including marsh, reedbeds and meadows. Everything on the site had to be recycled as part of the planning conditions. This major feat of civil engineering began in 1995.

A decision was made to plant much of the site rather than let it all regenerate naturally, as many of the plants required were no longer in the area and recolonisation is more difficult in such an urban area. Planting was carried out after researching plant communities. An example of this is the Cricklade type meadow with great burnet, *Sanguisorbia officinalis*, and snakeshead fritillary, *Fritillaria meleagris*.

Newer additions are three sustainable gardens including the Royal Bank of Canada Rain Garden opened in 2010 and designed by Nigel Dunnett. These gardens have a wide range of different habitats to encourage and benefit wildlife, including rain gardens, a stream, a dry stream and green-roof planting.

A true haven for wildlife and its observers has been created only four miles from Westminster.

Sustainability can be a guiding principle in design whatever the scale of garden or landscape. There are features that can be incorporated on a small scale for a private back garden or adapted for a large landscape, such as composting, creating habitats and rainwater harvesting. Chapter 13 describes features to include for creating a wildlife garden.

At the heart of sustainable design is the basic premise of working with the site. This includes working with existing contours.

Should it be deemed necessary to change the contours, e.g. to provide a level area for a lawn or patio, care needs to be taken to minimise damage to the topsoil. The best option for changing contours is cut and fill levelling, in order to keep the subsoil and topsoil in their correct relative positions. The aim should be to balance the amount of soil cut out with that used to fill in, thus avoiding the need to import topsoil, or remove excess topsoil from site. Once topsoil is cut away from areas that need to be lowered, the subsoil is then graded, avoiding excessive work on the topsoil.

However, cut and fill levelling is not only an expensive landscape procedure, but it can cause damage to topsoil and may even be illegal under the Land Drainage Act 1991 if it diverts the flow of water.

Green roofs

The idea of using roof space as growing space is a good example of one that combines modern technology with ancient practice, for example in Ancient Rome. Roof gardens may involve a few containers of plants to brighten a small area, or may extend to the scale of the roof garden created in the 1930s at Derry and Toms in Kensington, London. One-and-a-half acres of themed gardens were created on the roof of the department store. These types of roof greening are referred to as intensive systems, with relatively deep substrates to support a wide variety of plants. The history of planting roof spaces has mainly been for large buildings or wealthier clients in the UK. Other northern and central European countries, such as Norway as illustrated in Figure 18.5, have a long tradition of using green roofs for everyday buildings. These types of green roofs are referred to as extensive systems with lighter weight and shallower substrates. Lower growing plants are used, which are hardy and drought tolerant. They are not intended for use as conventional gardens.

Today, green roofs are increasing in popularity in a range of situations, for example on top of office buildings. On a smaller scale they can be used in domestic gardens on features such as garden sheds and garage roofs.

The benefits of green roofs are immense and varied. In urban areas, the vast, untapped horizontal space of roof tops can provide an aesthetically pleasing green environment. Biodiversity can be increased by the provision of extra habitats. There are benefits in terms of ameliorating the climate and managing water run-off. The amount of energy used to heat the building can be

Figure 18.5 Norway has a long tradition of green roofs.

Figure 18.6 A modern green roof.

reduced by the insulation provided, and conversely, the planting can have a cooling effect in hot temperatures, thus reducing the carbon footprint of air conditioning. On a large scale, recreational space can be provided in urban areas, and also food production.

Much research has been carried out by NASA on the therapeutic effects of indoor plants on office workers. This also applies to offices or high-rise buildings being surrounded by green spaces on roofs.

Figure 18.7 The Permaculture Association is a national charity and was established in 1983.

The Green Roof Centre, the University of Sheffield

The University of Sheffield is the leading research establishment in the UK for green roofs.

The RHS Chelsea Flower Show of 2011 saw an exhibit in the main pavilion by the Green Roof Centre, showcasing the research work and possibilities for green roofs. The research work centres on plant trials and selection, sustainable water management and the Green Roof Code. The Green Roof Code promotes best practice in the design, specification, installation and maintenance of green roofs.

Permaculture

The term 'permaculture' was originally derived from 'permanent agriculture' later becoming related more generally to 'permanent culture'. The term was coined by two Australians, Bill Mollison and David Holmgreen in the 1970s. However, the ideas and systems are recognised as not being new, but based on many sustainable, traditional growing systems in the tropics.

Permaculture is a design system that models growing and living systems on ecosystems. The system relies on diversity rather than monoculture. It is a holistic and ethical approach and looks for connections and communities. The Earth is regarded as a single living organism. It also uses a low-energy approach. The forest garden is a typical permaculture design, but not the only approach. As it looks to mimic natural plant and animal communities, it can also include ecosystems such as meadows.

Forest gardens – A forest garden mimics the natural ecosystem by having layers of planting. If carefully designed, this can be far more productive than one single layer of planting in the same area. Care must be taken not to use plants that create too dense a shade for the lower layers. A typical edible forest garden could have the upper layer as a canopy of fruit and nut trees; a shrub layer of fruit bushes; and a ground layer of perennial herbs and vegetables. Climbers can be incorporated to further increase productivity and exploitation of an ecological niche.

The no-dig system is a crucial part of permaculture, with the system relying on ground-cover planting and mulching. A typical design would use a system of keyhole-shaped beds.

SELECTION OF HARD LANDSCAPING MATERIALS

A range of factors need to be considered when selecting hard landscape materials for a site, including function, aesthetic appearance, durability and cost. A further factor to be considered is the environmental impact of the product. This may involve quarrying from a site that is environmentally sensitive, the carbon footprint in transportation and any manufacturing process, and human exploitation involved.

The use of local materials is both in keeping with the wider site and generally has less environmental impact. Transportation of local products is less costly, both for economics and carbon footprint. It may be a requirement in conservation areas to use local materials.

To reduce economic costs, imported stone is often used, such as Indian sandstone. This has been found to involve widespread exploitation of children, with child labour used for quarrying. Children as young as six have been involved in working in quarries. Ethical sourcing of landscape materials therefore needs to be a prime consideration for selection of materials.

The Ethical Trading Initiative (ETI) was set up to prevent the exploitation of workers and involves companies, trade unions and voluntary organisations. The ETI Base Code outlaws the use of child labour and sets out conditions for a safe working environment. Bonded employment is also outlawed, a practice that involves the worker borrowing money in return for promised labour, with accrued debts commonly being passed on to children. Discrimination and abuse are not allowed, and living wages and acceptable working hours set out.

The permeability of paving, as discussed under SUDS, is also another important consideration.

Timber products also vary in their impact on the environment. The Forest Stewardship Council (FSC) is an international, non-governmental organisation that promotes sustainable forestry on a worldwide scale. The FSC logo is used on timber products that have been sustainably produced and have not contributed to deforestation.

SELECTION OF SOFT LANDSCAPING MATERIALS

Also see Chapter 13 for a discussion on site assessment.

The phrase 'right plant, right place' has almost become a gardening mantra and with great justification. A site assessment will provide much information to aid decisions over plant selection. When the right plant is chosen, to suit the soil, climate and aspect, it is far more likely to thrive with the minimum of inputs. Plants growing in the wild do so without artificial fertilisers and pesticides.

Many leading garden authorities, such as Beth Chatto, have promoted this commonsense approach. An early experiment of finding the right plants for a seemingly hostile environment was carried out by Sir Frederick Stern in the early twentieth century.

Plants have evolved to adapt to a multitude of situations, from windy cliff sides with salt-laden air, to tropical rainforests; from acid heathland to chalk downland; from desert to pond.

A knowledge and understanding of the suitability of plants to the range of situations that are presented by gardens and landscapes is invaluable for sustainable design.

The provenance of plants and seeds is also important, particularly when planting or sowing native species.

CASE STUDY Highdown Gardens, Worthing, Sussex

The gardens at Highdown were an early experiment in growing the right plant in the right place. They are situated on a south-facing slope of Highdown Hill on the chalk of the South Downs. Sir Frederick Stern started to think about creating the gardens in 1909.

At the time, Stern found it difficult to find advice on the type of plants that would grow there, and was generally advised not to create a garden in such conditions. However, he decided to experiment as to which plants would grow. The garden was created at the time of exciting plant-hunting expeditions, providing much new material for experimentation. Stern subscribed to several of the plant-hunting expeditions.

Stern studied geological maps of temperate areas to try and discover which plants would tolerate the alkaline conditions. In general, he found that plants introduced from China and some parts of Europe fared better. Those from Japan and North America fared less well. This, he discovered, correlated with the geological maps of the areas, with Japan and North America having few areas of lime.

Through a mixture of this more scientific approach, and trial and error with large numbers of plants, Stern succeeded in creating a thriving garden. It is still full of many exciting plant introductions from the expeditions of hunters such as Ernest Wilson, Reginald Farrer, George Forrest and Frank Kingdon-Ward.

Plants found to thrive there include: *Acer griseum*, *Hydrangea villosa*, *Kolkwitzia amabilis*, *Lilium regale*, *Prunus serrula*, *Rosa moyesii* and *Viburnum rhytidophyllum*.

After Stern died in 1967, the garden was looked after by his wife, Lady Stern. She then gave the garden to Worthing Corporation, now Worthing Borough Council, and it is now open to the public. Please check with the garden before you visit to confirm the opening dates and times.

MANAGEMENT TECHNIQUES

Following sound horticultural practice forms a mainstay for sustainable gardening. This should be linked with scientific understanding and knowledge. Aspects to consider include an understanding of suitable soil management techniques carried out at the right time for both the soil type and conditions, and the plant concerned. For example, the rough digging of a heavy clay soil in the autumn to allow the winter frosts to carry out the work for you in breaking up the clods, compared to lighter spring cultivation of a sandy soil to avoid erosion and compaction over winter. Please see the discussion of this topic in Chapter 6.

Sowing or propagating at the correct time, and correct methods and timing for pruning, are essential. An understanding of the essential times for irrigation as discussed in the section 'Conserving water' is also important. Weed control is more important at specific times, for example seedling carrot, *Daucus carota*, and onion, *Allium cepa*, are not likely to recover if swamped by vigorous weeds at this early stage.

Knowledge of the life cycle of pests and diseases enables far more efficient prevention and control, please also see Chapter 12.

Organic growing systems

Organic growing systems are based on sustainable and holistic principles. They centre on looking after the soil rather than relying on inputs of fertilisers and pesticides, and aim to work with nature rather than against. A long-term view is taken of soil management, taking regard of future generations and building up natural soil fertility. As a wider view, recycling is promoted to reduce pollution, and the provision of habitats to benefit wildlife is encouraged. Alongside these aims lies the recognition of the value of scientific discoveries in achieving a more sustainable method of growing.

Two major organic organisations in the UK are the Soil Association and Garden Organic.

CASE STUDY Toxic manure

Reports came in from 2008 from many gardeners and allotment holders that their crops of potatoes, beans, peas, carrots and salad vegetables were severely withered and deformed. This was traced back to the use of manure contaminated by the hormone-based herbicide, aminopyralid. Aminopyralid is not licensed for food crops. The chemical entered the food chain as the treated grass was used for silage, fed to horses or cattle and then ended up in the manure. As it stays tightly bound to the plant material, it persists until the manure is fully decomposed.

Approval for the herbicide was suspended while the Chemicals Regulation Directorate (CRD) investigated the matter. Approval was subsequently granted again with extra restrictions. The importance of following label precautions was highlighted and training is now required for potential purchasers.

This case study highlights several factors: the importance of checking the provenance of organic mulches; the importance of following label recommendations for any pesticide; and the ease at which pesticides can unwittingly enter the food chain.

The bed system

The bed system, as described in Chapter 12, preserves soil structure. As long as a suitable width is selected, all of the bed can be reached without ever treading on the soil. It is worth remembering that a suitable length will also help. If the beds are too long there is always the temptation to take a short cut! Initial cultivation is generally necessary, but after that minimal cultivation or no-dig systems can be used.

The bed system also facilitates crop rotation, as one bed can be given over to one group of crops.

Conservation of topsoil

Topsoil is a valuable resource and every effort made to conserve it. Consideration needs to take place at the planning and design stage whether it is necessary to move topsoil in the first place.

During landscape works, valuable topsoil is often damaged or, at worst, removed from the site altogether. To avoid this degradation of topsoil, various procedures need to be followed. Any compaction of the topsoil should be avoided, which is more likely when heavy machinery is involved, such as diggers and scrapers. To avoid

Green manure crop	Properties	Soil type	Time of sowing
Lupinus angustifolius (bitter lupin)	Nitrogen fixing. Deep rooting, which improves soil structure and brings up nutrients washed through the soil.	Light acid soil	Early spring to early summer
Medicago lupulina (trefoil)	Nitrogen fixer. Low growing and tolerant of some shade, so it is suitable for undersowing.	Light soils, avoid acid conditions	Early spring to late summer
Medicago sativa (alfalfa)	Deep rooting, resulting in foliage that is rich in nutrients, including micro-nutrients. Will overwinter.	Not in acid or badly drained soils	Spring to mid summer
Phacelia tanacetifolia (phacelia)	Will overwinter when sown later. Foliage is a good weed suppressant. Extensive root system improves soil structure. Attractive to bees and hoverflies.	Most soil types	Early spring to late summer
Secale cereale (grazing rye)	Good weed suppression over winter. Prevents leaching of nutrients. Extensive root system benefits soil structure.	All soil types, especially useful for heavy soils	Late summer to late autumn
Sinapsis alba (mustard)	Quick growing. May help prevent wireworm, but is susceptible to clubroot.	Avoid poor soil	Spring to late summer
Trifolium incarnatum (crimson clover)	Nitrogen fixing. Weed suppressant.	Light soils	Early spring to late summer
Vicia faba (winter field beans)	Nitrogen fixing.	Useful for heavy soils but not too wet or dry	Late summer to autumn
Vicia sativa (winter tares)	Nitrogen fixing.	Avoid dry or acid soils	Early spring to late summer

Table 18.2 Green manure crops

unnecessary compaction, notice needs to be taken of weather conditions, particularly not working on the soil when too wet.

Should topsoil need to be moved for levelling and contouring of ground, it needs to be stored correctly. On no account should the subsoil and topsoil be mixed. Topsoil should be stored in stacks or 'bunds' with a maximum height of 1.25 m (4 ft). Storage should be as short as possible, ideally no longer than three to four months, to prevent the breakdown of the soil micro-organisms. Weeds need to be controlled during storage. When replacing the topsoil on site, this should be done in layers of 50 to 75 mm (20 to 30 inches), lightly consolidating each layer before adding the next. As with other stages, it is vital to avoid overcompaction.

Prevention of pests and diseases

The widespread use of pesticides has caused much harm to the environment in the past. Pesticide use increased greatly in the post-war period in Britain. Whereas it is important to recognise the role of pesticides in providing a much needed food supply, this widespread use did have many damaging effects on the environment. There is now strict regulation on the use of pesticides, including withdrawals of many pesticides. (Please see Chapter 17.)

Sustainable methods of preventing and controlling pests and diseases include:

- ✿ plant selection – an ongoing theme of choosing the right plant for the conditions and situation
- ✿ correct cultural techniques
- ✿ barrier methods, such as horticultural fleece or Enviromesh to prevent pests such as carrot flies and cabbage butterflies from laying their eggs
- ✿ hardening-off plants thoroughly before planting out
- ✿ avoiding excess fertiliser – too much nitrogen can cause soft, sappy growth, which is far more susceptible to pests and diseases, e.g. more palatable for aphids
- ✿ encouraging natural predators
- ✿ selecting resistant types and cultivars of plants.

If pesticides are required, then there are options that are less damaging to the environment. (Please see Chapter 17 for a detailed discussion.)

Integrated pest management (IPM) systems opt for the least harmful method first, and look for ways of combining pest-control methods for both efficiency in dealing with the problem and causing least harm to the environment.

Encouraging beneficial predators

The encouragement of beneficial insects and other natural predators goes hand in hand with the creation of habitats and preservation of the balance of nature. Monocultures tend to encourage the build-up of pests, and thus often lead to the need for large-scale use of pesticides. Beneficial predators to encourage

include ladybirds, *Coccinella* spp., lacewings, *Chrysoperla* spp., beetles, *Carabus violaceus*, thrushes, *Turdus* spp., frogs, *Rana* spp., hedgehogs, *Erinaceus europaeus*, and slow worms, *Anguis fragilis*.

Coppicing

Coppicing is an ancient practice that maintains diversity of habitats. Long practised in woodlands for production of wood for items such as chestnut paling fencing, hurdles, firewood, thatching spars and walking sticks, trees are cut back to near ground level in a cyclical manner, every 12 to 15 years. Trees such as hazel, *Corylus avellana*, sweet chestnut, *Castanea sativa*, beech, *Fagus sylvatica*, common oak, *Quercus robur*, and ash, *Fraxinus excelsior*, are grown in this way as stooled plants, with many, relatively straight stems growing from ground level, thus providing suitable material. The cyclical nature results in a variety of habitats being created, from open newly planted or newly coppiced areas through to maturing woodland, and more dense shade. Although not practised as widely now, the method is being introduced again. It can also work well in relatively small gardens. Plants such as hazel, *Corylus avellana*, can be coppiced to provide bean poles and pea sticks, thus keeping production entirely within the garden site. From an aesthetic and practical point of view, many trees can be grown in a smaller garden that would otherwise become too large, such as *Paulownia tomentosa*.

Much shrubby material that has been cut back as part of regular garden maintenance can be used to provide pea sticks for plant support, either for vegetables, or as attractive support in an herbaceous border, providing a sustainable approach.

Laying hedges

Figure 18.8 A laid hedge of *Crataegus monogyna*.

A further traditional practice, which enhances wildlife value, is laying hedges. The aim of hedge laying is to provide a dense, stock-proof hedge. Hedges are thinned out, generally in late autumn or winter, and selected branches are partly cut through close to their base using a billhook. These branches are then laid diagonally or flat, starting as close to the ground as possible. Ash or hazel poles are then used to stake the hedge. The process encourages dense bushy growth as the apical dominance maintained by auxins is disturbed. (Please see Chapter 8.)

Although primarily a practical agricultural feature, laid hedges have great conservation and aesthetic value. The dense corridor of growth provides valuable shelter for wildlife. The skill of craftsmen in producing a well-laid hedge provides an attractive boundary feature for a garden in a country style.

Timing of hedge trimming

Under the Wildlife and Countryside Act 1981, it is an offence to knowingly trim hedges while birds are nesting. The main bird nesting and breeding season is from the beginning of March to the end of July, during which time hedge trimming should be avoided. Obviously, these times vary from season to season, and different hedges and situations are more likely to have birds nesting.

Meadows

Meadow gardening may be exclusively for native plants, or may be extended to include suitable exotics. It is becoming increasingly popular in urban parks as an alternative to high-maintenance lawn areas, with associated high energy inputs and carbon footprint. It also provides a more colourful and exciting alternative.

Many large gardens are leaving larger areas of grass uncut and encouraging or planting wild flowers in the grass. The University of Sheffield has carried out extensive research into suitable methods and plant mixes for meadows, developing the pictorial meadows seed mixes. This was showcased with the spectacular annual meadow planting for the London 2012 Olympics.

Christopher Lloyd was a great exponent of meadow planting, creating stunning permanent meadows at Great Dixter, Sussex.

CONSERVATION BODIES AND LEGISLATION

To achieve sustainability in horticulture, the wider picture needs to be considered. Gardens and parks should be viewed as part of a wider landscape. Some conservation bodies exist for gardens and landscapes, or garden plants in particular. Others exist for the conservation of wild habitats and native plants. They all have a role to play in creating a more sustainable environment. Many often now work together on projects, such as the Wildlife Trusts and the Royal Horticultural Society (RHS) on their 'Wild About Gardens' website and 'The Big Wildlife Garden', encouraging gardeners and community gardens to turn their gardens into wildlife habitats. This project was founded by Natural England.

The important role of botanic gardens and Botanic Gardens Conservation International (BGCI) is discussed in Chapters 1 and 15.

Plantlife

The charity Plantlife was founded in 1989 with the aim of protecting and conserving native plants. Campaigns that Plantlife has run include invasive non-native plants; protecting peat habitats; preventing species decline with a 'Back from the Brink' campaign; saving meadows; and connecting people and plants through a 'Flora locale' and a County Flowers campaign.

Wildlife Trusts

The origin of the Wildlife Trusts dates back to 1912 when the Society for the Preservation of Nature Reserves was founded by the banker and naturalist Charles Rothschild. The society acquired nature reserves and identified sites needing protection. Many local conservation branches later developed. Rothschild's society eventually became the Royal Society of Wildlife Trusts, and is now the national representative of all the local Wildlife Trusts.

There are now 47 Wildlife Trusts organised on a county-wide basis. These local groups take on the day-to-day management and habitat restoration of conservation sites, such as nature reserves. They promote the management of gardens for wildlife and highlight the importance of the garden habitat in forming a patchwork linking green spaces in urban areas to larger parks and the wider countryside.

Plant Heritage

Plant Heritage is a UK organisation formed for the conservation of garden plants and gardens. (Please see Chapters 1 and 15.)

The Royal Horticultural Society (RHS)

The RHS was founded in 1804 by Sir Joseph Banks and John Wedgewood, as the Horticultural Society of London, with the aim of collecting and exchanging information on plants and encouraging the improvement of horticultural practices. This role still continues today for the RHS, and it has also taken on wider issues such as advising and educating the public on gardening in a changing climate. The RHS combines scientific research, education and the all-important practical skills for horticulture. These are showcased in the stunning RHS gardens at Wisley, Hyde Hall, Rosemoor and Harlow Carr.

The RHS Biological Records Centre keeps records of all naturalised and native flora and fauna on its properties. These date back to 1909 thus providing valuable historical data.

National Trust

The National Trust was founded in 1895 by Octavia Hill, Robert Hunter and Hardwicke Rawnsley. It is now a major conservation charity, concerned with landscapes, gardens, historic houses and threatened coastlines. There are over 200 National Trust gardens and landscape parks open to the public in England, Wales and Northern Ireland. Scotland has its own organisation: the National Trust for Scotland.

English Heritage

English Heritage is the government's advisor on the historic environment. Part of this involves the conservation of parks and gardens. English Heritage manages several historic parks and gardens such as Audley End, Chiswick House and Wrest Park, including their restoration. On a wider scale, English Heritage set up the Register of Historic Parks and Gardens to protect a wide range of designed landscapes. Any planning decisions concerning these sites have to take into account the fact that they are listed on the register, and it is termed a 'material consideration'.

The Botanical Society of the British Isles (BSBI)

The BSBI is a non-governmental organisation that provides the major source of information on the status and distribution of British and Irish flowering plants and ferns. Information provided by the society forms the basis of plant atlases and publications on rare species, and is necessary for targeting species that are priorities for conservation and sustainable use. The society's database includes information on the taxonomy, history, exploration, collection, biography, literature, distribution, breeding behaviour and autoecology (ecology of individual species) of native and alien plants found in the wild in the British Isles. The society also organises plant distribution surveys and field meetings.

Specialist plant societies

There are many specialist plant societies performing valuable conservation tasks. Table 18.3 lists a selection of specialist societies. This list is by no means exhaustive and whatever your particular interest, there is likely to be a society to represent it.

The Tree Register

The Tree Register is a charity concerned with Britain's notable and ancient trees. Records of champion trees are kept, a champion tree being the best specimen of its type in Britain. This generally refers to the largest or tallest specimen with the thickest trunk, but other criteria may be used such as the most spreading, oldest, heaviest or most beautiful (Johnson, 2011). It has produced a comprehensive database of champion trees, including valuable historical records. Volunteers record, measure and update the records, providing information on size and growth rates.

European Landscape Convention

The European Landscape Convention was formed by the Council of Europe and came into force in the UK in 2007. It is concerned with the overall protection of landscapes, recognising their importance for the general quality of life, and is based on principles of sustainability.

Society	Web address
British Cactus and Succulent Society	www.bcss.org.uk/
British Clematis Society	www.britishclematis.org.uk/
British National Carnation Society	www.britishnationalcarnation society.co.uk/
British Fuchsia Society	www.thebfs.org.uk/
Cyclamen Society	www.cyclamen.org/
Hardy Plant Society	www.hardy-plant.org.uk/
National Dahlia Society	www.dahlia-nds.co.uk/
National Vegetable Society	www.nvsuk.org.uk/
Royal National Rose Society	www.rnrs.org/
The Alpine Garden Society	www.alpinegardensociety.net/
The Cottage Garden Society	www.thecottagegardensociety.org.uk/
The Daffodil Society	www.thedaffodilsociety.com/
The Heather Society	www.heathersociety.org/
The National Auricula and Primula Society	www.auriculaandprimula.org.uk/
The Orchid Society of Great Britain	www.osgb.org.uk/
The Wildflower Society	www.thewildflowersociety.com/
Walled Kitchen Gardens Network	www.walledgardens.net/

Table 18.3 Specialist societies.

Natural England

Natural England is the government-funded body whose purpose is to promote the conservation of England's wildlife and natural features. The governing council is appointed by the Secretary of State for the Environment. The Acts of Parliament that give it power are:

National Parks and Access to the Countryside Act 1949
Countryside Act 1968
Nature Conservancy Council 1977
Wildlife and Countryside Act 1981 (amended 1985)
Environmental Protection Act 1990
Countryside and Rights of Way Act (CRoW) 2000

Figure 18.9 Level 3 arboriculture students from Sparsholt College, Hampshire, climbing and measuring the tallest of several very large *Platanus × hispanica* at Bryanston School, Dorset. At 49.85 metres it is the tallest broadleaved tree in the UK, and probably Northern Europe.

CASE STUDY | A landscape-changing event?

Introduction

Ash dieback is also known as chalara dieback, *Chalara fraxinea*, owing to there already being a disease in the UK with the common name of ash dieback but unrelated to the strain causing problems in 2012.

It is believed the disease came into Europe from Asia where it appears that ash trees have some immunity to it. It was first discovered in Poland in 1992 and spread westwards and northwards over the following 20 years. In Poland and Denmark over 80% of ash trees have been infected and killed by the disease causing devastation to the ash population. It has killed large numbers of ash trees throughout Europe. The disease was first discovered in Denmark in 2002 and had spread across the entire country by 2005 affecting up to 95% of the ash trees.

It was first identified in Britain in February 2012 in a nursery on young trees imported from the Netherlands. There had been warnings in the nursery industry during the previous three years that the disease may spread to this country on imports.

By the end of November 2012 it had been discovered on 237 sites throughout Great Britain and Northern Ireland.

It is possible that the disease has been here for a number of years but as a less aggressive strain, although research is currently being carried out to ascertain if this is correct.

Life cycle

Chalara fraxinea is an ascomycete fungus and is the asexual stage of the disease. It was first described in 2006; four years later it was discovered that it was the asexual stage of *Hymenoscyphus pseudoalbidus*, which is closely related to *H. albidus*, which is a distinct fungus.

Chalara is spread by ascospores mainly by wind and it is possible it may have come into Eastern England from Europe by this method.

As of November 2012 our current understanding of the disease is:

- ❀ the spores only survive a few days
- ❀ spore dispersal is mainly by wind and could have come from Europe by this method
- ❀ trees need a high dose of spores to be infected
- ❀ spores are produced on infested leaves from June to September
- ❀ it can attack any ash species, *Fraxinus* species
- ❀ disease symptoms show within a month of infection
- ❀ at present there is no cure
- ❀ not all trees die of infection some show genetic resistance
- ❀ if timber is treated there should be no spread by this method
- ❀ there is a possibility of spread by animals, including birds, and on clothing and boots.

(Forestry Commission website, 2012)

Symptoms

It is usually first noticed as wilting leaves – these turn a dark brown to black colour fairly quickly and if inspected closely the veins of the leaves have turned black before the leaf. The leaf rachis will also be black. All the leaves on infected stems will turn black and the stem will die back. Many of the leaves in the tree crown will turn black; this is more noticeable in young trees as they are easier to see than on mature large trees.

The shoots will then die back (hence it common name) and lesions (cankers) can be found at the base of young shoots on the main stem. If the stem is cut open the wood will be stained black where the fungus has blocked the xylem and caused the tree to wilt and die back.

Mature trees will show large amounts of dieback in the crown although it may take a few years for the tree to fully succumb to the disease. Where stems have died back the tree will produce large amounts of epicormic growth below the point of dieback.

Controls

At this point in time, December 2012, *Chalara* is being treated as a quarantine pest under the national emergency measures and any suspected cases should be reported to the Forestry Commission or FERA. This may change as our knowledge improves on the disease and other options can be considered.

Young ash trees are being removed, cut up, double bagged and buried deeply in landfill sites.

Imports of ash trees into the UK were banned from 29 October 2012.

At present mature trees are being monitored. It is not felt that a wholesale felling and burning of infected trees would either control the disease or prevent its spread as it is now widespread throughout Great Britain. Environmentally the felling of a large number of trees would deprive wildlife from much of its habitat and little would be achieved.

No chemical control is available at present (December 2012) but some chemical companies believe they have chemicals currently available but not approved for the control of *Chalara*. They are doing research to see if they will give good control and whether it will be possible to inject the chemical into the tree rather than spraying, which would be environmentally unacceptable.

Research in Denmark has shown that between 2% to 5% of ash trees have some genetic resistance to *Chalara* disease. Seed has been collected from these trees and work is being carried out at Copenhagen University to try and develop a resistant strain of ash tree.

Scientists in this country are also looking for resistant strains of ash tree to develop resistant trees.

LEGISLATION AND GOVERNMENT ACTION

Agenda 21

The first international Earth Summit was held in 1992 in Rio de Janeiro, Brazil. Over a hundred heads of state adopted Agenda 21, a comprehensive plan of action for sustainable development in the twenty-first century. This plan of action is to be taken globally, nationally and locally by organisations of the United Nations, governments and major groups in every area in which there is human impact on the environment.

To monitor and report on implementation of the Earth Summit agreements, the Commission on Sustainable Development (CSD) was created. Five-year reviews are held.

Convention on Biological Diversity

Signed at the Earth Summit, this commits governments to a planned approach to conserve and enhance biological diversity (biodiversity). Please see Chapter 1 for further details of the impact of this development.

UK Biodiversity Action Plan (UK BAP)

The UK BAP was the government's response to signing the Convention on Biological Diversity. The action plans are aimed at protecting threatened species and habitats within the UK and aiding their recovery.

Convention on International Trade in Endangered Species of Wild Flora and Fauna (CITES)

CITES is an international agreement between 175 countries providing protection against international trade and exploitation of wild plants and animals. Documentation for export and import is required for plants listed under CITES. There are three categories of protection as outlined in Table 18.4.

Appendix	Description	Examples
1	Plants threatened by extinction – trade is only allowed in exceptional circumstances.	*Araucaria araucana*. Many cacti, orchids, euphorbia.
2	Plants that require control in trade to avoid overexploitation.	*Cyclamen* spp. *Dicksonia* spp. *Galanthus* spp. *Lewisia serrata Sternbergia* spp. Many cacti, orchids, euphorbia.
3	Species already protected in a country asking for assistance in controlling trade.	*Diospyros* spp. *Meconopsis regia*

Table 18.4 CITES Appendices.

Town and Country Planning Act, 1990

Local authorities are empowered under planning legislation, to preserve and enhance pleasant features of both urban and country environments by making and enforcing policies for their conservation.

Town and Country Planning (General Permitted Development) Order 1995

This order sets out which alterations are allowed to houses and gardens without requiring planning permission. An amendment in 2008 placed restrictions on the amount of non-permeable paving for front gardens, as discussed under the section on sustainable urban drainage (SUDS).

Tree Preservation Orders

Tree Preservation Orders, commonly referred to as TPOs, play an important role in conserving trees. Trees are given protection that benefit the local environment and provide public enjoyment. Individual trees may be protected, groups of trees, areas of woodland or trees within defined areas.

It is against the law to cut down, uproot, top, lop, wilfully damage or destroy a tree with a TPO. Permission from the local planning authority needs to be granted for any work on the tree. When permission is given for felling of trees, provision is generally made within the consent for the replanting of new trees.

Information on local TPOs is readily available from local planning authorities, either on websites or by contacting the local office.

Conservation Areas

Local planning authorities determine if parts of their areas should be treated as conservation areas to preserve or enhance them. Extra restrictions then apply to these areas. Trees are given greater protection, and six weeks' notice needs to be given before work is carried out on any trees over 75 mm (3 inches) in diameter, not protected by TPOs. In this time, the local authority can prepare a provisional TPO if necessary. Tree replacement can be enforced with a tree replacement notice.

Wildlife and Countryside Act, 1981

Under this act, protection is provided for wild birds, their nests and eggs. All snakes are protected and endangered species such as red squirrels and water voles. It is illegal to pick, uproot or destroy scheduled wild plants, e.g. bluebell, *Hyacinthoides non scripta*, alpine catchfly, *Lychnis alpina*, rock cinquefoil, *Potentilla rupestris*, spring gentian, *Gentiana vernis*, and pennyroyal, *Mentha pulegium*.

The Hedgerows Regulations, 1997

The important wildlife habitat of hedges is given protection under the Hedgerows Regulations. Criteria are set out to determine whether the hedgerow is 'important', including the length, age, diversity and archaeological, historical, wildlife or landscape value. 'Important' hedgerows are not allowed to be removed, but management such as coppicing, laying or cutting out dead and damaged material is.

Natural Environment White Paper June 2011

The government White Paper of June 2011, as set out in 'The Natural Choice: securing the value of nature', recognises the importance of nature in all spheres of life, both on a local and global scale. The importance of an integrated approach for ecosystems and biodiversity is stressed and the threat of fragmentation of wildlife areas highlighted.

The role of gardeners is specifically mentioned, stating the importance of making well-informed choices and improving wildlife habitats. Gardeners are encouraged to adopt environmentally friendly practices, including using alternatives to peat. The role of pollinators is stressed in maintaining biodiversity.

Management of soils is a further important theme, concerning the value of safeguarding soils to prevent erosion, flooding and the emission of carbon into the atmosphere. The loss of organic matter from soils and compaction are highlighted as problems.

The role of sustainable urban drainage systems (SUDS) is a further important topic. Such systems are to be encouraged in a variety of environments, such as public wetlands, rain gardens and community ponds.

Summary

Conservation and sustainability are wide-ranging topics that affect all spheres of life and cannot be looked at in isolation. In a similar way, this chapter brings together all the chapters in the book. For conservation and sustainability to successfully underpin horticulture, all aspects need to be taken into account. Science plays a key role, for example in understanding how plants work and the role of the soil. Traditional horticultural practices play a fundamental role, for example in knowing the best times for cultivation and the benefits of crop rotation. Modern technology is crucial, for example in breeding plants resistant to pests and diseases, and providing sustainable methods of heating and lighting.

Practical skills are fundamental, for example planting and pruning techniques. As interesting as theoretical knowledge is, it needs to be applied to practical horticultural skills to be of any value. A good knowledge of plants, including their origins and growth patterns, provides a basis for suitable plant selection that is most likely to succeed.

Combined with scientific knowledge and practical skills, the artistic side of horticulture needs to be incorporated. The value of horticulture is both practical, for example in terms of food production, and aesthetic. An understanding of design and skill in combining plants and designing with hard landscape materials provides gardens and parks that give great pleasure to many people.

Many conservation bodies exist to protect habitats, native plants, cultivated plants, gardens, parks and landscapes. These range from small local organisations to large landowners and national or international bodies. Each plays a crucial role, from the local to the global. The realisation of the impressive sounding aims of larger organisations and government initiatives often depends on many individuals and communities implementing them on a practical level.

Although it does need to be recognised that much of horticulture concerns the manipulation of nature to grow the crops needed and provide pleasant environments for amenity, the general aim is to work in harmony with nature rather than battling against it. This can be achieved by selecting and using resources wisely, recycling as much as possible and careful plant selection. Many traditional horticultural methods fit well with sustainable practices, being concerned with fundamental principles such as maintenance of soil structure and plant health. It is important to look back to the past for much valuable information on traditional techniques and combine this with scientific knowledge and technological advances.

Sustainable design and sound horticultural techniques can create gardens and horticultural enterprises that balance inputs and outputs to not only limit damage to the environment, but positively enhance it.

Review questions

1. Describe the **role** of conservation bodies in the UK.
2. Suggest **five** methods of conserving water in a garden.
3. Describe **two** types of hard landscaping materials that can be used to provide a permeable surface.
4. Describe the **role** of certification bodies in the UK for organic growing.
5. Explain how sustainability can be **incorporated** into garden design.

REFERENCES

Blake, J. (1999). *Introduction to Landscape Design and Construction*. London: Gower Publishing Ltd.

Johnson, O. (2011). *Champion Trees of Britain and Ireland: The Tree Register Handbook*. Richmond: Kew Publishing.

Oldfield, S. (2007). *Great Botanic Gardens of the World*. London: New Holland Publishers (UK) Ltd.

Thompson, I. (2000). *Ecology, Community and Delight: Sources of Values in Landscape Architecture*. London: E & FN Spon.

FURTHER READING

Blake, J. (1999). *Introduction to Landscape Design and Construction*. London: Gower Publishing Ltd.
A down-to-earth introduction to the vast subject of landscape design and construction.

Dunnett, N. & Clayden, A. (2007). *Rain Gardens: Managing Water Sustainably in the Garden and Designed Landscape*. Cambridge: Timber Press.
Nigel Dunnett and Andy Clayden have provided a thorough and comprehensive guide to a wide range of methods for capturing rainwater. Innovative approaches make the most of available water in an ecologically sound and aesthetically pleasing manner.

Dunnett, N. & Kingsbury, N. (2004). *Planting Green Roofs and Living Walls*. Cambridge: Timber Press.
A detailed and inspiring guide to the greening of walls and roofs by Nigel Dunnett and Noel Kingsbury. The history of green roofs and living walls is described followed by comprehensive detail on their construction and planting.

Hall, J. & Tolhurst, I. (2009). *Growing Green: Organic Techniques for a Sustainable Future*. Cheshire: The Vegan Organic Network.
This book introduces stock-free organic systems based on extended crop rotation and building soil fertility.

Lavelle, M. (2011) *Sustainable Gardening*. Marlborough: The Crowood Press Ltd.
A practical approach to gardening sustainably with sound horticultural practice described.

Lloyd, C. (2004) *Meadows*. London: Cassell Illustrated.
An inspiring account of creating and managing a variety of meadow types by the great plantsman Christopher Lloyd. Excellent detail is given on a wide range of garden-worthy plants to include.

Shepherd, A. (2007). *The Organic Garden: Green Gardening for a Healthy Planet*. London: Collins.
Allan Shepherd covers a range of organic methods with many ecological alternatives and ethical choices.

Thompson, I. (2000) *Ecology, Community and Delight: Sources of Values in Landscape Architecture*. London: E & F N Spon.
This book examines the three principal value systems that influence landscape architectural practice: the aesthetic, the social and the environmental.

RHS level	Section heading	Page no.
2 2.5	Mulches and ground cover and the no-dig system	391
2 5.10	Green manures	391
2 5.11	Green manures	391
2 8.3	Selection of hard landscaping materials	388
2 8.4	Selection of hard landscaping materials	388
2 8.5	Selection of hard landscaping materials	388
3 4.1	Green roofs	387
3 5.1	Rainwater harvesting	384
	Green roofs	387

APPENDIX 1

Nutrient	Chemical symbol	Major sources	Major losses	Disorders developed
Carbon	C	Atmosphere	Burning of fossil fuels	Stunted and twisted growth
Oxygen	O	Atmosphere and green plants	Burning of fossil fuels and animal respiration	Death
Hydrogen	H	Atmosphere	Poor drainage due to lack of soil structure	Intolerance to drought
Nitrogen	N	Atmosphere	Crop removal, soil erosion and leaching	Slow and reduced growth with pale green leaves
Phosphorus	P	Rock particles	Crop removal, fixation in soil and reversion to a form unavailable to plants	Bright purple-red lower foliage with reduced yield due to stunted growth
Potassium	K	Rock particles	Crop removal, fixation in soil and leaching	Magnesium deficiency caused by oversupply of K, especially common in tomato, *Lycopersicon esculentum*, production
Calcium	Ca	Rock particles and animal deposits	Crop removal, soil erosion via weathering and leaching	Bitter pit in apples, *Malus domestica*, and blossom end rot in tomato, *Lycopersicon esculentum*
Magnesium	Mg	Rock particles	Crop removal and leaching	Interveinal yellowing (chlorosis) on wilting older leaves
Sulphur	S	Rock particles	Crop removal, soil erosion and leaching	Weak stems with foliage signs similar to nitrogen deficiency; acid rain

Table A 1.1 **Major (macro) nutrients.**

Nutrient	Chemical symbol	Major sources	Major losses	Disorders developed
Boron	B	Animal manure and superphosphate fertiliser	Crop removal and leaching	Top growth dies back and lower branches form a rosette. Star crack in beetroot, *Beta vulgaris* subsp. *vulgaris*
Cobalt	Co	Found naturally as a mineral in copper and nickel ores	Crop removal, soil erosion and leaching	Not known to be definitely essential for higher plants but found in the foliage of leafy plants, e.g. lettuce, *Lactuca sativa*
Copper	Cu	Soils with deposits of copper salts	Crop removal, soil erosion and leaching	Chlorotic foliage and stunted growth lead to death of top leaves and buds
Chlorine	Cl	Chloride salts, e.g. coastal areas	Crop removal but not usually deficient under field conditions	Wilted, bronzed foliage, leading to necrosis (dead sections)
Iron	Fe	Soils and animal manures	Crop removal, soil erosion and leaching. Less available in alkaline conditions	Chlorosis, the yellowing of foliage, commonly seen on fruit trees and roses, *Rosa* spp., on soils of high pH
Manganese	Mn	Soils and animal manures	Crop removal, soil erosion and leaching. May become toxic in acid situations	On chalky soils peas, *Pisum sativum*, develop marsh spot: shown as a brown mark inside the pea on the cotyledon. Also foliage speckling will show on a wide range of crops, e.g. carrots, *Daucus carota*, and onions, *Allium cepa*
Molybdenum	Mo	Animal manures	Crop removal, soil erosion and leaching	On acid soils 'whiptail' often develops on cauliflowers, *Brassica oleracea* var. Botrytis Group – shown as reduced leaf blades and uneven curding
Silicon	Si	Any sandy soils from the quartz particles (silica and silicates)	Soil erosion	Bamboo plants, e.g. *Phyllostachys nigra*, contain up to 6%
Sodium	Na	Sandy soils	Crop removal, soil erosion and leaching	Movement of solutes around some plant species
Zinc	Zn	Soil deposits and animal manures	Crop removal and leaching. May become toxic in acid situations and unavailable in alkaline soils	Foliage shows as mottled or interveinal chlorosis followed by bronzing. Misshapen roots also noted

Table A 1.2 Minor (micro) nutrients or trace elements.

Nutrient	Chemical symbol	Major functions	Growth stage	Notes
Carbon	C	Base building block	All	Enhanced in commercial tomato, *Lycopersicon esculentum*, production
Oxygen	O	All	All	Reduced in fruit storage
Hydrogen	H	All	All	Required for growth and development
Nitrogen	N	Synthesis of amino acids, proteins, chlorophyll, nucleic acids and co-enzymes	All	Required for growth and development
Phosphorus	P	Used in proteins, nucleo-proteins, metabolic transfer processes, e.g. ATP and ADP. Photosynthesis and respiration. Also a component of phospholipids	All	Energy transfer used heavily during propagational stages
Potassium	K	Sugar and starch formation, synthesis of proteins. Catalyst for enzyme reactions, neutralises organic acids, growth of meristematic tissue	All	Quickens maturity for flower and fruit production
Calcium	Ca	Production of cell walls, cell division and growth. Nitrogen assimilation	All	Also a co-factor for some enzymes
Magnesium	Mg	Essential in chlorophyll production, formation of amino acids and vitamins. Neutralises organic acids. Essential in the formation of fats and sugars	All	Also aids in seed germination
Sulphur	S	Essential ingredient in amino acids and vitamins	All	Flavours brassica and onion, *Allium cepa*, plants.
Boron	B	Affects flowering, pollen germination, fruiting, cell division and nitrogen metabolism	Young growth stages	Also plays a role in water relations and hormone movement
Cobalt	Co	May have a role in seed formation and longevity of storage. Aids in the fixation of nitrogen in the root nodules of legumes, e.g. peas, *Pisum sativum*	Mature growth stages	Not known to be definitely essential for higher plants but found in the foliage of leafy plants, e.g. Lettuce, *Lactuca sativa*
Copper	Cu	Constituent in enzymes, chlorophyll synthesis, catalyst for respiration, carbohydrate and protein metabolism	Active growth stages	Very narrow band of availability; can become toxic or deficient very quickly
Chlorine	Cl	Not too much known except that it aids in root and shoot growth	Active growth stages	Required for growth and development
Iron	Fe	Catalyst in synthesis of chlorophyll. Component in many enzymes	Active growth stages	Involved in the formation of many compounds. Provides physical disease resistance via ripening of tissues

(cont.)

Manganese	Mn	Chlorophyll synthesis and acts as a co-enzyme	Active growth stages	Promotion of rapid photosynthesis: stimulates the growth of several crops substantially, e.g. rice, *Oryza sativa*, pea, *Pisum sativum*, and cabbage, *Brassica oleracea* Capitata Group
Molybdenum	Mo	Essential in some enzyme systems that reduce nitrogen	Seed and young growth stages	Also required in protein synthesis
Silicon (silica and silicates)	Si	Cell strength and flexibility for some specialist plant groups	Active growth stages	Bamboo plants, e.g. *Phyllostachys nigra*, contain up to 6%
Sodium	Na	Nutrient movement around the plant: stimulates the growth of Joseph's coat, *Amaranthus tricolor*	Active growth stages	Required by plants with the C-4 pathway of photosynthesis
Zinc	Zn	Used in the formation of auxins, chloroplasts and starch	Seed and young growth stages	The legume plant group need zinc for seed production

Table A 1.3 Functions in the plant of both major and minor nutrients.

pH range >6.5	pH range 6.0–6.5	pH range 5.5–6.0	pH range 5.0–5.5
Broad beans, *Vicia faba*, and peas, *Pisum sativum*	Carrots, *Daucus carota*, and parsnips, *Pastinaca sativa*, and other roots	Potatoes, *Solanum tuberosum*, and tomatoes, *Lycopersicon esculentum*	Raspberries, *Rubus idaeus*, and strawberries, *Fragaria* × *ananassa*
Beets, e.g. *Beta vulgaris* subsp. *vulgaris*, and onions, *Allium cepa*	Sweet corn, *Zea mays*, and bush fruits, e.g. *Ribes nigrum*	Turnips, *Brassica rapa* Rapifera Group, and marrows, *Cucurbita pepo*	Rhododendrons, e.g. *R. lutea* and *R. arboreum*
Brassicas, e.g. *Brassica oleracea* var. Botrytis Group, and lettuce, *Lactuca sativa*	Apples, *Malus domestica*, and pears, *Pyrus communis*	Melons, *Cucumis melo*, and cucumbers, *Cucumis sativus*	Heaths, *Erica carnea*, and heathers, *Calluna vulgaris*
Spinach, *Spinacia oleracea*, and celery, *Apium graveolens* var. *dulce*	Bulbs, e.g. *Lilium regale* and *Crocus chrysanthus*	Lawn grasses, e.g. brown top bent, *Agrostis tenuis*, and chewings fescue, *Festuca rubra* var. *commutata*	Holly, *Ilex aquifolium*, and laurel, *Prunus laurocerasus*
Sweet peas, *Lathyrus odorata*, and stone fruits, e.g. *Prunus domestica*	Many evergreen shrubs, e.g. *Escallonia rubra* and *Viburnum tinus*	Many conifers, e.g. *Thuja plicata* and *Chamaecyparis pisifera* cultivars	Blueberries, *Vaccinium corymbosum*, and cranberries, *Vaccinium macrocarpon*

Table A 1.4 Preferred pH values of selected plants.

Type of mulch	Major sources	Notes	Safety notes	Extra details
Bullock manure	Animals	Can contain weed seeds, e.g. common dock, *Rumex obtusifolius*	Use gloves and wash exposed skin after mulching	Once composted the smell is lost or considerably reduced
Chipped bark	Pine, *Pinus* spp., and spruce, *Picea* spp.	Requires composting to break down the resin and avoid locking up of nitrogen in the short term	A longer lasting product but do request a sample to check the bark content	A range of sizes is available: chunky-chip is best with pieces measuring up to 7.5 cm (3 inches) long
Shredded bark	Pine, *Pinus* spp., and spruce, *Picea* spp.	Still requires composting and has a quicker breakdown time	Therefore not as long lasting as chipped bark	Cheaper than chipped bark
Grass clippings	Home lawns	Can be used fresh or composted with woodier material for a more stable breakdown	Do not use clippings that have been chemically treated with hormonal weedkillers, e.g. 2, 4 D.	Can also be used fresh on top of a sheet of newspaper, especially with onions, *Allium cepa*, or leeks, *Allium porrum*
Green waste	Local authorities (green compost bag schemes)	When fresh will have a pH level around 8, due to the sterilisation process	Fully sterilised with all contamination removed	Can be obtained loose or bagged. The very dark colour frames the plant structures
Newspaper	Daily papers	The broad sheets provide greater coverage and can be topped with a 5 cm (2 inch) layer of fresh lawn clippings	Do not use the colour supplements as prolonged use can lead to the excess application of heavy metals	Shredded paper can also be added to compost heaps to stabilise excess lawn clippings
Sewage sludge	Processed via earthworms	Can take the direct route and call it night soil	A certificate of chemical analysis, including the heavy metal contents, is provided when purchased	Excellent use of a common waste product; once composted the smell is reduced
Seaweed	Coastal areas	High salt levels, use as a slug and snail control or wash it before use	Can contain small pieces of plastic and raw sewage sludge, so use gloves and wash exposed skin after mulching	Add to your home compost, as an additive, which will keep the flies away from the borders
Leaf mould	Deciduous tree species	Requires composting to break down the leaves and avoid locking up of nitrogen in the short term	Ensure you select a clean sample	Common oak, *Quercus robur*, provides the longest lasting mould. Avoid horse chestnut, *Aesculus hippocastanum*, as the midribs do not break down quickly
Pine needles	*Pinus* spp.	Acid in action, which will reduce soil pH	Ensure you select a clean sample	Requires composting to break down the resin and avoid locking up of nitrogen in the short term

(cont.)

Garden compost	Home garden; shred the materials for a quicker and more even breakdown	Excellent for recycling of nutrients and use when you cannot identify the components	Check your compost for rodents and do not compost meat products	Turn the heap regularly and do not allow it to become too dry or too wet as this will extend the time for the breakdown process
Non-organic:				
Sea shells	Coastal areas	Historical use as path covering. Use approved local sources to reduce the carbon footprint	Wash before use around plants to reduce the high salt covering	Do not self-collect from beaches as this will cause erosion and reduction of the specialist habitat
Gravel and sand	Coastal, river areas and quarries	Historical use as path covering. Use approved local sources to reduce the carbon footprint	Wash before use around plants to reduce the high salt covering	Do not self-collect from beaches as this will cause erosion and reduction of the specialist habitat
Slate chips	Approved quarries and mine workings	Use local sources to reduce the carbon footprint	Wash before use around plants to reduce the covering of fine dust and debris	Do not self-collect from natural deposits as this will cause erosion and reduction of the specialist habitat
Pebbles	Coastal and river areas	Use approved local sources to reduce the carbon footprint	Wash before use around plants to reduce the high salt covering	Do not self-collect from beaches as this will cause erosion and reduction of the specialist habitat
Recycled aluminium	Spoil from industrial processes and manufacturing	Use approved local sources to reduce the carbon footprint	Only use on ornamental features as continued use can lead to a toxic build-up	Can reach high temperatures causing plant damage
Recycled stainless steel	Spoil from industrial processes and manufacturing	Use approved local sources to reduce the carbon footprint	Only use on ornamental features as the shards can physically contaminate edible crops	Can reach high temperatures causing plant damage
Recycled glass	Recycling and industrial reclamation	Use approved local sources to reduce the carbon footprint	Must be fully tumble ground to remove the sharp edges – check a sample before buying	Can reach high temperatures causing plant damage

Table A 1.5 Organic and non-organic materials used for mulching.

Botanical name	Common name	High water levels	Low water levels	Notes
Aronia arbutifolia	Red chokeberry	Good in well-drained soils, including clay	Improved seasonal foliage colour in full sun	Deciduous three-season shrub: flowers, berries and autumn foliage colour
Butomus umbellatus	Flowering rush	Can be grown in a pond or in moist soils as a marginal	Not recommended, but a site in full sun means maximum flowering	Excellent high-summer feature; attracts insects and also has edible roots
Chaenomeles speciosa 'Snow'	Flowering quince	Excellent for well-drained soils, including clays	Can be positioned as a wall shrub in a drier position	Attractive pure white flowers in early spring with red-tipped young foliage
Dierama pulcherrimum	Angel's fishing rod	Enjoys moist fertile soils, including during the summer	Also a position in full sun encourages maximum flowering	The hanging deep pink flowers dance in the wind
Escallonia 'Iveyi'	White escallonia	Excellent for well-drained soils, including clays	Excellent for coastal sites in full sun	Dark glossy evergreen foliage topped by luminous white flowers
Festuca glauca	Blue fescue	Not recommended as it results in dieback	Tolerant of dry soils in full sun	Excellent for use as an edging feature or in a container
Galium odoratum	Sweet woodruff	Moist soil in woodland sites	Tolerant of summer dryness	Can be used as a flowering living mulch
Hebe 'Red Edge'	Hebe	Moist but well-drained soils	Reduced water need due to waxy grey leaves	Good for coastal sites
Iris laevigata	Japanese iris	Can be grown as a shallow pond marginal or in moist soils	Also a position in full sun encourages maximum flowering	Wide range of flower colour forms available
Juniperus communis	Common juniper	Excellent for well-drained soils, including clays	Tolerant of dry soils in full sun due to reduced leaf size	Seasonal berries
Kerria japonica	Kerria	Excellent for clay soils	Also a position in full sun encourages maximum flowering	Also available with double flowers or variegated foliage
Lychnis flos-cuculi	Ragged robin	Can be grown as a pond marginal or in moist soils	Also a position in full sun encourages maximum flowering	Excellent for wildlife attraction
Macleaya cordata	Plume poppy	Excellent for well-drained soils, including clays	Reduced water need due to waxy leaves	Also sap used as a dye: be careful of your clothing
Nepeta subsessilis	Catmint	Excellent for any well-drained soil	Enjoys the dry area next to paving	As the common name suggests very attractive to cats
Osmunda regalis	Royal fern	Can be grown as a pond marginal or in moist soils	Reduced size is the only major effect of drier situations	Excellent autumn colour

(cont.)

APPENDIX 2

Type:	Spring: M A M	Summer: J J A	Autumn: S O N	Winter: D J F	All year round
Beetroot, *Beta vulgaris* subsp. *vulgaris*	▲ ▲ ▲	▲ ▲ ▲			
Broad bean, *Vicia faba*	▲ ▲ ▲		▲ ▲		
Carrot, *Daucus carota*	▲ ▲	▲ ▲ ▲			
Cabbage, *Brassica oleracea* Capitata Group					▲ ▲ ▲ Select seasonal cultivars
Cauliflower, *Brassica oleracea* Botrytis Group	▲ ▲ ▲	▲	▲		
Celery, *Apium graveolens* var. *dulce*	▲ ▲ ▲	▲			
Chard, *Beta vulgaris* subsp. *cicla* var. *flavescens*	▲ ▲	▲ ▲ ▲			
Courgette, *Cucurbita pepo*	▲ ▲	▲			
Leek, *Allium porrum*	▲ ▲ ▲			▲ ▲ ▲	
Lettuce, *Lactuca sativa*					▲ ▲ ▲ Select seasonal cultivars
Onion, *Allium cepa*					▲ ▲ ▲ Select seasonal cultivars
Mizuna, *Brassica rapa* var. *nipposinica*		▲ ▲ ▲	▲		
Pak choi and su choi, *Brassica rapa* var. *chinensis*		▲ ▲ ▲	▲ ▲ ▲		
Parsnip, *Pastinaca sativa*	▲ ▲	▲			
Radish, *Raphanus sativa*					▲ ▲ ▲ Select seasonal cultivars
Salsify, *Tragopogon porrifolius*	▲ ▲ ▲	▲			
Sweet potato, *Ipomoea batatas*	▲ ▲ ▲				
Sweet corn, *Zea mays*	▲ ▲	▲ ▲			
Tomato, *Lycopersicon esculentum* (outdoor)	▲ ▲ ▲			▲	

Table A 2.1 Recommended sowing times for seasonal vegetables.

Type	Spring: M A M	Summer: J J A	Autumn: S O N	Winter: D J F	All year round
Asparagus, *Asparagus officinalis*	▲▲	▲▲▲			
Beetroot, *Beta vulgaris* subsp. *vulgaris*					▲▲▲ Select seasonal cultivars
Broad bean, *Vicia faba*	▲▲▲		▲▲	▲▲▲	
Carrot, *Daucus carota*					▲▲▲ Select seasonal cultivars
Cabbage, *Brassica oleracea* Capitata Group					▲▲▲ Select seasonal cultivars
Cauliflower, *Brassica oleracea* Botrytis Group					▲▲▲ Select seasonal cultivars
Celery, *Apium graveolens* var. *dulce*	▲▲	▲▲▲			
Chard, *Beta vulgaris* subsp. *cicla* var. *flavescens*		▲▲▲	▲▲▲		
Courgette, *Cucurbita pepo*	▲	▲▲▲	▲▲		
Leek, *Allium porrum*	▲	▲▲▲	▲▲▲	▲	
Lettuce, *Lactuca sativa*					▲▲▲ Select seasonal cultivars
Onion, *Allium cepa*					▲▲▲ Select seasonal cultivars
Mizuna, *Brassica rapa* var. *nipposinica*			▲▲▲	▲▲	
Pak choi and su choi, *Brassica rapa* var. *chinensis*			▲▲▲	▲▲▲	
Parsnip, *Pastinaca sativa*	▲	▲▲▲	▲▲▲	▲▲	
Potato, *Solanum tuberosum*	▲▲	▲▲▲	▲▲▲		
Radish, *Raphanus sativa*					▲▲▲ Select seasonal cultivars
Salsify, *Tragopogon porrifolius*		▲▲▲	▲▲▲		
Sweet potato, *Ipomoea batatas*	▲	▲▲▲	▲		
Sweet corn, *Zea mays*		▲▲▲	▲▲▲		
Tomato, *Lycopersicon esculentum* (outdoor)	▲	▲▲▲	▲▲		

Table A 2.2 Recommended growing periods for seasonal vegetables.

Type:	Spring: M A M	Summer: J J A	Autumn: S O N	Winter: D J F	All year round:
Asparagus, *Asparagus officinalis*	▲ ▲				
Beetroot, *Beta vulgaris* subsp. *vulgaris*	▲	▲ ▲ ▲	▲ ▲ ▲		
Broad bean, *Vicia faba*	▲ ▲	▲ ▲ ▲			
Carrot, *Daucus carota*					▲ ▲ ▲ Select seasonal cultivars
Cabbage, *Brassica oleracea* Capitata Group					▲ ▲ ▲ Select seasonal cultivars
Cauliflower, *Brassica oleracea* Botrytis Group					▲ ▲ ▲ Select seasonal cultivars
Celery, *Apium graveolens* var. *dulce*					▲ ▲ ▲ Select seasonal cultivars
Chard, *Beta vulgaris* subsp. *cicla* var. *flavescens*					▲ ▲ ▲ Select seasonal cultivars
Courgette, *Cucurbita pepo*		▲ ▲ ▲	▲ ▲		
Leek, *Allium porrum*	▲		▲ ▲ ▲	▲ ▲ ▲	
Lettuce, *Lactuca sativa*					▲ ▲ ▲ Select seasonal cultivars
Onion, *Allium cepa*					▲ ▲ ▲ Select seasonal cultivars
Mizuna, *Brassica rapa* var. *nipposinica*			▲ ▲	▲ ▲ ▲	
Pak choi and su choi, *Brassica rapa* var. *chinensis*	▲		▲ ▲	▲ ▲ ▲	
Parsnip, *Pastinaca sativa*	▲	▲	▲ ▲ ▲	▲ ▲ ▲	
Potato, *Solanum tuberosum*		▲ ▲ ▲	▲ ▲ ▲	▲ ▲ ▲	
Radish, *Raphanus sativa*					▲ ▲ ▲ Select seasonal cultivars
Salsify, *Tragopogon porrifolius*			▲ ▲ ▲	▲ ▲ ▲	
Sweet potato, *Ipomoea batatas*			▲ ▲ ▲		
Sweet corn, *Zea mays*		▲ ▲	▲ ▲ ▲		
Tomato *Lycopersicon esculentum* (Outdoor)		▲ ▲ ▲	▲ ▲ ▲		

Table A 2.3 Recommended use periods for seasonal vegetables.

APPENDIX 3

Trade and common name (a.i.) Resistance code number	Unit sizes	Pests controlled and rates of use	Effects on beneficials

Aphox®: pirimicarb
  Also several other trade names
  IRAC code 1A
  Max 2 to 6 sprays per crop per year
	1 kg bottle	Mainly aphids, 5 g per 10 L	Harmful to *Aphidoletes*, *Chrysoperla*, *Cryptolaemus* and *Macrolophus* for 7–10 days. Moderately harmful to *Amblyseius*, *Dacnusa* and *Diglyphus*.

AQ10®: *Ampelomyces quisqualis*	10 g and 30 g sachet	Powdery mildew	Safe to all beneficials although some fungicides can be harmful to AQ10.

Chess WG®: pymetrozine
  IRAC code 9B
  Max 4 applications per crop per year or max 400 g per ha
	200 g Card box	Sucking pests such as aphids, leaf hopper, mealybug, psyllids (suckers), soft scale insect, whitefly. 2 to 6 g per 10 L (check approval notices).	Safe to the majority of beneficials except *Macrolophus* nymphs for up to 2 weeks.

Conserve®: spinosad
  Also Tracer
  IRAC code 5
  Max 6 applications per structure per year
	1 L bottle	Caterpillar, leaf miner, saw fly, scatella, sciarid, thrips. 7.5 ml per 10 L.	Safe to predatory mites and insects. Harmful to parasitoid wasps for 7 to 10 days.

Contans®: *Coniothyrium minitans*
  No FRAC code given.
  No frequency restrictions
	4 kg box	*Sclerotinia* in all edible and non-edible crops, protected crops. 2 to 8 kg/ha, depending on depth of incorporation.	Safe to all beneficials, specific to *Sclerotinia*.

Dynamec®: abamectin IRAC code 6, max 6 applications for tomato, unrestricted for ornamentals.	250 ml and 1 L bottle	Spider mite adults at 2.5 ml per 10 L. Leaf miner larvae and thrips at 5 ml per 10 L (on label rates).	Harmful to most beneficials but short persistence of 7 to 14 days allows good integration.

Lepinox®: *Bacillus thuringiensis*
  IRAC code 11
  No frequency restrictions
	500 g bottle	Most caterpillars (moth larvae), better on young stages, 7.5 to 10 g per 10 L.	Safe to all beneficials; specific to caterpillars.

Majestik®: plant extracts
  No IRAC code given
  No frequency restrictions
	5 L bottle	Physical control of many small insects and mites 250 ml per 10 L.	Harmful to young stages of most beneficials until dry.

(cont.)

Met52 Granular Bioinsecticide®: *Metarhizium anisopliae* No IRAC code given	1 kg and 10 kg bag	Vine weevil larvae, thrips pupal stages in soil/compost. 500 g per m^3 of compost.	Harmful to *Aphidoletes* pupae, safe to most other beneficials.
Mycotal®: *Lecanicillium muscarium* No IRAC code given No frequency restrictions	500 g box	Whitefly 10 g per 10 L.	Harmful to *Aphidoletes* larvae, safe to most other beneficials
Naturalis-L®: *Beauveria bassiana* No IRAC code given Max 5 applications per crop.	1 L bottle	Wide range of insect and mite pests. 15 to 30 ml per 10 L.	Harmful to *Aphidoletes* larvae, safe to most other beneficials.
Prestop®: *Gliocladium catenulatum* No IRAC code given, max applications variable; check approvals	1 kg	Wide range of plant pathogens	Safe to all beneficials although some fungicides can be harmful.
Savona®: fatty acids No IRAC code given No frequency restrictions	4 L bottle	Physical control of many small insects and mites 100 to 200 ml per 10 L.	Harmful to young stages of most beneficials until dry.
SB Plant Invigorator®: foliar lattice, linear sulphanate + iron chelate & nitrogen. No IRAC code given; physical mode of action. No frequency restrictions	1 L and 5 L (250 ml and 500 ml ready to use bottles for amateur use).	Physical control of many small insects and mites (20 ml per 10 L professional formulation; 100 ml per 10 L amateur formulation). Also control of powdery mildew.	As above.
Serenade ASO®: *Bacillus subtilis* FRAC code 44 Max 20 applications per crop per year	10 L	*Botrytis* and many other plant pathogens 10 ml per L (10 L/ha).	Safe to all beneficials.
T34 Biocontrol®: *Trichoderma asperellum* No FRAC code.	100 g and 500 g	Many root diseases	Safe to all beneficials.

Integration of biological control agents with pesticides is a critical part of most IPM programmes. The above table lists just a few products and their compatibility with some beneficials.

Table A 3.1 IPM-compatible pesticides. Some products are approved for use in production houses and as such may not be approved for use in amenity situations or display houses. Always check off-label approvals and EAMU/SOLA notices before use of any pesticide. The rates of use given below are those commonly used on a range of horticultural crops and should be checked before use.

Biological control agents	Pack sizes	Rates of use	Description of commercial packs	Comments
Aphelinus abdominalis Parasitoid wasp for larger elliptical species: *A. solani* and *M. euphorbiae*	250 adults	$1/m^2$/week or $2/m^2$/fortnight (preventive). $2/m^2$/week or $5/m^2$/fortnight (curative).	Adults and emerging mummies in vial or bottle.	Spring onwards min 12°C. Also in mix with other parasitoids.
Aphidius colemani Parasitoid wasp for smaller round-bodied species: *A. gossypii* and *M. persicae*.	500, 1,000 and 5,000 adults	$0.5/m^2$/week or $1/m^2$/fortnight (preventive). $1/m^2$/week or up to $5/m^2$/fortnight (curative).	Adults and emerging mummies in vial or bottle, larger units have sawdust as carrier material to aid distribution.	Early spring onwards min 12°C. Also in mix with other parasitoids.
Aphidius ervi Parasitoid wasp for larger elliptical species: *A. solani* and *M. euphorbiae*	250 adults	$0.5/m^2$/week or $1/m^2$/fortnight (preventive). $2/m^2$/week or up to $5/m^2$/fortnight (curative).	Adults and emerging mummies in vial or bottle.	Spring onwards min 12°C. Also in mix with other parasitoids.
ACE mix is all three of the above species, **CE** mix is a 50:50 mix of *A. colemani* and *A. ervi*.	500 adults	$1/m^2$/week or up to $5/m^2$/fortnight (preventive) or $2/m^2$/week or up to $5/m^2$/fortnight (curative).	Adults and emerging mummies in vial or bottle.	Easy to use with mixed aphid species or when unsure of species.
MACE mix contains the above species plus *A. matricariae* and *Ephedrus cerasicola*. **PACE** mix, as above plus *Praon* spp.	240	$1/m^2$/fortnight (preventive) or early stage curative.	Adults and emerging mummies in small tube.	Ideal for mixed cropping situations.
Aphidoletes aphidimyza Predatory midge suitable for most species of aphids.	250 and 1,000 midge cocoon/adults	$1/m^2$/fortnight (preventive) or $1/m^2$/week to $10/m^2$/week for heavier infestations.	Midge cocoons in vermiculite, blister pack of 250 cocoons or 1,000 in bottle.	Very good broad-spectrum predator, adults fly to find aphid colonies.
Chrysoperla carnea Predatory lacewing larvae suitable for most aphids and other soft-bodied prey.	500, 1,000 and 2,500 young larvae	$10/m^2$ up to $20/m^2$ as required, useful as curative. Can be used inside and outside.	Young larvae in tube with buck-wheat husk carrier.	Useful generalist predator, can clean up quite large infestations.
Beauveria bassiana Naturalis-L ® Insect pathogenic fungus, wide range of activity.	1 L bottle MAPP 14655	15 to 30 ml per 10 L.	Vegetable oil suspension.	Keep in fridge, mix in water, apply as spray, maintain humidity.

Table A 3.2 Aphid biological control agents.

Biological control agents	Pack sizes	Rates of use	Description of commercial packs	Comments
Amblyseius montdorensis Predatory mite attacks all stages of whitefly eggs, first and second instar whitefly larvae of most species including *Bemisia tabaci*.	25,000 & 125,000 loose mites & CRS sachets.	15 to 75 per m^2 of loose material 1 CRS per 1 to 2 m^2 per 6 to 8 weeks, 1 CRS per 1 to 2 plants (e.g. cucumber).	Tubes and bulk bags with bran and vermiculite, CRS mini sachets with hanging hooks or Gemini CRS sachets.	Introduce as preventive or at first sign of pest, or allow specified time after spraying.
Amblyseius swirskii Predatory mite of thrips and whitefly, attacks younger stages of both pests.	25,000 and 125,000 loose mites & CRS sachets	15 to 75/m^2 twice at 14-day interval as preventive or early stage control.	Tubes and bulk bags with bran and vermiculite, CRS sachets with or without hanging hooks or Gemini CRS	Summer or heated plants needs min 20°C for any development.
Encarsia formosa Small parasitoid wasp, better for glasshouse whitefly *T. vaporariorum*.	3,000, 6,000, 10,000 and 15,000 black scales	2 to 5/m^2/week as preventive; 5 to 10 or more weekly for 6 to 8 weeks as curative.	Parasitised black scales on hanging cards.	Spring onwards, needs min 18°C for a few hours each day for activity.
Macrolophus pygmaeus Licensed predatory bug for most species of whitefly, will also feed on leaf miner and spider mite.	250 and 500 adults and nymphs	0.25 to 0.5/m^2 usually a single introduction per year.	Adults and nymphs in bottle with inert carrier.	Late winter for heated crops to spring for cooler crops.
Beauveria bassiana Naturalis-L® Insect pathogenic fungus, wide range of activity, including spider mite.	1 L bottle MAPP 14655	15 to 30 ml per 10 L, apply as high-volume spray to contact target pest species.	Vegetable oil suspension	Keep in fridge, mix in water, apply as spray, maintain 60 to 80% humidity.
Lecanicillium muscarium Mycotal® Insect pathogenic fungus, can infect other pests including; mealybug and thrips.	500 g MAPP 04782	1 g per L as HV contact spray, good spray coverage is essential, spray weekly as curative.	Dry white powder in foil bag within card box.	Can be kept in fridge for several weeks, requires high humidity for activity.

Table A 3.3 Whitefly biological control agents.

Biological control agents	Pack sizes	Rates of use	Description of commercial packs	Comments
Amblyseius andersoni Native (European) predatory mite for most mite species and thrips.	25,000 and 125,000 mites and CRS sachets	1 per small plant; 10 to 20 per m^2 light infestation; 20 to 50 per m^2 as curative. 1 CRS per 1 to 2 plants or 1 per m^2.	Tube or 5 L paper bag	Introduce at first sign of pest, or allow specified time after spraying.
Amblyseius californicus Licensed predatory mite for spider mites.	2,000 and 25,000 mites	10 to 20/m^2 as preventive or early infestation.	Vial or bottle with corn grit carrier.	Introduce early or when hot weather predicted.
Feltiella acarisuga Predatory midge for spider mites	250 midge cocoon/adults	1 per 2 to 5 m^2 at early stage or 1 per m^2 for higher pest numbers.	White cocoons on cut leaf sections in tub with wood shavings.	Open in house place in shade, keep dry.
Phytoseiulus persimilis Predatory mite specifically for spider mites.	1,000, 2,000 and 10,000 mites	1 per small plant; 10 to 20 per m^2 for light infestations; up to 50 per m^2 as a curative.	Vial, tube or small bottle, sawdust or vermiculite carrier.	Introduce at first sign of pest, or allow specified time after spraying.
Beauveria bassiana Naturalis-L ® Insect pathogenic fungus, wide range of activity.	1 L bottle MAPP 14655	15 to 30 ml per 10 L. Apply as high-volume spray to contact target pest species.	Vegetable oil suspension	Keep in fridge, mix in water, apply as HV spray, maintain 60 to 80% humidity.

Table A 3.4 Spider mite biological control agents.

Biological control agents	Pack sizes	Rates of use	Description of commercial packs	Comments
Amblyseius cucumeris Predatory mite attacks young thrips larvae	10,000 to 250,000 loose mites and various units of CRS sachets.	75 to 200 per m^2 of loose material, 1 CRS sachet per 1 to 2 m^2 per 6 to 8 weeks, 1 sachet per 1 to 2 plants (e.g. cucumber).	Tubes and bulk bags with bran & vermiculite or bran free, CRS sachets with or without hanging hooks or Gemini CRS sachets.	Introduce as preventive or at first sign of pest, or allow specified time after spraying.
Amblyseius montdorensis Predatory mite attacks young thrips larvae and young whitefly stages.	25,000 and 125,000 loose mites and various units of CRS.	35 to 80 per m^2 of loose material, 1 mini sachet per L – 2 m^2 per 6 to 8 weeks, 1 sachet per 1 to 2 plants (e.g. cucumber).	Tubes and bulk bags with bran and vermiculite, CRS mini sachets with hanging hooks or Gemini CRS sachets.	Introduce as preventive or at first sign of pest, or allow specified time after spraying.
Amblyseius swirskii Predatory mite for young thrips and young whitefly.	10,000, 25,000 and 125,000 loose mites and CRS sachets.	5 to 25 per m^2 twice at 14-day interval as preventive or early stage control.	Tubes and bulk bags with bran and vermiculite, CRS sachets with or without hanging hooks or Gemini CRS sachets.	Summer or heated plants, needs min 20°C for development.

(cont.)

Orius laevigatus Predatory bug attacks all active stages of thrips.	500 and 1,000 adults, 2,000 nymphs.	Use as curative at 1 to 10 per m^2 or as preventive to plants with good supply of pollen.	Small bottle with buckwheat husk carrier.	Active late spring to autumn.
Beauveria bassiana Naturalis-L ® Insect pathogenic fungus, with wide range of activity.	1 L bottle MAPP 14655	15 to 30 ml per 10 L apply as high-volume spray to contact target pest species.	Vegetable oil suspension	Keep in fridge, mix in water, apply as spray, and maintain 60 to 80% humidity.
Metarhizium anisopliae Met52 Insect pathogenic fungus, with wide range of activity.	1 and 10 kg bags. MAPP 15168	0.5 kg per m^3 compost as an incorporation, or 122 kg per ha broadcast rate.	Foil bags containing rice grains infected with fungus.	Incorporate (pre-mix) in compost before planting or apply as mulch.

Table A 3.5 Thrips biological control agents.

Biological control agents	Pack sizes	Rates of use	Description of commercial packs	Comments
Dacnusa sibirica Parasitoid wasp, lays single eggs inside leaf miner larva, pupates within leaf miner puparium.	250 adults	1 per 2 to 5 m^2 weekly as preventive or early stage of pest, up to 1 per m^2 weekly as curative.	250 adult wasps in small bottle.	Leaf miner damage will continue for a while as parasitoid develops within pest.
Diglyphus isaea Parasitoid wasp 'stings' leaf miner larva to paralyse it, lays one or more eggs next to the larva.	250 adults	As above	As above	Leaf miner damage stops quickly after wasp paralysed the leaf miner larva.
Steinernema feltiae Insect pathogenic nematode, swims along mine tunnel to kill leaf miner larvae within leaf tissue.	50 million, 250 million and 1.25 billion	50 million per 50 m^2 as curative or per 100 m^2 as preventive or early stage curative.	Trays of dry carrier with nematodes in suspension.	Nematodes require a period with free water on leaf surface for maximum mobility and penetration.

Table A 3.6 Leaf miner biological control agents.

Biological control agents	Pack sizes	Rates of use	Description of commercial packs	Comments
Chrysoperla carnea Predatory lacewing larvae suitable for most aphids and other soft-bodied prey.	500, 1,000 and 2,500 young larvae	10/m^2 up to 20/m^2 as required, useful as curative. Can be used inside and outside.	Young larvae with buck-wheat husk carrier.	Useful generalist predator, can clean up quite large infestations. Dead bodies remain on plants.
Cryptolaemus montrouzeri Predatory ladybird mainly for mealybug but will feed on other prey.	25 adults or 25 larvae	2 to 10 per m^2 of infested plants depending on level of infestation; every 2 to 4 weeks.	Small vial with folded honey/water soaked filter paper.	Needs temperature above 20°C for activity, may be very noticeable and easily mistaken for mealybug.
Various parasitoid wasps and predatory ladybirds. Contact supplier for details.	25 and 50 adults	2 to 10 per m^2 of infested plants every 2 to 4 weeks, until wasp established.	As above.	Needs temperature above 20°C to check pest species.
Steinernema feltiae Insect pathogenic nematode for scale insects; swim in film of water to enter insect body.	50 million, 250 million and 1.25 billion	50 million per 50 m^2 as curative or per 100 m^2 as preventive or early stage curative.	Trays of dry carrier with nematodes in suspension.	Nematodes require a period with free water on leaf surface for maximum mobility and penetration.

Table A 3.7 Mealybug and scale insect biological controls.

Common and example trade name: plant mode of action	Class of pesticide and IRAC mode of action code	Target pest(s)	Targeted system or biological process	Specific mode of action on organism
Abamectin as **Dynamec**: Contact and translaminar, spray.	Avermectin IRAC 6	Spider mites, leaf miner, thrips.	Nervous system	Chloride channel activator.
Bacillus thuringiensis as **Lepinox**: Spray must be ingested for activity.	Microbial pathogen IRAC 11	Lepidopterous caterpillar.	Metabolic/feeding processes	Insect mid-gut membrane disrupter.
Deltamethrin as **Decis**: Contact with long residual activity, spray.	Pyrethroid (synthetic) IRAC 3	Broad-spectrum insecticide.	Nervous system	Sodium channel modulator.
Pyrethrins as **Pyrethrum 5EC** or **Spruzit**: Contact with short persistence, spray.	Pyrethrum (natural) IRAC 3	Broad-spectrum insecticide.	Nervous system	Sodium channel modulator.

(cont.)

Imidacloprid as **Imidasect 5G** or **Intercept 70WG**: Systemic with long persistence, incorporated in growing media.	Neonicotinoid IRAC 4A	Vine weevil larvae in compost and sucking pests on foliage.	Nervous system	Acetylcholine agonist (mimic/blocker).
Metarhizium anisopliae as **Met52**: Contact with long persistence, incorporated in growing media.	Fungal pathogen Not IRAC classified (biopesticide)	Vine weevil larvae and thrips in compost and soil.	Whole body	Fungus enters cuticle and invades body.
Pymetrozine as **Chess WG**: Systemic and translaminar, spray.	Azomethine IRAC 9B	Most sap-sucking pests, e.g. aphids.	Nervous system	Selective feeding blocker.
Spinosad as **Conserve** or **Tracer**: Systemic and translaminar, spray.	Spinosyn IRAC 5	Thrips, leaf miner, sciarid, caterpillar.	Nervous system	Acetylcholine agonist (mimic/blocker).
Thiacloprid as **Calypso**: Systemic and translaminar, spray.	Chloronicotinyl IRAC 4A	Most sap-sucking pests.	Nervous system	Nicotinic acetyl-choline receptor agonist (mimic).

Table A 3.8 A range of useful horticultural insecticides, showing their IRAC mode of action code number, target pests and specific site of activity. (NB Some have the same IRAC code number, indicating the same mode of activity.)

GLOSSARY

2-chloroethyl phosphonic acid (CEPA) A chemical, which when applied as a spray, releases the gaseous hormone **ethylene** within plants.

abscisic acid (ABA) A terpenoid plant hormone produced throughout the plant, but especially in the roots, leaves, fruits and seeds, frequently in response to stress. It has growth-inhibiting properties and is involved in leaf **abscission**, senescence, the development and dormancy of seeds and possibly the dormancy of buds. It induces stomatal closure and is therefore central to the water economy of plants. It also increases the tolerance of plants to salinity, low temperature and drought.

abscission The breaking away of leaves, flowers or fruits from the plant at the abscission zone. This is commonly seen in the autumn when deciduous plants lose their leaves.

acaricide Substance that is used to kill mites such as red spider mite.

accession A unique number or record given to an individual plant in a botanic garden or living collection.

accession policy A policy and procedure that ensures all plants in a botanic garden are recorded accurately on a database and that accurate records are kept.

acid rain Rain that has a low pH and can be harmful to plants, soil and aquatic life. It is caused by the emissions of sulphur dioxide and nitrogen oxides from industrial processes, which react with water molecules in the air to produce the acid rain.

actinomorphic Flowers that are regular in the arrangement of their petals. If the flowers are cut through vertically they will always produce two mirror images of each other.

active ingredients (a.i.) The active component (chemical) of a pesticide, which is toxic to the pest, disease or weed it is intended to control.

adventitious Of roots, shoots or buds that grow from an unusual position on a plant as, for example, when roots grow from a stem **cutting**.

aerenchyma A plant tissue with long files of gas-filled spaces that allow oxygen to diffuse rapidly through the plant. Such tissues are normally found in plants adapted to grow in water or in waterlogged soils.

aeroponics Plants are grown in a closed environment and the roots and lower parts of the plant are sprayed with a mist solution containing nutrients. The plants are supported by a frame or mesh support. It is considered to be a very sustainable method of growing as it has a low water and energy input.

Agricultural and Horticultural Development Board (AHDB) Levy A levy that growers are required to pay the AHDB if they grow and sell horticultural products with a value of more than £60,000 a year under the AHDB order 2008.

air filled porosity (AFP) This is the volume of soil filled with air after any **gravitational water** has drained away. It is made up of the **macropores** and **mesopores** in the soil and these should be maintained to a high level without restricting the supply of water to the plant.

alate Means winged: as in aphids with wings; or *Euonymus alatus*, which has winged stems.

allelopathic Substances that are allelopathic are naturally produced in the plant and work like a pesticide and inhibit or prevent a pest, disease or weed growing around, on or in the plant. An example being the black walnut tree, *Juglans nigra*, which produces juglone that inhibits other plants growing below it.

ambient The environmental conditions prevailing in the immediate vicinity of an organism or group of organisms at any given time. The term is usually used by plant scientists and horticulturists with reference to temperature.

ameliorate To improve, usually, the soil by the addition of organic matter.

amelioration In horticulture, reduction of the damaging effects of particular environmental conditions such as water stress by, for example, the shedding of leaves and/or the development of summer **dormancy**.

androecium The collective name for the male parts of the flower; the stamens that provide the pollen grains for pollination.

angiosperms Plants that have their seeds contained in a pericarp. They are commonly referred to as the flowering plants and are also called the higher plants.

anholocyclic When higher ambient winter conditions permit the survival and continued reproduction of aphids, there is no sexual cycle in the autumn (see **monoecious holocycle**). For example in aphids males are produced as a result of adverse environmental conditions, such as the onset of autumn, these mate with females that lay resistant or winter eggs. In spring and summer all aphids will be **viviparous parthenogenetic** females. This type of life cycle can happen in warm greenhouses and warmer parts of the world where the aphid does not need to hibernate over winter.

antheridium (plural antheridia) A **haploid** structure producing male **gametes** that are present in the **gametophyte** phase of bryophytes (mosses) and ferns as well as many algae and some fungi.

anthocyanins Water-soluble, coloured plant pigments belonging to the flavonoid group of chemicals. Anthocyanins usually accumulate in **vacuoles** and are especially important in determining the colours of flowers and the autumn colours of the leaves of deciduous shrubs and trees, although they may also occur in stems and fruit.

anti-gibberellins (growth-retardants) Compounds that retard the growth of plants by reducing the amounts of natural **gibberellins** in the tissues.

apetulous Flowers without petals.

apical dominance The term used for the process in which the apical bud (= the **terminal bud**) of a plant stem or branch prevents or inhibits the growth of the lateral buds below it.

apomixes The production of seed without pollination and fertilisation.

apterous A wingless insect.

aquifers An underground source of water usually held in permeable rock. This is the source of water for wells and boreholes.

arboretum (plural arboreta) A botanical collection of trees and shrubs cultivated for scientific research. Derived from the Greek word *arbour* – meaning tree.

arboriculturalist A person who specialises in the maintenance of trees.

arbour A shady enclosed area that is surrounded by trees, shrubs or trellis covered with climbing plants. They are often at the ends of pergolas or long paths lined with trees.

arbuscular mycorrhiza This is a type of symbiotic mycorrhiza fungus that penetrates the cortex in the root and improves the plant's uptake of nutrients. It is estimated that 85% to 90% of plants have a relationship with arbuscular mycorrhiza. These plants **range from** clover, *Trifolium* spp. to *Populus tremuloides*.

archegonium (plural archegonia) An organ in the **gametophyte** phase of certain plants (ferns and mosses), which produces and contains the female **gametes**.

arisings The grass or pruned wood removed when an area is mown or plants pruned.

artificially modified organism Also referred to as artificial selection and selective breeding. This is where humans deliberately breed for certain traits or groups of traits. Examples could be higher yields, pest or disease resistance, or bigger flowers.

ascomycetes A class of fungi (the other main class being basidiomycetes) in which the spores are held in a sac or ascus, which usually contain eight spores. This class includes a number of plant diseases including ash dieback, *Chalara fraxinea*, and brown rot of apple, *Monilinia fructigena*. It also includes the *Penicillium* moulds, which have proved to be very useful to humans.

ascospores Spores from a fungus in the **ascomycetes** class that are borne in an ascus.

aspirated humidity sensor An electronic device for measuring the relative humidity of the atmosphere, in a plant propagation unit or elsewhere, in which a fan is used to pass the air over the sensor(s) in order to increase the accuracy and consistency of the measurements by reducing the risk of a build-up of stagnant air in the measuring chamber and condensation on the sensor(s). See also **humidistat**.

asymmetry An asymmetrical design does not form a mirror image along a central axis. If divided in half the two halves will be different in a number of ways.

auxin A plant hormone, the first to be recognised as such, produced mainly in leaf primordia, young leaves and developing seeds and involved in promoting (at relatively low concentrations) and inhibiting (at relatively high concentrations) the elongation of plant cells. Auxins are also involved in vascular tissue differentiation and fruit growth. Recent evidence suggests that they are not, however, directly involved in apical dominance. Exogenously applied auxins promote ethylene synthesis, induce the formation of adventitious roots and delay senescence. The major natural auxin is **indole-3-acetic acid (IAA)**, although several chemically related synthetic auxins are used in horticulture, such as 2, 4-dichlorophenoxy acetic acid (2, 4-D), a selective herbicide, and 4-indole-3-butyric acid (IBA), which promotes the rooting of cuttings.

axillary bud A bud growing in the upper angle between the leaf stalk and stem.

bacterium vector The vector is a DNA molecule that is used to transfer foreign genetic material from one cell to another, possibly in a different organism. In the case of a bacterium vector it is bacteria that act as the vector.

bare root Plants grown in beds at the nursery and then lifted for selling. There is little or no soil on the roots and they are not sold in a container such as a pot. The roots should be kept moist by covering with polythene.

basal plate The compressed stem that forms the basal part of a **bulb** and from which the bulb scales arise.

BASIS (British Agrochemicals Safety Inspection Scheme) An independent registration scheme that organises training courses with examinations to allow staff to obtain a certificate of competence. All agricultural and horticultural advisors and merchants giving advice or selling pesticides must be qualified and registered with BASIS. A qualification is available for the safe storage and handling of pesticides. There is also a Fertiliser Advisers

Certification and Training Scheme (FACTS) for people advising and selling plant nutritional fertilisers.

bed system The growing of vegetables in, usually, raised beds with paths between each bed. The bed is a maximum of 1.2 m wide, so that the centre of the bed can be reached from each side without treading on the bed. Commercially the beds are set at the width of the tractor wheels. The height of the bed varies from 75 mm to 300 mm for low beds but up to 900 mm if for disabled gardeners. The length of the bed depends on the size of the garden or allotment. Once constructed the beds are not cultivated and are mulched annually with organic matter.

beneficial organism A living organism that aids in the control of pests also called **natural enemies**. They form part of biological control methods.

beneficial pathogen A pathogen that is beneficial to the grower in that it will help to control a pest or disease, which it attacks and therefore prevents or reduces damage to the crop plant.

beta-carotene A strongly coloured red-orange pigment found in some plants and fruits. Carotene is the substance that gives carrots their orange colour.

BG-Base (Botanic Garden Database) Established in 1985, it is the most widely used plant records and botanic garden database with 184 users across 28 countries.

biocide A substance that is used to control harmful organisms by chemical or biological means.

biodiversity The variety of living organisms and the genetic variability of these organisms in a particular area.

biofuel A type of fuel that is derived from a biological source. This can include growing crops to provide the fuel to heat a greenhouse or the extracting of gases by anaerobic digestion and using these to provide heat and power. There are a number of methods of producing biofuels being investigated at present to find a cheap and sustainable supply of energy.

biological control The control of pests, diseases and weeds by the use of other living organisms such as predators, parasites or pathogens. This can include the use of natural native predators or introducing predators and parasites.

biomass A biological material that is used for energy production, this could be woodchip from trees grown for this purpose or the burning of *Miscanthus* canes. It also includes the converting of the biomass material to a gas or liquid to then produce energy as in **biofuel**.

biopesticide Organisms applied to control or manage a pest population as a living pesticide in biological control programmes.

bioswales Open channels that have plants growing in them and in heavy rain or floods will fill with water. They are designed into the landscape to control the flow of surface water. Their purpose is to slow down the water flow and reduce flooding. They do this by holding the water until after most water has flowed away. They can be used to divert the flow of water as well as clean some of the silt from the water.

biota The complete fauna and flora of a region or period of time in history. It is the total plant and animal species that make up the living organisms of a particular region or ecosystem.

biotechnology Defined by the UN as 'any technological application that uses biological systems, living organisms, or derivatives thereof, to make or modify products or processes for specific use'. It involves a wide range of sciences including genetics, microbiology and tissue culture.

bishop Also called a thumper or whacker. It is a block of wood approximately 30 cm × 30 cm fixed to the end of a 1.8 m broom handle. Used to firm newly laid turf to encourage it to knit to the soil below.

bolting The premature flowering of a plant, usually a vegetable; an example being lettuce producing a flower head before the crop can be harvested.

bonsai A method of root and shoot pruning of trees and shrubs to restrict their size and produce a certain shape. The plants are usually grown in a decorative container and trained using wire ties. They are widely grown in Far Eastern countries like Japan and China.

borehole A deep shaft drilled into the ground down to water-bearing rocks in order to extract water for irrigation. Permission to drill a borehole needs to be applied for and there can be restrictions on how much water can be extracted.

borrowed landscape The landscape or view that is not part of the garden but is included in the design. It could be a view, clump of trees in the distance or a feature that can be viewed from the garden and is included as part of the design. It adds to the features of the garden.

borrowing money Money that is lent to the company to help set it up at the beginning or to expand or purchase expensive equipment. The money is lent or loaned to the company by investors like banks, venture capital trusts and investment companies.

boundary layer The layer of air, of variable thickness, depending on weather conditions, between the surface of a plant leaf and the surrounding air. It normally contains more water vapour than the surrounding air.

boxing off The collecting of grass clippings when mowing the lawn or sports surface as opposed to allowing them to fall on the turf surface.

broad-spectrum pesticide A pesticide that will kill a wide range of pests or weeds; e.g. a broad-spectrum herbicide will kill many species of weeds and is not specific to a limited range.

Bt toxin A toxin produced by the bacteria *Bacillus thuringiensis*. *Bacillus* is used as a biological control of various pests. There are several commercial variants or subspecies; Btk (*kurstaki*) is specific for lepidopterous caterpillars whereas Bti (*israelensis*) is active against dipterous pests such as mosquitoes and Btj (*japonensis*) is active against certain coleopteran beetles. The Bt spore contains a Cry protein toxin that kills the target organisms.

budding A form of **grafting**, used with a wide range of woody plants, in which a small piece of **scion**, tissue with a bud attached is applied, above soil level, to a cut surface beneath the lifted bark on a rootstock and secured there so that the tissues fuse and grow together as one plant. Two types of budding are commonly employed: in chip budding, used mainly with fruit trees and some ornamental shrubs, a small sliver of scion wood is inserted into a slit in the bark of the rootstock; and in shield- or T-budding, used mainly with roses, a shield-shaped piece of scion bark is inserted into a T-shaped cut in the bark of the rootstock.

bud union The point at which the plant was budded; see above. It usually shows as a slight swelling on the stem.

bulb An underground structure functioning as an organ for storage and survival during unfavourable periods, comprising a central bud and a series of modified leaf bases lacking chlorophyll, the scales, attached to a compressed stem, the **basal plate**, from which roots arise. **Scaly** and **tunicate** are the two main forms of bulbs.

bulb scaling A form of **vegetative propagation** used especially with *Lilium* and *Fritillaria* species, in which scales, with a small piece of the **basal plate** attached, are removed from the mother **bulb** and placed in a polythene bag with moist compost or inserted direct into a tray of compost. After a six-week period at around 21°C, the small bulblets, which by then will have developed at the base of the scale, may be detached and potted on.

bulb scoring A form of **vegetative propagation**, used principally with *Narcissus* and *Hyacinthus* spp. At the end of the dormant period the **basal plate** of the **bulb** is either scored in a cross shape to a depth of 5.0 mm or scooped out to a similar depth and the bulb placed upside-down in a dry medium or on a wire tray at 21°C, with occasional misting to prevent drying out. After a period of 8 to12 weeks small **bulblets** form on the injured surfaces and may be removed and grown on.

bulbils Small bulbs that develop above ground on some species of bulbous plants, e.g. in the leaf axils of the flower stem of *Lilium tigrinum*.

bulblets Small offsets that develop below ground in both **scaly** and **tunicate** bulbs.

bundle-sheath A single layer of cells surrounding a vascular bundle and completely enclosing it except at the tip. It effectively provides a seal around the bundle, preventing the entry of intercellular air into the xylem vessels, which would otherwise cause the water column to break. Other functions include the transfer of mineral ions from the non-living xylem to the surrounding living cells and the storage of waste products. In some plants the bundle-sheath cells may be involved in a specialised form of photosynthesis, called **C-4 metabolism**.

business plan A business plan is usually the initial plan for setting up a business. The banks or other investors will require it before lending any money. It is a formal statement of the business goals and how they will be achieved. It will set out the background to the company and financial information, customer profile, staffing levels and many other details.

C-4 photosynthesis (C-4 metabolism) A modification of normal photosynthesis in which the first products of carbon dioxide fixation, produced in the **palisade mesophyll** cells of the leaf, are organic acids (usually malic and/or aspartic) containing four carbon atoms. By this means carbon dioxide at low levels in the atmosphere is concentrated and stored. The organic acids are subsequently transferred to modified **bundle-sheath** cells containing chloroplasts (**Krantz anatomy**) and broken down to release carbon dioxide, which is then fixed by normal, C-3 photosynthesis. C-4 photosynthesis is an adaptation to environments with high light intensities, high temperatures and periods of water stress and is usually found in plants from arid and tropical/subtropical regions.

callus tissue The soft tissue produced at the site of a wound to form a protective layer over the area. It is seen when pruning wounds heal or at the base of cuttings just prior to rooting.

cambium A specialised layer of cells, located between the **xylem** and **phloem** of stem and roots, which divide to form secondary xylem and phloem, thereby increasing the transport and support capacity of the vascular tissue and the girth of the organ. A specialised cambium, the **cork cambium** (or **phellogen**), forms in the outer tissues of older stems and roots and produces a waterproof, protective layer of cells, their walls being thickened with **suberin**.

capital expenditure Is expenditure incurred when a business spends money to purchase fixed assets like new greenhouses or buy expensive equipment that is required for the business. It also has tax implications as the cost is not deducted in one year but spread over the lifetime of the equipment. Capital expenditure is not part of the ongoing costs of running the business.

carbon footprint Usually used to show the amount of carbon (greenhouse gases) produced by an individual, family or organisation. It is a measurement of how much carbon is used in carrying out daily life. The aim is to reduce this to the lowest level possible but still maintain a good standard of living.

carbon neutral Also called zero carbon footprint. The aim of being carbon neutral is to have a net zero carbon footprint whether for a person or organisation. This can be achieved by using renewable energy, reducing energy use and offsetting the carbon used by planting trees. Basic living will require the use of some carbon but being carbon neutral should keep this to a minimum.

carbon sink Any natural or artificial reservoir that stores carbon-containing chemical compounds such as carbon dioxide. A forest is a carbon sink as during photosynthesis the trees absorb large amounts of carbon dioxide and store it for the future. The oceans and soil are also large carbon sinks.

cash flow The movement of money into and out of the business, this could be payments for fuel, seeds, plants or other items; or income from the sale of plants, crops or other products. Businesses need a good cash flow to stay solvent and in business. If dealing with customers that only pay after 60 to 90 days, like many of the supermarkets, this delays the income coming into the business and makes paying wages and other costs difficult.

Casparian strip A strip of cells in the endodermis of the root that have walls thickened with **suberin**. It controls the flow of water and minerals into the centre of the root.

cation exchange capacity (CEC) The capacity of the soil to exchange cations between the soil water and surfaces of active materials like clay, organic matter and living roots. Cations have a positive charge and are therefore attracted to particles with a negative charge such as clay.

chelated The nutrient that the plant requires is chemically bonded to a metallic ion. This is a method of ensuring the nutrient is available for the plant and is not 'locked' in the soil. An example is iron, which is often 'locked' in alkaline soils resulting in lime-induced chlorosis, is applied to the soil and the plant can absorb it.

chemical scissors A chemical that can be used to cut **chromosomes** without causing any unwanted damage.

chimaera (chimera) A tissue made up of two or more genetically distinct cell types or a plant made up of two or more genetically distinct tissues. Chimaeras may be created by somatic mutation or by certain types of grafting that result in the mixing of the cells of two or more distinct individuals.

chlorophyll The green colouring in plants that absorbs the energy from sunlight during the first stages of photosynthesis. It is made up of four pigments, chlorophyll a, chlorophyll b, carotin and xanthophyll and is normally contained in the **chloroplasts**.

chloroplasts The **organelles** in plant cells that contain the chlorophyll pigments and where **photosynthesis** takes place.

chromophore The light-absorbing region of a molecule of **phytochrome** or of other light-sensitive pigments.

chromosome One of several lengths of DNA found in each nucleus. Each chromosome can hold many genes. The number of chromosomes found in each species varies but is a set number for that particular species. Chromosomes are located in the nuclei of all plant cells and occur in pairs.

circadian clock An internal metabolic clock that measures time by passing through a cycle that returns to the same point at intervals of approximately 24 hours. The resulting **circadian rhythm** may persist, even when the plant is removed from the conditions that induced it. The term 'circadian' is derived from the Latin words *circa* (about) and *dies* (day) and is used because the cycles of a circadian clock or rhythm are not exactly 24 hours long.

circadian rhythm See **circadian clock.**

clone A group of plants of identical genetic constitution derived from one original plant. They have been propagated by non-sexual means such as vegetative propagation.

cold-units Arbitrary units used by growers for recording the exposure of crop and ornamental plants or storage organs to specified periods of low temperature important in their development. For example, in the breaking of dormancy to produce 'forced' rhubarb, cold-units are calculated on the basis of exposure to temperatures of less than 10°C, taken at a soil depth of 10 cm, at 09.00 GMT, beginning on 1 October. All degrees below 10°C for each day are added together to give the cold-unit total. Some cultivars of rhubarb require exposure to up to 400 cold-units to break the dormancy of the crown, thereby allowing leaves to be produced. See also **thermal time.**

coloniser Plants that spread and populate (colonise) a new area. This is useful when bare earth is colonised by plants and helps to prevent erosion. Weeds and wild flowers are typical plants that will colonise disturbed earth quickly.

colour wheel A circle or wheel showing the arrangement of the primary colours, secondary colours and complementary colours. It shows the relationship between the colours and how they react with each other. When designing gardens and borders it is important to get the right colour combinations.

companion cell See **phloem**.

computer aided design (CAD) The use of computers and the correct software to create designs for gardens, landscaped areas and borders. Many of them are capable of designing the area, calculating the amount of materials and plants required, and writing a specification to complete the work.

conservation The protection and preservation of plants, fungi, animals including insects and their habitats to ensure their survival.

conserve Maintain and looking after a garden to retain its significance or value.

continuing professional development (CPD) The way that staff maintain and develop their skills and knowledge. It involves keeping up to date with the latest developments and methods and also learning new skills. It is a structured approach to learning.

Convention on Biological Diversity An international treaty that protects the Earth's biodiversity and places a number of governmental targets on how this will be achieved.

conveyor belt A continuous loop of material that travels around a number of pulleys and is used to move materials quickly and efficiently. It reduces carrying as the items are placed onto the conveyor belt, which moves them to the point they are required.

coppicing The hard pruning back to nearly soil level of trees and shrubs in order to promote fast vigorous growth. It is often used on shrubs grown for coloured bark to maintain a good colour.

cork cambium See **cambium**.

corm A plant storage organ made up of a compressed stem covered with dried scale-leaves.

cotyledon A seed leaf attached to the embryo or young plant. In most plants cotyledons have a storage function and in some species may contain chlorophyll and act as the first photosynthetic organs following germination. In the *Monocotyledonae* only one cotyledon is present, while in the *Dicotyledonae* there are usually two or more.

crassulacean acid metabolism (CAM) A modification of the normal process of **photosynthesis** in certain drought-tolerant plants, usually members of the family *Crassulaceae*. These open their **stomata** at night and fix carbon dioxide into malic acid, in which form it is stored temporarily. In the day-time the stomata are closed, to conserve water, and the carbon dioxide released from the malic acid and refixed into the usual products of photosynthesis, namely sugars.

critical day-length The duration of the daily light period above which **short-day plants**, or below which **long-day plants**, do not flower or exhibit delayed flowering. The critical **day-length (photoperiod)** differs among groups of plants but is always constant within a single species or cultivar.

crosiers The unrolling fern frond.

cryptochrome A group of plant pigments sensitive to blue light and involved in light-induced movement of **stomata**, **phototropism** and responses to shade caused by buildings and other physical structures (but not shade caused by other plants, these responses being controlled by the ratio of red to far-red light).

cultivar An abbreviation of the words cultivated variety. A term used for a plant that has been raised by humans in cultivation, to distinguish it from a natural variety.

cultivation The operations carried out on soil to produce conditions ideal for sowing or planting. These operations are broken down into two main phases, the first are **primary cultivations**, such as digging or ploughing, and these are followed by **secondary cultivations**, like raking.

cultural techniques Also called cultural controls. It is the control or suppression of pests, diseases or weeds using good husbandry methods. These include good soil management, use of resistant cultivars, clearing debris and correct disposal and regular inspection of plants and crops to promptly identify any problems so that action can be taken quickly to prevent an epidemic arising.

cut and fill levelling An operation used to level sloping ground. The topsoil is removed and stacked to one side, the higher area is then cut into using machinery and the material removed and placed on the low area to raise it. The upper part of the slope is reduced and the material used to raise the lower area, resulting in a level area. The topsoil is then replaced.

cuticle A relatively thin, non-cellular, waterproof layer, containing the hydrophobic, fatty-acid polymer, **cutin**, that covers the outer surfaces of plant stems and leaves.

cutin See **cuticle**.

cuttings The most widely used type of **vegetative propagation** of plants in which a portion of a stem, petiole, leaf lamina or bud is removed by cutting with a sharp knife and inserted into a suitable growing medium and maintained in a suitable (usually warm and moist) environment to induce the formation of **adventitious roots** and eventually, development into a new plant. Stem material may be soft-wood (herbaceous), semi-ripe or hard-wood, with a **node** and **terminal bud** included, and may be left with a small 'heel' or larger 'mallet' of tissue from a main stem attached at the base. An artificial rooting compound, usually a synthetic **auxin**, may be applied to the cut surface of the cutting to be inserted into the growing medium, to induce the formation of roots.

cytokinins A group of plant hormones usually produced in the roots and transported up the plant in the xylem. They appear to stimulate the production of cells in the meristems and therefore help to control plant growth.

damping down The wetting of glasshouse floor, staging and other surfaces to increase humidity, especially at higher temperatures.

damping-off diseases A complex of root diseases sometimes found in propagation, commonly species of *Alternaria*, *Fusarium*, *Phoma*, *Pythium*, *Phytophthora*, *Rhizoctonia* etc. Usually associated with sciarid fly (fungus gnat).

dark-grey water Laundry and dish-washing water, which usually contains pollutants, including bleaching and whitening agents and boron from washing powders and liquids. Dark grey water may only be used in the garden if 'eco-friendly' washing agents have been used.

day-length (photoperiod) The duration of exposure of a plant to light in a 24-hour period. See also **critical day-length**.

day-neutral plant A plant in which flowering is not controlled by **day-length**.

de-accession To remove an accession from a plant collection because it is surplus to requirements or is no longer required.

deadheading The removal of dying blooms from the plant. It is done to improve the appearance of the plant and encourage it to keep flowering as it prevents the production of seed.

dehiscent A fruit that splits open to release its seeds when ripe such as pea, *Pisum*, and bean.

dementia A collective term for certain symptoms that affect the brain. People with dementia may experience cognitive difficulties, changes in personality and behavioural problems. Dementia typically affects the elderly; however, it is not a normal part of ageing.

deoxyribonucleic acid (DNA) Linear (string-like) chemical that is found within the nuclei of cells. The codes for an organism's characteristics are held in the sequence of chemicals that make DNA.

desire lines Lines made by pedestrians or cyclists. They are usually a shortcut across an area and not part of the original design. On grass areas they show up as well-worn grass or muddy lines.

deutonymph the second nymphal stage of mites, after egg and **protonymph**.

de-vernalisation Reversal of **vernalisation** by exposure to a period of higher temperatures, notably day-time temperatures above 15°C. De-vernalisation progressively decreases as vernalisation proceeds and is finally lost when vernalisation is complete.

diapause the hibernation stage of an insect's life cycle that is induced by the lower temperatures and shorter days in the autumn. This allows the organism to overwinter.

dibber A bluntly pointed tool that is used to make a hole in the soil or growing media that a plant, seed, tuber, bulb or corm could be planted into. Dibbers can vary in size depending on what is being planted.

dicotyledons Plants that have two cotyledons or seed leaves, whereas monocotyledons only have one. Dicotyledons are often abbreviated to dicots and consist of most of the plants with broad leaves.

dioecious holocycle Where an insect completes its life cycle using two host plants. An example being some aphids have a summer host, usually an herbaceous plant, and a winter host, usually a woody plant. The black bean aphid, *Aphis fabae*, overwinters on *Euonymus europeaus* and migrates to the summer host of bean in late spring.

diploid (2n) Cells that contain two complementary sets of chromosomes, which is the normal state.

diversity The range of plants and animals in a particular habitat, area, region or country. The greater the range the better the diversity. It is also referred to as **biodiversity**.

division A type of **vegetative propagation** in which the crown of clump-forming herbaceous plants is divided into two or more pieces, each including one or more intact shoots with roots.

dormancy The temporary inability of seeds, buds or storage organs to grow, even though the prevailing environmental conditions may be favourable for growth. Such structures are said to be dormant and may be released from this condition (the 'breaking' of dormancy) by exposure to particular environmental conditions, especially low or high temperatures.

dormant See **dormancy**.

Dutch roll In **vegetative propagation**, a number of **cuttings** bound together as a bundle and rolled up in polythene or inserted into a single hole in the rooting medium. Once rooting has occurred, the cuttings are separated and potted-on. Dutch rolls are used when space for propagation is limited.

EAMU (Extension of Authorisation for Minor Uses) This is a programme run by the Chemicals Regulation Directorate (CRD) that supports the approval for pesticides to be used on crops or in situations not covered by the pesticide label. They were formally known as **SOLA**s (Specific Off-Label Approvals).

earthing up see **ridging**.

earthworms A cylindrical-shaped segmented animal that lives in the soil. Its digestive system runs straight through its body and its main food is organic matter remains. They help to breakdown **organic matter**, create burrows in the soil and help to maintain good **soil structure**.

ecological niche Is an area with particular characteristics that favour certain species. For example a north-facing area would be shady and cooler than a south-facing one; and would be more suitable for plants that prefer a cool shady environment. Ferns, hostas and climbing *Hydrangea* would thrive in such an area.

ecosystems A community of living organisms such as plants, animals and insects and their environment, which is made up of the soil particles, water and air. It forms an interacting system between organisms and their environment, for example animals eating plants, excreting waste that feeds the other plants.

efficacy The capacity of the product to produce the desired effect. How effective the product is at controlling the problem pest, disease or weed.

endangered plants These are plants classified as an endangered species on the IUCN Red List system. Plants on this list are in danger of becoming extinct owing to their population being very low and easily being lost or destroyed.

endogenous In relation to plant development, the determination of the timing of an event such as flowering from within the plant, rather than as a result of triggering by an environmental signal such as low temperature.

entomopathogenic nematodes a nematode worm that carries a bacterial pathogen that attacks and kills various, usually specific, insects.

environmental Referring to the environment. This could be an environmental policy on waste disposal to recycle as much as possible, even to aim for zero waste.

ephemeral Plants that can complete their life cycle in less than a year and are therefore capable of having more than one generation a year. It is common in weeds like annual meadow grass, *Poa annua*.

epicormic These are shoots that grow from epicormic buds situated on old stems or the trunks of trees; particularly common on some species of lime tree. They are sometimes referred to as water shoots.

epicotyl The seedling shoot above the seed leaves; the emerging growing point.

epidermis The outermost layer of cells of a plant organ.

epigeal The seed leaves come above ground during **germination**.

epinasty The curving down of the edges of the leaves as the upper surface layers have made more growth; this can be caused by **ethylene**.

epizootic A disease that appears as new cases in an animal or insect population, during a given period. It usually indicates a rapid population explosion and the start of an epidemic.

ericaceous Plants in the family *Ericaceae*, e.g. *Calluna vulgaris*. Usually these only grow well in acidic soils.

ethylene A gaseous plant hormone especially important in fruit ripening, **senescence** and **epinasty**.

ethyl methanesulphonate Is an organic compound capable of producing mutations in genetic material by nucleotide substitution.

etiolated The state of plants grown in the dark. They lack chlorophyll, have elongated internodes with few fibres and exhibit limited leaf expansion.

etiolation The condition developed by seedlings or plants grown in the dark. Chlorophyll is usually absent, the stem is elongated with reduced numbers of fibres; and leaf expansion is much reduced.

eukarya An organism whose cells contain complex structures that are contained within a membrane. The cells will contain a nucleus, and include all the animals, plants and fungi in the world.

eutrophication The result of large amounts of nutrients like nitrogen or phosphate getting into streams, rivers and oceans and causing the growth of plankton or algae blooms that can turn the water bright green or red. It can also be caused by large amounts of untreated sewage getting into rivers or oceans. It is a form of water pollution and can result in the death of large numbers of fish.

EVA (ethylene-vinyl acetate) This is an additive that is included to improve the clarity of polythene, it also helps to reduce its stiffness and improves heat retention.

evapo-transpiration The combination of evaporation of water from the walls of the cells surrounding the **sub-stomatal cavity** and its **transpiration** into the atmosphere via the **stomata**. Evaporation from leaf surfaces is sometimes also included in evapo-transpiration.

exosmosis Where water flows or is drawn out of the plant cell or organism into the surrounding solution. This can happen if high amounts of fertilisers are applied often accidentally, and the soil solution salt concentration is higher than the plant cells.

explant See **micropropagation**.

ex situ Of a plant, where work, e.g. propagation, is undertaken away from, the position or place in which it is currently growing.

F^2 generation The second filial generation from a cross of the F^1 generation. The F^2 generation are the progeny of the F^1 generation.

fallow An area of ground that is rested between crops. If going to be vacant for a long period it can be sown with **green manure** to improve the soil.

family tree A fruit tree comprising two or more cultivars grafted onto a common **rootstock**.

far-red light Light in the near infra-red region of the spectrum, between 700 and 750 nm. It is an important source of information for plants since it converts the pigment **phytochochrome** from the active to the inactive form and by this means is involved in the control of various physiological and developmental processes including seed germination, growth, flowering and **circadian rhythms**.

feathered maiden A one-year-old tree with a single upright trunk that has a number of lateral growths, which are referred to as feathers.

fertilisation The fusion of the male gamete from the pollen grain with the female gamete in the ovule, which results in the production of seed.

fertiliser A natural (organic) or synthetic (chemical) material that is applied to the soil or added to growing media to increase the amount of nutrients available for plant growth. Fertilisers consist of nutrients (mainly nitrogen, phosphate and potassium) and a carrier and are applied in small amounts as indicated on the packaging. The percentage of each type of nutrient contained in the fertiliser has to be shown on the packaging.

festooning The tying down of young vigorous nearly vertical growth to a near-horizontal position. This is done to encourage maximum fruiting and reduce the vigour of the shoot.

fibre See **xylem**.

field capacity The point when the soil is holding its maximum amount of water after rainfall or irrigation and any gravitational water has drained away.

financial records The company accounts that show the income, expenditure and profit or loss. These are completed each year and show the financial position of the company.

fissures Cracks or gaps that appear in the soil, they can be shallow or deep and improve the drainage of turf areas. They appear naturally in clay soils when they dry out, or can be made by using heavy aeration equipment to lift the turf to create fissures.

fluid drilling A method of sowing seeds in a gel solution. The seeds are pre-germinated (chitted) and then mixed with the fluid gel. They are sown using specialist equipment or the gardener can use a polythene bag with the corner cut off and squeeze the gel out as a continuous stream.

fluoresce To emit light using a fluorescent material.

fogging The creation of a mist of ultra-small droplets of water, delivered by very fine nozzles, to increase the **relative humidity** of the atmosphere surrounding plants. Fogging is usually controlled by an electronic device (an 'artificial leaf') that detects changes in the relative humidity of the atmosphere.

folia lattice surfactants A series of long-chain molecules that smother and suffocate small insects and mites, usually soap based.

food web A complex web made up of food chains. It involves all living things including plants, animals and insects and how they inter-react. It shows the feeding connections from the herbivores up to the carnivores and how they link together. It includes the plant life cycle from germinating, growing, dying and decomposition and the flora and fauna that are involved in this process.

formal A garden of geometric, usually symmetrical, design that often includes clipped plants, architectural features and water. The area will be high maintenance as it is highly manicured with low-mown lawns, clipped hedges, topiary and dense planting.

formative pruning This is carried out within the first two years of planting and is used to form the basis of the tree's branches and shape. It forms the structure of the tree and is important with top fruit to ensure a high-cropping tree. Unwanted growth is removed and the remaining growth pruned back to strategically placed buds to encourage the correct type of growth.

frost-hardening The induction of tolerance of freezing temperatures in some woody plants by prior exposure to autumn days and/or relatively low, but not sub-zero, temperatures. See also **hardening-off.**

frost pocket A small low-lying area that is prone to frost. Cold air flows downhill and will collect behind a barrier or at the bottom of the slope and form a frost pocket. This is a pocket of cold air with a temperature well below the surrounding area and will damage tender plants.

fruit buds Lateral buds, or buds on spurs that will flower and go on to produce fruit. They are usually larger and fatter than wood buds.

gamete The male and female **haploid** sex cells that fuse during fertilisation to form the **diploid zygote**; that will then form the seed.

gametophyte The haploid generation of plants and algae. The male and female **gametes** fuse to produce the **diploid zygote**, which develops into the **sporophyte**. This phase of the generation takes place in the flower and ovary where the pollen and embryo sac fuse to form the seed.

gene A portion of DNA that holds information pertaining to a characteristic (e.g. plant height, flower colour).

genetic engineering The transference of specific desirable gene(s) from one organism into another, using technological means. The organisms involved may not have a close taxonomic relationship.

genetic modification See **genetic engineering.**

genetically modified organism (GMO) An organism that has received genes from another organism that has modified some of its characteristics. See **genetic engineering.**

genome All the genes carried by a single set of **chromosomes** of an individual plant or animal.

genotype All the genetic information contained within an organism. See also **phenotype.**

geophyte Any species of land plant adapted to survive unfavourable climatic conditions, such as cold or dry periods, by means of underground storage structures such as bulbs, corms or tubers.

germination The starting of growth from a seed. The first visible signs are the radicle exiting the testa. Prior to that the seed will have imbibed water and enzymes will have started respiration.

gibberellins A major group of terpenoid plant hormones. Important functions include the promotion of cell division and cell elongation and, therefore, the control of stem elongation. Gibberellins also promote flowering in some **long-day plants**, promote seed germination in many species, cause reversion to a **juvenile** form in some species and inhibit tuber formation in some other species. See also **anti-gibberellins.**

global positioning systems (GPS) Are space-based satellite navigation systems that can be used to locate positions anywhere on Earth. All that is required is a GPS receiver and an unobstructed line of sight to four or more satellites.

golden rectangle A golden rectangle is one whose side lengths are in the golden ratio of 1:1.618... It is very useful in forming the basis for a design. When a square section is removed, the remainder forms another golden rectangle. It has been widely used by artists over the years.

grafting A type of **vegetative propagation** in which a union is created between two or more different plants by inserting a piece of stem tissue (the **scion**), with the end cut to expose the cambium, into a cut in a **rootstock**, in which the cambium is also exposed, binding them into position and allowing them to grow together and eventually function as a single plant. Many different grafting techniques are employed (see index), including whip/splice, whip and tongue, apical wedge, side veneer and **budding.**

gravel garden A garden, bed or border of free-draining soil, top-dressed with gravel or pebbles and used especially for growing Mediterranean or coastal plants.

gravitational water Water that drains out of the soil by the force of gravity. This occurs after heavy rain and the soil is saturated, the surplus water drains out of the macropores, which allows air in for the roots. Once the gravitational water has drained out the soil is at **field capacity.**

Green Flag Award A national standard that parks and open spaces are judged against. It is awarded to areas that achieve the standard each year following judging; they receive a green flag to fly at that particular park or open space.

green gene A gene that promotes drought resistance, usually including the retention of chlorophyll, in, for example, lawn grasses.

green manures A crop that is grown for the sole purpose of digging into the soil to improve the organic matter level of the soil, or to be cut and composted. Typical green manures include mustard, *Sinapsis alba*, clovers, *Trifolium* spp., and winter tares, *Vicia sativa.*

greenwood cuttings Are taken from the soft tip of a stem after the spring flush of artificial rooting compound applied (or the stem further distressed by twisting) and the treated section of the stem enclosed in an opaque plastic sleeve filled with moist packing material, such as sphagnum moss or peat, and sealed at both ends.

leader The main central shoot of a woody plant or the main shoot of a branch. The **terminal bud** of a leader exhibits **apical dominance**, discouraging or reducing the rate of growth of side shoots.

lichens Are composite plants consisting of a fungus and algae. They form a symbiotic relationship, the fungus taking water and nutrients from the soil and the algae absorbing carbon dioxide from the air and carrying out the process of photosynthesis.

light-grey water Water used in the home for personal washing and preparing and cooking vegetables. Being relatively free of pollutants, it may be collected and used for watering plants. See also **dark-grey water**.

light integral The total amount of light received by a given area of plant tissue during a specified period of time.

light spectrum This is the range of wavelengths of light that affect plants. The leaves on the plant do not use green light for photosynthesis, this is reflected, which gives leaves their green colour to the human eye. Plants use mainly blue light at 430 nm and red at 662 nm wavelengths for photosynthesis. The visible spectrum to the human eye of light is from 380 nm to 750 nm

light-saturation point The rate of photosynthesis increases with increasing light levels, until a specific value is reached, the light-saturation point, above which there is no further increase in rate.

lignin A tough hardening material in the cell walls of woody tissues in plants, especially in the xylem. It is the main component of wood as seen in the trunks of trees.

limiting factor A factor that limits the growth of plants compared to what the plant is capable of achieving. In drought conditions water is the limiting factor; in heavy shade light will be.

local authorities Also called local councils. A public body that administrates government in a local area. They include county or metropolitan councils as well as district, borough and parish councils.

locus This is the position of a specific gene on a **chromosome**.

long-day plant A plant that can only flower, or flowers earlier, when the **day-length** is longer than a specific duration known as the **critical day-length**.

long-persistence pesticide A pesticide that stays in the environment for a long period; possibly months or even years. An example being DDT that was banned many years ago but is still being found in the environment and food chain many years later. Currently several synthetic pesticides show long persistence against the pests they control as well as several non-target organisms such as biological control agents.

lowland bogs Are peatbogs that have developed in low-lying areas usually as a result of poor drainage that has resulted in the formation of peat. Owing to the area being waterlogged there is a lack of oxygen that results in the slow partial decomposition that produces peat.

lute A tool for spreading and working top dressing into the turf surface. It consists of a long handle with a frame at the base that spreads and levels the top dressing.

lux The SI unit of **illuminance**. One lux equals the illumination produced by a luminous flux of 1 lumen, distributed evenly over a surface of one square metre.

macropores Pores in the soil that are greater than 5 mm. These are the largest pores and allow good drainage and aeration.

manufacturers safety data sheet Also called material or product safety data sheets (see **safety data sheets**). Information supplied by the manufacturer about the chemical supplied by them. The information gives details of the chemical, first aid treatment, flash point, storage, potential hazards and protective equipment to be worn. They should be used when carrying out a risk and COSHH assessment.

MAPP (Ministerially Approved Pesticide Product) This used to be referred to as the MAFF number. It is a unique product registration number issued by the Chemicals Regulation Directorate (CRD) and the Health and Safety Executive (HSE). This indicates the product has been approved for use in the UK and any pesticide used in the UK must have a MAPP number on the packaging.

marginal planting Plants growing in shallow water or in moist soil round the edges of ponds or watercourses.

maturation The completion of the process. In the case of compost the decomposition reaches a point where the material can be used on the garden. In fruit it is the point where the fruit is ripe and ready for use.

melanise Where invertebrates darken during their life cycle usually after moulting but also after being parasitised.

Mendelian genetics The principles of genetic inheritance initially proposed by the nineteenth-century monk, Gregor Mendel.

mesophyll See **palisade mesophyll**.

mesopores Pores in the soil that are between **macropores** and **micropores** in size. They can contain water that is available to plants, or air if the water has been used or evaporated.

metamorphic A rock that has been altered in some way from its previous form through the action of heat and pressure. Examples include marble formed from limestone and slate formed from shale.

microclimate The climate in a specific area of garden, which may vary from the surrounding area. It could be a particularly sheltered or sunny area or exposed to wind; or affected by being at a high altitude. The microclimate could have a higher or lower temperature or higher or lower rainfall.

micromesh A very fine plastic mesh that is used to prevent access to crops from various flying insects. The holes in the mesh are small enough to stop most insect pests getting access to the crop. It will prevent access to root flies, aphids, moths and butterflies.

micropores The smallest pores in the soil that are of minute size and any water in the pores is unavailable to the plant. They have little if any effect on drainage or aeration.

micropropagation A form of **vegetative propagation** in which the formation of new plantlets is induced in culture from small **explants** of parental tissue.

micropyle A pore in the ovule through which the pollen tube enters during **fertilisation**. Also when the seed is about to germinate water will enter to start the process.

modified Lorette system This is the most common method of summer pruning apple trees and is carried out once the young shoots have become woody at the base. New laterals over 22 cm long, growing directly from the main stem, are reduced back to three leaves from the basal cluster. Side shoots are pruned back to one leaf above the basal cluster. This is carried out to maintain the trees shape and encourage fruiting.

mole drain Narrow drainage tunnel made in heavy clay soil by using a flat-bladed plough to draw a narrow cylinder through the soil, smearing the sides to keep the channel open for a number of years, running below the cultivation depth.

molluscicide A substance used to control molluscs (slugs and snails).

monocotyledons Plants that have only one cotyledon or seed leaf as opposed to **dicotyledons**, which have two. Monocotyledons include all the grasses, lilies, orchids and palms.

monoculture The growing of a single crop on an area; a field of carrots is a monoculture.

monoecious holocycle The most frequent and simplest type of life cycle in which all generations develop on a single host plant. For example the green

apple aphid, *Aphis pomi*, and some conifer aphids. In spring and summer all aphids will be **viviparous parthenogenetic** females; males are produced in the autumn that mate with females, which lay resistant or winter eggs.

morphogenesis Progressive changes in the three-dimensional shape and form of an organism during development or in response to changes in the environment.

morphological Relating to the three-dimensional shape and form (the **morphology**) of an organism.

morphology The three-dimensional shape and form of an organism.

Morse code A method of sending text information by a series of dots and dashes or clicks that can be understood by a skilled user. It is not used now owing to the many modern forms of communications.

mulch A covering of bulky organic material (e.g. rotted manure, compost or shredded bark) or non-organic material (e.g. pea gravel, stone chippings, polythene sheeting or old carpet) placed on the surface of cultivated soil for such purposes as the conservation of water by reducing evaporation, the suppression of weed growth and the improvement of soil temperatures.

mutation A change in the genetic code of a plant resulting in changes to the plant's appearance or other characteristic. It can happen naturally or be induced by chemicals or radiation. Natural mutations are known as sports and are often a change in flower colour, but can be other characteristics.

mycorrhizae A fungus that forms a symbiotic relationship with some plants roots. The plant passes carbohydrates to the fungus species; in return the fungus absorbs water and nutrients (particularly phosphates) that it passes onto the plant roots for use by the plant. There are a number of types of mycorrhiza, two of the main ones being endomycorrhiza and ectomycorrhiza, they differ in how they enter the root structure.

national minimum wage A national minimum wage set by the government starting on 1 October each year. It varies depending on your age and whether an apprentice.

natural enemies The parasites, predators, pathogens and other antagonists that attack an animal, insect or plant and cause debility or death.

nectaries (singular nectary) The glandular organs in flowers that secrete nectar, which attracts insects for pollination.

nematicide A chemical that will control nematodes, at present there are few available but more may be developed in the future, particularly to control nematodes in turf.

neutralising value (NV) This indicates the strength of the lime, the higher the number the greater the effect on neutralising the acidity of the soil. Too high a value can result in plants being scorched if it gets onto the leaves. The most commonly used lime is calcium carbonate with a NV of 45 to 55, approximately half that of calcium oxide.

New Zealand bin A traditional wooden compost bin made from planks of wood fixed to four corner posts. The front is made so that the planks can be removed easily by sliding them out. There are usually two or three bins constructed side by side, one is full of nearly decomposed material, one is partly decomposed and the other is being filled. When the material is ready to be applied to the soil the bin is emptied and the adjacent bin turned into the empty bin. This gives better decomposition and can help speed up the process. The empty bin is then filled with new material.

niche A particular place or habitat that has conditions that allow particular species to thrive. This could be a certain soil type or pH or other environmental conditions that a plant needs to grow and complete its life cycle.

nicking and notching Nicking is used to weaken a lateral bud's growth. A small nick is made with a sharp knife *below* the bud into the cambium layer to reduce the sap flow to the bud, which reduces the bud's growth. Notching is used to stimulate a bud's growth. A small notch is made just *above* the bud into the cambium layer to reduce sap flow past the bud and encourage the bud to break and grow.

no-dig technique A method of growing plants, usually vegetables, without any cultivation of the soil. The vegetables are usually grown in a **bed system** that may have been dug when first constructed but thereafter is not dug. A layer of **organic matter** is added each year, which soil fauna break down. Seeds are sown into drills made into the surface and plants are planted using a trowel.

node The point of attachment of a leaf or leaves to a plant stem.

nomenclatural standards A herbarium specimen or other device to which the name of a cultivar is attached.

nomenclatural type specimens A herbarium specimen to which a taxonomic name is permanently attached.

non-indigenous An alien non-native introduced species of plant or animal, including insects. A species that has been introduced to an area by humans either deliberately or accidentally. They can sometime cause considerable problems like Japanese knotweed, *Fallopia japonica*, or ash dieback disease, *Chalara fraxinea*.

nursery level The level at which newly planted trees and shrubs should be planted. It is the level at which the plants were originally grown at the nursery and can be identified by the soil mark on the plant's stem.

nursery stock The production of trees and shrubs, including fruit trees, for sale in the UK and export if the grower has an overseas market. The plants can be grown in containers or in the soil.

nutrient film technique (NFT) A hydroponic system where the plants are grown in covered troughs and a nutrient solution is circulated past the roots. The plants are not grown in any media but some may be started in

rockwool blocks to get them established.

octoploid Having eight sets of chromosomes (8n).

offsets As in surveying. A lateral measurement taken from the base line to fix the position of an object in relation to the base line. They are usually at right angles to the base line and are called perpendicular offsets. It can be necessary to take oblique offsets to certain points to establish their position and tie in lines. This will make use of **triangulation**.

offspring The next generation of a plant or animal.

organelles Are structures in the plant cell that have a specialised function, examples include **chloroplasts** and mitochondria.

organic matter A term for plant and animal material in or on the soil, in all stages of decomposition. It breaks down to form **humus**. Organic matter is applied to the soil surface and can be left as a mulch or incorporated into the soil by digging. It holds water and nutrients as well as improving the structure of the soil.

osmosis The movement of water (solvent) molecules across a **semi-permeable** or differentially permeable membrane from a region of a relatively low concentration of dissolved substances (solutes) and therefore of a high concentration of water molecules (a region of high **water-potential**), to a region of a relatively high concentration of solutes, and therefore of a relatively low concentration of water molecules (a region of low **water-potential**).

outcrossing Is the introduction of new or unrelated genetic material into a breeding line. This will increase genetic diversity and reduces the chances of all the progeny being prone to the same disease. The breeder will outcross a breeding line to introduce new traits that can reduce disease susceptibility or even give resistance to diseases.

outdoor room A term sometimes used to describe a garden or part of a garden that can be used like an outdoor room.

This could include the patio area, barbecue and other areas used for sitting out and entertaining. The garden is considered as an extension to the house – an extra 'outdoor room'.

oxygenators Plants growing submerged or partly submerged in water that produce oxygen, an example being Canadian pondweed, *Elodea canadensis*.

parenchyma Is made up of thin-walled cells that form the bulk of the non-woody tissue in plants and is found in roots, stems and leaves carrying out a number of functions in the plant. It forms the ground tissue of plants and makes up the cortex of roots and stems and the mesophyll of leaves as well as many other tissues.

parterre A level area of garden usually adjacent to the house and often separated from the rest of the garden by stone balustrades or hedges. Flower beds are planted up with seasonal bedding surrounded by gravel paths.

parthenocarpy The production of fruits without pollination or fertilisation. This can happen in greenhouse cucumber, *Cucumis* spp., and banana, *Musa* spp.

parthenogenesis Reproduction that does not involve fertilisation, examples being the vine weevil, which only has female species, and aphids that give birth without males being present usually in the summer.

parthenogenetic Pertaining to **parthenogenesis.**

partial de-horning This technique is used on old fruit trees that have become tall and straggly. It consists of cutting back the main branches in the winter to a smaller branch with a good shoot of the previous season's growth. This reduces the height and spread of the tree over a period of two or three years but still keeps a framework of branches.

palisade mesophyll The tissue usually on the upper side of the leaf made up of columnar cells set closely together like a palisade fence. These cells contain many chloroplasts and are the main area of photosynthesis in most plants.

pans Also called soil pans. These are a hard layer that is difficult for water and roots to penetrate. They can be caused by poor cultivation that smears or compacts the soil at the same depth each year; or by iron oxides that cement the soil particles together that cause a pan.

pathogen A parasite that causes disease in its host. They can be bacteria, fungi or viruses.

peds A number of individual soil particles that have naturally formed a unit such as an aggregate, crumb, block or granule. The peds make up the tilth that is prepared for sowing seeds outdoors.

perennating organs Organs in the plant that enable it to survive from one year to the next, such as bulbs, corms or tubers.

permaculture This term was derived from the words permanent and culture. It describes a philosophy and practice of a holistic approach to growing. Permaculture involves working with nature with a minimal labour input and makes greater use of permanent planting such as trees and shrubs than conventional horticulture.

permanent wilting point When plants wilt and do not recover overnight as they can no longer extract water from the soil. Any water remaining in the soil is too tightly held for the plant to absorb.

permeability A windbreak should be approximately 50% permeable to give the best protection. That is half of the windbreak is solid and the other half open to allow through some of the wind. This helps to prevent turbulence when wind hits a solid object, goes over the object and creates turbulence immediately behind it. Artificial windbreaks can be purchased with different levels of permeability for different uses.

persistence The ability of some pesticides to remain in the environment for a long time. Some pesticides are broken down by bacteria or light, but others have a stronger chemical structure, which takes time to break down.

personal protective equipment (PPE) This is protective equipment that must be provided free of charge by the employer to the employee who should

wear it when required. It includes such items as ear defenders, goggles, gloves, hard hats and other specialist equipment for when using chainsaws or other hazardous equipment or materials.

pesticide A substance that will kill insects, rodents, slug, snails, mites, diseases and weeds. At one time it referred to just insecticides but now covers all chemicals that control a living problem that attacks plants. They include both man-made and natural chemicals; some are approved for use by organic gardeners but most are not. Pesticides include acaricides, algicides, bactericides, fungicides, insecticides, molluscicides and nematicides.

petiole The stalk that attaches the leaf to the stem.

pH Is a logarithmic scale measuring the percentage of hydrogen ions in the soil water. The scale runs from 1 to 14 but most soils in the UK have a pH of between 4.0 to 8.0. A pH of below 7 is acid – the lower the figure the greater the acidity. A pH above 7 is alkaline – the higher the figure the more alkaline the soil. The pH affects the range of plants that can be grown in the soil and their ability to absorb nutrients.

phellogen See **cambium**.

phenology The study of the timing of periodic biological indicators of the seasons, such as the times of flowering in plants, or breeding and migration in animals and birds, in relation to climate.

phenotype The physical manifestation of the genetic information (**genotype**) contained within a plant (i.e. the actual appearance of a plant). The phenotype results from the interaction of the genetic information with the environment in which the plant is growing.

phenotypic variation The variation of a **phenotype** that has evolved over time. Plants of the same species growing in a different environment may develop slightly different characteristics like plant size or leaf size. The genotype + environment → **phenotype**. Phenotypic variation leads to evolution by natural selection.

pheromone A chemical produced by an insect to get a specific response from another insect. Most commonly they are sex pheromones that attract the opposite sex for mating. Synthetic pheromones are now being produced that are used in **integrated pest management** to control specific pests. By attracting males or females to a trap this prevents mating and reduces the damage caused by that particular insect.

phloem A living tissue composed mainly of elongated **sieve cells**, joined end to end by perforated end walls to form **sieve tubes**, which are involved in the transport of dissolved sugars, certain hormones, some minerals and other substances within the plant. The sieve cells are associated with metabolically active **companion cells**. **Fibres** may also be present, their walls thickened with additional cellulose.

photon A **quantum** of electromagnetic energy in the visible spectrum, between wavelengths 400 and 700 nm ('visible light').

photoperiod See **day-length**.

photoperiodism The triggering of a particular phase of growth of plants, such as flowering, dormancy or the formation of underground storage organs, in response to day-length.

photorespiration A form of respiration that takes place in the chloroplasts of C-3 plants that is induced by light. It is thought it may prevent the build-up of substances in the chloroplasts during periods of high photosynthetic activity.

photosynthesis The process of the plant using sunlight, carbon dioxide and water to produce sugars and other carbohydrates. It is carried out in the green parts of the plant in the **chloroplasts** that contain **chlorophyll**.

phototropism A process in which a plant's growth is stimulated by a source of light. The plant's reaction can be negative, away, or positive, towards.

phytochemicals Are chemical compounds found within plants. The chemicals that give many of the 'superfoods' their positive effects are phytochemicals, and many drugs are derived from phytochemicals. An example of a phytochemical is lycopene, which occurs in the tomato giving the fruit its red colour.

phytochrome A light-sensitive, proteinaceous plant pigment that exists in two interchangeable forms: an active form (in red light) or an inactive (in **far-red light**). This switch, linked to environmental signals, is utilised in plants to control a diversity of physiological and developmental processes, including seed germination, vegetative growth, flowering and **circadian rhythms**.

phytosanitary certificate – A certificate of inspection attached to a shipment of plants that will cross international boundaries.

pinetum A specialised collection of conifers or cone-bearing gymnosperms and their close relatives. The plural of pinetum is pineta.

pits The point in the plant cell wall through which the **plasmodesmata** pass that link adjacent cells.

plant communities A collection of plant species (including fungi) in a designated area. They form a group of plants that grow and thrive in the soil conditions, climate and factors that affect the growth of plants in the area. Plant communities include woodland, heathland, grassland and others.

plant pathogen A micro-organism that is harmful to its host and causes disease.

plasmodesmata (singular plasmodesma) Fine strands of cytoplasm that connect adjacent cells and allow the flow of water and minerals between cells.

pleaching The training of trees such as lime, *Tilia* spp., or hornbeam, *Carpinus betulus*, along wires to form a hedge-like structure. Pleached trees look like a hedge on stilts! The plants require annual training and pruning to maintain their shape and keep them narrow. They are often used alongside paths and to create a tall screen.

pleasure garden A public park or garden in the eighteenth or nineteenth centuries, used for various forms of entertainment such as concerts, firework displays and parties.

pollarding The hard pruning of a tree back to the trunk at intervals (removing all the branches); this encourages the production of fast growth. It is often seen used as a method of street tree pruning in order to maintain the size of the tree, but is not a recommended method of tree pruning.

pollination The transfer of compatible male pollen from the anther to the receptive stigma. This can be carried out by insects, wind, birds or water.

polycarbonate A transparent plastic often used on the roof of large greenhouses and conservatories. It can be double- or triple-walled to give it better heat-retaining properties and greater strength. Light transmission is close to glass when new but deteriorates over time.

polypropylene also called polypropene, is a polymer plastic that is widely used in the horticultural industry and for packaging.

polyurethane A polymer that can be made into a number of different types of materials like foams and hard plastics; these have a number of uses in the garden and landscaped areas.

potager A kitchen garden containing a mixture of flowers, vegetables and fruit laid out for ornamental effect. The beds are often edged with low hedges.

pot bound The plant's roots will be densely packed within the growing media in the container. It is a sign that the plant has been in the pot too long and should have been planted out or potted on earlier.

predators An insect or animal that preys upon others for food. They are a useful method of controlling some pests. Examples include ladybirds, frogs and grass snakes.

pressure potential In a plant, the hydrostatic pressure to which water in a liquid state is subjected. The pressure potential in a turgid plant cell (see **turgor**) has a positive value, but in the xylem of a transpiring plant, where the water column is under tension, it has a negative value.

prevailing wind The direction that the most common wind for the area comes from. Usually in the UK it is south westerly.

prill Are like granules of fertiliser but with a very uniform spherical shape, which tend to flow better through fertiliser distributors.

primary cultivations The first cultivations carried when preparing ground for planting or sowing. These can include clearing the ground, sub-soiling, ploughing and discing on a large scale. On a smaller scale they include digging (single or double) and forking over. These are carried out well before any planting or sowing to allow the soil to settle and are often done over winter. They will be followed by the **secondary cultivations**.

profit margin Is a measurement of the profitability of a particular product or operation (e.g. landscaping a garden). A low profit margin indicates that this product or operation should be looked at otherwise it could soon start to lose money.

propagule A name given to any part of a plant that is used to establish another plant. The term may be used to describe, for example, a seed, **cutting**, off-set, **bulb** or **corm**, or a piece of tissue used for **micropropagation** (also called an **explant**).

prothallus The **gametophyte** phase of ferns, mosses and liverworts. It is usually a heart-shaped structure with **rhizoids** growing from the base. They also have the **archegonia** and **antheridia**, which are fertilised to form the **diploid zygote** that grows into the **sporophyte** generation.

protists These are a group of micro-organisms that used to be called *Protista*. They are now part of a group of distantly related phyla, consisting mainly of unicellular organisms that do not fit into other kingdoms.

protonymph the first nymphal stage in mites after the egg stage. The eggs hatch into the protonymph and move on from this stage to the **deutonymph**.

protoplast The living contents of a plant cell contained within the plasmalemma.

provenance Refers to the area from which the plants were originally derived. An example would be seeds collected from plants in Scotland would be from that provenance.

pulhamite Is a patented human-made rock-like material that was used to build rock gardens or used in other hard landscape type projects like building a grotto. It was developed in the ninteenth century by James Pulham. It looks like a dark gritty sandstone rock.

puparium Is a barrel-shaped skin remaining from the last larval skin in which the pupa is hidden as it develops to adulthood; this occurs mainly in the fly families.

quantum (plural **quanta**) A discrete unit of electromagnetic energy (including, in order of decreasing wavelength, radio waves, infra-red waves, visible light, ultra-violet waves, X-rays and gamma rays). A quantum within the wavelengths of the visible spectrum ('visible light') is known as a **photon**.

quincunx A planting layout of five trees with one at each corner and one in the centre.

rachis (1) The main stalk of a fern frond or compound leaf. The leaflets are attached to the rachis.

rachis (2) The main stem of a grass flower.

rain shadow The area adjacent to buildings or walls that does not receive the full amount of rain. This is usually owing to the eaves projecting out or it is on the leeward side of a wall or building that prevents the rain falling on the area.

recessive A gene whose characteristics are hidden by a dominant gene. The characteristics of the recessive gene are masked by the dominant gene's characteristics. They will only show up when two recessive genes combine and no dominant gene is present.

recreate Returning a garden to a known earlier state by using new materials.

regulated water deficit The application of water to plants only during critical periods of growth. Up to 50% water savings have been achieved by this means in some ornamental species under protected cultivation, without impairing flowering.

relative humidity The amount of water vapour present in a given sample of air, expressed as a percentage of the amount that would be present if that volume of the same air was saturated (see also **vapour-pressure deficit**).

research and development Work carried out by scientists to improve on the current ways of producing crops or equipment. This includes producing or discovering new chemicals or breeding new cultivars of plants that are higher cropping or have a resistance to a pest or disease. Once something new is discovered it needs to be developed so that it can be used by growers and any problems sorted out before it is available for use.

resistance The ability of some plant species or cultivars to resist some pests or diseases. This reduces the chances of infection and therefore reduces the use of **pesticides** and crop losses. Some **cultivars** may have thicker **cuticles** that prevent the infection of the leaf by fungal spores.

response period With regard to irrigation it is the period when the plants will give the best response to water applied by irrigation. When the water will have its maximum effect on the crop and yield.

restore Returning the garden to a known earlier state, by removing newer additions or re-assembling existing materials.

rhizoid A very small tubular structure that grows from the lower epidermis of bryophytes and absorbs water and minerals and anchors mosses and liverworts to their growing 'media', which could be tree bark, tile roof or soil. They function in a similar way to root hairs.

rhizome A specialised stem, formed completely or partially underground, enabling a perennial plant to spread laterally and establish a **clone** of genetically uniform plants. Rhizomes are usually adapted for storage and as organs for survival during periods of dormancy

rhizosphere The zone of soil immediately surrounding the plant's roots.

RIDDOR Reporting of Injuries, Diseases and Dangerous Occurrences Regulations 1995. These regulations state that employers must report certain types of injuries, diseases or dangerous occurrences to the Health and Safety Executive. They can now be reported online, by phone or by completing a form and sending by post.

ridging (1) During digging the ground is left in ridges to allow the maximum surface area of the soil to be exposed to weathering by frost and rain over winter.

ridging (2) It is also used to refer to when a potato crop is **earthed up**; this is when soil is pulled over the new young potato shoots during the early stages of the crop.

risk assessment An assessment of any risks in your workplace to people who may be affected. The risks should be identified, who may be affected, what precautions you have in place and whether any further action is required to reduce the risks. It is a legal requirement in the UK for employers to carry out a risk assessment.

rodenticide A substance used to control rodents such as rats and mice.

root pruning Is carried out when the tree is dormant and is used to encourage fruit trees to flower and then produce fruit. It is a drastic measure and should be used as a last resort. A trench is dug a measured distance from the trunk around the tree and the main roots cut and a section removed. Normally half the root system is done one year and half the next to help maintain the tree's stability and reduce the shock to the tree.

rootstock In **grafting**, the rooted plant onto which is grafted a shoot or bud (the **scion**). Also known as an **understock**.

rootzone The growing medium for high-standard turf areas such as golf and bowling greens and also many of the top football pitches. It is a mixture of specific-sized sand particles, screened topsoil and possibly a small amount of organic matter. They are developed to have very good drainage but provide sufficient water and nutrient for the grass growth if irrigated and fertiliser is applied regularly.

safety data sheet (SDS) Also called material or product safety data sheets. Information supplied by the manufacturer about the chemical supplied by them. They should be used when preparing a COSHH risk assessment. Current SDSs must have 16 sections including measures to be taken in the event of fire, first aid, accidental release, handling and storage, the product's physical and chemical properties and information on toxicology, stability and reactivity, disposal considerations, transport, ecological and regulatory details.

saprophyte An organism, usually a fungus or bacterium, growing on and deriving its nourishment from dead or decaying organic matter.

scaly bulbs Bulbs that consist of overlapping fleshy leaf bases that give it a scaly appearance. Scaly bulbs do not have a protective tunic and therefore are more easily damaged and desiccated. *Lilium* spp. is an example of scaly bulbs.

scarification (1) An operation carried out on lawns or sports fields to remove thatch and moss from the area but leaving most of the grass intact. It is done using a springbok rake or scarifier machine.

scarification (2) A mechanical or chemical treatment to reduce or break the testa of a seed to improve **imbibition** and **germination**.

Scientific National Plant Collection A National Plant Collection status awarded to an existing National Plant Collection because it has attained a higher standard due to taxonomic research linked to the National Collection.

scion In **grafting**, a piece of a stem or a bud detached from a plant and placed in contact with, and ultimately growing together with, a root system (**rootstock**) or other tissue of another plant.

seasonality The period of time when a crop is in season in the UK for example runner bean, *Phaseolus coccineus*, used to be available from late July to the end of September; but now that they can be imported from all around the world they can be seen in supermarkets nearly all the year round. Therefore consumers have forgotten the season they were traditionally available.

secondary cultivations Carried out after the primary cultivations and include operations like light forking, using a three-tined prong, raking and treading. The aim is to produce a **tilth** for sowing or a reasonably fine soil for planting or turfing. They are carried out just prior to the sowing or planting.

sedimentary Rock that has formed from particles deposited from suspension in water. The main sedimentary rocks are sandstones, shale and limestones.

selective pesticides A pesticide with the ability to select out the pest, disease or weed and control it but leave other insects or plants unharmed. For example selective herbicides that will kill broad-leaved weeds but not the grasses when used on turf.

semi-permeable membrane In a plant cell, a membrane whose structure allows only the passage of solvent (water) molecules and that therefore prevents the passage of dissolved substances.

senescence The degenerative processes leading, eventually, to the death of a plant cell or tissue, or the whole plant.

sensory garden A garden designed and planted to involve the senses. As well as colour to appeal to the visual sense there will be scents to appeal to the olfactory sense (smell). Leaves and materials of different textures can be felt to involve the sense of touch. Some plants such as grasses and bamboos move in the wind to create sound that involves the sense of hearing.

sessile (1) A leaf joined directly to the stem without a **petiole**.

sessile (2) A flower joined directly to the stem without a pedicel.

shade avoider A plant that elongates rapidly in shade conditions, thereby enabling it to outgrow its neighbours and reach better light conditions. Many weeds are typical shade avoiders.

shade tolerator (shade plant) A plant that does not become elongated in shade conditions. Such plants, which have physiological and structural adaptations that enhance their ability to collect light in shady conditions, usually flourish in woodland habitats.

short-day plant A plant that can only flower, or flowers earlier, when the **day-length** is shorter than a specific duration known as the **critical day-length**.

short-persistence pesticide A pesticide that breaks down quickly and only persists in the environment for a short period of time.

sieve cell See **phloem**.

sieve tubes See **phloem**.

slit drainage A type of drainage used on sports grounds consisting of slits cut into the ground connecting the soil surface with the drainage pipes installed below ground. The slits are cut using machinery and back-filled with a coarse free-draining sand. The purpose is to remove water from the surface of the soil quickly to allow sports to be played even during wet periods.

smoke treatment A treatment given to certain seeds to mimic the effects of a bush fire and enhance **germination**. The chemicals in the smoke help to break the dormancy.

snag A jagged or misplaced (not close above a bud) wound in a stem or root pruned with a blunt or damaged blade of a knife or secateurs. Snagged tissues are vulnerable to infections causing die-back or rot.

soft fruit This group of fruits consist of the cane fruits including raspberries, blackberries and hybrid berries; and the bush fruits like blackcurrants, red-currants and gooseberries. They are referred to as soft fruit as they are soft and easily damaged.

softwood cuttings These are cuttings taken from the very soft fast early-spring growth and can include growth that has been forced early in a polytunnel or glasshouse. They are taken in the spring up until late June, after this time the speed of growth slows down and they are classed as **greenwood cuttings**.

Soil Association A charity that campaigns for healthy and sustainable food production. It supports the production of organic food and also runs a certification scheme for growers who produce organic food.

soil-based composts These are composts (growing media) that are based on soil, the main example is John Innes seed and potting composts. They are a mixture of loam, peat and coarse sand. They were very widely used before the peat-based composts were developed, which are now being replaced by peat-free composts.

soil creep The movement of soil on sloping ground owing to it being too wet. The weight of the water and the fact that it lubricates the soil enables it to slide down the slope. On steeper slopes it can result in landslides.

soil structure This is the arrangement of the primary soil particles into crumbs, aggregates or **peds**. This helps to form the drainage and aeration channels in the soil and prevents it becoming a muddy unstructured quagmire.

soil texture The relative proportions of sand, silt and clay in the soil. The amount of each constituent dictates into which class of soil it would be categorised. Examples would be sandy loams or clay loams.

SOLA (Specific Off-Label Approval) This was a system for getting pesticides approved for certain specific uses not stated on the product label. See **EAMU**.

solarisation The inhibition of photosynthesis by photo-oxidation of certain of the compounds involved, followed by the bleaching of leaves, resulting from the transfer of a plant from shady conditions to bright sunlight. The phenomenon may also be termed '**heliosis**'.

solute A substance, such as a salt or sugar, that will dissolve in a solvent, such as water, to form a solution.

spongy mesophyll A tissue, located between the **palisade mesophyll** and lower **epidermis** of the leaf, comprising large, irregular cells with wide air spaces between them, the latter allowing the free diffusion of gases (principally carbon dioxide, oxygen and water vapour) within the leaf. Its functions include photosynthesis (utilising any light not trapped by the cells of the palisade layer), the storage of nutrients, water and secondary metabolites, and the provision of support for the palisade layer.

sporophyte Is the diploid generation of a plant or algae and is the most prominent phase of the flowering plant's growth. It consists of the main plant parts of leaves, flowers, stems and roots.

staff recruitment The advertising, interviewing and selection of staff for a company.

standard A plant grown with a bare stem or trunk surmounted with a head of branches. Many trees are bought and grown as standard trees with a trunk of 1.8 m. It is also possible to grow roses and fuchsia as standard plants.

stock control Controlling the stock (plants, materials or equipment) so that the owner/company knows what they currently have in stock. This could be what chemicals and how much they have in store or the number of plants and what species and cultivars are in the nursery. Good stock control is important to maintain the profitability of a business.

stock plants In **vegetative propagation**, a plant used as a source of **cuttings** or other material for propagation.

stoma (plural **stomata**) A pore, in the **epidermis** of a leaf or stem, whose diameter may be varied by changes in the **turgor** of the two **guard cells** that bound it. By this means the exchange of gases (principally carbon dioxide, oxygen and water vapour) between the plant and the atmosphere is controlled.

stooled see **stooling**.

stooling In **vegetative propagation**, the repeated cutting back of parent plants to be used as a source of hardwood **cuttings** or for **layering**, to maintain their **juvenility**.

stratifying Layering seeds in moist soil or sand usually in a container, placed outdoors or in a cold frame and maintaining them at low temperatures in order to break **dormancy**.

strigolactones A type of plant hormone that reduces the branching in plants but stimulates the growth and branching of **arbuscular mycorrhiza** growth on the roots.

strobili (singular **strobilum**) These are commonly called cones as in conifers but also include the reproductive structures of *Equisetums* (horsetails), cycads, ginkgoes and some other primitive plants.

suberin A mixture of waxy substances present in the thickened dead cell walls of some plants. It is the basis of cork as in *Quercus suber*. It forms an outer protective layer to the plant.

sub-lethal dose of pesticide This applies when a pesticide has been applied at a dose insufficient to kill either the beneficial insects or possibly the pest. The pesticide may have an effect on the insect but not kill it and may be a contributory factor to pesticide resistance. This is one of the possible causes of the large drop in pollinator insects, such as bees, that have been affected by pesticides.

sub-stomatal cavity The large air space, delimited by **mesophyll** or cortical cells, immediately adjacent to a stomatal pore.

super weeds Weeds that have developed resistance to some herbicides. The widespread use of glyphosate herbicide in GMO crops has resulted in some weeds becoming resistant to glyphosate.

supply route The route from the grower to the retailer sometimes via a wholesaler, but can be direct or via a co-operative supplier. Shrubs are usually supplied directly from the nursery to a garden centre. But some crops will go from a grower to a processor and then a retailer.

sustainable A method of growing and/or using resources so that the resource is not permanently damaged and can be re-used again in the future. The growing of **biomass** is a sustainable energy method.

sustainability Creates and maintains the conditions under which humans and nature can exist in productive harmony. It is important in making sure that we have and continue to have the water, materials and resources to protect human health and our environment.

symmetry A symmetrical design is one that forms a mirror image along a central line or axis. A structure or garden that divided in half will have two identical images. They are balanced and symmetrical.

synchronous In plants, a developmental phase, such as flowering or leaf-fall, occurring at approximately the same time all over an individual or within a group of individuals. The term is usually used to describe a situation where all the flowers on an individual or within a group of individuals develop together, usually in response to an environmental signal such as day-length. Synchrony may facilitate cross-fertilisation or the avoidance of the damaging effects of adverse environmental conditions such as drought or very low temperatures.

systematic collections A type of living collection displayed around a particular plant family or plants that are closely related. These are often used to demonstrate plant evolution or systemic orders. In university botanic gardens these are often displayed as order beds.

systematics The study of the evolution of plants and other living organisms and their relationships through time. It includes the naming of plants using botanical Latin: describing, classifying

and preserving collections of them for future research.

target breeding Plant breeding to address a specific objective such as water-use efficiency or drought tolerance.

taxa (singular taxon) A taxonomic unit in the biological system of classification of organisms, for example: a phylum, order, family, genus or species. They consist of a group of organisms that a taxonomist places in units. Taxa are arranged in an order from kingdom to subspecies of related organisms.

taxonomic key A method used to identify plants or other organisms. The keys are usually dichotomous and are followed in a strict and logical order. The user carefully studies the plant and works through the key answering the statements, which should lead to the correct identification of the plant.

temperate plants Plants that originate and grow in the temperate zones of the world. These are areas where the climate is not subject to extremes of weather and temperature. A rough guide would be the zone between the tropic of Cancer and Arctic Circle and tropic of Capricorn and Antarctic Circle.

terminal bud The end bud of a shoot, which usually exerts **apical dominance**, preventing or inhibiting the growth of lateral shoots.

terrestrial ecosystems The interaction between living organisms and non-living material on the land mass of the Earth's surface. How plants and animals interact with soil, rocks and water in the environment.

thatch A build-up of organic matter on the soil surface of turf, it is formed from dead material produced by the turf including grass clippings and old plant material. It results in a spongy turf when walked on and encourages worms, disease and moss. The grasses also tend to be shallow rooting and so becomes stressed in drought conditions.

thematic collections Living plant collections based around a theme such as North American trees, culinary or

medicinal herbs, Mediterranean plants, or plants used in industry.

thermal time A concept whereby certain stages in a plant's development are related to particular periods of exposure to particular temperatures, in the absence of other controlling factors such as responses to **day-length** or **vernalisation**. The concept indicates that within specified limits, the rate of biochemical reactions and, consequently, the rate of growth of an organism increases with increasing temperatures.

tilth When the soil has a fine crumbly surface ready for seed sowing. All large lumps will have been broken down or removed as will large stones or debris. It is achieved by roughly raking the area, treading it to break up the lumps of soil and then giving a final rake to get the correct levels and remove stones or debris.

tonoplast A membrane enclosing a **vacuole** in a plant cell.

top dressing (1) The application of a soil mixture to a lawn or sports area to help level it and improve root growth.

top dressing (2) The application of a fertiliser to stimulate growth. It is normally applied in the spring and is a high nitrogen fertiliser on crops such as spring cabbage.

top fruit Also called tree fruit and consisting of two main groups: the pome fruits, which are the apples and pears, and the stone fruits, such as cherries, plums, apricots and other related fruits. Top fruit are usually grown as trees or on a dwarfing rootstock in highly trained shapes like cordons and espaliers.

topiary An old art of training plants into geometric shapes, birds or animals. They are commonly grown over wire frames and trained and trimmed to shape. Yew, *Taxus baccata*, and box, *Buxus* spps., are widely used for this purpose.

topography The shape of the ground surface showing the hills and plains (rising and falling ground). A steep topography indicates steep slopes or hilly ground; flat topography indicates flat or slightly undulating ground.

totipotent A term used to describe a plant cell that is capable of undergoing division to produce a complete new individual, if suitably induced.

trace elements Also called minor and micro-nutrients. They are chemical elements that are essential for plant growth but required in very small amounts. For example iron, boron, copper, zinc and manganese.

tracheid A dead, elongated, spindle-shaped plant cell, lacking contents and with its walls thickened with bands of **lignin**, that forms part of the **xylem** tissue. Tracheids, linked together by pits, may be involved in the transport of water and dissolved substances such as mineral ions and hormones. They also provide mechanical support.

Trade Description Act The Trade Descriptions Act 1968 is an Act of the Parliament of the UK, which prevents manufacturers, retailers or service industry providers from misleading consumers as to what they are purchasing. This law can punish companies or individuals who make false claims about their products or services.

transgenic organisms Organisms that have been altered in some way using **genetic modification**. The organism contains a gene or genes transferred from another species.

translaminer The movement of a pesticide within a leaf from the point of application. The pesticide will stay within the leaf, not move out into the stem as a systemic pesticide would.

transpiration The flow of water vapour from a plant to the atmosphere, principally through the **stomata** of the leaves and stems.

transposon Fragments of DNA that can move around the genome by removing themselves from one position and slotting themselves into a new position. This occurs naturally and will change the characteristics of the plant.

triangulation A method to determine a particular position. Measurements are

taken from two selected fixed points to a third point to establish its position.

triploid A plant with three sets (3n) of chromosomes rather than the normal two. They can often be sterile. A good example of a triploid plant is the apple cultivar Malus Bramley's Seedling.

trompe-l'oeil A feature in the garden that is intended to deceive the eye to create a certain effect. Examples are mirrors, ha-ha and wall paintings.

tropical plants Plants that originate from the tropical zones of the world. This is the area between the tropic of Cancer and tropic of Capricorn. It is characterised by a hot climate.

tropism A response to an external stimulus that controls the direction of plant growth such as phototropism (light) and geotropism (gravity). A tropism can be both positive and negative; a root is positively geotropic as it grows towards gravity, but negatively phototropic as it grows away from light.

tuber, root and stem An underground, swollen portion of a root (e.g. *Dahlia* spp.) or stem (e.g. potato, *Solanum tuberosum*). Tubers normally have a storage function and provide the means for a plant to survive during periods unfavourable for growth, such as winter or drought conditions.

tunicate bulbs Have the fleshy leaf bases arranged in concentric rings around the growing point. They are called tunicate bulbs as they have an outer layer of dry leaves that protect the bulb, a 'tunic'. Examples include *Narcissi, Alliums, Tulipa* and *Galanthus* spp.

turgor The rigidity of plant cells and, as a consequence, tissues and organs, resulting from the **hydrostatic pressure** (the turgor pressure) exerted on the cell walls by the protoplasts. When the amount of water lost from a plant's leaves is greater than can be supplied from the roots and adjacent soil, turgor is lost and **wilting** occurs. Also see **pressure potential** and **wall pressure**.

understock See **rootstock**.

unit costs The cost to produce, store and sell one unit of a product or crop. It includes all fixed costs and variable costs. A unit could be the cost to produce a large tree or the cost to produce a kilogram of tomatoes.

unity The unifying of the design to give an attractive appearance. Unity can be achieved by the use of rhythm, balance, proportion, scale, colour, texture and form. This can be done by the repeating of particular plants or colours in the garden or the use of similar hard landscaping materials.

UVI (ultra-violet light inhibitor) This is an additive that is put into polythene sheets used on polytunnels that reduces the ultra-violet light damage to the polythene extending its life.

vacuole A sac, filled with liquid and bounded by a membrane (the **tonoplast**), contained within the **protoplast** of a plant cell.

vapour pressure The absolute amount of water vapour in air.

vapour-pressure deficit The difference between the amount of water vapour present in a given volume of air and the amount that would be present at saturation (see also **relative humidity**).

vapour-pressure gradient The concentration gradient along which water vapour moves, from a region of higher **vapour pressure** to one of lower vapour pressure.

vascular bundle One of a number of long strands in stems, leaves and roots, comprised principally of **xylem**, **phloem**, **cambium** and a **bundle-sheath**. Vascular bundles are chiefly responsible for mechanical support and the transport of water, sugars, hormones and other dissolved substances within the plant.

vegetative propagation The propagation of a plant by non-sexual means, such as by **division**, taking **cuttings**, **layering**, **grafting** or planting **bulbs**, **corms** or **tubers**. Plants propagated vegetatively from a single parent form a **clone**.

vernalisation The triggering of flower development by exposure of seeds, seedlings or young plants to a period of low temperature. The term is sometimes used, incorrectly, to describe the triggering of seed germination by exposure to low temperatures. See also **de-vernalisation**.

viviparous The embryo develops within the female's body leading to live birth, for example the summer form of aphids.

wall pressure In a turgid plant cell, the inwardly directed pressure exerted by the wall that balances the **turgor pressure (hydrostatic pressure)**.

water potential In a plant system, the potential energy of water (free energy per molecule); in other words, its potential to do 'work'. By definition, pure water has a potential of zero, so the water potential of plant cells and of unsaturated soils have negative values. In any plant system, water molecules will flow from a region of relatively high water potential to one of relatively low potential. Thus in a fully saturated soil, water molecules will flow from the soil, across the **semi-permeable membrane** of a root hair cell and into the protoplast, thereby increasing the **turgor pressure** of the cell.

weaning The acclimatisation of a newly rooted cutting or other **propagule** to a natural growing environment, usually involving a reduction in temperature and humidity.

weathering The physical and chemical changes produced in rocks at the Earth's surface caused by the weather conditions like wind, rain, freezing and thawing. This is how the original soils were formed.

whorl An arrangement of plant parts attached to a stem usually at a single point. These include a whorl of sepals or petals, but some plant species have a whorl of leaves, *Galium aparine* (cleavers) being an example.

wildlife corridors Areas that link together to form a corridor along which wildlife can pass. This could be as simple as hedges or a number of gardens through which wildlife can pass.

wilting Occurs when the foliage or soft plant parts wilt (droop) owing to a drop in turgor pressure as the plant cells are short of water. With insufficient water in the cells there is nothing to keep the soft plant parts firm and erect so the plant wilts. This can occur in drought conditions or owing to pest or disease attack.

windbreak A line of trees, bushes or material placed in a position to reduce the force of the wind. These are used to prevent damage to plants and reduce heat loss from greenhouses and other structures. A permeable windbreak will give protection for about ten times its height.

wind rock Caused by strong winds moving and rocking plants, which disturbs the roots restricting growth; and can also create a hollow around the stem that can collect water, which if it freezes can damage the stem.

wood buds Buds on fruit trees that when they grow will produce stem and leaves, as opposed to fruit buds that produce flowers and fruit.

wounding In **vegetative propagation** of woody plants, severely injuring the tissues of a **layered** stem or the base of a **cutting** in order to improve rooting.

xeromorphic Plants (**xerophytes**) that have adaptations that help them grow in extreme conditions including deserts or drought-prone areas. The plants usually develop thick (often waxy) **cuticles**, sunken stomata or thick-walled tissues to reduce water loss.

xerophytes see **xeromorphic**.

xylem A plant tissue in which most of the cells are dead, lack contents and have walls thickened with **lignin**. These cells may be elongate, lacking end walls and joined end to end to form vessels; or they may be spindle-shaped, dead **tracheids**, linked by pits; or they may be long, structural fibres. Vessels and tracheids, when first formed, are involved in the transport of water and dissolved minerals, some hormones and other substances towards the apex of the plant and into the leaves, but as they age, during secondary thickening, they may become occluded with gums, resins and phenolic substances. Vessels, **tracheids** and **fibres** are all involved in the structural support of the plant. Wood is chiefly composed of xylem.

zygomorphic Flowers that are irregular in their petal arrangement. If the flower is cut through vertically there is only one plane that would produce a mirror image, any other plane would produce differing halves. Examples of zygomorphic flowers are the deadnettle, *Lamium album*, and sweet pea, *Lathyrus odoratus*.

zygote A cell resulting from the fusion of two **haploid gametes** during **fertilisation** from which the seed will develop. The cell will have a **diploid** nucleus (2n).

FIGURE CREDITS

The authors and publisher acknowledge the sources of copyright material below and are grateful for the permissions granted. While every effort has been made, it has not always been possible to identify the sources of all material used, or to trace all copyright holders. If any omissions are brought to our notice, we will be happy to include the appropriate acknowledgements on reprinting.

All images in this manuscript have been provided by the contributing authors, except for the ones in the list below.

CHAPTER 1

1.1 Illustration by Simon Tegg
1.2 Illustration by Simon Tegg
1.4 Cambridge University Botanic Garden/Howard Rice
1.5 Cambridge University Botanic Garden/Howard Rice
1.6 Cambridge University Botanic Garden/Juliet Day
1.7 Cambridge University Botanic Garden/Howard Rice
1.8 Cambridge University Botanic Garden/Juliet Day

CHAPTER 2

Chapter Opener, image kindly provided by Rosie Yeomans

2.24 Illustration by Simon Tegg
2.25 Illustration by Simon Tegg
2.32 Illustration by Simon Tegg
2.33 Illustration by Simon Tegg
2.34 Illustration by Simon Tegg
2.35 Illustration by Simon Tegg
2.36 Illustration by Simon Tegg
2.37 Illustration by Simon Tegg
2.38 Illustration by Simon Tegg
2.39 Illustration by Simon Tegg
2.40 Illustration by Simon Tegg
2.41 Illustration by Simon Tegg

CHAPTER 3

Chapter Opener, image kindly provided by Janet Prescott. Reproduced with permission.

3.1 Illustration by Simon Tegg
3.2 Illustration by Simon Tegg. Redrawn from the image kindly provided by Professor Brian Thomas, University of Warwick
3.3 Illustration by Simon Tegg. Redrawn from F. Salisbury and C. Ross, *Plant Physiology* 1991, with permission from Thomson Higher Education, Belmont, CA.

3.4 Image redrawn from David S. Ingram, Daphne Vince-Prue and Peter J. Gregory, Figure 8.8, *Science and* the Garden, 2nd edition. Copyright © 2008, John Wiley and Sons

3.5 Image redrawn from David S. Ingram, Daphne Vince-Prue and Peter J. Gregory, Figure 14.3, *Science and the Garden*, 2nd edition, Copyright © 2008, John Wiley and Sons

3.6 Illustration by Simon Tegg

3.0 Illustration by Simon Tegg. Redrawn from David S. Ingram, Daphne Vince-Prue and Peter J. Gregory, Figure 8.2, *Science and the Garden*, 2nd edition, Copyright © 2008, John Wiley and Sons

3.9 Image © Professor Brian Thomas, University of Warwick

CHAPTER 4

Chapter Opener, image © Artazum/Shutterstock

4.1 Illustration by Simon Tegg. Redrawn from F. Salisbury and C. Ross, *Plant Physiology* 1991, with permission from Thomson Higher Education, Belmont, CA.

4.2 Image redrawn from David S. Ingram, Daphne Vince-Prue and Peter J. Gregory, Figure 8.1, *Science and the Garden*, 2nd edition, Copyright © 2008, John Wiley and Sons

CHAPTER 5

Chapter Opener, image © Freer/Shutterstock

5.1 Image redrawn from David S. Ingram, Daphne Vince-Prue and Peter J. Gregory, Figure 13.1, *Science and the Garden*, 2nd edition, Copyright © 2008, John Wiley and Sons

5.2 Illustration by Simon Tegg

5.3 Image redrawn from Garner, W. W. & Allard, H. A. (1923), *Journal of Agricultural Research* 23, 871–920. Image courtesy of Professor Brian Thomas, University of Warwick

5.6 Illustration by Simon Tegg

5.7 Illustration by Simon Tegg

5.9 Image © Professor Royal Heins, Michigan State University

CHAPTER 6

Chapter Opener, image © Fotogr/Shutterstock

6.4 Image modified by ADAS

6.5 Illustration by Simon Tegg

6.16 Image kindly provided by Ray Broughton

CHAPTER 7

7.1 Illustration by Simon Tegg

7.3 Illustration by Simon Tegg

7.4 Illustration by Simon Tegg

7.5 Illustration by Simon Tegg

7.6 Illustration by Simon Tegg

7.11 Illustration by Simon Tegg

7.20 Illustration by Simon Tegg

7.28 Illustration by Simon Tegg

7.29 Illustration by Simon Tegg

7.31 Illustration by Simon Tegg

7.33 Illustration by Simon Tegg

7.35 Illustration by Simon Tegg

CHAPTER 8

8.1 Image reproduced with permission of Roy Niblett/Slide from Thermo Scientific, (6B 17 *Ligustrum* Stem Apex L.S. X40)
8.3 Image redrawn with permission of Peter Gregory
8.5(a) Image reproduced with permission of Ray Broughton
8.11 Illustration by Simon Tegg
8.12(a–d) Image reproduced with permission Ray Broughton
8.13 Illustration by Simon Tegg
8.16 Image reproduced with permission of Ray Broughton

CHAPTER 9

Chapter Opener, image courtesy of the Department of Energy Genome Programs

9.2 Illustration by Simon Tegg. Redrawn from the image kindly provided by the National Human Genome Research Institute
9.8 Illustration by Simon Tegg

CHAPTER 10

Chapter Opener, image © Andrew Roland/Shutterstock

10.14(c) Image reproduced with permission of Anneli Salo
10.16 Image reproduced with permission of Martin Burr

CHAPTER 12

Chapter Opener, © Zigzag Mountain Art/Shutterstock

12.8(a, b, c) Illustration by Simon Tegg

CHAPTER 13

13.1 Illustration by Simon Tegg
13.10 Illustration by Simon Tegg
13.20 Illustration by Simon Tegg

CHAPTER 14

14.4 Image reproduced with permission of Jamie Cryer
14.7(a,b) Image reproduced with permission of Ray Broughton
14.8 Image reproduced with permission of Ray Broughton
14.16 Illustration by Simon Tegg. Reproduced with permission of Stephen J White
14.17 Illustration by Simon Tegg. Reproduced with permission of Stephen J White

CHAPTER 15

15.1 Image reproduced with permission of Aaron Mills. Courtesy of the University of Reading
15.2 Image courtesy of Bayer CropScience. Reproduced with permission
15.3 Image courtesy of Plant Heritage. Reproduced with permission
15.4 Image courtesy of BGCI. Reproduced with permission
15.5 Image courtesy of RBG Kew. Reproduced with permission
15.6 Image courtesy of RBG Kew. Reproduced with permission
15.7 Image courtesy of Bayer CropScience. Reproduced with permission

CHAPTER 16

16.2 Image courtesy of BALI

CHAPTER 17

Chapter Opener photo © Fun/Shutterstock

17.1 Image courtesy of Syngenta Bioline UK. Reproduced with permission
17.2 Image courtesy of Syngenta Bioline UK. Reproduced with permission
17.3 Image courtesy of Syngenta Bioline UK. Reproduced with permission
17.4 Image courtesy of Syngenta Bioline UK. Reproduced with permission
17.5 Image courtesy of Syngenta Bioline UK. Reproduced with permission
17.6 Image courtesy of Syngenta Bioline UK. Reproduced with permission
17.7 Image courtesy of Syngenta Bioline UK. Reproduced with permission
17.8 Image courtesy of Syngenta Bioline UK. Reproduced with permission
17.9 Image courtesy of Syngenta Bioline UK. Reproduced with permission
17.10 Image courtesy of Syngenta Bioline UK. Reproduced with permission
17.11 Image courtesy of Syngenta Bioline UK. Reproduced with permission
17.12 Image courtesy of Syngenta Bioline UK. Reproduced with permission
17.14 Image courtesy of Syngenta Bioline UK. Reproduced with permission
17.15 Image courtesy of Syngenta Bioline UK. Reproduced with permission

CHAPTER 18

Chapter Opener, image kindly provided by Harry Mycock

18.1 Image courtesy of Veolia Environmental Services. Reproduced with permission
18.2 Image courtesy of Veolia Environmental Services. Reproduced with permission
18.3 Image courtesy of Veolia Environmental Services. Reproduced with permission
18.4 Image courtesy of Veolia Environmental Services. Reproduced with permission
18.7 Image reproduced with permission of Janet Prescott
18.8 Image reproduced with permission of Janet Prescott

INDEX